Probability and Mathematical Statistics (Continued)

SEBER • Linear Regression Analysis
SEBER • Multivariate Observations
SEN • Sequential Nonparametrics: Invariance Principles and Statistical Inference
SERFLING • Approximation Theorems of Mathematical Statistics
TJUR • Probability Based on Radon Measures
WILLIAMS • Diffusions, Markov Processes, and M
I: Foundations
ZACKS • Theory of Statistical Inference

Applied Probability and Statistics

ABRAHAM and LEDOLTER • Statistical Method
AGRESTI • Analysis of Ordinal Categorical Data
AICKIN • Linear Statistical Analysis of Discrete Data
ANDERSON, AUQUIER, HAUCK, OAKES, VANDAELE, and WEISBERG • Statistical Methods for Comparative Studies
ARTHANARI and DODGE • Mathematical Programming in Statistics
BAILEY • The Elements of Stochastic Processes with Applications to the Natural Sciences
BAILEY • Mathematics, Statistics and Systems for Health
BARNETT • Interpreting Multivariate Data
BARNETT and LEWIS • Outliers in Statistical Data, *Second Edition*
BARTHOLOMEW • Stochastic Models for Social Processes, *Third Edition*
BARTHOLOMEW and FORBES • Statistical Techniques for Manpower Planning
BECK and ARNOLD • Parameter Estimation in Engineering and Science
BELSLEY, KUH, and WELSCH • Regression Diagnostics: Identifying Influential Data and Sources of Collinearity
BHAT • Elements of Applied Stochastic Processes, *Second Edition*
BLOOMFIELD • Fourier Analysis of Time Series: An Introduction
BOX • R. A. Fisher, The Life of a Scientist
BOX and DRAPER • Evolutionary Operation: A Statistical Method for Process Improvement
BOX, HUNTER, and HUNTER • Statistics for Experimenters: An Introduction to Design, Data Analysis, and Model Building
BROWN and HOLLANDER • Statistics: A Biomedical Introduction
BROWNLEE • Statistical Theory and Methodology in Science and Engineering, *Second Edition*
CHAMBERS • Computational Methods for Data Analysis
CHATTERJEE and PRICE • Regression Analysis by Example
CHOW • Analysis and Control of Dynamic Economic Systems
CHOW • Econometric Analysis by Control Methods
CLARKE and DISNEY • Probability and Random Processes: A First Course with Applications
COCHRAN • Sampling Techniques, *Third Edition*
COCHRAN and COX • Experimental Designs, *Second Edition*
CONOVER • Practical Nonparametric Statistics, *Second Edition*
CONOVER and IMAN • Introduction to Modern Business Statistics
CORNELL • Experiments with Mixtures: Designs, Models and The Analysis of Mixture Data
COX • Planning of Experiments
DANIEL • Biostatistics: A Foundation for Analysis in the Health Sciences, *Third Edition*
DANIEL • Applications of Statistics to Industrial Experimentation
DANIEL and WOOD • Fitting Equations to Data: Computer Analysis of Multifactor Data, *Second Edition*
DAVID • Order Statistics, *Second Edition*
DAVISON • Multidimensional Scaling
DEMING • Sample Design in Business Research
DILLON and GOLDSTEIN • Multivariate Analysis: Methods and Applications
DODGE and ROMIG • Sampling Inspection Tables, *Second Edition*
DOWDY and WEARDEN • Statistics for Research

continued on back

Finite Algorithms in Optimization and Data Analysis

Finite Algorithms in Optimization and Data Analysis

M. R. Osborne
Department of Statistics,
Australian National University,
Canberra

JOHN WILEY & SONS
Chichester · New York · Brisbane · Toronto · Singapore

Library of Congress Cataloging in Publication Data:

Osborne, M. R. (Michael Robert)
Finite algorithms in optimization and data analysis.

(Wiley series in probability and mathematical
statistics. Applied probability and statistics)
Includes bibliographies and index.
1. Mathematical optimization—Data processing.
2. Least squares—Data processing. 3. Approximation
theory—Data processing. 4. Algorithms. I. Title.
II. Series.
QA402.5.08 1984 519 84-11841

ISBN 0 471 90539 9

British Library Cataloguing in Publication Data:

Osborne, M. R.
Finite algorithms in optimization and data
analysis.—(Wiley series in probability and
mathematical statistics. Applied
probability and statistics)
1. Algorithms
I. Title
511′.8 QA9.58

ISBN 0 471 90539 9

Printed and bound in Great Britain.

Contents

Preface

Finite algorithms are calculation sequences which involve at most a strictly limited number of steps. They have a fundamental place in the theory and practice of computation both as objects of importance in their own right and as building blocks for more elaborate (iterative or recursive) procedures. Our purpose is to study a class of such algorithms – a class defined in part by the qualifications 'optimization and data analysis', which suggest that the algorithms minimize an objective or cost function subject to conditions on the acceptability of trial solution points, but ultimately by the taste, interests, and limited knowledge of the author. While it is hoped that they are not too obtrusive, taste and interests are important. They serve to delimit the algorithms presented in the sense that all were met first in fitting models to data, in many cases in nonlinear situations where the algorithm was applied to a 'linear subproblem' as part of an iterative solution process. This establishes a context in which methods of continuous rather than discrete analysis are important. However, finiteness would seem to imply an inevitable combinatorial element, and this will usually enter here through the problem possessing a reformulation as a linear complementarity problem so that, at least in principle, the solution can be found by an exhaustive search of a finite number of possibilities. In general, such an approach is impractical, and in order to direct the search the idea of minimizing an objective or cost function in a feasible descent direction is exploited systematically and in a manner typical of the strategy used in solving continuous optimization problems.

The implied problem context has another important consequence for, after possible reformulation, none of the problems considered is large by the standard of much contemporary work in mathematical programming. To be more specific, although one of the key problem descriptors n (the number of observations in the input data set) is not limited in size explicitly, it is assumed that the second key descriptor p (the number of parameters to be determined in the model) is typically fixed and small relative to n, and that the matrix representation of the model to be fitted to the data is dense. Immediate consequences are:

(a) That estimates of the cost in carrying out basic steps of the algorithms are quoted for fixed p as n increases without limit; and

(b) that the important and rapidly developing technology for manipulating sparse matrices is irrelevant to our considerations.

These assumptions are accepted here without further question and must be borne in mind in assessing the value of the material presented in other contexts.

A striking feature of many of the problems treated is that they lead to the minimization of nondifferential objective functions (for example polyhedral convex functions). For this reason methods of convex analysis are used extensively in algorithmic development as they simplify the problems of characterizing optimality and finding descent directions. The necessary results are developed in Chapter 1 which treats basic properties of convex sets and functions, theories of the alternative, differential properties of convex functions, necessary conditions for optima in convex programming, and duality.

The best-known optimization problem for which a finite algorithm is available is the linear programming problem. This is discussed in Chapter 2 where the main methods are developed. The treatment is somewhat unusual in that the simplex algorithm is not given its customary prominence but is dealt with as one among a range of possibilities in which the descent methods based on active set strategies are at least as important for later applications. Because much of this material is relevant to subsequent considerations, advantage is taken of the comparatively simple problem setting to treat the main computational questions in some detail. These include:

(a) problem scaling, testing for termination, and the provision of scale free tests;
(b) methods for rank-deficient problems;
(c) resolution of degeneracy problems; and
(d) implementation of stable algorithms (in particular in tableau form).

Problems in discrete approximation that conventionally have been solved by linear programming methods (in particular, problems in l_1 and l_∞ approximation) are considered in Chapter 3. It is shown that descent algorithms also can be developed by working with the approximation problem formulation. As a result of these possible different approaches there is a surprisingly large literature of equivalent algorithms, and an attempt is made to collect these together and to reconcile apparent differences. Also, some insight into the behaviour of algorithms for discrete approximation problems can be gained by considering algorithms for the corresponding continuous problem. For this reason a selection of results for the continuous problems is given in Appendix 2.

Chapter 4 considers the general problem of minimizing a polyhedral convex function. This problem subsumes the problems of both Chapters 2 and 3. In a sense this involves nothing new as the problem of minimizing a polyhedral convex function can always be posed as a linear programming problem. However, in many cases the resulting formulation is not practical (in one of our examples the linear program has up to $2^n n!$ constraints), and this has led to several interesting and otherwise valuable estimation procedures being largely ignored. Thus it is significant that our approach leads to methods that are both compact and effective. The idea is to formulate the problem in such a way that key structural elements are emphasized. The appropriate analytical tool for this purpose turns out to be the directional derivative of a convex function. This is used to construct a compact parametrization of the subdifferential of a polyhedral convex function in which the structural elements are made explicit. This property makes this form of the subdifferential readily accessible in many cases, and subsequent development of the standard forms of descent algorithm becomes routine.

If the norm is smooth then the most important discrete approximation problem is just least squares fitting. This problem is treated in Chapter 5, and certain ramifications are developed. Thus constrained least squares problems (and in particular the restricted least squares problem), and iteratively reweighted least squares methods are considered. The idea of the resistance of an algorithm to outliers in the data is developed as a method for comparing estimation procedures and used to introduce the M-estimator as a way of increasing resistance in procedures based on least squares. Properties of this estimator are then considered in some detail and a finite algorithm for calculating it presented using continuation with respect to a parameter as a basic algorithmic idea. Intriguingly, this idea surfaces in other applications including the Lemke algorithm for linear complementarity problems considered in Chapters 2 and 5, and in an approach to the development of projected gradient algorithms considered in Chapter 4.

In Chapter 6, two data analysis applications which differ from previous problems in being non-convex are considered. These are the total approximation problem which occurs when there are errors in the model specification as well as in the data (for algorithmic purposes consideration is specialized to the l_1 norm), and the problem of computing centres in cluster analysis using the Jaccard metric. Again finite descent algorithms are possible. However, these problems require for their analysis tools more appropriate to general nondifferentiable optimization problems, and it is necessary to develop an appropriate theory. Thus a nondifferentiable form of the implicit function theorem is used in establishing first-order necessary conditions and in characterizing directional sequences, and second-order conditions are given to provide sufficient conditions for an isolated local minimum.

Chapter 7 draws together theoretical and experimental results on the complexity and general performance of selected algorithms. Thus far almost all studies have centred on the simplex algorithm of linear programming, and the challenge has been to reconcile the observed satisfactory performance in practice with the existence of specially constructed examples which require an exponentially large number of steps for their solution. That linear programs can be solved in polynomial time has been proved by Khachiyan in the case that the data can be represented exactly by a finite number of digits. His proof can be adapted readily to show the existence of polynomial algorithms for minimizing polyhedral convex functions when the number of structure functionals is polynomial (this does not follow from the linear programming formulation when this involves exponentially large numbers of constraints). However, the ellipsoidal algorithm used by Khachiyan does not appear to provide a practical alternative (highlighting the difference between average and worst-case performance), and the challenge remains to explain the generally good behaviour of the simplex algorithm. One approach has been to estimate expected execution times for problems generated by random sampling from a suitable distribution. In particular, the expected complexity of Lemke's algorithm applied to the linear complementarity problem derived from the linear programming problem has been estimated in this way. The experimental evidence derives in the main from Monte Carlo studies of algorithm performance (so that again an important consideration is selection of a suitable sampling distribution). This serves to suggest the deeper analytical results, but it also provides a convenient method for checking out particular computing strategies (for example unnormalized versus scale-free tests), and for comparing competing algorithms.

The impetus to put this book together stems from visits to the Australian National University by Bill Steiger in 1979 and 1981. This proved a period of particularly stimulating and fruitful collaboration shared also by David Clark, whose PhD thesis provided important material for Chapter 5, and by David Anderson. I was able to consolidate the early chapters and provide a first draft of Chapter 4 while a Visiting Professor at the University of Utah during the academic year 1981/82 and was privileged to be able to try out the material on a lively seminar. Chapter 6 results from work done while Alistair Watson was visiting the Australian National University from January 1983. Several people contributed to the preparation of the typescript, and mention should be made of Chris Jirasek who perhaps had the largest share of the initial load, and Debbie Spencer who looked after the several stages of correction and assembly. Di Cook converted programs developed on an HP9845 so that larger-scale evaluations could be carried out on a Univac 1100–80, and persevered with the graphics system in order to prepare the diagrams.

Table of notation

$\forall$	all (or any)
$\exists$	there exists
$\ni$	such that
$\to$	goes to
$\Rightarrow$	implies
$\Leftrightarrow$	if and only if
$:=$	becomes
$\in$	is an element of
$\notin$	is not an element of
$\subset$	is a subset of (is contained in)
$\supset$	contains
$\not\subset$	is not contained in
$\subseteq$	is contained in or equals
$\supseteq$	contains or equals
$\{x; P(x)\}$	set of elements x having property P
$\{x\}$	set consisting of a single element
X^c	set $\{\mathbf{x}; \mathbf{x} \notin X\}$
$\varnothing$	empty set
$\cup$	set union
$\cap$	set intersection
$\backslash$	set difference ($A \backslash B$ is the set of elements in A but not in B)
A	matrix (implied by context as is dimensions – matrix notation is discussed further in Section A1.1 of Appendix 1)
$\mathbf{x}$	vector (dimension implied by context)
A^T	transpose of A (similarly $\mathbf{x}^T$ and $(A^{-1})^T = A^{-T}$)
ν	index set $\{i, j, k, \ldots\}$ or $\{\nu(1), \nu(2), \ldots\}$ pointing to rows (usual case) of A
A_ν	submatrix pointed to by ν
$\mathbf{b}_\nu$	subvector pointed to by ν
$\|\nu\|$	number of elements in ν
$\mathbf{e}_i$	ith coordinate vector
$\mathbf{e}^{(p)}$	vector of dimension p each component of which is 1
I_p	unit matrix of dimension p
$\boldsymbol{\rho}_i(A)$	ith row of A
$\boldsymbol{\kappa}_i(A)$	ith column of A

$[\mathbf{x}]_j^i$	subvector consisting of the ith to the jth components of $\mathbf{x}$ inclusively (cases $i=1$, $j=\dim x$ are suppressed)
$E(\mathbf{u}, \mathbf{v}, \alpha)$	elementary matrix (Section A1.2)
P_{ij}	elementary permutation matrix
$L(i, \mathbf{x})$	elementary lower triangular matrix
$H(i, \mathbf{x})$	elementary Householder matrix
$R(i, j, (\alpha, \beta))$	elementary rotation matrix zeroing element in the (α, β) positions of the object matrix
$\times$	product (of sets of vectors) $\left(X \times Y = \left\{\begin{bmatrix}\mathbf{x}\\ \mathbf{y}\end{bmatrix}; \mathbf{x} \in X, \mathbf{y} \in Y\right\}\right)$
R	$\{x; -\infty < x < \infty\}$
R^p	$R \times R \times \ldots \times R$ (p terms)
R_p^+	$\{\mathbf{x} \in R^p; \mathbf{x} \geqslant 0\}$
$L^\perp$	$\{\mathbf{x}; \mathbf{x}^T\mathbf{y} = 0, \forall \mathbf{y} \in L\}$
$H(\mathbf{u}, \nu)$	hyperplane (set $\{\mathbf{x}; \mathbf{u}^T\mathbf{x} = \nu\}$)
$H^+(\mathbf{u}, \nu)$	$\{\mathbf{x}; \mathbf{u}^T\mathbf{x} > \nu\}$
$H^-(\mathbf{u}, \nu)$	$\{\mathbf{x}; \mathbf{u}^T\mathbf{x} \leqslant \nu\}$
C	cone, $\mathbf{x} \in C \Rightarrow \lambda\mathbf{x} \in C, \forall \lambda \geqslant 0$.
C^*	polar cone $= \{\mathbf{y}; \mathbf{y}^T\mathbf{x} \leqslant 0, \forall \mathbf{x} \in C\}$ (particular case of the polar to a set S, $S^* = \{\mathbf{y}; \mathbf{y}^T\mathbf{x} \leqslant 1, \forall \mathbf{x} \in S\}$
$\circ$	composition operation ($A \circ B\mathbf{x}$ is the image of $\mathbf{x}$ under the mapping in which first B is applied, followed by A).
$\lVert\cdot\rVert$	norm (Example 2.1 of Chapter 1)
$\lVert\cdot\rVert_D$	norm on domain space
$\lVert\cdot\rVert_R$	norm on range space
$\lVert\cdot\rVert_\alpha$	$\alpha = 1$, l_1 norm; $\alpha = 2$, l_2 or Euclidean norm; $\alpha = \infty$, l_∞ or maximum norm.
$\lVert\cdot\rVert_P$	polyhedral norm (Definition 1.2 of Chapter 3)
$\lVert\cdot\rVert_F$	Frobenius or Euclidean norm of a matrix (see equation (5.50) of Chapter 2)
$\delta(\mathbf{x} \mid S)$	indicator function of the set S (Section 1.6)
$\delta^*(\mathbf{x} \mid S)$	support function of the set S (Section 1.2)
$\Delta(\mathbf{r})$	dispersion (Definition 1.2 of Chapter 4)
∂	subdifferential of convex function (Definition 4.4 of Chapter 1) (∂_x indicates that subdifferential is taken with respect to $\mathbf{x}$)
∂_ε	ε-subdifferential (Definition 7.1 of Chapter 1)
∇	gradient operator (∇_x indicates that the gradient is taken with respect to $\mathbf{x}$)
$\lfloor z \rfloor$	largest integer $\leqslant z$
$f'(\mathbf{x} : \mathbf{t})$	directional derivative of f in direction $\mathbf{t}$ at $\mathbf{x}$ (Definition 4.5 of Chapter 1)
$f^{(k)}(\mathbf{x} : \cdot, \ldots \cdot)$	tensor formed by the partial derivatives of order k
$\mathcal{T}(S, \mathbf{x})$	tangent cone to S at $\mathbf{x}$ (Definition 5.2 of Chapter 1)
$\mathcal{L}(\mathbf{x}, \mathbf{u})$	Lagrangian of CPP (Definition 6.2 of Chapter 1)

$\mathscr{J}(\mathbf{f}, \mathbf{w})$	generalized Jacobian of $\mathbf{f}$ (Definition 2.1 of Chapter 6)
$\mathscr{E}$	mathematical expectation
max, min, sup, inf	maximum, minimum, supremum, and infimum of indicated quantities
aff	affine hull of set (Definition 2.9 of Chapter 1)
arg	argument for which the indicated condition is satisfied
cl	closure of set
cond	condition number of a matrix (cond_F is the condition number in the Frobenius norm)
conv	convex hull of set (Definition 2.1 of Chapter 1)
det	determinant of a matrix
diag	diagonal matrix formed from indicated elements
dim	dimension (of a set of vectors)
dom	essential domain of argument function (Definition 4.1 of Chapter 1)
epi	epigraph of argument (Definition 4.2 of Chapter 1)
int	interior of set
lim	limit of indicated sequence
med	median of indicated set
rank	rank of a matrix
ri	relative interior of set (Definition 2.10 of Chapter 1)
sgn	sign of indicated quantity
span	linear hull of the elements of S
trace	trace of matrix
LP1	basic form of linear programming problem (Section 2.1)
LP2	program dual to LP1
IP	interval program (Section 2.7)
DIP	dual program to IP
RIP	restricted interval program
QP	quadratic programming problem (Section 5.3)
EQP	equality constrained QP
CPP	convex programming problem (Section 1.5)
USE	unnormalized steepest edge test (Definition 4.1 of Chapter 2)
NSE	normalized steepest edge test (Definition 4.1 of Chapter 2)
RND	uniform random number generators (Section 7.4)
FNRND	another random number generator
IRLS	iteratively reweighted least squares (Section 5.4)

CHAPTER 1

Some results from convex analysis

1.1 INTRODUCTION

This chapter summarizes primarily material relating to convex optimization problems in order to provide a suitable underpinning for subsequent algorithmic developments. The problem domain is one in which convexity is important (for example there is a sense in which linear programming is our archetypal problem), and which can be characterized formally as

$$\min_{\mathbf{x} \in S} f(\mathbf{x}) \tag{1.1}$$

where S is a prescribed convex set (the constraint set), and f is convex on $S \subseteq R^p$.

To motivate the kind of results to be considered, note that the structure of S is clearly of relevance, and this leads directly to consideration of representation theorems for convex sets. These results have a direct and elegant application to linear programming problems. The second class of results concerns separation theorems and their direct relation to the characterization of optimizing points. To see this, note that if the minimizing value in (1.1) is $f = \bar{f}$ then

$$S \cap \{\mathbf{x}; f(\mathbf{x}) < \bar{f}\} = \emptyset. \tag{1.2}$$

As both sets involved are convex there exists a separating hyperplane, and from the equation of this hyperplane we can deduce necessary conditions for an optimum (Kuhn–Tucker conditions). An interesting feature of this approach to characterizing optima is that it does not require f to be differentiable.

If $\mathbf{x} \in S$, $f(\mathbf{x}) > \bar{f}$ then there exists a direction $\mathbf{t}$ at $\mathbf{x}$ such that $\mathbf{x} + \gamma \mathbf{t} \in S$ provided $\gamma > 0$ is small enough, and such that f decreases in the direction $\mathbf{t}$. Such a direction is called a *direction of descent* for f at $\mathbf{x}$. Theorems of the alternative arise naturally in this context as either a downhill direction exists or $\mathbf{x}$ is an optimum. Convex functions need not be differentiable (an example is $|x|$ at $x = 0$), but a generalized set-valued derivative called the subdifferential can be defined if the epigraph of f, epi f, the convex set lying above the graph of f in R^{p+1}, possesses a nonvertical supporting hyperplane at $\mathbf{x}$. The subdifferential is important in characterizing descent directions and in developing multiplier relations giving necessary conditions for $\mathbf{x}$ to be an optimum. The related concept of the directional derivative of a convex function will prove to be an important tool in analysing the structure of polyhedral convex functions. Also, we note that considering the supporting hyperplanes to epi f leads naturally to the conjugate convex function $f^*(\mathbf{u})$ and to the idea of duality which greatly enhances the structural richness of the problem setting. Finally, a brief description of descent methods for minimizing a convex function is given.

1.2 CONVEX SETS

In this section necessary material on the representation of convex sets and on separation theorems is developed. The key result on the existence of a hyperplane separating two disjoint convex sets is used extensively in subsequent sections. It has direct application to the development of alternative theorems and of necessary conditions for solutions of optimization problems. The representation theorem states that a convex set, under certain mild conditions, can be described completely in terms of quantities which have a natural geometric importance (extreme points and directions of recession). In the linear programming context this theorem has the direct interpretation that the optimum must be obtained at a vertex of the feasible region (an extreme point of this convex set), and that the problem can have a bounded solution only if the directions of recession bear a particular relationship to the objective function. Linear programming (discussed in Chapter 2) and the problems discussed in Chapters 3 and 4 are all particular cases of problems involving polyhedral convex functions. Polyhedral convexity occurs when there are only a finite number of extreme points and directions of recession, and the opportunity is taken here to develop some of the basic ideas.

Definition 2.1 The set $S \subseteq R^p$ is *convex* if

$$\mathbf{x}, \mathbf{y} \in S \Rightarrow \theta\mathbf{x} + (1-\theta)\mathbf{y} \in S \quad \text{for} \quad 0 \leq \theta \leq 1. \tag{2.1}$$

Equivalently, S is convex if all finite convex combinations of points in S is again in S. That is,

$$\mathbf{x}_i \in S, i = 1, 2, \ldots, m, \quad \sum_{i=1}^{m} \lambda_i = 1, \quad \lambda_i \geq 0, 1 \leq m < \infty \Rightarrow \sum_{i=1}^{m} \lambda_i \mathbf{x}_i \in S \tag{2.2}$$

Association with any set S is the set obtained by taking all convex combinations of points of S in the sense expressed by (2.2). This set is called the *convex hull* of S and is written conv S.

Example 2.1 Any function $\|\cdot\|$ taking bounded values for finite $\mathbf{x}$ is a *norm* if it satisfies the conditions:

(i) $\|\mathbf{x}\|>0, \quad \mathbf{x}\neq 0,$
(ii) $\|\mathbf{x}+\mathbf{y}\|\leqslant\|\mathbf{x}\|+\|\mathbf{y}\|,$ the triangle inequality, and
(iii) $\|\sigma\mathbf{x}\|=|\sigma|\,\|\mathbf{x}\|$.

Define $S=\{\mathbf{x};\|\mathbf{x}\|\leqslant 1\}$. Then it follows from (ii), (iii) that S is convex. Note that S is *balanced* ($\mathbf{x}\in S\Rightarrow -\mathbf{x}\in S$), and has a proper interior as $\theta\mathbf{x}/\|\mathbf{x}\|\in S$, $-1\leqslant\theta\leqslant 1$ for any $\mathbf{x}$. It is instructive to sketch S when $p=2$ for the particular cases

$$\|\mathbf{x}\|=\max|x_i|, \quad \left\{\sum_{i=1}^{2} x_i^2\right\}^{1/2}, \quad \sum_{i=1}^{2}|x_i|,$$

corresponding to the maximum, Euclidean, and l_1 norms respectively.

Remark 2.1 Alternatively, given S satisfying the above requirements, a norm can be defined by

$$\|\mathbf{x}\|_S=\inf\lambda, \qquad \mathbf{x}\in\lambda S. \tag{2.3}$$

If S is not balanced then the resulting function does not satisfy (iii) but is still convex. It is called a gauge function.

Definition 2.2 A *hyperplane* is the set of points

$$H(\mathbf{u},\nu)=\{\mathbf{x};\mathbf{u}^{\mathrm{T}}\mathbf{x}=\nu\}. \tag{2.4}$$

It should be noted that $H(\mathbf{u},\nu)$ *separates* R^p into two distinct half-spaces

$$H^{+}(\mathbf{u},\nu)=\{\mathbf{x};\mathbf{u}^{\mathrm{T}}\mathbf{x}>\nu\}, \tag{2.4a}$$

and

$$H^{-}(\mathbf{u},\nu)=\{\mathbf{x};\mathbf{u}^{\mathrm{T}}\mathbf{x}\leqslant\nu\}. \tag{2.4b}$$

Lemma 2.1 (*Lemma of the separating hyperplane – simplest case*). *Let S be a closed convex set in R^p and $\mathbf{x}_0$ a point not in S. Then there exists a hyperplane H separating $\mathbf{x}_0$ and S in the sense that $S\subset H^{+}$, $\mathbf{x}_0\in H^{-}$.*

Proof Let $\mathbf{x}_1$ be any point in S. Then in the Euclidean vector norm

$$\inf_{\mathbf{x}\in S}\|\mathbf{x}-\mathbf{x}_0\|_2^2\leqslant\|\mathbf{x}_1-\mathbf{x}_0\|_2^2=r^2$$

The function $\|\mathbf{x}-\mathbf{x}_0\|_2^2$ is continuous on the closed set $S\cap\{\mathbf{x};\|\mathbf{x}-\mathbf{x}_0\|_2\leqslant r\}$ so the minimum value for $\mathbf{x}\in S$ is attained for $\mathbf{x}=\mathbf{x}^*$ (say). Let $\mathbf{y}\in S$ and consider $\mathbf{x}^*+\gamma\mathbf{z}$ where $\mathbf{z}=\mathbf{y}-\mathbf{x}^*$ and $\gamma>0$. Then

$$\|\mathbf{x}^*+\gamma\mathbf{z}-\mathbf{x}_0\|_2^2=\|\mathbf{x}^*-\mathbf{x}_0\|_2^2+2\gamma\mathbf{z}^{\mathrm{T}}(\mathbf{x}^*-\mathbf{x}_0)+\gamma^2\|\mathbf{z}\|_2^2.$$

Letting $\gamma\to 0$ gives

$$\mathbf{z}^{\mathrm{T}}(\mathbf{x}^*-\mathbf{x}_0)\geqslant 0, \qquad \mathbf{y}\in S. \tag{2.5}$$

It follows that $H(\mathbf{x}^* - \mathbf{x}_0, \mathbf{x}^{*T}(\mathbf{x}^* - \mathbf{x}_0) - \varepsilon)$ is a suitable hyperplane for all $\varepsilon > 0$ small enough (ε must be chosen smaller than $\|\mathbf{x}_0 - \mathbf{x}^*\|_2^2$).

Remark 2.2 Separation in this form is called *strong* separation, and there is even a small enough ball about $\mathbf{x}^*$ which excludes $\mathbf{x}_0$. A more subtle result is also important. It can be proved by induction in finite-dimensional spaces, but requires the Hahn–Banach theorem in more general situations.

Theorem 2.1 *Let S, T be convex sets, and $S \cap T = \varnothing$. Then there exists a hyperplane such that $S \subset H^+$, $T \subset H^-$. Equivalently we can find $\mathbf{u}$ such that*

$$\inf_{\mathbf{x} \in S} \mathbf{u}^T\mathbf{x} \geqslant \sup_{\mathbf{x} \in T} \mathbf{u}^T\mathbf{x} \tag{2.6}$$

Remark 2.3 An important application corresponds to S an *open* convex set, and $\mathbf{x}_0 \in \text{cl } S \backslash S$. The theorem shows there is a hyperplane through $\mathbf{x}_0$ containing S in H^+. It follows that H contains only boundary points of S.

Definition 2.3 The hyperplane H in Remark 2.3 *supports* S at $\mathbf{x}_0$. Note that there exists a supporting hyperplane at every finite boundary point of S.

Definition 2.4 The function

$$\delta^*(\mathbf{u} \mid S) = \sup_{\mathbf{x} \in S} \mathbf{u}^T\mathbf{x} \tag{2.7}$$

is called the *support* function for S.

If the supremum in (2.7) is attained at $\mathbf{x}_0$, then $\mathbf{x}_0$ is a boundary point of S, $H(\mathbf{u}, \mathbf{u}^T\mathbf{x}_0)$ supports S at $\mathbf{x}_0$, and $S \subseteq H^-$. For compatibility with the support function, unless otherwise indicated, the convention will be followed that if H supports S then $S \subseteq H^-$.

Definition 2.5 The point $\mathbf{x}_0$ is an *extreme point* of S if and only if it cannot be expressed as a point properly in the interior of the line segment joining two distinct points of cl S. An extreme point which has the property that there exists a hyperplane supporting S at $\mathbf{x}_0$ such that

$$H \cap \text{cl } S = \{\mathbf{x}_0\} \tag{2.8}$$

is called an *exposed point.* Exposed points are equivalent to extreme points if the set of extreme points is finite.

Example 2.2 Consider $S = \{\mathbf{x};\ \|\mathbf{x}\| \leqslant 1\}$.

(a) If $\|\mathbf{x}\| = \left\{\sum_{i=1}^{p} x_i^2\right\}^{1/2}$ then every point in cl $S \backslash \text{int } S$ is an extreme point.

(b) If $\|\mathbf{x}\| = \max_{1 \leqslant i \leqslant p} |x_i|$ then the extreme points of S have the form

$$\mathbf{x} = \sum_{i=1}^{p} \theta_i \mathbf{e}_i, \qquad \theta_i = \pm 1, \ i = 1, 2, \ldots, p.$$

(c) If $\|\mathbf{x}\| = \sum_{i=1}^{p} |x_i|$ then the extreme points of S have the form

$$\mathbf{x} = \pm \mathbf{e}_1, \pm \mathbf{e}_2, \ldots, \pm \mathbf{e}_p.$$

In each case note that any $\mathbf{x} \in S$ can be written as a convex combination of extreme points. These are particular examples of a general result which will now be developed.

Lemma 2.2 *Let H support S at $\mathbf{x}$. If $\mathbf{y}$ is an extreme point of $H \cap S$ then $\mathbf{y}$ is an extreme point of S.*

Proof If $H \cap S = \{\mathbf{y}\}$ then $\mathbf{y}$ is an extreme point of S. Thus we assume that $H \cap S = T$, not a singleton, and that $\mathbf{y}$ is an extreme point of T but not of S. Then we can find $\mathbf{x}, \mathbf{z}$ in S but not in T such that

$$\mathbf{y} = \theta \mathbf{x} + (1 - \theta)\mathbf{z}, \qquad 0 < \theta < 1.$$

Now $\mathbf{y} \in H$, but $\mathbf{x}, \mathbf{z} \in \operatorname{int} H^-$. Thus

$$0 = \mathbf{u}^T\mathbf{y} - \nu = \theta(\mathbf{u}^T\mathbf{x} - \nu) + (1 - \theta)(\mathbf{u}^T\mathbf{z} - \nu) < 0$$

This gives a contradiction.

Lemma 2.3 *Let S be a closed, bounded, convex set. Then S has extreme points.*

Proof This is by induction with respect to dimension. If S is a singleton then the result is immediate. Now assume $S \subset R^p$, $\mathbf{x}_0$ is a boundary point of S, and H is a hyperplane supporting S at $\mathbf{x}_0$. It follows that $H \cap S$ is bounded, so that by the induction hypothesis it has extreme points. But then, by Lemma 2.2, these points are extreme points of S.

The representation theorem for bounded sets can now be given.

Theorem 2.2 *A closed bounded convex set S in R^p is the closed convex hull of its extreme points.*

Proof Let $\hat{S}$ be the closed convex hull of the extreme points of S. Then $\hat{S} \subseteq S$. Assume that $\hat{S} \subset S$ properly so that there exists a point $\mathbf{x} \in S$ strongly separated from $\hat{S}$. The separating hyperplane theorem now gives $H(\mathbf{u}, \nu)$

such that $\hat{S} \subset H^-, \mathbf{x} \in H^+$. Consider

$$\nu_0 = \delta^*(\mathbf{u} \mid S) > \nu$$

It follows from the definition of the support function that $H(\mathbf{u}, \nu_0)$ supports S. Also $T = H(\mathbf{u}, \nu_0) \cap S$ is closed and bounded. Thus, by Lemma 2.3, it has an extreme point. By Lemma 2.2 this is also an extreme point of S. But by construction it is strongly separated from $\hat{S}$, giving a contradiction.

To extend this result to unbounded sets a further concept is needed in order to describe the property that points can be arbitrarily far apart.

Definition 2.6 Let S be a convex set, and let $\mathbf{t}$ have the property that

$$\mathbf{x} \in S \Rightarrow \mathbf{x} + \lambda \mathbf{t} \in S, \qquad \lambda \geqslant 0 \tag{2.9}$$

then $\mathbf{t}$ is a *direction of recession* for S. It is an *extreme direction* if it cannot be represented as a convex combination of other directions (so that there do not exist directions $\mathbf{t}_i$ and constants $\lambda_i \geqslant 0$, $i = 1, 2, \ldots, k$ such that $\mathbf{t} = \sum_{i=1}^{k} \lambda_i \mathbf{t}_i$ for any finite k). S is said to *contain a line* if both $\mathbf{t}$ and $-\mathbf{t}$ are directions of recession.

It is not difficult to give examples of important convex sets containing directions of recession.

Definition 2.7 Let $\mathbf{b} \in R^m$, $A : R^p \rightarrow R^m$, then

$$M = \{\mathbf{x};\ A\mathbf{x} = \mathbf{b}\} \tag{2.10}$$

is an *affine set* or *flat* in R^p.

Thus every hyperplane is an affine set, and every affine set is an intersection of hyperplanes. An affine set can be represented as a translated subspace

$$M = \mathbf{x}_0 + L \tag{2.11}$$

where $\mathbf{x}_0$ is any point such that

$$A\mathbf{x}_0 = \mathbf{b} \tag{2.12}$$

and

$$L = \{\mathbf{y};\ A\mathbf{y} = 0\} \tag{2.13}$$

is a linear space. The dimension of M is the dimension of $L = p - \operatorname{rank}(A)$.

Remark 2.4 A form of the separation theorem that will be required is that if M is an affine set, S convex, and $S \cap M = \varnothing$, then there exists a separating hyperplane containing M.

Example 2.3 (a) Any element of $L \neq 0$ defines a direction of recession for M. Any set of the form $S+L$ where S is convex and L a subspace has no extreme points. (b) A convex set $C \subseteq R^p$ is a *cone* pointed at $\mathbf{x}_0$ if

$$\mathbf{y} \in C \Rightarrow \mathbf{x}_0 + \lambda(\mathbf{y}-\mathbf{x}_0) \in C \qquad \forall \lambda \geq 0. \tag{2.14}$$

Thus $\mathbf{y}-\mathbf{x}_0$ is a direction of recession for $\mathbf{y}$ in C, $\mathbf{y} \neq \mathbf{x}_0$, and $\mathbf{x}_0$ is the only point in C that can be an extreme point. In particular, it is contained in every supporting hyperplane to C (either $\delta^*(\mathbf{u} \mid C) = \mathbf{u}^T\mathbf{x}_0$ if $\mathbf{u}^T(\mathbf{y}-\mathbf{x}_0) \leq 0$, $\forall \mathbf{y} \in C$, or $\delta^*(\mathbf{u} \mid C) = +\infty$). An important example is $R_p^+ = \{\mathbf{x};\, x_i \geq 0\}$ which is a cone pointed at 0.

We now state the extended representation theorem. The standard proofs of this result use induction explicitly with respect to the dimension (in the proof of Theorem 2.2 this was hidden in the appeal to Lemma 2.3). This extended result is particularly important in providing a theoretical basis for linear programming.

Theorem (*Klee*) *A closed convex set in R^p containing no lines is the convex hull of its extreme points and directions of recession. That is, given $\mathbf{x} \in S$, then there exist extreme points of S, $\mathbf{s}_1, \mathbf{s}_2, \ldots, \mathbf{s}_k$, and extreme directions of recession, $\mathbf{t}_1, \mathbf{t}_2, \ldots, \mathbf{t}_l$, such that*

$$\mathbf{x} = \sum_{i=1}^{k} \theta_i \mathbf{s}_i + \sum_{i=1}^{l} \lambda_i \mathbf{t}_i \tag{2.15}$$

where

$$\sum_{i=1}^{k} \theta_i = 1, \quad \theta_i \geq 0, i = 1, 2, \ldots, k, \quad \lambda_i \geq 0, i = 1, 2, \ldots, l.$$

Definition 2.8 If the set of extreme points and extreme directions is finite then S is *polyhedral*. If S is bounded then it is a *convex polyhedron* or a bounded *convex polytope*.

A convex polyhedron with a proper interior possesses a representation as the intersection of a finite number of closed half-spaces. These are determined by the subsets of $\geq p$ extreme points which determine a hyperplane such that the convex polyhedron is contained in one or the other of the closed half-spaces so generated.

A convex set is a *convex body* if it has an interior point. If it does not contain interior points then it is contained in some affine set.

Definition 2.9 The intersection of all affine sets containing S is called the *affine hull* of S and is denoted aff S. It is the smallest affine set containing S.

Definition 2.10 The *relative interior* of the convex set S (ri S) is

$$\text{ri}\, S = \{\mathbf{x};\, \mathbf{x} \in \text{aff}\, S, \exists \varepsilon > 0 \ni \varepsilon B(\mathbf{x}) \cap \text{aff}\, S \subset S\} \tag{2.16}$$

where $B(\mathbf{x})$ is the unit ball centered at $\mathbf{x}-B(\mathbf{x})=\{\mathbf{y};\ \|\mathbf{y}-\mathbf{x}\|<1\}$. This permits us to talk about ri S when S is (say) a line segment in R^p. Such a set has no proper interior in the usual sense when $p>1$. The *relative boundary* of S is the set of points in $S\backslash\text{ri}\, S$. If $S=\text{ri}\, S$ then S is *relatively open.*

In Figure 2.1 the relations between various possible convex sets are developed.

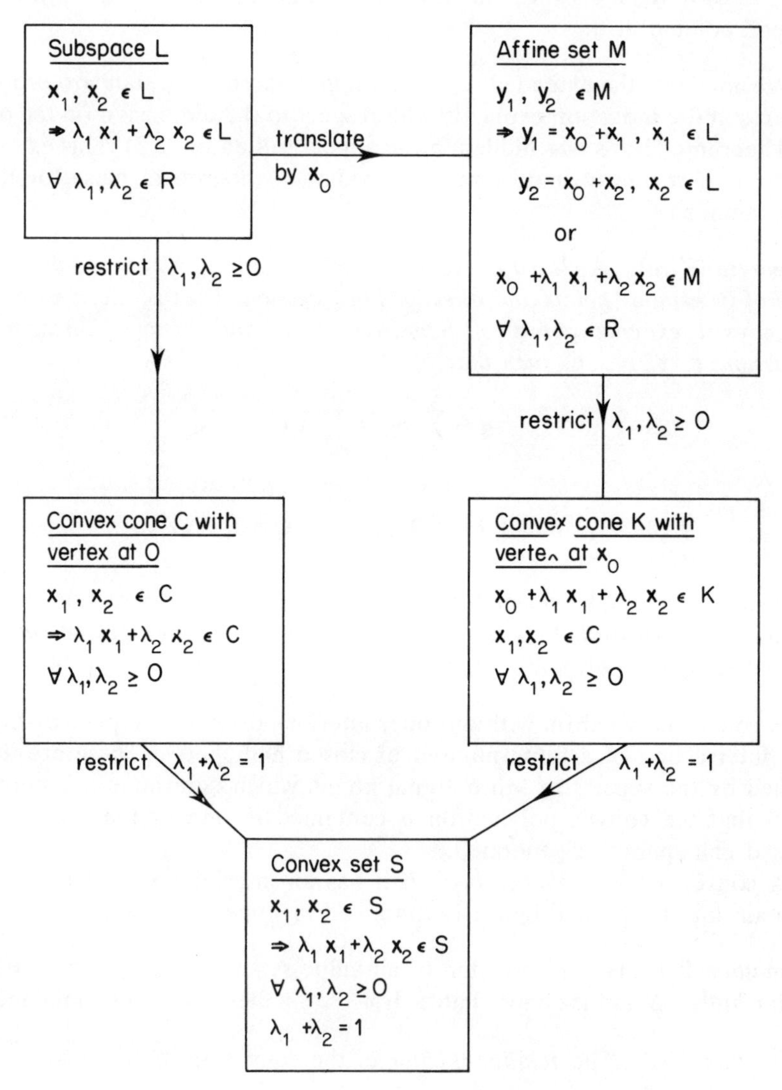

Figure 2.1 Relationships between sets, cones, and subspaces in R^p

Exercise 2.1

(i) A closed convex set is the intersection of the closed half-spaces containing it.

(ii) Let C_1, C_2 be convex sets, then

$$\text{cl}\, C_1 \subseteq \text{cl}\, C_2 \quad \text{if and only if} \quad \delta^*(\mathbf{u} \mid C_1) \leqslant \delta^*(\mathbf{u} \mid C_2).$$

(iii) Let S be the set determined by the linear inequalities

$$x + y \geqslant 1,$$
$$10y - x \geqslant -1$$
$$-y + 10x \geqslant -1$$

Determine its extreme points and directions of recession. Also determine the hyperplanes supporting S at each extreme point.

1.3 THEOREMS OF THE ALTERNATIVE

Alternative theorems are important in function minimization as they permit the formalization of the alternatives that either the current point is optimal or there exists a direction in which the function can be reduced.

Separation theorems are the main analytic device used in this section. This is well illustrated by the first result which is perhaps the simplest form of alternative theorem and gives a criterion for the consistency of a system of linear inequalities.

Theorem 3.1 *Let S be a closed convex set. Then either the system of inequalities*

$$\mathbf{u}^T\mathbf{t} < 0, \qquad \mathbf{t} \in S, \tag{3.1}$$

is consistent or

$$0 \in S.$$

Proof The inequality (3.1) states that $\delta^*(\mathbf{u} \mid S) < 0$ so that 0 is strongly separated from S.

Remark 3.1 This result has meaning when S is not convex. Either (3.1) holds or $0 \in \text{conv}\, S$.

Perhaps the most celebrated theorem of the alternative is known as Farkas' lemma, and it plays a key role in developing multiplier conditions characterizing optimality in mathematical programming problems. Here the separation theorem is applied to the particular case in which S is the cone generated by convex combinations of the rows of a matrix A. The resulting cone is polyhedral as the set of extreme directions is clearly no larger than the row dimension of A. Such a cone is also called *finitely generated.* A key preliminary result is that a finitely generated cone is closed.

Lemma 3.1 *Let the cone C be finitely generated by $\mathbf{a}_1, \mathbf{a}_2, \ldots, \mathbf{a}_m$. That is,*

$$C = \left\{\mathbf{x}; \exists \lambda_i \geq 0,\ i = 1, 2, \ldots, m \ni \mathbf{x} = \sum_{i=1}^{m} \lambda_i \mathbf{a}_i\right\} \tag{3.2}$$

Then C is closed.

Proof C is convex by definition. However, the closure result is complicated if C contains lines. To exclude this case let L be the linear space generated by the lines in C, and introduce coordinates into L, $L^\perp$ such that

$$C = L \times L^\perp \cap C \tag{3.3}$$

As L is closed the result entails showing that the cone $L^\perp \cap C$ (which by construction does not contain lines) is closed. Thus there is no restriction in considering only this case ($L^\perp \cap C$ is clearly finitely generated).

If C does not contain lines then

$$\sum_{i=1}^{m} \lambda_i \mathbf{a}_i = 0 \Rightarrow \lambda_i = 0, \qquad i = 1, 2, \ldots, m$$

for if this is not the case then $\lambda_k > 0$ for some k so that both $\mathbf{a}_k$ and

$$-\mathbf{a}_k = \sum_{i \neq k} \frac{\lambda_i}{\lambda_k} \mathbf{a}_i$$

are in C, showing that C contains a line. It follows that $\mathbf{x}$ is bounded if and only if the λ_i in every representation of $\mathbf{x}$ are bounded, and closure is an immediate consequence.

Theorem 3.2 (*Farkas' lemma*) *Let $A : R^p \to R^n$, $\mathbf{a}$ be such that*

$$A\mathbf{x} \geq 0 \Rightarrow \mathbf{a}^T\mathbf{x} \geq 0. \tag{3.4}$$

Then there exists $\mathbf{y} \geq 0 \in R^n$ ($\mathbf{y} \in R_n^+$) such that

$$\mathbf{a}^T = \mathbf{y}^T A. \tag{3.5}$$

Proof This proceeds by constructing the cone

$$C = \left\{\mathbf{z};\ \mathbf{z} = \sum_{i=1}^{n} \lambda_i \mathbf{a}_i,\ \lambda_i \geq 0,\ \mathbf{a}_i = \boldsymbol{\rho}_i(A),\ i = 1, 2, \ldots, n\right\} \tag{3.6}$$

and assumes that $\mathbf{a} \notin C$. The separating hyperplane theorem is then used to give a contradiction, for by Lemma 3.1 C is closed so that $\mathbf{a}$ must be strongly separated from C. Choose the separating hyperplane to support C

so that it passes through $\mathbf{z}=0$. Then

$$C\subset H^{-}(\mathbf{u},0),\qquad \mathbf{a}\in H^{+}$$

But then choosing $\mathbf{x}=-\mathbf{u}$ in (3.4) gives a contradiction.

Remark 3.2 The idea of the alternative may be clearer if the result is put slightly differently. It says that exactly one of the pairs of systems

(i) $$A^{\mathrm{T}}\mathbf{y}=\mathbf{b},\qquad \mathbf{y}\geqslant 0,$$

and

(i) $$A\mathbf{x}\geqslant 0,\qquad \mathbf{b}^{\mathrm{T}}\mathbf{x}<0,$$

can have a solution.

An alternative route to this result makes use of the *polar* to C. This is the cone

$$C^{*}=\{\mathbf{y};\,\mathbf{y}^{\mathrm{T}}\mathbf{x}\leqslant 0,\,\forall\,\mathbf{x}\in C\}\tag{3.7}$$

The chief properties of the polar cone are summarized below.

(a) C^{*} is a closed convex cone.
(b) If $C_1\subseteq C_2$ then $C_2^{*}\subseteq C_1^{*}$.
(c) $C^{**}=C$ if and only if C is a closed convex cone.
(d) $C^{*}=(\mathrm{cl\ conv}\ C)^{*}$, the polar cone of the closure of the convex hull of C.
(e) If A is a linear space then $A^{\perp}=A^{*}$.
(f) If C_1 and C_2 are convex cones then

$$C_1^{*}\cap C_2^{*}=(C_1+C_2)^{*}\tag{3.8}$$

Farkas lemma follows from (3.8) for if C_1 is given by (3.6), and C_2 is generated by $\mathbf{a}$, then (3.4) gives

$$C_1^{*}\subseteq C_2^{*}$$

so that

$$C_1^{*}=C_1^{*}\cap C_2^{*}=(C_1+C_2)^{*}$$

and the result follows by taking polars and using (c).

Theorem 3.2 can be generalized substantially to partial orders defined on functions taking values in cones ($a \underset{C}{\geqslant} b$ if $a-b\in C$), to operators on normed linear spaces, and to convex (not only linear) operators. The following result is typical.

Theorem 3.3 *Let C be a convex cone in R^n, $S\subset R^p$ convex, and $A:R^p\to R^n$. Then exactly one of the systems*

(i) $$A\mathbf{x}\in\mathrm{int}\ C,\qquad \mathbf{x}\in S,\tag{3.9}$$

and

(ii) $$(\mathbf{v}^{\mathrm{T}}A)S \subset R_1^+, \qquad 0 \neq \mathbf{v} \in C^* \tag{3.10}$$

has a solution.

Proof It is clear that (i) and (ii) cannot both have a solution. If (i) has no solution then AS and int C are disjoint. But then the separating hyperplane theorem guarantees an $H(\mathbf{u}, 0)$ such that $AS \subset H^-$, int $C \subset H^+$. But this implies that

$$(\mathbf{u}^{\mathrm{T}}A)S \leqslant 0, \mathbf{u}^{\mathrm{T}}\mathbf{w} > 0, \forall\, \mathbf{w} \in \text{int } C$$

so that $\mathbf{u} \in C^*$ and $\mathbf{u}$ solves (ii). This completes the proof of the theorem as either (i) has a solution or not, and if it does not then (ii) has a solution by the above argument.

Exercise 3.1

(i) Prove the properties (a)–(f) for polar cones.
(ii) Restate Farkas' lemma for cone-valued operators.
(iii) Prove Motzkin's theorem: Let $A: R^p \to R^n$, $B: R^p \to R^m$, $T \subset R^n$ a closed convex cone, $S \subset R^m$ a convex cone with int $S \neq \varnothing$. Then exactly one of the following systems has a solution

(a) $$-A\mathbf{x} \in T, \qquad -B\mathbf{x} \in \text{int } S \tag{3.11}$$

(b) $$\mathbf{v}^{\mathrm{T}}B + \mathbf{w}^{\mathrm{T}}A = 0, \qquad \mathbf{w} \in T^*, \qquad 0 \neq \mathbf{v} \in S^*. \tag{3.12}$$

(iv) Deduce Farkas' lemma as a special case of Motzkin's theorem.
(v) Show that either

$$\exists\, \mathbf{x} \ni A\mathbf{x} = \mathbf{c},$$

or

$$\exists\, \mathbf{v} \ni A^{\mathrm{T}}\mathbf{v} = 0, \qquad \mathbf{v}^{\mathrm{T}}\mathbf{c} = 1.$$

Exercise 3.2 Let $A: R^p \to R^n$ and $X = \{\mathbf{x}; A\mathbf{x} \geqslant \mathbf{b}\}$. The constraint $\boldsymbol{\rho}_i(A)\mathbf{x} \geqslant b_i$ is *redundant* if X is unchanged by deletion of this inequality.

(i) What is the maximum possible value of dim $(X \cap \{\boldsymbol{\rho}_i(A)\mathbf{x} = b_i\})$.
(ii) Use Farkas lemma to determine conditions for a constraint to be redundant.

Distinguish between the cases that the constraint hyperplane contains or does not contain points of the feasible region X.

1.4 CONVEX FUNCTIONS

All the problems considered in Chapters 2 to 5 can be reduced to that of minimizing particular convex functions (frequently even polyhedral convex functions). Convexity is a strong assumption (for example, points at which a convex function is not continuous are unusual). However, convex functions are not necessarily differentiable; but if a nonvertical supporting hyperplane can be constructed at the point $\mathbf{x}$ to the convex set lying above the graph of f then the normals to the set of possible hyperplanes at $\mathbf{x}$ have certain of the properties of the gradient vector of a smooth (convex) function. Suitably normalized, this set is called the subdifferential, and if it is nonempty then the function is said to be stable at $\mathbf{x}$. The properties of the subdifferential and, in particular, its relationship to the directional derivative of f will be important subsequently. One important application is made in Chapter 4 where a general approach is developed to minimizing polyhedral convex functions based on explicit knowledge of a compact parametrization of the subdifferential.

In this section basic properties of convex functions are summarized.

Definition 4.1 Let X be a convex set in R^p. Then $f: R^p \to R$ is *convex* on X if

$$f(\theta\mathbf{x}+(1-\theta)\mathbf{y}) \leqslant \theta f(\mathbf{x})+(1-\theta)f(\mathbf{y}), \qquad \mathbf{x}, \mathbf{y} \in X, \qquad 0 \leqslant \theta \leqslant 1. \quad (4.1)$$

If strict inequality holds for $0<\theta<1$ and $\mathbf{x} \neq \mathbf{y}$, then f is *strictly convex.*

It is convenient to make the assumption that f is bounded below on X except that f may tend to $-\infty$ if $\|\mathbf{x}\| \to \infty$ in case X has directions of recession (so that only regular convex functions in the sense of Rockafellar are considered). Also f can be extended to all R^p by setting $f=+\infty$, $\mathbf{x} \in X^c$, and in this case there is no lack of generality in assuming that

$$X = \{\mathbf{x};\ f(\mathbf{x}) < \infty\}$$

in which case X is called the *effective domain* of f and written dom f.

Definition 4.2 The set

$$\operatorname{epi} f = \left\{ \begin{bmatrix} \mathbf{x} \\ \mu \end{bmatrix};\ \mathbf{x} \in X,\ \mu \geqslant f(\mathbf{x}) \right\} \quad (4.2)$$

is called the *epigraph* of f. It is a convex set in R^{p+1} if and only if f is convex. The convex function f is *closed* if epi f is closed.

If epi f is closed then it is the intersection of the closed half-spaces containing it. It follows that the supremum of an affine family is a closed convex function, and any closed convex function can be represented as the supremum of an affine family

$$f = \sup_{i \in \alpha} \mathbf{a}_i^T \mathbf{x} - b_i \quad (4.3)$$

where α is a given index set. If $|\alpha|$ is finite then f is *polyhedral.*

Example 4.1 A convex function is not necessarily closed. Consider

$$f=0, \qquad 0 \leqslant x < 1,$$
$$f=1, \qquad x=1.$$

However, it is intuitively clear that (4.1) must make this kind of behaviour possible only at the boundary of dom f. In fact convexity forces quite strong regularity conditions in the interior of X.

Lemma 4.1 *Let $\delta > 0$ be sufficiently small for*

$$N = \delta B(\mathbf{x}) \cap \operatorname{dom} f \subset \operatorname{ri} \operatorname{dom} f$$

where $B(\mathbf{x})$ is the unit ball centred on $\mathbf{x}$. Then f is continuous on N.

Proof Note that f is bounded on N so that there exists K such that $|f(\mathbf{x})| \leqslant K$, $\mathbf{x} \in N$, by the definition of dom f. Now choose $\mathbf{t}$, $\|\mathbf{t}\| < \delta$, such that $\mathbf{x} \pm \mathbf{t} \in N$. Let $0 < \gamma < 1$. Then the convexity inequality gives

$$f(\mathbf{x}+\gamma\mathbf{t}) - f(\mathbf{x}) \leqslant \gamma\{f(\mathbf{x}+\mathbf{t}) - f(\mathbf{x})\}$$

and

$$f(\mathbf{x}+\gamma\mathbf{t}) - f(\mathbf{x}) \geqslant \gamma\{f(\mathbf{x}) - f(\mathbf{x}-\mathbf{t})\}$$

so that

$$|f(\mathbf{x}+\gamma\mathbf{t}) - f(\mathbf{x})| \leqslant 2\gamma K. \tag{4.4}$$

Remark 4.1 Continuity of f on dom f implies closure of epi f. However, a formally weaker condition suffices. It is readily verified that the following conditions are equivalent:

(a) f is lower semicontinuous for $\mathbf{x} \in \operatorname{dom} f$ so that

$$f(\mathbf{x}_0) = f\left(\lim_{i\to\infty} \mathbf{x}_i\right) \leqslant \liminf_{i\to\infty} f(\mathbf{x}_i) \tag{4.5}$$

(b) $\{\mathbf{x};\, f(\mathbf{x}) \leqslant \alpha\}$ is closed for all α.

(c) epi f is a closed set in R^{p+1}.

Example 4.2 Lower semicontinuity is actually a weaker condition. An example is given by

$$f = \begin{cases} x_2^2/x_1, & x_1 > 0 \\ 0, & x_1 = x_2 = 0 \\ +\infty & \text{otherwise} \end{cases}$$

Then f is convex, continuous for $x > 0$, lower semicontinuous at $(0, 0)$.

Definition 4.3 The convex function f is *stable* at $\mathbf{x}_0 \in \operatorname{dom} f$ if there exists K, $0 < K < \infty$, such that

$$f(\mathbf{x}_0) - f(\mathbf{x}) \leqslant K\, \|\mathbf{x} - \mathbf{x}_0\|_2. \tag{4.6}$$

Stability implies lower semicontinuity at $\mathbf{x}_0$. For let $\{\mathbf{x}_i\} \to \mathbf{x}_0$. Then (4.6) gives

$$f\left(\lim_{i\to\infty} \mathbf{x}_i\right) - \lim_{i\to\infty} f(\mathbf{x}_i) \leq 0.$$

Also the first of the inequalities leading to (4.4) gives

$$f(\mathbf{x}) - f(\mathbf{x} + \gamma \mathbf{t}) \geq \gamma\{f(\mathbf{x}) - f(\mathbf{x} + \mathbf{t})\}$$

showing that stability for γ small enough (that is for $\|\mathbf{x} - \mathbf{x}_0\|$ small enough in (4.6)) implies stability.

At points in ri dom f, stability is a consequence of (4.4) so that the interesting points are boundary points. In this case the stability constant gives information about the rate of change of f as $\mathbf{x}$ moves off the boundary of dom f. This is illustrated by the following examples which show also that continuity need not imply stability at boundary points of dom f.

Example 4.3 (i) Let $f = e^{-x}$, dom $f = 0 \leq x < \infty$. Then

$$|e^{-x_0} - e^{-x}| = |e^{-\xi}(x - x_0)| \leq \|x - x_0\|_2,$$

where ξ is a mean value between x_0 and x, so that f is stable on dom f with stability constant 1.

(ii) Let $f = -x^{1/2}$, dom $f = 0 \leq x < \infty$. It is a consequence of the inequalities

$$\frac{1}{2x^{1/2}}(x - x_0) \leq -x_0^{1/2} + x^{1/2} \leq \frac{1}{x^{1/2}}(x - x_0)$$

valid if $x_0 \leq x$, that f is not stable at the boundary point $x_0 = 0$.

Both examples are sketched in Figure 4.1. In the first case there is a nonvertical supporting hyperplane to epi f at

$$\begin{bmatrix} \mathbf{x}_0 \\ f(\mathbf{x}_0) \end{bmatrix} = \begin{bmatrix} 0 \\ 1 \end{bmatrix}$$

separating epi f and the inverted cone pointed at $\mathbf{x}_0$ having semivertical angle

$$\tan^{-1} \frac{\|\mathbf{x} - \mathbf{x}_0\|_2}{f(\mathbf{x}_0) - z} = \tan^{-1} \frac{1}{K} > 0. \tag{4.7}$$

In the second case the rate of decrease of f is infinite at $\mathbf{x}_0$ and this construction is not possible. The connection between stability and the existence of a nonvertical supporting hyperplane is fundamental. It is formalized in the following result.

Lemma 4.2 *Let f be convex and $\mathbf{x}_0 \in$ dom f. Then there exists a nonvertical supporting hyperplane to epi f at $\mathbf{x}_0$ if and only if f is stable at $\mathbf{x}_0$.*

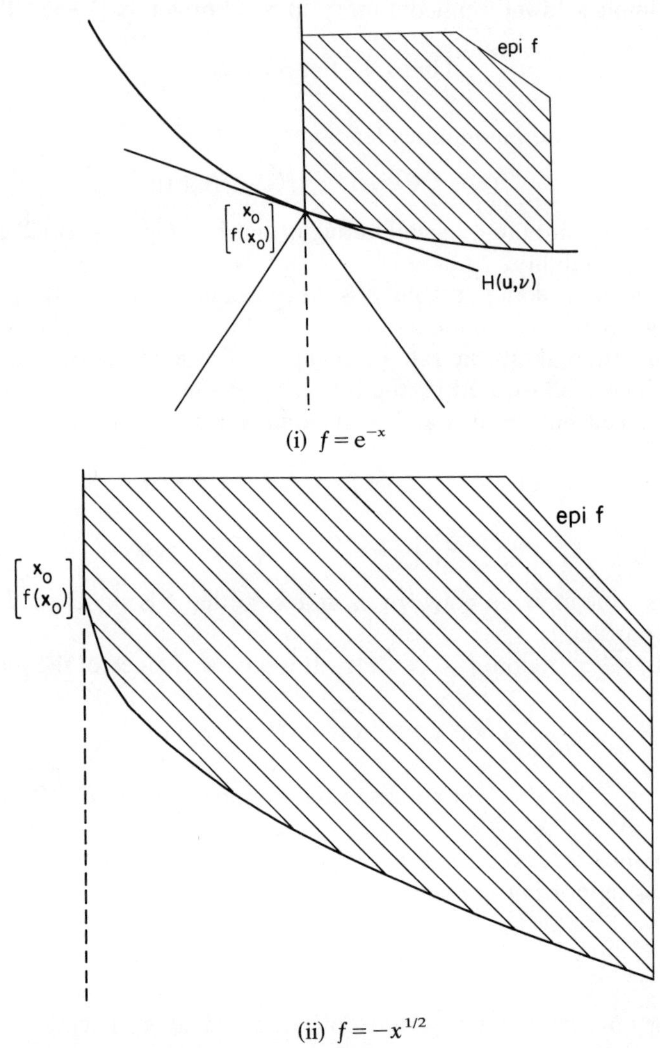

(i) $f = e^{-x}$

(ii) $f = -x^{1/2}$

Figure 4.1 The stability property

Proof If there exists a nonvertical supporting hyperplane at $\mathbf{x}_0$ then there exists $\mathbf{d}$, $\|\mathbf{d}\|_2 < \infty$, such that for all $\mathbf{x} \in \operatorname{dom} f$

$$\left\{\begin{bmatrix}\mathbf{x}\\ z\end{bmatrix} - \begin{bmatrix}\mathbf{x}_0\\ f(\mathbf{x}_0)\end{bmatrix}\right\}^{\mathrm{T}}\begin{bmatrix}-\mathbf{d}\\ 1\end{bmatrix} = 0 \Rightarrow f(\mathbf{x}) \geqslant z.$$

Thus

$$\begin{aligned} f(\mathbf{x}_0) &= z + \mathbf{d}^{\mathrm{T}}(\mathbf{x}_0 - \mathbf{x}) \\ &\leqslant f(\mathbf{x}) + |\mathbf{d}^{\mathrm{T}}(\mathbf{x}_0 - \mathbf{x})| \\ &\leqslant f(\mathbf{x}) + \|\mathbf{d}\|_2 \, \|\mathbf{x}_0 - \mathbf{x}\|_2. \end{aligned}$$

On the other hand, if f is stable at $\mathbf{x}_0$ then there exists $K<\infty$ such that

$$\operatorname{epi} f \cap \operatorname{int}\left\{\begin{bmatrix}\mathbf{x}\\ z\end{bmatrix}; f(\mathbf{x}_0)-z>K\,\|\mathbf{x}-\mathbf{x}_0\|_2\right\}=\varnothing$$

But this says that epi f is disjoint from an open cone pointed at $\mathbf{x}_0$ with semivertical angle $\tan^{-1} 1/K$. Thus there exists a separating hyperplane which must support epi f at $\mathbf{x}_0$ and which is necessarily nonvertical because the cone has finite angle (as, for example, in Figure 4.1(i)).

Definition 4.4 Let $\mathbf{x}\in\operatorname{dom} f$. Then $\mathbf{d}$ is a *subgradient* of f at $\mathbf{x}$ if and only if

$$f(\mathbf{y})\geqslant f(\mathbf{x})+\mathbf{d}^{\mathrm{T}}(\mathbf{y}-\mathbf{x}),\qquad \forall\,\mathbf{y}\in\operatorname{dom} f \tag{4.8}$$

(and hence for all $\mathbf{y}$ as $f(\mathbf{y})=+\infty$, $\mathbf{y}\notin\operatorname{dom} f$ by assumption). The set of all subgradients at $\mathbf{x}$ form the *subdifferential* $\partial f(\mathbf{x})$ at $\mathbf{x}$.

The subdifferential is convex and closed. If it is a singleton then f is differentiable at $\mathbf{x}$ and $\partial f(\mathbf{x})=\nabla f(\mathbf{x})$. By convention the elements of the subdifferential are row vectors.

Remark 4.2 By (4.8), $\partial f(\mathbf{x})\neq\varnothing$ if and only if there exists a nonvertical supporting hyperplane to epi f at

$$\begin{bmatrix}\mathbf{x}\\ f(\mathbf{x})\end{bmatrix},$$

and each element of $\partial f(\mathbf{x})$ is associated with exactly one such support plane – in fact with the normalization

$$\begin{bmatrix}-\mathbf{d}\\ 1\end{bmatrix}$$

in the equation of the supporting hyperplane.

Combining this observation with Lemma 4.2 gives:

Lemma 4.3 $\partial f(\mathbf{x})\neq\varnothing$ *if and only if f is stable at* $\mathbf{x}$.

Remark 4.3 A convex function has a subgradient at each point of ri dom f. An example in which dom f is not p-dimensional is

$$f(x,y)=\begin{cases}x^2, & y=0\\ +\infty & y\neq 0\end{cases}$$

when

$$\partial f(x,0)=\{[2x,\gamma],\forall\,\gamma\}.$$

The stability cone is given by its restriction to aff epi f but the subgradient inequality (4.8) holds for all

$$\begin{bmatrix} x^* \\ y^* \end{bmatrix}$$

as a consequence of the convention that $f = +\infty$, $\mathbf{x} \notin \operatorname{dom} f$.

Lemma 4.4 *If* $\mathbf{x} \in \operatorname{int} \operatorname{dom} f$ *then* $\partial f(\mathbf{x})$ *is a bounded set.*

Proof Assume the contrary. Then $\partial f(\mathbf{x})$ has a direction of recession $\mathbf{t}$. Let $\mathbf{u}^{\mathrm{T}} \in \partial f(\mathbf{x})$. The subgraph inequality gives

$$f(\mathbf{x} + \varepsilon \mathbf{z}) \geqslant f(\mathbf{x}) + (\mathbf{u} + \lambda \mathbf{t})^{\mathrm{T}} \varepsilon \mathbf{z}$$

and the right-hand side can only be bounded by $+\infty$ for all $\varepsilon > 0$ for any $\mathbf{z}$ such that $\mathbf{z}^{\mathrm{T}}\mathbf{t} > 0$ as λ can be chosen arbitrarily large. But if $\mathbf{x} \in \operatorname{int} \operatorname{dom} f$, then any direction $\mathbf{z}$ is allowed and f is bounded on $\varepsilon B(\mathbf{x})$ for some $\varepsilon > 0$, and this gives a contradiction.

Corollary 4.1 *If* $\partial f(\mathbf{x})$ *has a direction of recession then* $\mathbf{x}$ *is a boundary point of* $\operatorname{dom} f$.

Lemma 4.5 *Consider a sequence* $\{\mathbf{x}_i\} \to \mathbf{x}$ *such that* $\partial f(\mathbf{x}_i)$, *for each i, and* $\partial f(\mathbf{x}) \neq \emptyset$. *Select a corresponding (sub)sequence* $\{\mathbf{v}_i\} \to \mathbf{v}$ *where* $\mathbf{v}_i^{\mathrm{T}} \in \partial f(\mathbf{x}_i)$. *Then* $\mathbf{v}^{\mathrm{T}} \in \partial f(\mathbf{x})$.

Proof For all $\mathbf{y}$ the subgradient inequality gives

$$\begin{aligned} f(\mathbf{y}) &\geqslant f(\mathbf{x}_i) + \mathbf{v}_i^{\mathrm{T}}(\mathbf{y} - \mathbf{x}_i) \\ &= f(\mathbf{x}_i) + \mathbf{v}^{\mathrm{T}}(\mathbf{y} - \mathbf{x}_i) + (\mathbf{v}_i - \mathbf{v})^{\mathrm{T}}(\mathbf{y} - \mathbf{x}_i) \\ &= f(\mathbf{x}) + \mathbf{v}^{\mathrm{T}}(\mathbf{y} - \mathbf{x}) + f(\mathbf{x}_i) - f(\mathbf{x}) \\ &\qquad + \mathbf{v}^{\mathrm{T}}(\mathbf{x} - \mathbf{x}_i) + (\mathbf{v}_i - \mathbf{v})^{\mathrm{T}}(\mathbf{y} - \mathbf{x}_i). \end{aligned}$$

Using stability at $\mathbf{x}$ gives

$$f(\mathbf{y}) + K \|\mathbf{x} - \mathbf{x}_i\|_2 \geqslant f(\mathbf{x}) + \mathbf{v}^{\mathrm{T}}(\mathbf{y} - \mathbf{x}) + \varepsilon_i$$

where

$$\varepsilon_i = \mathbf{v}^{\mathrm{T}}(\mathbf{x} - \mathbf{x}_i) + (\mathbf{v}_i - \mathbf{v})^{\mathrm{T}}(\mathbf{y} - \mathbf{x}_i)$$

and, by assumption, $\varepsilon_i \to 0$, $i \to \infty$. Letting $i \to \infty$ gives the subgradient inequality as $\mathbf{y}$ is arbitrary showing that $\mathbf{v}^{\mathrm{T}} \in \partial f(\mathbf{x})$.

Definition 4.5 Let f be convex and $\mathbf{x} \in \operatorname{dom} f$. Then

$$f'(\mathbf{x} : \mathbf{t}) = \inf_{\lambda > 0} \frac{f(\mathbf{x} + \lambda \mathbf{t}) - f(\mathbf{x})}{\lambda} \tag{4.9}$$

is the *directional derivative* of f in the direction $\mathbf{t}$. Convexity gives

$$\frac{f(\mathbf{x}+\lambda_1\lambda_2\mathbf{t})-f(\mathbf{x})}{\lambda_1\lambda_2}\leqslant\frac{f(\mathbf{x}+\lambda_1\mathbf{t})-f(\mathbf{x})}{\lambda_1}$$

when $0<\lambda_1, 0<\lambda_2<1$ so that the difference quotient in (4.9) is a nondecreasing function of λ showing that the definition is sensible. It is convex and positively homogeneous as a function of $\mathbf{t}$.

The next results give the relationship between the directional derivative and the subdifferential.

Lemma 4.6 *Let f be stable at* $\mathbf{x}$ *and* $\mathbf{t}$ *be into* ri dom f, *then*

$$f'(\mathbf{x}:\mathbf{t})=\sup_{\mathbf{v}^T\in\partial f(\mathbf{x})}\mathbf{v}^T\mathbf{t} \tag{4.10}$$

Proof The subgradient inequality gives

$$f(\mathbf{x}+\lambda\mathbf{t})-f(\mathbf{x})\geqslant\lambda\mathbf{v}^T\mathbf{t},\qquad \mathbf{v}^T\in\partial f(\mathbf{x})$$

showing that

$$f'(\mathbf{x}:\mathbf{t})\geqslant\sup_{\mathbf{v}^T\in\partial f(\mathbf{x})}\mathbf{v}^T\mathbf{t}.$$

To obtain the opposite inequality note that

$$f(\mathbf{x})\geqslant f(\mathbf{x}+\lambda_i\mathbf{t})-\lambda_i\mathbf{v}_i^T\mathbf{t}$$

Now choose $\{\lambda_i\}\to 0$, $\{\mathbf{v}_i^T\in\partial f(\mathbf{x}+\lambda_i\mathbf{t})\}\to\mathbf{v}^T\in\partial f(\mathbf{x})$. Then

$$\mathbf{v}_i^T\mathbf{t}\geqslant\frac{f(\mathbf{x}+\lambda_i\mathbf{t})-f(\mathbf{x})}{\lambda_i}$$

so that letting $i\to\infty$ gives

$$\mathbf{v}^T\mathbf{t}\geqslant\inf_i\frac{f(\mathbf{x}+\lambda_i\mathbf{t})-f(\mathbf{x})}{\lambda_i}\geqslant f'(\mathbf{x}:\mathbf{t})$$

Corollary 4.2 *Let f be stable at* $\mathbf{x}$, *then* $\mathbf{v}^T\in\partial f(\mathbf{x})$ *if and only if*

$$f'(\mathbf{x}:\mathbf{y})\geqslant\mathbf{v}^T\mathbf{y},\qquad\forall\,\mathbf{y}$$

Proof This inequality implies the subgradient inequality because the difference quotients in (4.9) are nondecreasing functions of λ.

Corollary 4.3 *Let* $\mathbf{x}\in$ ri dom f. *Then*

$$f'(\mathbf{x}:\mathbf{t})=\delta^*(\mathbf{t}\mid\partial f(\mathbf{x})) \tag{4.11}$$

Proof This follows on noting that any direction in the orthogonal complement of the allowable $\mathbf{t}$ in (4.10) is a direction of recession for $\partial f(\mathbf{x})$ and gives $f'(\mathbf{x}:\mathbf{t}) = +\infty$.

This result can be refined slightly.

Corollary 4.4 *Let f be stable at $\mathbf{x}$, then*

$$\mathrm{cl}_y f'(\mathbf{x}:\mathbf{y}) = \delta^*(\mathbf{y} \mid \partial f(\mathbf{x})) \tag{4.12}$$

where the notation indicates the closure of the convex function $f'(\mathbf{x}:\mathbf{y})$ with respect to $\mathbf{y}$.

Proof By (4.3), $\mathrm{cl}_y f'(\mathbf{x}:\mathbf{y})$ must be expressible as the supremum of an affine family. As it is positively homogeneous, each member of the family must pass through the origin. But then it follows from Corollary 4.2 that the affine family must be generated by the elements of the subdifferential.

Remark 4.4 The set of $\mathbf{v} \in \partial f(\mathbf{x})$ maximizing (4.10) for each $\mathbf{t}$ are the points in the intersection of $\partial f(\mathbf{x})$ and $H(\mathbf{t}, \delta^*(\mathbf{t} \mid \partial f(\mathbf{x})))$ where the indicated hyperplane supports $\partial f(\mathbf{x})$. It follows from Lemma 2.2 that the extreme points of the intersection are extreme points of $\partial f(\mathbf{x})$, and clearly all extreme points can be obtained this way. This observation is useful in calculating the subdifferential of piecewise linear functions. For example, if $f = |x|$ then

$$\begin{aligned} \partial f(x) &= -1, \quad x<0, \\ &= 1, \quad x>0, \\ &= [-1, 1], \quad x = 0. \end{aligned}$$

It is an important result that the subdifferential of the sum of any finite number of convex functions is given by the sum of the subdifferentials at any point in the intersection of the relative interiors of the domains of definition. It is sufficient to consider the case of two functions.

Lemma 4.7 *Let f_1, f_2 be convex functions and set $g = f_1 + f_2$. If*

$$\mathbf{x} \in \mathrm{ri}\,\mathrm{dom}\, f_1 \cap \mathrm{ri}\,\mathrm{dom}\, f_2 \neq \varnothing$$

then

$$\partial g(\mathbf{x}) = \partial f_1(\mathbf{x}) + \partial f_2(\mathbf{x})$$

Proof Because the difference quotients are nondecreasing,

$$g'(\mathbf{x}:\mathbf{y}) = f_1'(\mathbf{x}:\mathbf{y}) + f_2'(\mathbf{x}:\mathbf{y}) \qquad \forall\, \mathbf{y}$$

But, by Corollary 4.3, this says the support function of $\partial g(\mathbf{x})$ is the sum of the support functions of $\partial f_1(\mathbf{x})$ and $\partial f_2(\mathbf{x})$ and the result follows.

Example 4.4 Consider $f(\mathbf{x}) = \|\mathbf{x}\|$. Then

$$f'(\mathbf{x}:\mathbf{x}) = \frac{(1+\lambda)\|\mathbf{x}\| - \|\mathbf{x}\|}{\lambda}$$

$$= \|\mathbf{x}\| = \sup_{\mathbf{v}^{\mathrm{T}} \in \partial \|\mathbf{x}\|} \mathbf{v}^{\mathrm{T}}\mathbf{x} \tag{4.13}$$

In fact equality holds for *every* $\mathbf{v}^{\mathrm{T}} \in \partial \|\mathbf{x}\|$. For assume the contrary. Then there exists $\mathbf{v} \in \partial \|\mathbf{x}\|$ with $\mathbf{v}^{\mathrm{T}}\mathbf{x} < \|\mathbf{x}\|$. The subgradient inequality gives for $\varepsilon > 0$ small enough,

$$(1-\varepsilon)\|\mathbf{x}\| \geqslant \|\mathbf{x}\| - \varepsilon \mathbf{v}^{\mathrm{T}}\mathbf{x}$$

so that

$$-\varepsilon \|\mathbf{x}\| \geqslant -\varepsilon \mathbf{v}^{\mathrm{T}}\mathbf{x}$$

which is a contradiction. Now consider

$$\|\mathbf{z}\|^* = \sup_{\|\mathbf{x}\| \leqslant 1} \mathbf{z}^{\mathrm{T}}\mathbf{x}. \tag{4.14}$$

It is straightforward to verify that $\|\cdot\|^*$ is in fact a norm. It is the polar or dual norm to $\|\cdot\|$. Also, it follows from (4.13) that $\|\mathbf{v}\|^* \leqslant 1$ if $\mathbf{v}^{\mathrm{T}} \in \partial \|\mathbf{x}\|$. Now the maximum in (4.14) is attained (maximum of a continuous function on a compact set). If a maximizer is $\mathbf{x}_z$, then

$$\|\mathbf{w}\|^* \geqslant \mathbf{x}_z^{\mathrm{T}}\mathbf{w}$$
$$\geqslant \mathbf{x}_z^{\mathrm{T}}\mathbf{z} + \mathbf{x}_z^{\mathrm{T}}(\mathbf{w}-\mathbf{z})$$

holds for all $\mathbf{w}$. Thus $\mathbf{x}_z^{\mathrm{T}} \in \partial \|\mathbf{z}\|^*$. Evidently the roles of the norm and its dual can be interchanged.

Exercise 4.1 (i) Let S be a convex set in R^{p+1}. Then

$$f(\mathbf{x}) = \inf\left(\mu; \begin{bmatrix}\mathbf{x}\\ \mu\end{bmatrix} \in S\right).$$

is convex (the convex function generated by S).

(ii) Associated with $S = \operatorname{epi} f$ is the cone

$$C = \{\lambda \mathbf{z};\ \lambda \geqslant 0, \mathbf{z} \in S\}.$$

Show that g, the convex function generated by C, is positively homogeneous. It is the largest positively homogeneous function satisfying $f \geqslant g$. Discuss the case $0 \in \operatorname{int} \operatorname{epi} f$.

Exercise 4.2 If f is twice continuously differentiable then $f^{(2)}(\mathbf{x}:\mathbf{t},\mathbf{t}) \geqslant 0$ and $f^{(2)}(\mathbf{x}:\mathbf{t},\mathbf{t}) > 0$ implies f strictly convex. Here $f^{(2)}(\mathbf{x}:\cdot,\cdot)$ is the matrix of second partial derivatives (Hessian) of f. The notation is consistent with that for the directional derivative.

Exercise 4.3 Definition of a steepest descent direction: Let $\mathbf{d}$, $\|\mathbf{d}\| \leq 1$ minimize $f'(\mathbf{x}:\cdot)$. Then $\nabla f = f'(\mathbf{x}:\mathbf{d})\mathbf{d}^T$. What additivity and continuity properties does ∇f have?

1.5 OPTIMALITY CONDITIONS

In this section the problem of characterizing the minimum of a convex function is considered. It is convenient to break the problem up into two cases:

(a) any point in dom f is allowable as a competitor in seeking a minimizer of f, and
(b) the set of points allowable in this sense is constrained a priori by the further condition that $\mathbf{x} \in S$ where S is a convex set having nonvoid intersection with dom f.

In the first case we say the problem is *unconstrained* while in the second case it is *constrained.* The distinction is exemplified by the problems $\min x^2$, x unconstrained and $\min x^2$, $x \geq 1$. In the constrained case the necessary conditions for an optimum follow from an elegant separation argument which is based on Figure 5.1. This leads to the familiar multiplier conditions in the important case that S is defined by a finite number of relations having the form $g(\mathbf{x}) \leq 0$, g convex, and that the problem data is differentiable. But the argument does not use differentiability. These necessary conditions will be important in subsequent developments.

The unconstrained case is considered first.

Definition 5.1 By assumption, $\inf_{\mathbf{x} \in \operatorname{dom} f} f(\mathbf{x})$ is bounded below $(= \bar{f}$ say). If this infimum is attained so that

$$\exists \mathbf{x}^* \in \operatorname{dom} f \ni f(\mathbf{x}^*) = \bar{f}$$

then $\mathbf{x}^*$ *minimizes* f and $\bar{f}$ is the *minimum* value of f.

Remark 5.1 For convex functions it is only necessary to test for a minimum in a small ball containing $\mathbf{x}^*$. For assume that
(i) $f(\mathbf{y}) \geq f(\mathbf{x}^*)$, $\forall\, \mathbf{y} \in \delta B(\mathbf{x}^*) \cap \operatorname{dom} f$, and
(ii) $\exists\, \mathbf{x}^{**} \in \operatorname{dom} f \ni f(\mathbf{x}^{**}) < f(\mathbf{x}^*)$,
and select $\mathbf{y} \in \delta B(\mathbf{x}^*)$ on the line segment joining $\mathbf{x}^*$ and $\mathbf{x}^{**}$. Then convexity gives

$$\begin{aligned} f(\mathbf{y}) &= f(\theta \mathbf{x}^{**} + (1-\theta)\mathbf{x}^*), \quad \text{with} \quad 0 < \theta < 1 \\ &\leq \theta f(\mathbf{x}^{**}) + (1-\theta) f(\mathbf{x}^*) \\ &< f(\mathbf{x}^*) \end{aligned}$$

and this is a contradiction. Thus a local minimum of a convex function is necessarily the global minimum. However, this minimum is not necessarily unique (a sufficient condition is strict convexity).

Theorem 5.1 *If* $\mathbf{x}^*$ *minimizes f then* $\partial f(\mathbf{x}^*) \neq \varnothing$ *and* $0 \in \partial f(\mathbf{x}^*)$. *If* $0 \in \partial f(\mathbf{x}^*)$ *then* $\mathbf{x}^*$ *minimizes f.*

Proof For any $\mathbf{y} \in \operatorname{dom} f$

$$f(\mathbf{x}^*) - f(\mathbf{y}) \leqslant 0 \leqslant K \|\mathbf{x}^* - \mathbf{y}\|_2$$

so that f is stable at $\mathbf{x}^*$ and $\partial f(\mathbf{x}^*) \neq \varnothing$. This inequality also shows that $0 \in \partial f(\mathbf{x}^*)$.

On the other hand, if $0 \in \partial f(\mathbf{x}^*)$ then the subgradient inequality gives

$$f(\mathbf{y}) \geqslant f(\mathbf{x}^*) + 0(\mathbf{y} - \mathbf{x}) \geqslant f(\mathbf{x}^*) \qquad \forall\, \mathbf{y}$$

so that $\mathbf{x}^*$ minizes f.

Remark 5.2 It is crucial that the minimum be attained, for consider

$$f = \begin{cases} -x, & -1 \leqslant x < 0 \\ 1, & x = 0 \\ +\infty & \text{otherwise} \end{cases}$$

Then $\inf_x f(x) = 0$ but is never attained, and $\partial f(0) = \varnothing$. However, it is not necessary for f to be differentiable at $\mathbf{x}^*$. Consider $f = |x|$. Then

$$\begin{aligned} \partial |x| &= -1, & x < 0, \\ &= [-1, 1], & x = 0, \\ &= 1, & x > 0. \end{aligned}$$

and $0 \in \partial f(0) = [-1, 1]$.

We turn now to the problem in which the range of allowable values of $\mathbf{x}$ is constrained to be a prescribed convex set $S \subseteq R^p$. Clearly there is no problem unless $S \cap \operatorname{dom} f \subset \operatorname{dom} f$, the inclusion being proper, and there is no loss of generality in identifying S with $S \cap \operatorname{dom} f$. However, we do make the important assumption (*Slater condition*) that

$$\operatorname{int} S \neq \varnothing \subset \operatorname{dom} f. \tag{5.1}$$

The minimization problem now becomes

$$\min_{\mathbf{x} \in S} f(\mathbf{x}) \tag{5.2}$$

where S is a convex body (which implies a nonvoid interior) contained in $\operatorname{dom} f$. The existence of a (finite) point $\mathbf{x}^*$ minimizing (5.2) is assumed, and we consider the problem of developing characterization results analogous to

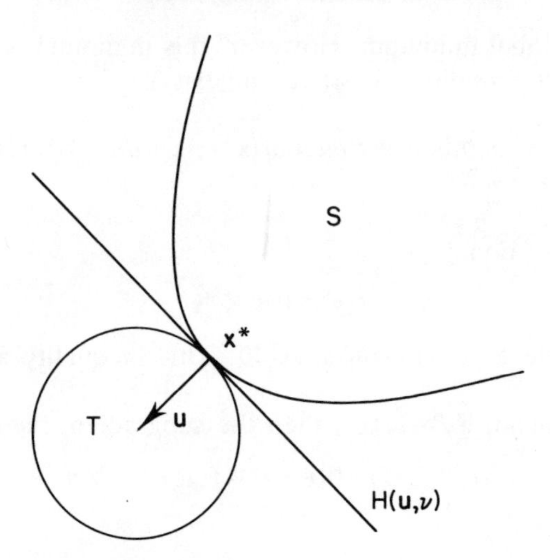

Figure 5.1 Constrained optimization implies a separation principle.

that of Theorem 5.1. The approach followed starts by introducing an associated separation principle. Let

$$T = \{\mathbf{x};\, f(\mathbf{x}) < f(\mathbf{x}^*)\}. \tag{5.3}$$

If $T = \varnothing$ then the minimization problem is unconstrained and Theorem 5.1 applies. Thus we assume $T \neq \varnothing$ and note that T is an open convex set, $\mathbf{x}^* \in \text{cl}\, T \cap S$ and S and T are disjoint. Thus there is a hyperplane $H(\mathbf{u}, \nu)$ separating S and T such that $S \subset H^-$, $T \subset H^+$, and which supports both S and T on the points of cl $T \cap S$. This is indicated diagrammatically in Figure 5.1.

To obtain a characterization theorem it is necessary to relate $\mathbf{u}$ to the elements of $\partial f(\mathbf{x}^*)$.

Definition 5.2 The *tangent cone* to a convex set S at $\mathbf{x} \in S$ is

$$\mathcal{T}(S, \mathbf{x}) = \text{cl}\, \{\mathbf{z};\, \mathbf{z} = \lambda(\mathbf{y} - \mathbf{x}),\, \forall\, \lambda > 0,\, \forall\, \mathbf{y} \in S\} \tag{5.4}$$

$\mathcal{T}(S, \mathbf{x})$ is a closed convex cone. $\mathcal{T} = R^p$ if $\mathbf{x}$ is an interior point of S.

The next result is an immediate consequence of this definition.

Lemma 5.1 *The hyperplane $H(\mathbf{u}, \nu)$ passing through $\mathbf{x}^*$ supports S at $\mathbf{x}^*$ if and only if*

$$\mathbf{u} \in \mathcal{T}(S, \mathbf{x}^*)^*. \tag{5.5}$$

Corollary 5.1 *A point $\mathbf{x}^*$ solves* (5.2) *if and only if*

$$\mathcal{T}(S, \mathbf{x}^*)^* \cap -\mathcal{T}(\text{cl}\, T, \mathbf{x}^*)^* \neq \varnothing. \tag{5.6}$$

Lemma 5.2 *If $\mathcal{T}(\text{cl } T, \mathbf{x}^*) \neq \varnothing$ so that there exist points in the immediate neighbourhood of $\mathbf{x}^*$ such that $f(\mathbf{x}) < f(\mathbf{x}^*)$ (this implies $0 \notin \partial f(\mathbf{x}^*)$) then*

$$\mathcal{T}(\text{cl } T, \mathbf{x}^*)^* = \{\lambda \mathbf{v};\ \lambda \geqslant 0,\ \mathbf{v}^{\mathrm{T}} \in \partial f(\mathbf{x}^*)\} \tag{5.7}$$

Proof

$$\begin{aligned}\mathcal{T}(\text{cl } T, \mathbf{x}^*) &= \text{cl}\,\{\lambda \mathbf{y};\ \lambda \geqslant 0,\ f(\mathbf{x}^* + \mathbf{y}) - f(\mathbf{x}^*) \leqslant 0\} \\ &= \{\mathbf{y};\ \text{cl}_y f'(\mathbf{x}^* : \mathbf{y}) \leqslant 0\} \\ &= \{\mathbf{y};\ \mathbf{y}^{\mathrm{T}}\mathbf{v} \leqslant 0,\ \forall\ \mathbf{v}^{\mathrm{T}} \in \partial f(\mathbf{x}^*)\}\end{aligned}$$

This shows that the elements of $\partial f(\mathbf{x}^*)$ generate the polar cone.

Corollary 5.2 *An application of Corollary* 5.1 *gives that $\mathbf{x}^*$ minimizes f if and only if*

$$-\partial f(\mathbf{x}^*) \cap \mathcal{T}(S, \mathbf{x}^*)^* \neq \varnothing. \tag{5.8}$$

This is about as far as we can go without being specific about the structure of S. Perhaps the case of most interest occurs when S is defined by a finite number of constraint conditions so that S has the form

$$S = \{\mathbf{x};\ g_i(\mathbf{x}) \leqslant 0,\ i = 1, 2, \ldots, n\} \tag{5.9}$$

where the g_i are convex. (Alternatively the $g_i(\mathbf{x})$ can be defined as concave, corresponding to multiplying by -1, so that the constraint inequalities are reversed. This second use is more conventional because it leads to positive multipliers.) The problem of minimizing f subject to $\mathbf{x} \in S$ given by (5.9) is called the *convex programming problem* or CPP. In the discussion of duality it will also be referred to as the *primal* problem. Let

$$\bar{g}(\mathbf{x}) = \max_{1 \leqslant i \leqslant n} g_i(\mathbf{x}). \tag{5.10}$$

Then $\bar{g}(\mathbf{x})$ is convex for, with $0 \leqslant \theta \leqslant 1$,

$$\begin{aligned}\bar{g}(\theta\mathbf{x} + (1-\theta)\mathbf{y}) &= g_i(\theta\mathbf{x} + (1-\theta)\mathbf{y}) \quad \text{for some } i \\ &\leqslant \theta g_i(\mathbf{x}) + (1-\theta) g_i(\mathbf{y}) \quad \text{by convexity} \\ &\leqslant \theta \bar{g}(\mathbf{x}) + (1-\theta)\bar{g}(\mathbf{y}).\end{aligned}$$

This means that the constraint functions can be summarized in the single convex constraint $\bar{g}(\mathbf{x}) \leqslant 0$.

Remark 5.3 Now the constraint set S is given by

$$S = \{\mathbf{x};\ \bar{g}(\mathbf{x}) \leqslant 0\}. \tag{5.11}$$

Thus the separation principle is equivalent to separating two sets defined in essentially similar ways. In particular, Lemma 5.2 can be applied to compute the elements of the tangent cones of both.

Lemma 5.3 *Let*

$$\sigma = \{i;\ g_i(\mathbf{x}^*) = 0\} \tag{5.12}$$

(*The index set* σ *is said to point to the constraints active at* $\mathbf{x}^*$). *Then*

$$\mathcal{T}(S, \mathbf{x}^*)^* = \{\gamma \mathbf{v};\ \gamma \geqslant 0,\ \mathbf{v}^{\mathrm{T}} \in \operatorname{conv}\{\partial g_i(\mathbf{x}^*),\ i \in \sigma\}\} \tag{5.13}$$

Proof Let

$$S_i = \{\mathbf{x};\ g_i(\mathbf{x}) \leqslant 0\}, \qquad i = 1, 2, \ldots, n,$$

Then, as a consequence of the definition of a tangent cone,

$$\mathcal{T}(S, \mathbf{x}^*) = \bigcap_{i \in \sigma} \mathcal{T}(S_i, \mathbf{x}^*)$$

These are closed convex cones so that property (f) for polar cones gives

$$\mathcal{T}(S, \mathbf{x}^*)^* = \sum_{i \in \sigma} \mathcal{T}(S_i, \mathbf{x}^*)^*.$$

The result now follows by an application of Lemma 5.2.

Remark 5.4 This result can be refined to give

$$\partial \bar{g}(\mathbf{x}) = \operatorname{conv}\{\partial g_i(\mathbf{x}),\ i \in \sigma = \{j;\ g_j(\mathbf{x}) = \bar{g}(\mathbf{x})\}\} \tag{5.14}$$

Combining Corollary 5.2 and Lemma 5.3 gives a form of characterization result in a more explicit form.

Theorem 5.2 *The point* $\mathbf{x}^*$ *solves* (5.2) *where* S *is given by* (5.9) *if and only if there is a* $\gamma > 0$ *such that*

$$-\partial f(\mathbf{x}^*) \cap \gamma \operatorname{conv}\{\partial g_i(\mathbf{x}^*),\ i \in \sigma\} \neq \emptyset. \tag{5.15}$$

Remark 5.5 In the particular case that f, g_i, $i = 1, 2, \ldots, n$, are smooth, this result gives the *multiplier conditions* or *Kuhn–Tucker* conditions that there exist $\lambda_i \geqslant 0$, $i = 1, 2, \ldots, n$, such that

(i)
$$\nabla f(\mathbf{x}^*) = -\sum_{i=1}^{n} \lambda_i \nabla g_i(\mathbf{x}^*), \tag{5.16}$$

and

(ii)
$$\lambda_i g_i(\mathbf{x}^*) = 0, \qquad i = 1, 2, \ldots, n. \tag{5.17}$$

The second condition is called a *complementarity* condition.

Example 5.1 Consider the problem

$$\min y$$

subject to the constraint

$$x^2+y^2 \leqslant 1.$$

This problem meets the conditions of the above remark and (5.16) gives

$$-[0, 1] = \lambda[2x, 2y].$$

This condition cannot be satisfied if $\lambda = 0$. Thus $x = 0$ at the minimum and $\lambda > 0$. The only possibility is thus $\lambda = \frac{1}{2}$, $y = -1$.

The condition (5.1) on S is stronger than necessary. One important case in which the separation argument leads to a multipliers result without this assumption is the case in which S is an affine set so that, by Remark 2.4, the separating hyperplane can be chosen to include S.

Lemma 5.4 *Let $S \neq \emptyset$ be the affine set*

$$S = \{\mathbf{x};\ A\mathbf{x} = \mathbf{b}\}$$

where $A: R^p \to R^m$. Then $\mathbf{x}^* \in \operatorname{cl} T \cap S$ minimizes $f(\mathbf{x})$ if and only if

$$\partial f(\mathbf{x}^*) \cap \operatorname{span}\ (\boldsymbol{\rho}_i(A),\ i = 1, 2, \ldots, m) \neq \emptyset. \tag{5.18}$$

Proof Here a direct computation shows that

$$\mathcal{T}(S, \mathbf{x}^*) = \{\mathbf{t};\ \boldsymbol{\rho}_i(A)\mathbf{t} = 0,\ i = 1, 2, \ldots, m\} \tag{5.19}$$

so that $\mathcal{T}$ is the subspace $L^\perp$ which is the orthogonal complement of the subspace L translated to form the flat S. Property (e) of polar cones gives

$$\mathcal{T}(S, \mathbf{x}^*)^* = L$$

and the result (5.18) follows from Corollary 5.2.

Remark 5.6 In the case that f is differentiable at $\mathbf{x}^*$, (5.18) leads to a multiplier condition in which the signs of the multipliers are no longer constrained to be positive.

An important application of this result is the following.

Lemma 5.5 *Let f be convex, $A: R^p \to R^m$, and*

$$g(\mathbf{t}) = f(A\mathbf{t} + \mathbf{b}).$$

Then g is convex,

$$\operatorname{dom} g = \{\mathbf{t};\ A\mathbf{t} + \mathbf{b} \in \operatorname{dom} f\}$$

and

$$\partial_t g(\mathbf{t}) = \partial_x f(A\mathbf{t} + \mathbf{b})A. \tag{5.20}$$

Proof Only the verification of (5.20) need be considered. If $\mathbf{x}=A\mathbf{t}+\mathbf{b}$, $\mathbf{v}^{\mathrm{T}}\in\partial f(\mathbf{x})$, then the subgradient inequality gives

$$f(\mathbf{y})\geqslant f(\mathbf{x})+\mathbf{v}^{\mathrm{T}}(\mathbf{y}-\mathbf{x}).$$

In particular, this is true when $\mathbf{y}=A\mathbf{s}+\mathbf{b}$, $\mathbf{s}\in\operatorname{dom} g$, so that

$$g(\mathbf{s})\geqslant g(\mathbf{t})+\mathbf{v}^{\mathrm{T}}A(\mathbf{s}-\mathbf{t})$$

showing that

$$\partial_x fA\subseteq\partial_t g.$$

On the other hand, if $\mathbf{w}\in\partial_t g$ then

$$f(A\mathbf{s}+\mathbf{b})\geqslant f(A\mathbf{t}+\mathbf{b})+\mathbf{w}^{\mathrm{T}}(\mathbf{s}-\mathbf{t})$$

so that $\mathbf{t}$ minimizes

$$G(\mathbf{y},\mathbf{s})=f(\mathbf{y})-\mathbf{w}^{\mathrm{T}}\mathbf{s}$$

subject to

$$\mathbf{y}=A\mathbf{s}+\mathbf{b}.$$

It follows that G is a convex function with effective domain dom $f\times R^m$, and that the constraint set is an affine set in $R^p\times R^m$. Lemma 5.4 can now be applied to give $\mathbf{u}^{\mathrm{T}}\in\partial f(\mathbf{u})$ and multiplier vector $\boldsymbol{\lambda}$ such that

$$[\mathbf{u}^{\mathrm{T}}\mid-\mathbf{w}^{\mathrm{T}}]=\boldsymbol{\lambda}^{\mathrm{T}}[I\mid-A]$$

Comparing terms yields

$$\mathbf{u}=\boldsymbol{\lambda},$$

and

$$\mathbf{w}^{\mathrm{T}}=\mathbf{u}^{\mathrm{T}}A$$

showing that

$$\partial_t g\subseteq\partial_x fA.$$

This inclusion completes the required proof.

Remark 5.7 Another case in which the restriction that S satisfy a Slater condition can be relaxed corresponds to S polyhedral. The following development represents a useful argument that will also be used subsequently. Let

$$S=\{\mathbf{x};\ A\mathbf{x}=\mathbf{b},\ C\mathbf{x}\leqslant\mathbf{d}\}$$

where $A:R^p\rightarrow R^m$ determines the affine hull of S and is assumed to have full row rank $m<p$, and $C:R^p\rightarrow R^n$. The aim is to use the information on aff S to reduce the problem to one in a smaller number of variables in which only inequality constraints occur by eliminating some of the variables using

the equality constraints. The assumptions on A ensure that it can be factorized (compare Appendix 1) to give

$$A^{\mathrm{T}}=[Q_1|Q_2]\begin{bmatrix}U\\0\end{bmatrix}$$

where $Q=[Q_1\,|\,Q_2]$ is a $p\times p$ orthogonal matrix and U is $m\times m$ upper triangular. Now $\mathbf{x}$ is in aff S provided

$$[U^{\mathrm{T}}\,|\,0]Q^{\mathrm{T}}\mathbf{x}=[U^{\mathrm{T}}\,|\,0]\begin{bmatrix}\mathbf{y}_1\\\mathbf{y}_2\end{bmatrix}=\mathbf{b}$$

where $\mathbf{y}_1,\mathbf{y}_2$ are defined by the obvious partitioning of $\begin{bmatrix}Q_1^{\mathrm{T}}\\Q_2^{\mathrm{T}}\end{bmatrix}\mathbf{x}$, that is, provided

$$U^{\mathrm{T}}\mathbf{y}_1=\mathbf{b}.$$

This system of equations fixes $\mathbf{y}_1$ so that $\mathbf{y}_2=Q_2^{\mathrm{T}}\mathbf{x}$ provides the free parameters in $\mathbf{x}$. In terms of the new variables the problem becomes

$$\min_{\mathbf{y}_2} f\left(Q\begin{bmatrix}\mathbf{y}_1\\\mathbf{y}_2\end{bmatrix}\right)$$

subject to the inequality constraints

$$CQQ^{\mathrm{T}}\mathbf{x}=CQ_2\mathbf{y}_2+CQ_1U^{\mathrm{T}}\mathbf{b}\leqslant\mathbf{d}.$$

An application of Theorem 5.2 gives multipliers $\boldsymbol{\lambda}\geqslant 0$ such that

$$\mathbf{v}^{\mathrm{T}}Q_2=-\sum_{i\in\sigma}\lambda_i\boldsymbol{\rho}_i(C)Q_2,\qquad \boldsymbol{\lambda}^{\mathrm{T}}(C\mathbf{x}^*-\mathbf{d})=0,$$

where $\mathbf{v}^{\mathrm{T}}\in\partial f(\mathbf{x}^*)$ and (5.20) is used to compute $\partial_{\mathbf{y}_2}g$, $g=f(Q\mathbf{y})$. This shows that

$$\mathbf{v}^{\mathrm{T}}+\sum_{i\in\sigma}\lambda_i\boldsymbol{\rho}_i(C)\in\operatorname{span}(\boldsymbol{\rho}_i(A),\ i=1,2,\ldots,m),$$

or equivalently, that there exist unconstrained multipliers μ_i, $i=1,2,\ldots,m$ such that there exists $\mathbf{v}^{\mathrm{T}}\in\partial f(\mathbf{x}^*)$ satisfying

$$\mathbf{v}^{\mathrm{T}}=-\sum_{i=1}^{n}\lambda_i\boldsymbol{\rho}_i(C)+\sum_{i=1}^{m}\mu_i\boldsymbol{\rho}_i(A)\tag{5.21}$$

where the multipliers $\lambda_i\geqslant 0$, $i=1,2,\ldots,n$ satisfy

$$\lambda_i(\boldsymbol{\rho}_i(C)\mathbf{x}^*-d_i)=0,\qquad i=1,2,\ldots,n.\tag{5.22}$$

Example 5.2 Consider the problem

$$\min x^2+y^2$$

subject to

$$x+y=-1,$$
$$y\leq\alpha.$$

This problem is in the form considered in Remark 5.7 and here (5.21) gives

$$-[2x, 2y]=\lambda[0, 1]+\mu[1, 1].$$

If the inequality constraint is inactive, then $\lambda=0$. In this case $x=y=-\mu/2$ and it follows from the equality constraint that $\mu=1$. If $\lambda>0$ then $y=\alpha\leq-\frac{1}{2}$. The equality constraint now gives $x=-1-\alpha$ and substituting in the multiplier condition gives $\mu=2(1+\alpha)$, $\lambda=-2-4\alpha$. Note that $\lambda>0$ if $\alpha<-\frac{1}{2}$ and that $\mu\geq0$ if $\alpha\geq-1$, but $\mu<0$ if $\alpha<-1$. This demonstrates that the multiplier associated with the equality constraint can have either sign.

Exercise 5.1 Consider the tangent cone to the set S at the point $\mathbf{x}_0$ where:

(i) $S=\{x^2+y^2\leq1\}$, $\mathbf{x}_0=0$,

(ii) $S=\{x^2+y^2\leq1\}$, $\mathbf{x}_0=\begin{pmatrix}0\\-1\end{pmatrix}$,

(iii) $S=\{x^2+(y+1)^2\leq2\}\cap\{y\geq0\}$, $\mathbf{x}_0=\begin{pmatrix}-1\\0\end{pmatrix}$.

Exercise 5.2 Use multiplier conditions to find the solution to the problem $\min|x|+|y|$ subject to $ax+by=c$, $x+y>0$ as a function of a, b, c.

Exercise 5.3 Use Farkas' lemma (Theorem 3.2) to deduce the multiplier conditions (5.16), (5.17) in the case that the data on the minimization problem is differentiable.

1.6 CONJUGATE CONVEX FUNCTIONS, DUALITY

The results of this section complete our general discussion of convexity. The basic idea is to seek a reciprocal relationship between the function f defined on the points $\mathbf{x}\in\operatorname{dom} f$ and a conjugate function f^* defined on the vectors $\mathbf{u}$ such that $\begin{bmatrix}\mathbf{u}\\-1\end{bmatrix}$ is the normal to a supporting hyperplane to epi f. By working on this relationship in the context of the CPP we are able to find a dual problem having the property that the supremum of this new problem coincides with the infimum of the CPP (the CPP is also called the primal problem). This duality theory is used in Chapter 2 to construct the dual linear programming problem but the results in this case can be obtained by more elementary means. Duality is also used in discussing quadratic programming in Chapter 5.

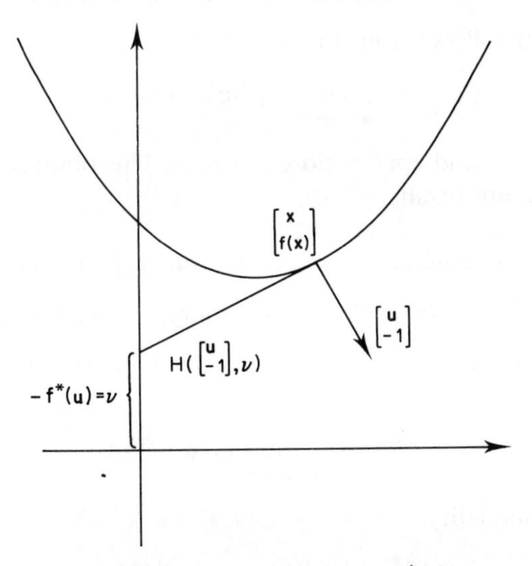

Figure 6.1 Construction of $f^*(\mathbf{u})$

The basic construction used in specifying the conjugate convex function to f is sketched in Figure 6.1.

If

$$H\left(\begin{bmatrix}\mathbf{u}\\-1\end{bmatrix}, \nu\right) \quad \text{supports} \quad \text{epi}\, f \quad \text{at} \quad \begin{bmatrix}\mathbf{x}\\f(\mathbf{x})\end{bmatrix}$$

then epi $f \subset H^-$ gives

$$\begin{bmatrix}\mathbf{y}\\\mu\end{bmatrix} \in \text{epi}\, f \Rightarrow \left\{\begin{bmatrix}\mathbf{y}\\\mu\end{bmatrix} - \begin{bmatrix}\mathbf{x}\\f(\mathbf{x})\end{bmatrix}\right\}^{\mathrm{T}} \begin{bmatrix}\mathbf{u}\\-1\end{bmatrix} \leqslant 0$$

so that

$$\mathbf{y}^{\mathrm{T}}\mathbf{u} - \mu \leqslant \mathbf{x}^{\mathrm{T}}\mathbf{u} - f(\mathbf{x}) = -\nu \tag{6.1}$$

This shows that $\nu = f(\mathbf{x}) - \mathbf{u}^{\mathrm{T}}\mathbf{x}$ is given by

$$\nu = \min_{\mathbf{y} \in \text{dom}\, f} (f(\mathbf{y}) - \mathbf{u}^{\mathrm{T}}\mathbf{y})$$

showing explicitly its dependence on $\mathbf{u}$. This motivates the following definition.

Definition 6.1 *The conjugate convex function* $f^*(\mathbf{u})$ to $f(\mathbf{x})$ is given by

$$f^*(\mathbf{u}) = \sup_{\mathbf{x} \in \text{dom} f} \{\mathbf{x}^{\mathrm{T}}\mathbf{u} - f(\mathbf{x})\} \tag{6.2}$$

It follows from (4.3) that $f^*(\mathbf{u})$ is a closed convex function of $\mathbf{u}$ and

$$\text{dom}\, f^* = \left\{\mathbf{u}; \begin{bmatrix}\mathbf{u}\\-1\end{bmatrix} \quad \text{supports} \quad \text{epi}\, f\right\}.$$

Remark 6.1 It follows from Figure 6.1 that

$$\sup_{\mathbf{u}\in \operatorname{dom} f^*} -f^*(\mathbf{u}) = f(0) \tag{6.3}$$

provided $0 \in \operatorname{dom} f$ and epi f is closed at zero. This observation will form the basis of our duality results.

An immediate consequence of the definition of f^* is *Fenchel's inequality*:

$$f^*(\mathbf{u}) + f(\mathbf{x}) \geq \mathbf{x}^{\mathrm{T}}\mathbf{u}, \qquad \mathbf{x} \in \operatorname{dom} f, \qquad \mathbf{u} \in \operatorname{dom} f^*. \tag{6.4}$$

The second conjugate function f^{**} is defined to be the conjugate of f^*; whence

$$f^{**}(\mathbf{x}) = \sup_{\mathbf{u}\in \operatorname{dom} f^*} (\mathbf{x}^{\mathrm{T}}\mathbf{u} - f^*(\mathbf{u})) \tag{6.5}$$

By Fenchel's inequality

$$f^{**}(\mathbf{x}) \leq f(\mathbf{x}), \qquad \mathbf{x} \in \operatorname{dom} f,$$

and equality holds at points at which epi f has a nonvertical hyperplane by the construction defining $f^*(\mathbf{u})$. As usual, problems can occur only at boundary points of dom f. In particular, f^{**} is closed. The precise result follows from these observations.

Theorem 6.1

$$f^{**}(\mathbf{x}) = f(\mathbf{x})$$

if and only if f is closed.

Example 6.1 (i) Let $f: R \to R$ be given by

$$f = \tfrac{1}{2}(x - x_0)^2 + \gamma\,|x|\,.$$

Then f is strictly convex and

$$\partial f(x) = \begin{cases} x - x_0 + \gamma \operatorname{sgn} x, & x \neq 0, \\ -x_0 + \gamma[-1, 1], & x = 0. \end{cases}$$

In particular, the supporting hyperplanes to epi f at $\begin{bmatrix} 0 \\ \frac{1}{2}x_0^2 \end{bmatrix}$ correspond to

$$u = -x_0 + \gamma[-1, 1]$$

so that

$$-\gamma \leq u + x_0 \leq \gamma$$

giving

$$f^*(u) = -\tfrac{1}{2}x_0^2, \qquad u + x_0 \leq \gamma.$$

If $|u+x_0|>\gamma$ then u defines a supporting hyperplane at $z\neq 0$ given by

$$z=u+x_0-\gamma \operatorname{sgn} z$$

whence

$$\operatorname{sgn} z=\operatorname{sgn}(u+x_0)$$

and

$$z=\operatorname{sgn}(u+x_0)(|u+x_0|-\gamma).$$

Thus

$$\begin{aligned} f^*(u) &= uz-\tfrac{1}{2}(z-x_0)^2-\gamma\,|z| \\ &= (|u+x_0|-\gamma)\,|z|-\tfrac{1}{2}z^2-\tfrac{1}{2}x_0^2 \\ &= \tfrac{1}{2}(|u+x_0|-\gamma)^2-\tfrac{1}{2}x_0^2,\ |u+x_0|>\gamma. \end{aligned}$$

(ii) Let X be a convex set in R^p. The function

$$\delta(\mathbf{x}\mid X)=\begin{cases} 0, & \mathbf{x}\in X, \\ +\infty & \cdot \text{ otherwise} \end{cases} \tag{6.6}$$

is called the *indicator function* for X. The conjugate convex function is

$$\delta^*(\mathbf{u}\mid X)=\sup_{\mathbf{x}\in X} \mathbf{x}^{\mathrm{T}}\mathbf{u}$$

which will be recognized as the support function for X. Note that $\delta^*(\mathbf{u}\mid\varnothing)=-\infty$ and is not a proper convex function.

Standard duality theory is developed under assumptions on the CPP that make the constraint functions concave (see the discussion following (5.9)). In this a central concept is that of the Lagrangian of the CPP.

Definition 6.2 Let f convex be the objective function of the CPP, and let the constraints be $g_i(\mathbf{x})\geqslant 0$, $i=1,2,\ldots,n$, where the g_i are concave. The *Lagrangian* of the CPP is

$$\mathscr{L}(\mathbf{x},\mathbf{u})=f(\mathbf{x})-\sum_{i=1}^{n} u_i g_i(\mathbf{x}) \tag{6.7}$$

Note that $\mathscr{L}$ is a convex function of $\mathbf{x}$ on

$$\operatorname{dom} f\cap\left\{\bigcap_{i=1}^{n}\operatorname{dom}(-g_i)\right\}\times\{\mathbf{u}\geqslant 0\}.$$

Let

$$S(\mathbf{y})=\{\mathbf{x};\,\mathbf{g}(\mathbf{x})\geqslant\mathbf{y}\} \tag{6.8}$$

so that $S(0)$ correspond to the constraint set for the CPP. The condition (5.15) for $\mathbf{x}^*$ to be a minimizer of the CPP can be restated in terms of $\mathscr{L}$. Thus $f(\mathbf{x})$ has a minimum at $\mathbf{x}^*\in S(0)$ if and only if there exists $\mathbf{u}^*\geqslant 0$ such

that
(i) $\mathbf{u}^{*T}\mathbf{g}(\mathbf{x}^*) = 0$, and
(ii) $0 \in \partial_x \mathscr{L}(\mathbf{x}^*, \mathbf{u}^*)$.
This follows because (ii) is already in the form of (5.15) provided the complementarity condition (i) holds. Because $\mathbf{u}^* \geqslant 0$, $\mathscr{L}(\mathbf{x}, \mathbf{u}^*)$ is convex so that (ii) ensures that $\mathbf{x}^*$ minimizes $\mathscr{L}(\mathbf{x}, \mathbf{u}^*)$.

This discussion is summarized in the next result.

Lemma 6.1 *Let* dom $f \cap S(0) \neq \varnothing$. *Then* $f(\mathbf{x})$ *has a minimum on* $S(0)$ *at* $\mathbf{x} = \mathbf{x}^*$ *if and only if*
(i) $\mathbf{x}^*$ *minimizes* $\mathscr{L}(\mathbf{x}, \mathbf{u}^*)$ *on* $S(0)$;
(ii) $\mathbf{u}^* \geqslant 0$;
(iii) $\mathbf{u}^{*T}\mathbf{g}(\mathbf{x}^*) = 0$; *and*
(iv) $\mathbf{g}(\mathbf{x}^*) \geqslant 0$.

Remark 6.2 The conditions on $(\mathbf{x}^*, \mathbf{u}^*)$ are called the *optimality conditions* for the CPP. They are equivalent to the validity of the inequalities

$$\mathscr{L}(\mathbf{x}^*, \mathbf{u}) \leqslant \mathscr{L}(\mathbf{x}^*, \mathbf{u}^*) \leqslant \mathscr{L}(\mathbf{x}, \mathbf{u}^*)$$

for all $\mathbf{u} \geqslant 0$ and $\mathbf{x} \in S(0)$. In this case we say that the Lagrangian has a *saddle point* at $(\mathbf{x}^*, \mathbf{u}^*)$.

The problem of minimizing $\mathscr{L}$ with respect to $\mathbf{x}$ looks a bit like the operation of generating a conjugate convex function. This turns out to be useful hint and some time will be spent on developing the relevant results. These in turn will lead quite directly to the quality theory.

Definition 6.3 The *perturbation function* for the CPP is given by

$$v(\mathbf{z}) = \inf_{\mathbf{x} \in S(\mathbf{z})} f(\mathbf{x}) \tag{6.9}$$

where $v(\mathbf{z}) = +\infty$ if $S(\mathbf{z}) = \varnothing$. Clearly $v(0)$ solves the CPP so that $v(\mathbf{z})$ gives information about the sensitivity of the CPP to certain constraint perturbations.

Lemma 6.2 *Let*

$$Y = \{\mathbf{y} \in R^n; \text{dom } f \cap S(\mathbf{y}) \neq \varnothing\}. \tag{6.10}$$

Then Y *is a convex set and* $v(\mathbf{z})$ *is a convex function with* dom $v = \text{cl } Y$.

Proof Let $\mathbf{y}_1, \mathbf{y}_2 \in Y$. Then there exist $\mathbf{x}_1, \mathbf{x}_2$ such that $\mathbf{g}(\mathbf{x}_1) \geqslant \mathbf{y}_1$, $\mathbf{g}(\mathbf{x}_2) \geqslant \mathbf{y}_2$. By concavity

$$\begin{aligned} \mathbf{g}(\theta\mathbf{x}_1 + (1-\theta)\mathbf{x}_2) &\geqslant \theta\mathbf{g}(\mathbf{x}_1) + (1-\theta)\mathbf{g}(\mathbf{x}_2), \qquad 0 \leqslant \theta \leqslant 1, \\ &\geqslant \theta\mathbf{y}_1 + (1-\theta)\mathbf{y}_2 \end{aligned}$$

showing that $\theta\mathbf{y}_1+(1-\theta)\mathbf{y}_2\in Y$. Now

$$v(\theta\mathbf{y}_1+(1-\theta)\mathbf{y}_2)=\inf_{\mathbf{x}\in S(\theta\mathbf{y}_1+(1-\theta)\mathbf{y}_2)} f(\mathbf{x})$$

$$\leqslant \inf_{\mathbf{x}_1\in S(\mathbf{y}_1),\mathbf{x}_2\in S(\mathbf{y}_2)} f(\theta\mathbf{x}_1+(1-\theta)\mathbf{x}_2),$$

as by the above argument such points are in $S(\theta\mathbf{y}_1+(1-\theta)\mathbf{y}_2)$,

$$\leqslant\theta \inf_{\mathbf{x}_1\in S(\mathbf{y}_1)} f(\mathbf{x}_1)+(1-\theta)\inf_{\mathbf{x}_2\in S(\mathbf{y}_2)} f(\mathbf{x}_2)$$
$$\leqslant\theta v(\mathbf{y}_1)+(1-\theta)v(\mathbf{y}_2).$$

Definition 6.4 The *dual* function of the CPP is

$$w(\mathbf{u})=\inf_{\mathbf{x}\in X}\mathscr{L}(\mathbf{x},\mathbf{u}),\qquad X=\operatorname{dom} f\cap\left\{\bigcap_{i=1}^{n}\operatorname{dom-g}_i\right\}\tag{6.11}$$

Lemma 6.3 *The dual function is concave.*

Proof This follows by the standard argument which interestingly does not use $\mathbf{u}\geqslant 0$. However, it is also an immediate corollary of the next result which expresses $-w(\mathbf{u})$ as the conjugate of $v(\mathbf{y})$.

Lemma 6.4

$$w(\mathbf{u})=\inf_{\mathbf{y}\in Y}(v(\mathbf{y})-\mathbf{y}^{\mathrm{T}}\mathbf{u})\tag{6.12}$$

Proof By (6.11),

$$w(\mathbf{u})=\inf_{\mathbf{x}\in X}\{f(\mathbf{x})-\mathbf{u}^{\mathrm{T}}\mathbf{g}(\mathbf{x})\}.$$

Let $\mathbf{y}\in Y$ and note that $\mathbf{g}(\mathbf{x})\geqslant\mathbf{y}$ for $\mathbf{x}\in S(\mathbf{y})$. Then as $f(\mathbf{x})=+\infty$, $\mathbf{x}\notin\operatorname{dom} f$,

$$w(\mathbf{u})\leqslant\inf_{\mathbf{x}\in S(\mathbf{y})}\{f(\mathbf{x})-\mathbf{y}^{\mathrm{T}}\mathbf{u}\}$$
$$=v(\mathbf{y})-\mathbf{y}^{\mathrm{T}}\mathbf{u}$$
$$\leqslant\inf_{\mathbf{y}\in Y}\{v(\mathbf{y})-\mathbf{y}^{\mathrm{T}}\mathbf{u}\}$$

as $\mathbf{y}$ was chosen arbitrarily. To obtain the opposite inequality, set $\mathbf{y}_1=\mathbf{g}(\mathbf{x}_1)$ for $\mathbf{x}_1\in X$. Then

$$f(\mathbf{x}_1)-\mathbf{g}(\mathbf{x}_1)^{\mathrm{T}}\mathbf{u}\leqslant\inf_{\mathbf{x}\in S(\mathbf{y}_1)}\{f(\mathbf{x})-\mathbf{y}_1^{\mathrm{T}}\mathbf{u}\}$$
$$\geqslant v(\mathbf{y}_1)-\mathbf{y}_1^{\mathrm{T}}\mathbf{u}$$
$$\geqslant\inf_{\mathbf{y}\in Y}\{v(\mathbf{y})-\mathbf{y}^{\mathrm{T}}\mathbf{u}\}.$$

But $\mathbf{x}_1$ was chosen arbitrarily; so

$$\inf_{\mathbf{x}\in X} \{f(\mathbf{x})-\mathbf{u}^{\mathrm{T}}\mathbf{g}(\mathbf{x})\} \geqslant \inf_{\mathbf{y}\in Y} \{v(\mathbf{y})-\mathbf{y}^{\mathrm{T}}\mathbf{u}\}$$

which gives the required inequality.

The relationship of conjugacy between $v(\mathbf{y})$ and $-w(\mathbf{u})$ permits certain results to be deduced immediately.

Theorem 6.2 *Let*

$$\tau = \sup_{\mathbf{u}\in \mathrm{dom}(-w)} w(\mathbf{u}) = \sup_{\mathbf{u}\in \mathrm{dom}(-w)} \{\mathbf{u}^{\mathrm{T}}0-(-w(\mathbf{u}))\} \tag{6.13}$$

Then

(i) *If τ is finite then $0\in \bar{Y}$, the closure of Y.*

(ii) *$\tau \leqslant v(0)$. This can be restated as* sup dual $\leqslant \inf_{\mathbf{x}\in S(0)} f(\mathbf{x}) =$ inf primal.

(iii) *If $v(0)$ is finite then $\tau = v(0)$ if and only if v is closed at 0. That is if and only if v is lower semicontinuous at 0.*

(iv) *If $v(\mathbf{y})$ is stable at zero then*

   (a) *the supremum of $w(\mathbf{u})$ is attained for $\mathbf{u}=\mathbf{u}^* \geqslant 0$,*

   (b) *$\mathbf{u}^*$ is a subgradient of $v(\mathbf{y})$ at $\mathbf{y}=0$, and*

   (c) *the set of optimal solutions of the CPP is characterized by the optimality conditions of Lemma 6.1.*

Proof It follows from (6.13) that

$$\tau = v^{**}(0)$$

where v^{**} is the second conjugate of v. Results (i)–(iii) are now a consequence of Theorem 6.1 and the discussion leading to it. To prove (iv) we have, as $v^{**}(0)=v(0)$, by (6.12)

$$\sup_{\mathbf{u}} w(\mathbf{u}) = v(0) \leqslant v(\mathbf{y})-\mathbf{y}^{\mathrm{T}}\mathbf{u} \quad \forall\, \mathbf{y} \Rightarrow \mathbf{u}^{\mathrm{T}} \in \partial v(0).$$

By exercise 6.1(iv) the maximum value of w is attained for any $\mathbf{u}^{\mathrm{T}}\in\partial v(0)$ which is non-null by assumption. To show that $\mathbf{u}^{\mathrm{T}}\in\partial v(0)\geqslant 0$, note that

$$\mathbf{y}_2 \leqslant \mathbf{y}_1 \Rightarrow S(\mathbf{y}_2) \supseteq S(\mathbf{y}_1) \Rightarrow v(\mathbf{y}_2) \leqslant v(\mathbf{y}_1).$$

The subgradient inequality applied at $\mathbf{y}_1=0$ with $\mathbf{y}_2=-\lambda\mathbf{e}_k$, $\lambda \geqslant 0$, gives

$$v(-\lambda\mathbf{e}_k) \geqslant v(0)-\lambda\mathbf{e}_k^{\mathrm{T}}\mathbf{u}$$

so that

$$\lambda u_k \geqslant v(0)-v(-\lambda\mathbf{e}_n) \geqslant 0, \qquad k=1,2,\ldots,m.$$

Example 6.2 Let the constraint set be defined by $g=-x$, $x \geqslant 0$. Then

$$S(y)=\begin{cases}\{y\leqslant -x\leqslant 0\}, & y\leqslant 0\\ \varnothing, & y>0.\end{cases}$$

Case (i) v not lower semicontinuous at 0. Let $f(0)=0$, $f(x)=-1$, $x>0$. Then $v(0)=f(0)$ as $S(0)=\{0\}$. However, $v(y)=-1$, $y<0$, so that $v^{**}(y)=-1$, $y\leqslant 0$. Here $v(0)=0\geqslant v^{**}(0)=\sup_{\text{dom}(-w)} w(u)=-1$. This is an example of a *duality gap*.

Case (ii) v closed at zero but not stable. Let $f(x)=-\sqrt{x}$, $x\geqslant 0$. Then $v(y)=-\sqrt{-y}$ and $\text{dom}\, v=(-\infty,0)$. This corresponds to example 4.3(ii) so that $v(y)$ is not stable at $y=0$ and $\partial v(0)=\varnothing$. Direct calculation gives

$$w(u)=\begin{cases}-1/4u, & u>0,\\ -\infty, & u\leqslant 0,\end{cases}$$

verifying that $\sup_{u>0} w(u)=0=v(0)$ but that the supremum is not attained for any finite u.

Case (iii) v is stable at zero. Let $f=e^{-x}$, $x\geqslant 0$. Then $v(y)=e^{y}$ and $\text{dom}\, v=(-\infty,0]$. The dual function is

$$w(u)=\begin{cases}-\infty, & u<0\\ u-u\log u, & 0\leqslant u\leqslant 1\\ 1 & u>1\end{cases}$$

so that $\sup_{u\geqslant 0} w(u)=1=v(0)$ and is attained for $1\leqslant u\leqslant\infty$.

Exercise 6.1 (i) Let

$$f(x)=\begin{cases}x^2/2, & |x|\leqslant c\\ c\,|x|-c^2/2, & |x|\geqslant c.\end{cases}$$

Compute f^*, f^{**}.

(ii) When does $f=f^*$ in the sense of having the same functional form?

(iii) Verify the properties

$$(f+\alpha)^*=f^*-\alpha,$$

$$(\alpha f)^*=\alpha f^*(1/\alpha), \qquad \alpha>0,$$

$$f^*(\mathbf{u})=g^*(\alpha\mathbf{u}) \quad \text{if} \quad f(\mathbf{x})=g(\mathbf{x}/\alpha), \qquad \alpha>0,$$

$$\left(\inf_{i\in\sigma} f_i\right)^*=\sup_{i\in\sigma} f_i^*.$$

(iv) $\partial f(\mathbf{x})=\{\mathbf{u}^{\mathrm{T}};\, f^*(\mathbf{u})+f(\mathbf{x})-\mathbf{u}^{\mathrm{T}}\mathbf{x}=0\}$.

Exercise 6.2 Show that if f is positively homogeneous and convex then f^*

is the indicator function of a closed convex set. What is this set in the case that f is a norm?

Exercise 6.3 (i) Let C be symmetric and positive definite, and consider the CPP

$$\min_{\mathbf{x}\in X} \tfrac{1}{2}\mathbf{x}^{\mathrm{T}}C\mathbf{x}-\mathbf{c}^{\mathrm{T}}\mathbf{x}; \qquad X=\{\mathbf{x};\, A\mathbf{x}\geqslant b\}$$

Show that the dual can be put in the form

$$\max_{\mathbf{u}\geqslant 0,\mathbf{x}} \mathbf{u}^{\mathrm{T}}\mathbf{b}-\tfrac{1}{2}\mathbf{x}^{\mathrm{T}}C\mathbf{x} \quad \text{where} \quad \mathbf{x}^{\mathrm{T}}C-\mathbf{u}^{\mathrm{T}}A-\mathbf{c}^{\mathrm{T}}=0.$$

What simplifications are possible when $A=I_p$, $\mathbf{b}=0$? This problem is called the quadratic programming problem.

1.7 DESCENT METHODS FOR MINIMIZING CONVEX FUNCTIONS

The main techniques applied to develop algorithms in subsequent chapters are direct adaptations of the descent methods that are commonly used in solving general optimization problems. Some aspects of this approach are described here to provide background for the later applications. We consider first the method of steepest descent and establish a convergence result for minimizing a convex differentiable function that is bounded below. It is noted that this method has a poor convergence rate which tends to make it impractical for general applications. This contrasts with the situation here, in which the method is used as a basis for deriving very satisfactory algorithms in which the finiteness of the procedure exorcizes the slow convergence problems. Steepest descent also can fail if the differentiability condition is relaxed (precisely the case of most interest here), but this problem can be removed by using a larger set, the ε-subdifferential, instead of the subdifferential as the set on which the descent direction is defined. In principle this additional information permits a convergent algorithm to be given. The significance of this result when the objective function is polyhedral convex is noted.

Let $f(\mathbf{x})$ convex with $\operatorname{dom} f=R^p$ be given. Each iteration of a descent algorithm has two essential components:

(i) It is necessary to provide a recipe which at each point $\mathbf{x}$ in $\operatorname{dom} f$ either determines that $0\in\partial f(\mathbf{x})$ so that $\mathbf{x}$ is optimal, or gives a vector $\mathbf{t}$ satisfying $f'(\mathbf{x}:\mathbf{t})<0$. If $\mathbf{t}$ satisfies this condition then we say that $\mathbf{t}$ is a *descent vector* or that it is downhill for minimizing f at $\mathbf{x}$; by which we mean that $f(\mathbf{x}+\gamma\mathbf{t})<f(\mathbf{x})$ for all $\gamma>0$ small enough.

(ii) The second component of the current iteration step must prescribe how to move in the direction determined by $\mathbf{t}$ to generate the new $\mathbf{x}$. If the new point is $\mathbf{x}+\gamma\mathbf{t}$ then an obvious requirement to impose on γ is that f must be decreased. However, the choice of γ must not be too conservative or there is some possibility of premature termination. Two possibilities are now developed under the assumptions:

(a) f is at least twice continuously differentiable,

(b) $f<K_1 \Rightarrow f^{(2)}(\mathbf{x}:\mathbf{t},\mathbf{t})/\|\mathbf{t}\|^2<K_2$, and

(c) the set $X=\{\mathbf{x}; f(\mathbf{x})<K_1\}$ is bounded (so that f has a bounded minimizer).

Rule A Assume $\mathbf{x}\in X$ and choose γ so that $\mathbf{x}+\gamma\mathbf{t}$ steps to the minimum of f in the direction determined by $\mathbf{t}$. Necessarily $\mathbf{x}+\gamma\mathbf{t}\in X$. In this case γ is characterized by

$$f'(\mathbf{x}+\gamma\mathbf{t}:\mathbf{t})=0 \tag{7.1}$$

so that our ability to carry out this step is predicated on our ability to solve this (in general) nonlinear equation for γ. Analytically,

$$\gamma=f'(\mathbf{x}:\mathbf{t})/f^{(2)}(\mathbf{x}+\bar{\gamma}\mathbf{t}:\mathbf{t},\mathbf{t}) \tag{7.2}$$

where $\bar{\gamma}$ is a mean value. As f is convex and X is bounded, it follows that $f^{(2)}(\mathbf{x}+\bar{\gamma}\mathbf{t}:\mathbf{t},\mathbf{t})>0$ so that a lower bound for γ is given by

$$\begin{aligned}\xi&=-f'(\mathbf{x}:\mathbf{t})/K_2\|\mathbf{t}\|^2.\\&=-f'(\mathbf{x}:\mathbf{t}/\|\mathbf{t}\|)/K_2\|\mathbf{t}\|.\end{aligned} \tag{7.3}$$

Using ξ to define the step leads to a procedure which is easy to analyse. Expanding $f(\mathbf{x}+\xi\mathbf{t})$ using Taylor's theorem with mean value form of the remainder gives

$$\begin{aligned}f(\mathbf{x}+\xi\mathbf{t})&=f(\mathbf{x})+\xi f'(\mathbf{x}:\mathbf{t})+\tfrac{1}{2}\xi^2 f^{(2)}(\mathbf{x}+\bar{\bar{\gamma}}\mathbf{t}:\mathbf{t},\mathbf{t})\\&=f(\mathbf{x})-\frac{f'(\mathbf{x}:\mathbf{t}/\|\mathbf{t}\|)^2}{K_2}\left\{1-\frac{1}{2K_2}f^{(2)}\left(\mathbf{x}+\bar{\bar{\gamma}}\mathbf{t}:\frac{\mathbf{t}}{\|\mathbf{t}\|},\frac{\mathbf{t}}{\|\mathbf{t}\|}\right)\right\}\\&\leqslant f(\mathbf{x})-\frac{1}{2K_2}f'(\mathbf{x}:\mathbf{t}/\|\mathbf{t}\|)^2\end{aligned} \tag{7.4}$$

This shows that the successive iterates $f(\mathbf{x}_i)$ are decreasing so that the sequence $\{f(\mathbf{x}_i)\}$ is convergent. Further,

$$|f'(\mathbf{x}:\mathbf{t}/\|\mathbf{t}\|)|/<\sqrt{2K_2(f(\mathbf{x})-f(\mathbf{x}+\xi\mathbf{t}))} \tag{7.5}$$

so that if $\mathbf{x}^*$ is a limit point of the sequence of iterates $\{\mathbf{x}_i\}$ then

$$f'(\mathbf{x}^*:\mathbf{t}/\|\mathbf{t}\|)=0.$$

In particular, if $\mathbf{t}=-\nabla f(\mathbf{x})$ corresponding to *steepest descent* then

$$f'(\mathbf{x}^*:\mathbf{t}/\|\mathbf{t}\|)=\|\nabla f(\mathbf{x}^*)\|=0 \tag{7.6}$$

so that $\mathbf{x}^*$ is a minimizer of f. If, in addition, f has a unique minimizer then

$\{\mathbf{x}_i\} \to \mathbf{x}^*$. These conclusions remain valid for other choices of $\mathbf{t}$ if the angle between $\mathbf{t}$ and $-\nabla f$ is uniformly bounded away from $\pi/2$.

Rule B Note that if f is linear in $\mathbf{x}$ then $\gamma f'(\mathbf{x}:\mathbf{t})$ is the change in f in the step γ. Now predictions of the linear theory are likely to be valid for nonlinear f only if the achieved reduction in f bears some relation to the predicted reduction. This suggests that γ be chosen to satisfy

$$0 < \sigma \leqslant \frac{f(\mathbf{x}+\gamma\mathbf{t}) - f(\mathbf{x})}{\gamma f'(\mathbf{x}:\mathbf{t})} \leqslant 1 - \sigma \tag{7.7}$$

for some prescribed $\sigma < .5$. This test can be satisfied always as a consequence of the continuity of f for as $\gamma \to 0$ the test quantity tends to 1 (so the right-hand inequality serves to prevent γ becoming too small), while for γ large, $f(\mathbf{x}+\gamma\mathbf{t}) > f(\mathbf{x})$ as X is bounded, so the test quantity is negative. An iterative scheme based on this rule for selecting γ is certainly convergent if $\inf_i \gamma_i \geqslant \gamma_0 > 0$, for then

$$f(\mathbf{x}_i) - f(\mathbf{x}_i + \gamma_i \mathbf{t}_i) \geqslant -\sigma\gamma_0 f'(\mathbf{x}_i : \mathbf{t}_i) > 0 \tag{7.8}$$

so that $\{f(\mathbf{x}_i)\}$ is convergent. Also, extracting a convergent subsequence with limit $\mathbf{x}^*$ from $\{\mathbf{x}_i\}$ gives

$$f'(\mathbf{x}^* : \mathbf{t}^*) = 0$$

as a direct consequence of (7.8). In the case $\mathbf{t} = \nabla f(\mathbf{x})$ we recover (7.6). It follows that $\{\mathbf{x}_i\}$ produced by the method of steepest descent can have a limit point which is not a minimizer of f only if $\inf_i \gamma_i = 0$. However, if this behaviour is assumed it leads immediately to the violation of the right-hand inequality in (7.7) and hence to a contradiction.

Renark 7.1 Rule B is a good example of the type of test now generally used in practical algorithms for nonlinear optimization problems. Another example is given in Exercise 7.1. One connection between the two rules is that if rule A is employed with exact line search (γ determined by (7.1)), and if $\{\mathbf{x}_i\}$ converges, then eventually (7.7) is satisfied for all $\sigma < .5$ (Exercise 7.2).

The above discussion shows that convergence results are readily obtained for the method of steepest descent. However, it must be stressed that it is not a satisfactory procedure in general because the rate of convergence is too slow. In the particular case $f = \frac{1}{2}\mathbf{x}^{\mathrm{T}}A\mathbf{x}$, A positive definite, a sharp upper bound for the rate of convergence can be given by a straightforward argument. Here $\nabla f(\mathbf{x})^T = A\mathbf{x}$ so the method of steepest descent gives

$$\mathbf{x}_{i+1} = \mathbf{x}_i - \gamma_i A \mathbf{x}_i = \frac{P_i(A)}{P_i(0)}\mathbf{x} \tag{7.9}$$

where choosing γ_i by rule A is equivalent to choosing $P_i(A)$ to minimize

$$f\left(\frac{P_i(A)}{P_i(0)}\mathbf{x}\right)$$

where the choice runs over all linear polynomials. Thus an upper bound for the reduction in f in the step (7.9) can be found by specializing P_i. A suitable choice is $Q(u)$ chosen to satisfy

$$Q(\lambda_1) = -1$$
$$Q(\lambda_p) = 1$$

where the eigenvalues of A satisfy $0 < \lambda_1 \leqslant \lambda_2 \leqslant \cdots \leqslant \lambda_p$. This gives

$$Q(u) = -(\lambda_p + \lambda_1)/(\lambda_p - \lambda_1) + 2u/(\lambda_p - \lambda_1).$$

If

$$\mathbf{x}_i = \sum_{j=1}^{p} \alpha_j^{(i)} \mathbf{v}_j,$$

where the $\mathbf{v}_j$ are the eigenvectors of A corresponding to the eigenvalues λ_j, then

$$\begin{aligned} f(\mathbf{x}_{i+1}) &= f\left(\frac{P_i(A)}{P_i(0)}\mathbf{x}_i\right) \leqslant f\left(\frac{Q(A)}{Q(0)}\mathbf{x}_i\right) \\ &\leqslant \frac{1}{2Q(0)^2}\sum_{j=1}^{p} \lambda_j Q(\lambda_j)^2 \alpha_j^{(i)2} \\ &\leqslant \frac{1}{2Q(0)^2}\sum_{j=1}^{p} \lambda_j \alpha_j^{(i)2} \\ &= \frac{1}{Q(0)^2} f(\mathbf{x}_i) \end{aligned}$$

as $|Q(\lambda_j)| \leqslant 1$ by construction. Thus

$$f(\mathbf{x}_{i+1}) \leqslant \left(\frac{\lambda_p - \lambda_1}{\lambda_p + \lambda_1}\right)^2 f(\mathbf{x}_i) \tag{7.10}$$

In the case $p = 2$ it is fairly straightforward to verify that this inequality is sharp (Exercise 7.3). However, it is sharp in an asymptotic sense for all p (this result appear to be rather difficult to prove). This predicts a very unsatisfactory rate of convergence when the spread of eigenvalues is large, and practical experience confirms this prediction.

The situation is perhaps even worse when the differentiability conditions are relaxed, as the method of steepest descent may converge to a point which does not minimize f (recall that ∇f is defined in Exercise 4.3 in the case that $\partial f(\mathbf{x})$ is more than a singleton). For example, let

$$f=\begin{cases}5(9x^2+16y^2)^{\frac{1}{2}}, & x>|y|,\\ 9x+16\,|y|, & x<|y|.\end{cases}$$

We consider steepest descent applied to this problem with starting values satisfying $\frac{256}{81}|y_1|>x_1>|y_1|$. In this regime

$$\nabla f=\frac{5}{\sqrt{9x^2+16y^2}}(9x,\ 16y)$$

The step to the minimum in the direction $-\nabla f$ is given by the solution of

$$[9x_i \quad 16y_i]\begin{pmatrix}9 & \\ & 16\end{pmatrix}\left\{\begin{pmatrix}x_i\\ y_i\end{pmatrix}-\alpha\begin{pmatrix}9x_i\\ 16y_i\end{pmatrix}\right\}=0$$

(this should be compared with equation (7.1)). It is readily verified that

$$\frac{x_{i+1}}{y_{i+1}}=-\frac{256y_i}{81x_i}=\left(-\frac{256}{81}\right)\left(-\frac{81}{256}\right)\frac{x_{i-1}}{y_{i-1}}=\frac{x_{i-1}}{y_{i-1}}.$$

This shows that provided the initial point satisfies the stated conditions then $x>|y|$ is satisfied by all subsequent iterates. Thus the iteration is equivalent to steepest descent applied to $f=9x^2+16y^2$, and successive iterates lie alternatively on the straight lines $y=\frac{y_1}{x_1}x$ and $y=\frac{y_2}{x_2}x$ and converge to the origin. However, $f\to-\infty$ as $x\to-\infty$ and a direct calculation gives

$$0\notin\partial f(0,0)=[9,15]\times[-16,16].$$

This example shows how nondifferentiability can decouple the different pieces of the function. In this case the iteration is confined to the quadratic section and never sees the linear section unbounded below.

One proposal for overcoming this problem computes the descent direction over a set which draws differential information from a neighbourhood of the current point rather than just at the point itself.

Definition 7.1 Let $\varepsilon>0$ be given. If

$$f(\mathbf{y})\geqslant f(\mathbf{x})+\mathbf{v}^{\mathrm{T}}(\mathbf{y}-\mathbf{x})-\varepsilon, \qquad \forall\,\mathbf{y}\in\operatorname{dom} f \tag{7.11}$$

then $\mathbf{v}^{\mathrm{T}}$ is an ε-*subgradient* at $\mathbf{x}$. The set of all ε-subgradients at $\mathbf{x}$ is called the ε-*subdifferential* $\partial_\varepsilon f(\mathbf{x})$. Clearly $\partial f(\mathbf{x})\subseteq\partial_\varepsilon f(\mathbf{x})$ for all $\varepsilon>0$. However, $\partial_\varepsilon f(\mathbf{x})$ also contains $\partial f(\mathbf{y})$ for $\mathbf{y}$ close enough to $\mathbf{x}$.

Lemma 7.1 *Let* $\mathbf{x} \in \operatorname{int} \operatorname{dom} f$. *Then for each* $\varepsilon > 0$ *there exists* $\delta = \delta(\varepsilon)$ *such that*

$$\|\mathbf{y}-\mathbf{x}\| < \delta \Rightarrow \partial f(\mathbf{y}) \subseteq \partial_\varepsilon f(\mathbf{x}). \tag{7.12}$$

Proof Let $\mathbf{v}^{\mathrm{T}} \in \partial f(\mathbf{y})$. Then the subgradient inequality gives

$$\begin{aligned}
f(\mathbf{z}) &\geqslant f(\mathbf{y}) + \mathbf{v}^{\mathrm{T}}(\mathbf{z}-\mathbf{y}) \\
&\geqslant f(\mathbf{x}) + \mathbf{v}^{\mathrm{T}}(\mathbf{z}-\mathbf{x}) + f(\mathbf{y}) - f(\mathbf{x}) - \mathbf{v}^{\mathrm{T}}(\mathbf{y}-\mathbf{x}) \\
&\geqslant f(\mathbf{x}) + \mathbf{v}^{\mathrm{T}}(\mathbf{z}-\mathbf{x}) + (\mathbf{v}_x - \mathbf{v})^{\mathrm{T}}(\mathbf{y}-\mathbf{x}) + f(\mathbf{y}) - f(\mathbf{x}) \\
&\quad - \mathbf{v}_x^{\mathrm{T}}(\mathbf{y}-\mathbf{x}) \\
&\geqslant f(\mathbf{x}) + \mathbf{v}^{\mathrm{T}}(\mathbf{z}-\mathbf{x}) + (\mathbf{v}_x - \mathbf{v})^{\mathrm{T}}(\mathbf{y}-\mathbf{x})
\end{aligned}$$

where $\mathbf{v}_x^{\mathrm{T}} \in \partial f(\mathbf{x})$ can be chosen to maximize $(\mathbf{v}_x - \mathbf{v})^{\mathrm{T}}(\mathbf{y}-\mathbf{x})$. Thus δ must be chosen so small that

$$\inf_{\substack{\mathbf{v}^{\mathrm{T}} \in \partial f(\mathbf{y}) \\ \|\mathbf{y}-\mathbf{x}\| < \delta}} \ \sup_{\mathbf{v}_x^{\mathrm{T}} \in \partial f(\mathbf{x})} (\mathbf{v}_x - \mathbf{v})^{\mathrm{T}}(\mathbf{y}-\mathbf{x}) > -\varepsilon, \tag{7.13}$$

and this is possible because $\partial f(\mathbf{y})$ is bounded for $\mathbf{y}$ close enough to $\mathbf{x}$ by Lemma 4.4.

The importance of the ε-subdifferential in setting up descent methods is a consequence of the following results.

Lemma 7.2

$$f(\mathbf{y}) \leqslant \inf_{\mathbf{x}} f(\mathbf{x}) + \varepsilon \Rightarrow 0 \in \partial_\varepsilon f(\mathbf{y}) \tag{7.14}$$

Proof This follows from the definition of ε-subgradients as

$$f(\mathbf{z}) \geqslant \inf_{\mathbf{x}} f(\mathbf{x}) \geqslant f(\mathbf{y}) - \varepsilon, \qquad \forall\, \mathbf{z}. \tag{7.15}$$

Lemma 7.3 *If* $\mathbf{t}$ *satisfies*

$$\mathbf{t}^{\mathrm{T}}\mathbf{v} < 0, \qquad \forall\, \mathbf{v}^{\mathrm{T}} \in \partial_\varepsilon f(\mathbf{x})$$

(*by Theorem* 3.1 *this is possible if and only if* $0 \notin \partial_\varepsilon f(x)$), *then*

$$\inf_{\gamma} f(\mathbf{x} + \gamma \mathbf{t}) < f(\mathbf{x}) - \varepsilon \tag{7.16}$$

Proof This follows from Lemma 7.2 by noting that
(i) $g(\gamma) = f(\mathbf{x} + \gamma\mathbf{t})$ is a convex function of γ, and
(ii) $\partial_\varepsilon g(\gamma) = \{\mathbf{t}^{\mathrm{T}}\mathbf{v}, \mathbf{v}^{\mathrm{T}} \in \partial_\varepsilon f(\mathbf{x} + \gamma\mathbf{t})\}$
so that, in particular,

$$0 \notin \partial_\varepsilon g(0).$$

One possible algorithm is now described. In it tol is a parameter used in testing for termination, and α is used to adjust the current value of ε.

Algorithm

(i) *Set* $\mathbf{x}=\mathbf{x}_0$, $\varepsilon=\varepsilon_0$, $\alpha<1$, tol.

(ii) *If* $0\in\partial_\varepsilon f(\mathbf{x})$ *then go to (iv);*
else find $\mathbf{t}\ni\mathbf{t}^{\mathrm{T}}\mathbf{v}<0, \forall\, \mathbf{v}^{\mathrm{T}}\in\partial_\varepsilon f(\mathbf{x})$.

(iii) $\gamma=\arg\min\limits_{\lambda} f(\mathbf{x}+\lambda\mathbf{t})$.
$\mathbf{x}:=\mathbf{x}+\gamma\mathbf{t}$
go to (ii).

(iv) *If* $\varepsilon<$tol *then stop;*
else while $0\in\partial_\varepsilon f(\mathbf{x})$ *do* $\varepsilon:=\alpha\varepsilon$
go to (ii).

It is an immediate consequence of the two previous lemmas that this algorithm terminates in a finite number of steps with a point $\mathbf{x}$ such that $f(\mathbf{x})-\inf\limits_{\mathbf{x}} f<$tol provided that the intermediate calculations necessary to test $0\in\partial_\varepsilon f(\mathbf{x})$ and to minimize $f(\mathbf{x}+\lambda\mathbf{t})$ with respect to λ can be carried out. Note that the finiteness of the procedure is not lost if instead of minimizing $f(\mathbf{x}+\lambda\mathbf{t})$ a search is made for a point $\mathbf{x}+\bar{\gamma}\mathbf{t}$ such that $f(\mathbf{x})-f(\mathbf{x}+\bar{\gamma}\mathbf{t})>\varepsilon$. This algorithm survives when applied to the previous example. Initially the steepest descent pattern can be followed exactly as before provided ε is small enough. But in at most $f(\mathbf{x}_0)/\varepsilon$ steps this pattern must be interrupted. It is easy to see that $\delta=O(\varepsilon^{\frac{1}{2}}f^{\frac{1}{2}})$ for small ε in (7.13). Thus δ decreases more slowly than the distance of the successive iterates from the origin while the iteration is in the quadratic regime (this distance is $O(f)$). As the width of the section satisfying $\frac{256}{81}|y|\geqslant x\geqslant|y|$ is also $O(f)$, equating these estimates shows that $\partial_\varepsilon f$ must see points in the linear region at the latest by the time the current iterates are distance $O(\varepsilon)$ from the origin.

For one class of nondifferentiable optimization problems, descent methods provide a valuable problem solving idea. These problems involve minimizing polyhedral convex functions defined as the maximum of a finite set of affine functions on a domain which is a polyhedral convex set which can be R^p. Here an immediate consequence of the locally linear behaviour is that for each $\mathbf{x}$ in dom f there exist δ, ε_0 (in general depending on $\mathbf{x}$) such that

$$\partial_\varepsilon f(\mathbf{y})\subseteq\partial f(\mathbf{x}),\qquad \|\mathbf{y}-\mathbf{x}\|<\delta,\qquad \varepsilon<\varepsilon_0. \tag{7.17}$$

It will prove possible (the details are in Chapter 4) to generate descent steps with the property that the successive iterates are points where δ and ε can be given a priori finite bounds in (7.17). Such points are the extreme points of epi f, and in this case the ε-subgradient algorithm and the descent method are effectively equivalent. Finiteness follows from the above discussion or from the observation that the number of extreme points of epi f is finite.

Exercise 7.1 Discuss the convergence of the basic descent algorithm when γ is chosen to satisfy

$$|\nabla f(\mathbf{x}+\gamma\mathbf{t})\mathbf{t}| \geqslant -\sigma\, \nabla f(\mathbf{x})\mathbf{t}.$$

Exercise 7.2 Let $\{\mathbf{x}_i\}$ be a convergent sequence produced by a descent algorithm using rule A to determine the step at each stage. Then for any $\sigma < .5$ there is an i_0 such that the step also satisfies rule B for $i > i_0$.

Exercise 7.3 Show that the inequality (7.10) is sharp by considering the particular case $A = \operatorname{diag}\{\lambda_1, \lambda_2\}$ with $\lambda_2 > \lambda_1 > 0$.

Exercise 7.4 Establish the result

$$\delta^*(\mathbf{t} \mid \partial_\varepsilon f(\mathbf{x})) = \inf_{\lambda>0} \frac{f(\mathbf{x}+\lambda\mathbf{t}) - f(\mathbf{x}) + \varepsilon}{\lambda}.$$

Use the results of Exercise 4.1 to construct the positively homogeneous convex function of $\mathbf{t}$ dominated by $f(\mathbf{x}+\mathbf{t}) - f(\mathbf{x})$ and then the results of Exercise 6.2.

NOTES ON CHAPTER 1

The basic reference for this chapter is Rockafellar (1970) and there is really very little more to be said. Points of presentation have been influenced by Craven (1978), Geoffrion's survey article on duality in Geoffrion (1972) which stresses the value of the concept of stability (attributed to Gale (1967)), and by Luenberger (1969) among others. Theorem 2.3 is of key importance for subsequent work. It is due to Klee (1957). Multiplier conditions for problems with inequality constraints were first obtained by John (1948). The contribution of Kuhn and Tucker (1951) was in large part to give a constraint qualification which was sufficient to eliminate certain degenerate cases. Their qualification is implied for convex problems by the Slater condition employed here. For a detailed account of descent methods Fletcher (1980) is recommended. That the inequality (7.10) cannot be improved for general p is due to Akaike (1959). The example of false convergence is due to Wolfe (1975). The ε-subgradient algorithm is given by Bertsekas and Mitter (1973).

CHAPTER 2

Linear programming

2.1 INTRODUCTION

The basic linear programming problem (LP1) considered here is the minimization of a linear function subject to a set of linear inequality constraints:

$$\min_{\mathbf{x}\in X} \mathbf{c}^{\mathrm{T}}\mathbf{x}\colon X=\{\mathbf{x};\ A\mathbf{x}\geqslant\mathbf{b}\} \qquad \text{LP1}$$

where $A: R^p \rightarrow R^n$, $p<n$. First a summary of the basic results relating to the geometric and algebraic properties of this problem is given which emphasizes the central role of duality. For this purpose, it is convenient to assume that $X \neq \emptyset$ and that A has full column rank p. However, in developing algorithms these are points which require checking, and an effort is made to present methods which are robust in this respect. Algorithmic developments make up the major part of the chapter, and three main approaches are presented:

(i) the simplex algorithm for solving the dual problem to LP1 (by far and away the most extensively used algorithm for solving linear programming problems);
(ii) descent methods applied to LP1 in both reduced and projected gradient form; and
(iii) an exact penalty function algorithm which is also developed for LP1.

In addition, the relation between the linear programming problem and a linear complementarity problem is developed by considering Lemke's algorithm, and the special case of interval constraints in LP1 is treated in detail. This latter problem is of interest because the structure can be exploited to improve the efficiency of the resulting algorithm for a significant class of problems. An application is made in Chapter 3 to the discrete approximation problem in the l_1 norm.

A feature of this chapter is the effort made to provide a general approach to the problem of degeneracy. Degeneracy occurs when the problem of moving to a better objective function value is overconstrained. It is resolved by interpreting the requirements of a suitable move geometrically, and showing that these can be satisfied by a recursive procedure without going outside the basic algorithmic framework already being employed. While the idea is implicit in some early work on degeneracy in the simplex algorithm, it has been made explicit in this context only comparatively recently. The other applications of the idea appear to be new. Attention is also given to the provision of scale-free tests in choosing descent directions and making decisions on termination. Again the advantages have been known for a long time. However, acceptable implementations for the simplex algorithm have become available only recently and still involve a cost penalty. Thus it is of interest that the extension to the other basic algorithms is relatively much cheaper.

The important question of how the algorithms actually perform is deferred until Chapter 7, where matters relating to complexity and performance are discussed. By this stage the basic descent algorithms will have been considered in a variety of contexts and some new implementation considerations will have emerged, but the general performance characteristics are sufficiently similar to justify a collective treatment.

2.2 GEOMETRIC THEORY, DUALITY

The explicit form for the feasible set X in LP1 makes it easy to characterize the extreme points and directions of recession. Here representation theory is applied to characterize the optimal solutions of LP1. Then the dual problem is introduced and the implications of uniqueness for both the primal and dual problems are discussed.

Lemma 2.1 *A point $\mathbf{x}^*$ is an extreme point of X if and only if there exists an index set σ pointing to the rows of a submatrix A_σ of A such that*

(i) $|\sigma| \geqslant p$,
(ii) $\operatorname{rank}(A_\sigma) = p$,
(iii) $A_\sigma \mathbf{x}^* = \mathbf{b}_\sigma$, *and*
(iv) $A_{\sigma^c}\mathbf{x}^* > \mathbf{b}_{\sigma^c}$

where σ^c indicates the set complement.

Proof If $A\mathbf{x}^* > \mathbf{b}$ then $\mathbf{x}^*$ is an interior point of X. Thus the set of constraints holding with equality must be nontrivial, and it is necessary to show that $\mathbf{x}^*$ is an extreme point if and only if $\operatorname{rank}(A_\sigma) = p$. But if $\operatorname{rank}(A_\sigma) < p$ then there exists a nontrivial $\mathbf{v}$ satisfying

$$A_\sigma \mathbf{v} = 0$$

and

$$A_{\sigma^c}(\mathbf{x}^* + \theta \mathbf{v}) \geqslant \mathbf{b}_{\sigma^c}$$

for all θ small enough so that $\mathbf{x}^*$ lies in the interior of a line segment in X, contradicting that $\mathbf{x}^*$ is an extreme point. Now assume that $\operatorname{rank}(A_\sigma) = p$

and that $\mathbf{x}^*$ is not an extreme point. Then there exist feasible points $\mathbf{y}$, $\mathbf{z}$ such that

$$\mathbf{x}^* = \theta\mathbf{y} + (1-\theta)\mathbf{z}$$

for $0 < \theta < 1$. Now $\mathbf{y}$, $\mathbf{z}$ must satisfy

$$A_\sigma \mathbf{y} \geqslant \mathbf{b}_\sigma, \qquad A_\sigma \mathbf{z} \geqslant \mathbf{b}_\sigma$$

with at least one strict inequality in each case (the contrary implies equality with $\mathbf{x}^*$ by the rank condition on A_σ). But than $\mathbf{x}^*$ cannot satisfy condition (iii).

Corollary 2.1 *The set of extreme points of X, $S(X) = \{\mathbf{s}_1, \mathbf{s}_2, \ldots, \mathbf{s}_{k_1}\}$, is finite.*

Lemma 2.2 *The feasible set has nontrivial directions of recession if and only if the system of inequalities*

$$A\mathbf{v} \geqslant 0$$

has a nontrivial solution.

If $\mathbf{t}^*$ is an extremal direction (of recession) then there exists an index set μ pointing to the rows of a submatrix A_μ of A such that

(i) $|\mu| \geqslant p-1$,
(ii) $\text{rank}\,(A_\mu) = p-1$,
(iii) $A_\mu \mathbf{t}^* = 0$, *and*
(iv) $A_{\mu^c} \mathbf{t}^* > 0$.

Proof This is an easy modification of the proof for Lemma 2.1.

Remark 2.1 The rank assumption on A implies that X cannot contain lines, for the alternative requires that there exist $\mathbf{v}$ such that both $\mathbf{v}$ and $-\mathbf{v}$ are directions of recession. This requires $A\mathbf{v} \geqslant 0$ and $-A\mathbf{v} \geqslant 0$ so that $A\mathbf{v} = 0$. But $\text{rank}\,(A) = p$ whence $\mathbf{v} = 0$.

Corollary 2.2 *The set of extreme directions of X, $C(X) = \{\mathbf{t}_1, \mathbf{t}_2, \ldots, \mathbf{t}_{k_2}\}$ is finite.*

Theorem 2.1 (*geometric form of the fundamental theorem*) *The primal problem (LP1) has a bounded solution if and only if*

$$\mathbf{c}^T\mathbf{t} \geqslant 0, \qquad \mathbf{t} \in C(X) \tag{2.1}$$

where $C(X)$ is the set of extreme directions for X. In this case

$$\min_{\mathbf{x} \in X} \mathbf{c}^T\mathbf{x} = \min_{\mathbf{s} \in S(X)} \mathbf{c}^T\mathbf{s} \tag{2.2}$$

where $S(X)$ is the set of extreme points for X.

Proof It follows from the Klee representation theorem (Theorem 2.3 of Chapter 1) that for each $\mathbf{x} \in X$ there exist multipliers θ_i, $i = 1, 2, \ldots, k_1$, λ_j, $j = 1, 2, \ldots, k_2$, such that $\theta_i, \lambda_j \geq 0$, $\sum_{i=1}^{k_1} \theta_i = 1$, and

$$\mathbf{x} = \sum_{i=1}^{k_1} \theta_i \mathbf{s}_i + \sum_{j=1}^{k_2} \lambda_j \mathbf{t}_j. \tag{2.3}$$

Taking the scalar product of (2.3) with $\mathbf{c}$ we see that (2.1) must hold as otherwise the λ_j can be chosen to make $\mathbf{c}^T\mathbf{x}$ negative and arbitrarily large in magnitude. Thus the smallest contribution the extreme directions can make to the objective function is zero when $\lambda_j = 0$, $j = 1, 2, \ldots, k_2$. In this case

$$\begin{aligned} \mathbf{c}^T\mathbf{x} &= \sum_{i=1}^{k_1} \theta_i \mathbf{c}^T\mathbf{s}_i \\ &\geq \left\{ \sum_{i=1}^{k_1} \theta_i \right\} \min_{\mathbf{s}_i \in S(X)} \mathbf{c}^T\mathbf{s}_i. \end{aligned}$$

Remark 2.2 *This theorem shows that the solution of LP1 can be found by considering only the extreme points of X.* This observation underlies the basic strategy of most linear programming algorithms. An immediate corollary is that there exist finite algorithms for linear programming, one such being provided by the exhaustive testing of all extreme points of X to find the smallest value of $\mathbf{c}^T\mathbf{s}_i$. The $\mathbf{s}_i$ can be found by considering all $p \times p$ submatrices of A and applying Lemma 2.1.

Example 2.1 Consider X defined by

$$A = \begin{bmatrix} 1 & 1 \\ -1 & 10 \\ 10 & -1 \end{bmatrix}, \qquad \mathbf{b} = \begin{bmatrix} 1 \\ -1 \\ -1 \end{bmatrix}.$$

To calculate the extreme points of X let

$$\nu = \{1, 2\}, \qquad A_\nu^{-1}\mathbf{b}_\nu = \begin{bmatrix} 1 \\ 0 \end{bmatrix}, \qquad \text{extreme point;}$$

$$\nu = \{1, 3\}, \qquad A_\nu^{-1}\mathbf{b}_\nu = \begin{bmatrix} 0 \\ 1 \end{bmatrix}, \qquad \text{extreme point;}$$

$$\nu = \{2, 3\}, \qquad A_\nu^{-1}\mathbf{b}_\nu = -\frac{1}{9}\begin{bmatrix} 1 \\ 1 \end{bmatrix}, \quad \notin X.$$

To calculate the extreme directions let

$$\mu=\{1\},\qquad A_\mu \mathbf{t}=0 \Rightarrow \mathbf{t}=\lambda\begin{bmatrix}1\\-1\end{bmatrix},$$

$$A\mathbf{t}=\lambda\begin{bmatrix}0\\-11\\11\end{bmatrix}\Rightarrow \mathbf{t}\quad \text{not extreme direction;}$$

$$\mu=\{2\},\qquad A_\mu \mathbf{t}=0 \Rightarrow \mathbf{t}=\lambda\begin{bmatrix}10\\1\end{bmatrix},$$

$$A\mathbf{t}=\lambda\begin{bmatrix}11\\0\\99\end{bmatrix}\Rightarrow \mathbf{t}\quad \text{is an extreme direction if } \lambda>0;$$

$$\mu=\{3\},\qquad A_\mu \mathbf{t}=0 \Rightarrow \mathbf{t}=\lambda\begin{bmatrix}1\\10\end{bmatrix},$$

$$A\mathbf{t}=\lambda\begin{bmatrix}11\\99\\0\end{bmatrix}\Rightarrow \mathbf{t}\quad \text{is an extreme direction if } \lambda>0.$$

Linear programming problems with X as feasible region can now be solved readily. For example consider the following possibilities where $\mathbf{c}$ defines the objective function and $\mathbf{x}^*$ minimizes LP1.

(i) $\mathbf{c}=\begin{bmatrix}0\\1\end{bmatrix}$, $\mathbf{x}^*=\begin{bmatrix}1\\0\end{bmatrix}$, (2.1) is satisfied, and the solution is unique.

(ii) $\mathbf{c}=\begin{bmatrix}1\\1\end{bmatrix}$, $\mathbf{x}^*=\theta\begin{bmatrix}1\\0\end{bmatrix}+(1-\theta)\begin{bmatrix}0\\1\end{bmatrix}$, $0\leqslant\theta\leqslant 1$, (2.1) is satisfied, and the solution is not unique.

(iii) $\mathbf{c}=\begin{bmatrix}-10\\1\end{bmatrix}$: in this case $\mathbf{c}^{\mathrm{T}}\begin{bmatrix}1\\10\end{bmatrix}=0$ but $\mathbf{c}^{\mathrm{T}}\begin{bmatrix}10\\1\end{bmatrix}=-99$ so that (2.1) is not satisfied and the linear program is unbounded.

At an optimum point the objective function must be nondecreasing for all directions into X. This is easily specified here because of the polyhedral nature of the constraints defining X. Thus a direction $\mathbf{t}$ is into X (or feasible) at $\mathbf{x}^*$ if $\mathbf{x}^*+\theta\mathbf{t}$ is feasible for all small enough θ. That is, provided $\boldsymbol{\rho}_i(A)\mathbf{t}\geqslant 0$ for all i such that $\boldsymbol{\rho}_i(A)\mathbf{x}^*=b_i$. It follows that $\mathbf{x}^*$ is optimal provided

$$\mathbf{c}^{\mathrm{T}}\mathbf{t}\geqslant 0,\qquad \forall\,\mathbf{t}\ni A_\sigma\mathbf{t}\geqslant 0 \tag{2.4}$$

where σ is the index set pointing to the constraints holding with equality at $\mathbf{x}^*$. An application of Farkas' lemma (Theorem 3.2 of Chapter 1) now gives

$$\mathbf{c}^{\mathrm{T}}=\mathbf{u}_\sigma^{\mathrm{T}}A_\sigma,\qquad \mathbf{u}_\sigma\geqslant 0. \tag{2.5}$$

Essentially (2.5) states the Kuhn–Tucker conditions for LP1 for it implies that there exists a vector of multipliers such that

$$\begin{aligned}&\mathbf{c}^{\mathrm{T}}=\mathbf{u}^{\mathrm{T}}A,\\&\mathbf{u}\geqslant 0,\quad \text{and}\\&u_i(\boldsymbol{\rho}_i(A)\mathbf{x}^*-b_i)=0,\quad i=1,2,\ldots,n,\end{aligned}\tag{2.6}$$

and (2.6) is identical with (5.16) and (5.17) of Chapter 1 in this case. Duality adds more here. The Lagrangian ((6.7) of Chapter 1) is

$$\mathscr{L}(\mathbf{x},\mathbf{u})=\mathbf{c}^{\mathrm{T}}\mathbf{x}-\mathbf{u}^{\mathrm{T}}(A\mathbf{x}-\mathbf{b})\tag{2.7}$$

and, because of the linearity in $\mathbf{x}$,

$$w(\mathbf{u})=\inf_{\mathbf{x}\in R^p}\mathscr{L}(\mathbf{x},\mathbf{u})=\begin{cases}\mathbf{u}^{\mathrm{T}}\mathbf{b}, & A^{\mathrm{T}}\mathbf{u}-\mathbf{c}=0\\-\infty & \text{otherwise}\end{cases}\tag{2.8}$$

Thus the dual problem (LP2) is

$$\max_{\mathbf{u}\in U}\mathbf{u}^{\mathrm{T}}\mathbf{b}: U=\{\mathbf{u};\, A^{\mathrm{T}}\mathbf{u}-\mathbf{c}=0,\mathbf{u}\geqslant 0\}\qquad \text{LP2}$$

where Theorem 6.2 of Chapter 1 has been used to impose the positivity conditions.

The equivalence between the solutions to LP2 and the Kuhn–Tucker multipliers for LP1 is summarized in the following theorem.

Theorem 2.2 (*duality theorem*) *If* $\mathbf{x}$ *and* $\mathbf{u}$ *are feasible for LP1 and LP2 respectively then*

$$\mathbf{c}^{\mathrm{T}}\mathbf{x}\geqslant\mathbf{b}^{\mathrm{T}}\mathbf{u}\tag{2.9}$$

so that min (LP1) ⩾ max (LP2). *If* $\mathbf{x}$ *solves LP1 then the associated Kuhn–Tucker multiplier vector provides an optimal* $\mathbf{u}$ *for LP2 for it is feasible, and equality holds in* (2.9).

Theorem 2.3 *If LP1 has an unbounded solution then the dual program LP2 has no solution.*

Proof If the primal has an unbounded solution then there exists a direction of recession $\mathbf{t}$ for X such that

$$\mathbf{c}^{\mathrm{T}}\mathbf{t}<0,\qquad A\mathbf{t}\geqslant 0\tag{2.10}$$

If the dual has a feasible point then there exists $\mathbf{u}\geqslant 0$ such that

$$A^{\mathrm{T}}\mathbf{u}-\mathbf{c}=0.$$

But then

$$\mathbf{t}^{\mathrm{T}}A^{\mathrm{T}}\mathbf{u}-\mathbf{t}^{\mathrm{T}}\mathbf{c}=0$$

contradicting (2.10).

Lemma 2.3 *Let* $\mathbf{x}^*$ *be an optimal extreme point of* X *and* σ *be the defining index set. Then there exists an index set* $\nu \subseteq \sigma$ *such that* $|\nu| = p$, $\text{rank}(A_\nu) = p$, *and there is a Kuhn–Tucker vector for LP1 with* $u_i = 0$, $i \notin \nu$.

Proof The complementarity conditions (2.6) give

$$u_i(\boldsymbol{\rho}_i(A)\mathbf{x}^* - b_i) = 0, \qquad i = 1, 2, \ldots, n$$

so that $u_i = 0$ if $i \in \sigma^c$. Thus there is nothing to prove unless $|\sigma| > p$. In this case consider the subset σ' of the rows of A_σ such that

(i) $u_i > 0 \Leftrightarrow i \in \sigma'$, and
(ii) $\exists\, \mathbf{v} \ni \mathbf{v}^T A_{\sigma'} = 0$.

If σ' exists we show that we can find an index set σ'' and new multipliers u_i' such that

(a) $|\sigma''| < |\sigma'|$, and
(b) $u_i' > 0 \Leftrightarrow i \in \sigma''$.

Consider

$$w_i(\theta) = u_i - \theta v_i, \qquad i \in \sigma'.$$

We have

$$w_i(\theta) \geq 0, \qquad \theta \leq \min_{j \in \sigma', v_j > 0} u_j / v_j = \hat{\theta}$$

where the set $v_j > 0$ can always be chosen to be nonempty as the defining equation is homogeneous. Let the minimizing index be k and define

$$\sigma'' = \{i;\ w_i(\hat{\theta}) > 0\},$$

$$u_i' = \begin{cases} w_i(\theta), & i \in \sigma'' \\ 0 & \text{otherwise.} \end{cases}$$

Clearly $\mathbf{u}'$ is a Kuhn–Tucker multiplier vector for LP1 and is optimal for LP2 (this also follows from the duality theorem which gives the additional information that $\mathbf{v}^T \mathbf{b}_{\sigma'} = 0$). By construction $|\sigma''| < |\sigma'|$. Thus we can repeat the process, and after at most a finite number of steps we must obtain an index set $\sigma^{(e)}$ and multiplier vector $\mathbf{u}^{(e)}$ such that

(a) $u_i^{(e)} > 0 \Leftrightarrow i \in \sigma^{(e)}$, and
(b) $\boldsymbol{\rho}_i(A)$, $i \in \sigma^{(e)}$, linearly independent.

Clearly $|\sigma^{(e)}| \leq p$. If $|\sigma^{(e)}| < p$ then we select an index set $\nu \supset \sigma^{(e)}$ such that $|\nu| = p$ and $\text{rank}(A_\nu) = p$. Otherwise set $\nu = \sigma^{(e)}$. That this imbedding is possible is a consequence of the following result.

Lemma 2.4 *Let* M *have rank* p. *If* M *contains a* $k \times k$ *nonsingular submatrix then this can be imbedded in a* $p \times p$ *nonsingular submatrix.*

Proof There is no restriction in reordering the elements of M so that the submatrix in question is in the leading position. Consider now the elimination factorization of M into the product of upper and lower triangular factors by the procedure considered in Appendix 1 but restricting the pivot choice to the leading submatrix for the first k steps and then using complete pivoting for the next $p-k$ steps. This pivoting sequence defines a suitable imbedding.

Definition 2.1 The set ν defined in Lemma 2.3 is an *optimal reference.* A reference is any index set pointing to p rows of A defining an extreme point of X, or, by extension, the corresponding rows of A themselves.

Remark 2.3 The proof of Lemma 2.3 shows that associated with an optimal reference ν there is an associated multiplier vector $\mathbf{u}_\nu \geqslant 0$, and $\mathbf{c} = A_\nu^T \mathbf{u}_\nu$.

Theorem 2.4 (*uniqueness*) *Let $\mathbf{x}^*$ be a solution to LP1. Then $\mathbf{x}^*$ is unique if there exists an optimal reference $\nu \subseteq \sigma$ such that the associated multiplier vector satisfies $u_\nu > 0$. If $|\nu| = |\sigma| = p$ then the condition is also necessary.*

Proof If $\mathbf{x}^*$ is not unique then $H(\mathbf{c}, \mathbf{c}^T\mathbf{x}^*)$ which supports X must do so on a subset of dimension >0. Thus there must exist nontrivial $\mathbf{t}$ such that $\mathbf{x}^* + \lambda\mathbf{t} \in X \cap H(\mathbf{c}, \mathbf{c}^T\mathbf{x}^*)$ for $\lambda \geqslant 0$, small enough. Optimality implies that

$$0 = \mathbf{c}^T\mathbf{t} = \mathbf{u}_\nu^T A_\nu \mathbf{t}$$

while feasibility for λ small enough requires that

$$A_\nu \mathbf{t} \geqslant 0.$$

Thus

$$\mathbf{u} > 0 \Rightarrow A_\nu \mathbf{t} = 0 \Rightarrow \mathbf{t} = 0$$

by the rank assumption on A. On the other hand, let $u_{\nu(i)} = 0$. Then if $|\sigma| = p$ there exists $\mathbf{t} \neq 0$ satisfying

$$0 = \mathbf{c}^T\mathbf{t} = \boldsymbol{\rho}_i(A_\nu)\mathbf{t}, \qquad j \neq i, \qquad \boldsymbol{\rho}_i(A_\nu)\mathbf{t} = 1.$$

Clearly $\mathbf{x}^* + \lambda\mathbf{t}$ is feasible and optimal for all small enough $\lambda > 0$.

Exercise 2.1 By finding the extreme points of X, solve the linear programming problem:

$$\begin{aligned} \text{minimize} \quad & 2x_1 - 3x_2 \quad \text{subject to} \\ & 2x_1 - x_2 \geqslant -2 \\ & -x_1 - 2x_2 \geqslant -8 \\ & x_1, x_2 \geqslant 0. \end{aligned}$$

Exercise 2.2 (i) Verify Theorem 2.2 by direct calculation. Also show that the dual of LP2 is LP1.

(ii) Show that the dual of

$$\min_{\mathbf{x}} \mathbf{c}^{\mathrm{T}}\mathbf{x}; \qquad A\mathbf{x} \geqslant \mathbf{b}, \qquad \mathbf{x} \geqslant 0$$

is

$$\max_{\mathbf{u}} \mathbf{b}^{\mathrm{T}}\mathbf{u}; \qquad A^{\mathrm{T}}\mathbf{u} \leqslant \mathbf{c}, \qquad \mathbf{u} \geqslant 0.$$

This is called the *symmetric dual problem.*

Exercise 2.3 (i) Construct an example of an extreme point with the property that for each reference the set of directions $\mathbf{t}$ satisfying $A_\nu \mathbf{t} \geqslant 0$, where ν is the index set specifying the particular reference, properly contains the set of feasible directions.

(ii) Verify that the solution set for the symmetric dual problem is identical with the solution set of the *linear complementarity problem*:

$$\begin{bmatrix} \mathbf{y} \\ \mathbf{v} \end{bmatrix} = \begin{bmatrix} 0 & A \\ -A^{\mathrm{T}} & 0 \end{bmatrix} \begin{bmatrix} \mathbf{u} \\ \mathbf{x} \end{bmatrix} + \begin{bmatrix} -\mathbf{b} \\ \mathbf{c} \end{bmatrix} \qquad \mathbf{u}, \mathbf{v}, \mathbf{x}, \mathbf{y} \geqslant 0, \qquad [\mathbf{y}^{\mathrm{T}}, \mathbf{v}^{\mathrm{T}}] \begin{bmatrix} \mathbf{u} \\ \mathbf{x} \end{bmatrix} = 0.$$

(iii) Prove the result dual to that of Theorem 2.3 that if LP2 is unbounded than LP1 has no solution.

Exercise 2.4 (i) An inequality constraint $\boldsymbol{\rho}_i(A)\mathbf{x} \geqslant b_i$ is an *implicit equality* if $X \subset \{\boldsymbol{\rho}_i(A)\mathbf{x} = b_i\}$. Show than an inequality constraint is an implicit equality if and only if the system

$$\begin{aligned} A^{\mathrm{T}}\mathbf{u} &= -\boldsymbol{\rho}_i(A)^{\mathrm{T}} \\ \mathbf{b}^{\mathrm{T}}\mathbf{u} &= -b_i \\ \mathbf{u} &\geqslant 0 \end{aligned}$$

is feasible.

(ii) If the specification of X in LP1 contains exactly one implicit equality then this constraint is also redundant (Exercise 3.2 of Chapter 1).

Exercise 2.5. Show that the condition $|\sigma| = p$ is essential in the necessity part of Theorem 2.4. A simple example can be constructed when $p = 2$ by adding a redundant constraint parallel to the objective function.

2.3 ALGEBRAIC THEORY

In the previous section we obtained information about the structure of the solution of LP2 using duality arguments. However, these results can be obtained directly by straightforward algebraic arguments and this approach is developed in this section. It is conventional to concentrate on the nonzero elements of $\mathbf{u}$.

Definition 3.1 $\mathbf{u}$ is a *basic solution* of LP2 if there exists an index set $\sigma, |\sigma| = p$, pointing to the columns of a submatrix A_σ^T of A^T having the properties that

(i) $A_\sigma^T \mathbf{u}_\sigma = \mathbf{c}$,
(i) $\mathbf{u}_{\sigma^c} = 0$,
(iii) rank $(A_\sigma^T) = p$.

If $\mathbf{u}_\sigma \geqslant 0$ then $\mathbf{u}$ is called a *basic feasible solution*. If not all the inequalities $\mathbf{u}_\sigma \geqslant 0$ are strict then the solution is *degenerate*. The columns of A_σ^T are said to form the *basis* (basis matrix).

Remark 3.1 It follows from the duality theorem (Theorem 2.2) that the index set σ pointing to an optimal reference set for LP1 points also to the elements of a basic feasible solution which is optimal for LP2. This suggests a structural result which is explored in the next theorem. Note that in the primal, σ points to the inequalities satisfied as equations (the active constraints), while in the dual it points to components of $\mathbf{u}$ (variables) which are positive in general and in this sense correspond to inactive constraints. This is no more than an expression of the complementarity condition. In general, solving the equations in LP1 pointed to by σ defining a nonoptimal basic feasible solution does not give a feasible point.

Fundamental Theorem (*algebraic version*) *If LP2 has a feasible solution then it has a basic feasible solution. If it has a feasible optimizing solution then it has a basic feasible optimizing solution.*

Proof Let $\mathbf{u}$ be a feasible solution with nonzero components pointed to by the index set μ, and let A_μ^T have less than full column rank. The argument proceeds as in Lemma 2.3. That is, another feasible solution is constructed with at least one less nonzero component (so that $|\mu'| \leqslant |\mu| - 1$), and the process is continued until at the lth step, rank $(A_{\mu'}^T) = |\mu'| \leqslant p$. The proof then follows by using Lemma 2.4 to imbed $A_{\mu'}^T$ in a $p \times p$ matrix having rank p. Let

$$A_\mu^T \mathbf{w}_\mu = 0$$

and consider

$$\mathbf{u}' = \mathbf{u} - \theta \mathbf{w}$$

where $w_i = 0$ if $i \notin \mu$. If θ is small enough $\mathbf{u}'$ is feasible. Choosing

$$\theta = \min_{i \in \mu, w_{\mu(i)} > 0} \frac{x_{\mu(i)}}{w_{\mu(i)}}$$

one extra component vanishes in $\mathbf{u}'$. As before, $\mathbf{w}_\mu$ can always have at least one positive component. If $\mathbf{u}$ is optimal then the same argument applies as

$$\mathbf{b}^T(\mathbf{u}-\theta\mathbf{w}) \leqslant \mathbf{b}^T\mathbf{u}$$

for θ small enough and of either sign. It follows that

$$\mathbf{b}^T\mathbf{w}=0$$

so that the reduction in the number of nonzero components can be carried out without affecting optimality.

Remark 3.2 This result shows that we need only consider the basic feasible solutions in seeking the optimum of LP2.

Any linear program can be written in the form LP2. For example, LP1 can be put in this form:

(a) by introducing additional nonnegative variables called *slack variables* to turn the inequalities into equalities, and
(b) by replacing the unconstrained variables by the difference of two nonnegative variables.

If this is done then LP1 can be written

$$\max\,[-\mathbf{c}^T\,|\mathbf{c}^T|\,0]\begin{bmatrix}\mathbf{y}\\ \hline \mathbf{z}\\ \hline \mathbf{w}\end{bmatrix}$$

subject to the constraints

$$[A\,|-A|-I]\begin{bmatrix}\mathbf{y}\\ \hline \mathbf{z}\\ \hline \mathbf{w}\end{bmatrix}=\mathbf{b}, \qquad \mathbf{y}\geqslant 0, \qquad \mathbf{z}\geqslant 0, \qquad \mathbf{w}\geqslant 0$$

where $\mathbf{w}$ is the vector of slack variables and $\mathbf{x}=\mathbf{y}-\mathbf{z}$. Note that no pair $\mathbf{y}_i$, $\mathbf{z}_i$ can be basic variables simultaneously (why?). The equivalence of this problem with LP1 can be verified by calculating the Kuhn–Tucker conditions.

Exercise 3.1 Consider the minimization of a linear objective function subject to a combination of linear equality and inequality constraints. Determine the dual to this problem. Show that if a variable is unconstrained in the primal then the corresponding inequality in the dual is actually an equation.

Exercise 3.2 A variable is a *null variable* if it is zero in every feasible solution.

(i) Show that if the feasible set is nonempty then u_i is a null variable if and only if there exists $\boldsymbol{\lambda} \geqslant 0$ such that

$$A\boldsymbol{\lambda} \geqslant 0, \qquad \boldsymbol{\rho}_i(A)\boldsymbol{\lambda} > 0,$$

and

$$\mathbf{c}^{\mathrm{T}}\boldsymbol{\lambda} = 0.$$

(ii) If any basic feasible solution is nondegenerate then the system contains no null variables.

Exercise 3.3 A variable u_i is *nonextremal* if the inequality $u_i \geqslant 0$ is redundant (compare Exercise 3.2 of Chapter 1). Show that if the feasible set is nonempty then u_i is a nonextremal variable if and only if there exists $\boldsymbol{\lambda}$, $\mathbf{d}$ such that

$$A\boldsymbol{\lambda} = \mathbf{d}, \qquad \mathbf{c}^{\mathrm{T}}\boldsymbol{\lambda} \leqslant 0$$

where

$$d_i = -1, \qquad d_j \geqslant 0, \qquad j \neq i.$$

2.4 THE SIMPLEX ALGORITHM

The simplex algorithm stands in a special position with regard to the development of linear programming as it provided the first satisfactory computational procedure for solving these problems at exactly the time that digital computers were becoming available. It thus made possible the applications of linear programming in government and commerce which have become so important in the last 35 years. A brief sketch of the early history is given in the notes at the end of the chapter.

The simplex algorithm provides a finite procedure for solving LP2

$$\max_{\mathbf{u} \in U} \mathbf{b}^{\mathrm{T}}\mathbf{u} : U = \{\mathbf{u};\, A^{\mathrm{T}}\mathbf{u} = \mathbf{c},\, \mathbf{u} \geqslant 0\}$$

starting from an initial basic feasible solution, and for purposes of presentation it is assumed that A has full column rank p. A step of the algorithm modifies the current basic feasible solution $\mathbf{u}$ determined by the index set σ pointing to the columns of A^{T} in the current basis by replacing one column by a nonbasic column to obtain a new basis pointed to by the index set $\bar{\sigma}$ in such a way that:

(i) the new basic solution $\bar{\mathbf{u}}$ is feasible, and
(ii) the objective function is increased unless $\mathbf{u}$ is optimal or degenerate.

Degeneracy represents an unusual situation, and it is conventional to ignore it in the initial development of the algorithm.

First the sequence of operations involved in modifying the basis is described. To simplify notation it is convenient to write

$$[A_{\sigma}^{\mathrm{T}} \mid A_{\sigma^c}^{\mathrm{T}}] = [B \mid N]. \tag{4.1}$$

The new basis is found by interchanging a pair of columns between B and N. Assume that the interchange is between column k of N and column s of B (that is, between columns $\sigma^c(k)$ and $\sigma(s)$ of A^T). Let

$$\mathbf{v}_k^\sigma = B^{-1}\boldsymbol{\kappa}_k(N) \tag{4.2}$$

and note that

$$[B \mid N]\begin{bmatrix} -\mathbf{v}_k^\sigma \\ \mathbf{e}_k \end{bmatrix} = 0. \tag{4.3}$$

It follows that

$$\bar{\mathbf{u}}(\lambda) = \begin{bmatrix} \mathbf{u}_\sigma \\ 0 \end{bmatrix} + \lambda \begin{bmatrix} -\mathbf{v}_k^\sigma \\ \mathbf{e}_k \end{bmatrix} \tag{4.4}$$

is a feasible, nonbasic solution for all $\lambda > 0$ small enough, and that the exchange between basic and nonbasic columns can be effected by choosing λ to make $(\bar{\mathbf{u}}(\lambda))_s = 0$. This gives

$$\lambda_s = (\mathbf{u}_\sigma)_s / (\mathbf{v}_k^\sigma)_s. \tag{4.5}$$

The resulting basic solution $\mathbf{u}_s$ is feasible provided λ_s is chosen as the smallest positive value of λ reducing an element of $\bar{\mathbf{u}}(\lambda)$ to zero. This is equivalent to choosing s as the index of the smallest element in the set

$$\Lambda_k = \{\lambda_j;\, j \in \sigma,\, (\mathbf{v}_k^\sigma)_j > 0\}. \tag{4.6}$$

Remark 4.1 If $\mathbf{v}_k^\sigma \leqslant 0$ so that $\Lambda_k = \varnothing$ then $\bar{\mathbf{u}}_s$ is feasible for all $\lambda \geqslant 0$. It follows that

$$\begin{bmatrix} -\mathbf{v}_k^\sigma \\ \mathbf{e}_k \end{bmatrix}$$

defines a direction of recession for U. LP2 will be unbounded if the objective function increases in this direction. The precise condition is given in Remark 4.2 below.

The new basis matrix obtained after the column interchange is

$$\begin{aligned} \bar{B} &= B + (\boldsymbol{\kappa}_k(N) - \boldsymbol{\kappa}_s(B))\mathbf{e}_s^T \\ &= B\{I + (B^{-1}\boldsymbol{\kappa}_k(N) - \mathbf{e}_s)\mathbf{e}_s^T\}. \end{aligned} \tag{4.7}$$

Setting

$$V_{k,s}^\sigma = \{I + (\mathbf{v}_k^\sigma - \mathbf{e}_s)\mathbf{e}_s^T\}^{-1} = \left\{I - \frac{\mathbf{v}_k^\sigma - \mathbf{e}_s}{(\mathbf{v}_k^\sigma)_s}\mathbf{e}_s^T\right\} \tag{4.8}$$

gives the nonzero part of the new basic solution as

$$\bar{\mathbf{u}}_{\bar{\sigma}} = V_{k,s}^\sigma B^{-1}\mathbf{c} = V_{k,s}\mathbf{u}_\sigma. \tag{4.9}$$

To verify directly that this is equivalent to (4.4) with λ given by (4.6), it suffices to expand (4.9) using (4.8) and identify terms.

It has been shown that each column of N can be exchanged with an appropriate column of B to give a new basic feasible solution provided the associated set Λ_k is nonempty. The next step is to select the column of N in such a way that the objective function is increased unless $\mathbf{u}$ is degenerate. If column k of N is chosen, then the new objective function value is

$$\mathbf{b}^{\mathrm{T}}\bar{\mathbf{u}}_s = \mathbf{b}^{\mathrm{T}}\mathbf{u} + \lambda\mathbf{b}^{\mathrm{T}}\begin{bmatrix} -\mathbf{v}_k^{\sigma} \\ \mathbf{e}_k \end{bmatrix}$$

$$= \mathbf{b}^{\mathrm{T}}\mathbf{u} + \lambda\{b_{\sigma^c(k)} - \mathbf{b}_{\sigma}^{\mathrm{T}}\mathbf{v}_k^{\sigma}\} \tag{4.10}$$

so that

$$\xi_k = b_{\sigma^c(k)} - \mathbf{b}_{\sigma}^{\mathrm{T}}\mathbf{v}_k^{\sigma} \tag{4.11}$$

gives the rate of increase of the objective function with λ. If $\xi_k > 0$ then

$$\begin{bmatrix} -\mathbf{v}_k^{\sigma} \\ \mathbf{e}_k \end{bmatrix}$$

defines a *direction of ascent* for LP2.

Remark 4.2 If $\xi_k > 0$ and the corresponding set $\Lambda_k = \varnothing$ then the objective function can be increased indefinitely in the direction

$$\begin{bmatrix} -\mathbf{v}_k^{\sigma} \\ \mathbf{e}_k \end{bmatrix}$$

as $\bar{\mathbf{u}}_s$ is feasible for all $\lambda \geqslant 0$ by Remark 4.1. It follows that LP2 is unbounded.

The actual sequence of operations that would be followed is the reverse of that described here. Each k for which $\xi_k > 0$ specifies a column that can be entered into the basis in order to provide an increase in the objective function value. Once k is determined the index of the column to leave the basis is specified by the smallest element in Λ_k. It is only in the selection of k that there is room to incorporate different strategies. The two considered here are perhaps the most important.

Definition 4.1 Let

$$\Xi^{\sigma} = \{\xi_j;\, \xi_j > 0,\, j = 1, 2, \ldots, n-p\}. \tag{4.12}$$

The USE (or *unnormalized steepest edge*) test for the variable to enter the basis is

$$k = \arg\max_j\, (\xi_j,\, \xi_j \in \Xi^{\sigma}). \tag{4.13}$$

The NSE (or *normalized steepest edge*) test (in the $\|\cdot\|$ norm) is

$$k = \arg\max_j\, \left(\xi_j \Big/ \left\| \begin{bmatrix} -\mathbf{v}_j^{\sigma} \\ \mathbf{e}_j \end{bmatrix} \right\|,\, \xi_j \in \Xi^{\sigma}\right). \tag{4.14}$$

Unless otherwise specified the norm is the Euclidean norm.

Remark 4.3 Much of our discussion will be in terms of USE tests for simplicity, but the NSE test has the advantages of comparing comparable quantities (directional derivatives) as well as the disadvantage of requiring extra work to compute the vector norms needed at each stage. It turns out that this extra computation can be reduced to acceptable levels in certain cases, and reported experience appears to show clearly enough that NSE tests perform better. This point is discussed further in Section 7.5.

The next result shows that in an important sense Ξ summarizes the set of available directions of ascent at the current point.

Lemma 4.1 *If* $\Xi^\sigma = \varnothing$ *then* $\mathbf{u}$ *is optimal.*

Proof If $\Xi^\sigma = \varnothing$ then it follows from (4.11) that

$$\mathbf{b}_{\sigma^c}^{\mathrm{T}} - \mathbf{b}_\sigma^{\mathrm{T}} B^{-1} N \leqslant 0.$$

Let

$$\mathbf{x} = B^{-\mathrm{T}} \mathbf{b}_\sigma = A_\sigma^{-1} \mathbf{b}_\sigma$$

then

$$A\mathbf{x} = \begin{bmatrix} B^{\mathrm{T}} \\ N^{\mathrm{T}} \end{bmatrix} \mathbf{x} = \begin{bmatrix} \mathbf{b}_\sigma \\ N^{\mathrm{T}} B^{-\mathrm{T}} \mathbf{b}_\sigma \end{bmatrix} \geqslant \begin{bmatrix} \mathbf{b}_\sigma \\ \mathbf{b}_{\sigma^c} \end{bmatrix}$$

so that $\mathbf{x}$ is feasible for LP1. Now

$$\mathbf{c}^{\mathrm{T}} \mathbf{x} = \mathbf{c}^{\mathrm{T}} A_\sigma^{-1} \mathbf{b}_\sigma = \mathbf{u}_\sigma^{\mathrm{T}} \mathbf{b}_\sigma = \mathbf{u}^{\mathrm{T}} \mathbf{b}$$

showing that the primal and dual objective functions are equal. This establishes optimality as a consequence of the duality theorem.

Remark 4.4 The appearance of the duality theorem suggests that the ξ_j should have an interpretation in terms of Kuhn–Tucker multipliers for LP2. The appropriate form of the Kuhn–Tucker conditions corresponds to (5.21) and (5.22) of Chapter 1 and gives, on taking account of the ordering implied by σ, and noting that nonbasic u_i correspond to active constraints,

$$\begin{aligned} [\mathbf{b}_\sigma^{\mathrm{T}} \mid \mathbf{b}_{\sigma^c}^{\mathrm{T}}] &= \boldsymbol{\zeta}^{\mathrm{T}} [B \mid N] - [0 \mid \boldsymbol{\eta}^{\mathrm{T}}] I_n, \qquad \boldsymbol{\eta} \geqslant 0, \\ &= [\boldsymbol{\zeta}^{\mathrm{T}} \mid \boldsymbol{\eta}^{\mathrm{T}}] \begin{bmatrix} B & N \\ 0 & -I_{n-p} \end{bmatrix}, \qquad \boldsymbol{\eta} \geqslant 0. \end{aligned}$$

Thus

$$\mathbf{b}_\sigma^{\mathrm{T}} = \boldsymbol{\zeta}^{\mathrm{T}} B,$$

and

$$\mathbf{b}_{\sigma^c}^{\mathrm{T}} = \mathbf{b}_\sigma^{\mathrm{T}} B^{-1} N - \boldsymbol{\eta}^{\mathrm{T}} \tag{4.15}$$

so that

$$\xi_j = -\eta_j \leqslant 0, \qquad j = 1, 2, \ldots, n-p,$$

showing that the ξ_j do indeed relate directly to the multipliers associated with the active inequality constraints at an optimal point. However, at any basic feasible $\mathbf{u}$ the $-\xi_j$, $j=1,2,\ldots,n-p$ can be considered as tentative multiplier estimates so that if $\Xi^\sigma \neq \emptyset$ then some multiplier estimates are infeasible. This permits a second point to be made, for a direction of ascent is defined by

$$\begin{bmatrix} \mathbf{v}_k^\sigma \\ -\mathbf{e}_k \end{bmatrix} = \boldsymbol{\kappa}_{p+k} \begin{bmatrix} B & N \\ 0 & -I_{n-p} \end{bmatrix}^{-1}, \tag{4.16}$$

that is, by the column of the inverse of the matrix associated with the active constraints corresponding to the infeasible multiplier estimate. This is a pattern of behaviour which will be encountered frequently. Note that it implies that there is a direction of ascent if and only if the current point is not optimal.

In outline the usual form of the simplex algorithm is as follows (a USE test is assumed).

Algorithm

(i) *Determine a basic feasible solution.*

(ii) *Compute Ξ^σ. If $\Xi^\sigma = \emptyset$ then stop. Otherwise determine the index k of the maximum element of Ξ^σ.*

(iii) *Compute Λ_k. If $\Lambda_k = \emptyset$ then LP2 is unbounded. Otherwise determine s as the index of the smallest element of Λ_k.*

(iv) *Exchange $\boldsymbol{\kappa}_k(N)$ with $\boldsymbol{\kappa}_s(B)$. $\sigma := \{\sigma \backslash \sigma(s)\} \cup \{\sigma^c(k)\}$. Repeat* (ii).

Example 4.1 Consider LP2 specified by the data

$$A^T = \begin{bmatrix} 1 & 0 & 0 & 1 & 0 & 0 \\ 20 & 1 & 0 & 0 & 1 & 0 \\ 200 & 20 & 1 & 0 & 0 & 1 \end{bmatrix},$$

$$\mathbf{b}^T = [100 \quad 10 \quad 1 \quad 0 \quad 0 \quad 0],$$

$$\mathbf{c}^T = [1 \quad 10 \quad 100].$$

An initial basic feasible solution is obtained by choosing $\sigma = \{4, 5, 6\}$. This gives

$$\mathbf{b}_\sigma = \begin{bmatrix} 0 \\ 0 \\ 0 \end{bmatrix}, \quad \mathbf{u}_\sigma = \begin{bmatrix} 1 \\ 100 \\ 10000 \end{bmatrix}, \quad B^{-1}N = \begin{bmatrix} 1 & 0 & 0 \\ 20 & 1 & 0 \\ 200 & 20 & 1 \end{bmatrix}.$$

The quantities required for the first step of the simplex algorithm are

$$\{\xi_k, k \in \sigma^c\} = \{100, 10, 1\} = \Xi^1 \Rightarrow k = 1$$

and

$$\{\lambda_s, s \in \sigma\} = \{1, 5, 50\} = \Lambda_1 \Rightarrow s = 4.$$

It follows that $\bar{\sigma} = \{1, 5, 6\}$ and that the increase achieved in the objective function is $\xi_k \lambda_s = 100$.

In the second iteration of the algorithm

$$\mathbf{b}_\sigma = \begin{bmatrix} 100 \\ 0 \\ 0 \end{bmatrix}, \qquad \mathbf{u}_\sigma = \begin{bmatrix} 1 \\ 80 \\ 9800 \end{bmatrix}, \qquad B^{-1}N = \begin{bmatrix} 0 & 0 & 1 \\ 1 & 0 & -20 \\ 20 & 1 & -200 \end{bmatrix}.$$

Thus

$$\{\xi_k, k \in \sigma^c\} = \{10, 1, -1000\}, \qquad \Xi^2 = \{10, 1\}, \qquad k = 2,$$

and

$$\{\lambda_s, s \in \sigma\} = \{+\infty, 80, 490\} = \Lambda_2 \Rightarrow s = 5,$$

so that $\bar{\sigma} = \{1, 2, 6\}$ and the increase in the objective function is $\xi_k \lambda_s = 800$.

The full calculation is summarized in Table 4.1. This problem is a special case of an example which is designed to show that the number of iterations cannot always be bounded by a polynomial in the problem dimensions. This point is discussed further in Chapter 7. The point to notice is that the columns of A are effectively paired and in the successive bases all possible combinations occur in which one member of each pair is chosen. The advantage of this example is that it provides sufficient steps to give a good illustration of the mechanics of the simplex algorithm. Although the dimensions in this problem are too small to be meaningful, the number of iterations taken is already rather larger than experience suggests is usual.

Table 4.1 Successive iterations of the simplex algorithm summarized.

Iteration	σ	$\lambda_s \xi_k$	Objective function
1	4, 5, 6	100	0
2	1, 5, 6	800	100
3	1, 2, 6	100	900
4	4, 2, 6	8000	1 000
5	4, 2, 3	100	9 000
6	1, 2, 3	800	9 100
7	1, 5, 3	100	9 900
8	4, 5, 3	—	10 000

Points relating to the implementation of this algorithm are now considered in greater detail.

(1) Computing a basic feasible solution

The requirement to compute an initial basic feasible solution can be solved within the overall framework of the simplex algorithm by considering an augmented constraint set for which a basic feasible solution is immediate. Consider the constraint conditions

$$A^{\mathrm{T}}\mathbf{u}+\mathbf{y}=\mathbf{c}, \qquad \mathbf{u}\geq 0, \qquad \mathbf{y}\geq 0. \tag{4.17}$$

There is no restriction in assuming $\mathbf{c}\geq 0$, and this ensures that $\mathbf{u}=0$, $\mathbf{y}=\mathbf{c}$ is a basic feasible solution, and that the associated basis matrix is $B=I_p$. Here the $\mathbf{y}$ are called *artificial variables*. To obtain a basic feasible solution to the original problem it is necessary to make $\mathbf{y}$ nonbasic, and this can be done by maximizing $-\mathbf{e}^{(p)\mathrm{T}}\mathbf{y}$ which has a maximum of zero when all components of $\mathbf{y}$ have been reduced to zero level (and in the nondegenerate case been swapped out of the basis). At this stage a basic feasible solution to LP2 is available. This form of the algorithm is called the *two-phase simplex method.*

An alternative approach considers the objective function $\mathbf{b}^{\mathrm{T}}\mathbf{u}-\gamma\mathbf{e}^{(p)\mathrm{T}}\mathbf{y}$. We will show that choosing $\gamma>0$ large enough will force $\mathbf{y}=0$ at the maximum in general, thus avoiding the need to split the computation into two distinct phases. However, usually γ is not known a priori, so that if the optimum solution contains nonzero components of $\mathbf{y}$ it will be necessary to increase γ and resolve. An idea of how large γ has to be can be obtained from the Kuhn–Tucker conditions for the augmented problem. These give

$$[\mathbf{b}^{\mathrm{T}} \mid -\gamma\mathbf{e}^{(p)\mathrm{T}}]=\boldsymbol{\zeta}^{\mathrm{T}}[A^{\mathrm{T}} \mid I_p]-[0, \boldsymbol{\eta}_u^{\mathrm{T}}, \boldsymbol{\eta}_y^{\mathrm{T}}]I_{n+p}, \qquad \boldsymbol{\eta}_u\geq 0, \qquad \boldsymbol{\eta}_y\geq 0.$$

In particular,

$$\gamma\mathbf{e}^{(p)\mathrm{T}}=(\boldsymbol{\eta}_y^{\mathrm{T}}-\boldsymbol{\zeta}^{\mathrm{T}})I_p.$$

Now $(\boldsymbol{\eta}_y)_i>0\Rightarrow y_i=0$ by complementarity, so the requirement reduces to

$$\gamma>-\min_i \zeta_i, \qquad i=1,2,\ldots,p. \tag{4.18}$$

Note that the matrix of (4.17) has rank p irrespective of the rank of A. Thus *this second procedure can be applied to rank-deficient problems.* If it produces an optimum then either:

(a) One or more artificial variables remain in the basis but at zero level. In this case a solution of the rank-deficient problem is obtained; or
(b) One or more artificial variables remain in the basis at a positive level as $\gamma\to\infty$. It follows from the Kuhn–Tucker conditions that $\boldsymbol{\zeta}$ is feasible for LP1 with objective function value $\mathbf{c}^{\mathrm{T}}\boldsymbol{\zeta}=-\gamma\mathbf{c}^{\mathrm{T}}\mathbf{e}^{(p)}-\mathbf{c}^{\mathrm{T}}\boldsymbol{\eta}_y\to-\infty$, $\gamma\to\infty$ as $\mathbf{c}\geq 0$ by assumption. Thus LP1 is unbounded. It follows from Theorem 2.3 that the feasible set for LP2 is empty. This means that the constraint equations are inconsistent.

(2) Organization of the calculation

Two different approaches with merits appropriate to different classes of problems have been widely used in implementing the simplex algorithm.

Product form of the inverse

This method keeps the basis matrix explicitly in a factored form which makes solution of linear equations easy and which is updated at each iteration. Thus at the ith stage

$$B = R_i U_i. \tag{4.19}$$

Modifications to the columns of B imply modifications to the columns of U_i. These are written

$$U_i \rightarrow \bar{U}_i = S_i U_{i+1} \tag{4.20}$$

where S_i^{-1} represents the operations applied to $\bar{U}_i$ to bring it back to the form appropriate to the factorization. It follows that the updated basis matrix has the factored form

$$\bar{B} = R_i S_i U_{i+1} = R_{i+1} U_{i+1} \tag{4.21}$$

and R_i possesses the decomposition into the product form

$$R_i = R_0 \prod_{j=0}^{i-1} S_j \tag{4.22}$$

which gives the method its name. This method has the advantage that inverse matrices and the results of applying inverse matrices are not kept explicitly. In solving large problems, exploitation of sparsity in the data is important, and this tends to be destroyed by computation. Thus it is in such problems that the product form is seen to greatest advantage. A possible disadvantage is that as the number of iterations is not known in advance, neither is the number of update matrices S_j that must be stored if R_i is not formed explicitly.

Referring to the outline of the algorithm, it will be seen that the calculations that have to be carried out are

(i) computation of the ξ_j, $\xi_j > 0$, to determine the variable to enter the basis;
(ii) computation of Λ_k to determine the variable to leave the basis;
(iii) updating $\mathbf{u}_\sigma$ and the factors of B.

Note that only the computation of the ξ_j is defined on the full array and so involves a cost proportional to n. To reduce the work in this step a recurrence is used to update these values from iteration to iteration. By (4.11) the work involved is in calculating $\mathbf{b}_\sigma^{\mathrm{T}}\mathbf{v}_j^\sigma$ for $j = 1, 2, \ldots, n-p$. In updating the basis from σ to $\bar{\sigma}$ the change in these quantities when $j \neq k$ is given by

$$\begin{aligned}\mathbf{b}_{\bar{\sigma}}^{\mathrm{T}}\mathbf{v}_j^{\bar{\sigma}} &= \{\mathbf{b}_\sigma^{\mathrm{T}} + (b_{\sigma^c(k)} - b_{\sigma(s)})\mathbf{e}_s^{\mathrm{T}}\} V_{k,s}^\sigma \mathbf{v}_j^\sigma \\ &= \mathbf{b}_\sigma^{\mathrm{T}}\mathbf{v}_j^\sigma + \alpha_j \xi_k,\end{aligned}$$

where

$$\alpha_j = (\mathbf{v}_j^\sigma)_s / (\mathbf{v}_k^\sigma)_s, \tag{4.23}$$

while when $j = k$ we must take account of the column interchange which alters the partitioning of $\mathbf{b}$ and obtain

$$\begin{aligned}\mathbf{b}_{\bar{\sigma}}^{\mathrm{T}} \mathbf{v}_k^{\bar{\sigma}} &= \{\mathbf{b}_\sigma^{\mathrm{T}} + (b_{\sigma^c(k)} - b_{\sigma(s)})\mathbf{e}_s^{\mathrm{T}}\} V_{k,s}^\sigma \mathbf{e}_s \\ &= b_{\sigma(s)} + \xi_k / (\mathbf{v}_k^\sigma)_s.\end{aligned}$$

Thus

$$\bar{\xi}_j = \begin{cases} \xi_j - \alpha_j \xi_k, & j \neq k \\ -\xi_k / (\mathbf{v}_k^\sigma)_s, & j = k. \end{cases} \tag{4.24}$$

This relation requires α_j given by (4.23). However, explicit calculation of $\mathbf{v}_j^\sigma$, $j \neq k$, can be avoided ($\mathbf{v}_k^\sigma$ is required to compute Λ_k) by computing

$$\mathbf{y}^{\mathrm{T}} = \mathbf{e}_s^{\mathrm{T}} B^{-1} \tag{4.25}$$

so that

$$\alpha_j = \mathbf{y}^{\mathrm{T}} \boldsymbol{\kappa}_j(N) / (\mathbf{v}_k^\sigma)_s, \qquad j = 1, 2, \ldots, n - p. \tag{4.26}$$

For $n \gg p$, computation of the $(n - p)$ scalar products needed to give the α_j dominates the work done in an iteration. What is involved is $p(n - p)$ basic computational steps which can be equated with the multiplications involved as a crude approximation. Asymptotic estimates with leading term np for large n are about as good as we might expect (equivalent to each element of A entering one nontrivial computing step per iteration).

Tableau form

In small dense problems it is possible to avoid keeping explicit information on the factors of B by updating the matrix

$$W_i = R_i^{-1}[A^{\mathrm{T}} | I_p | \mathbf{c}] \tag{4.27}$$

where U_i is available in the columns of W_i pointed to by σ, and the unit matrix can be deleted if R_i^{-1} is not required. Now both the updating of W_i and the calculation of α_j involve the full array A^{T}, and it is necessary to use information on the form of the factorization (4.19) to keep the asymptotic complexity estimate to np multiplications. To compute α_j we note that

$$\begin{aligned}\mathbf{e}_s^{\mathrm{T}} B^{-1} \boldsymbol{\kappa}_j(N) &= \mathbf{e}_s^{\mathrm{T}} U_i^{-1} R_i^{-1} \boldsymbol{\kappa}_j(N) \\ &= \mathbf{e}_s^{\mathrm{T}} U_i^{-1} \boldsymbol{\kappa}_{\sigma^c(j)}(W_i)\end{aligned} \tag{4.28}$$

so that it is appropriate to compute $\mathbf{e}_s^{\mathrm{T}} U_i^{-1}$ instead of $\mathbf{y}$, and savings are possible by exploiting the zeros in $\mathbf{e}_s$. Updating the tableau involves computing

$$W_{i+1} = S_i^{-1} W_i \tag{4.29}$$

and it is easily verified that the other quantities required are readily available.

We note briefly two possible factorizations which have been used extensively. In the early developments which were based on the use of tableau methods, the Jordan elimination procedure held a central position. This corresponds to

$$U_i = I_p, \qquad S_i^{-1} = V_{k,s}^{\sigma}. \tag{4.30}$$

In this case (4.28) has a particularly simple form so that the bulk of the computation in our asymptotic sense takes place in updating W_i and it is straightforward to develop the np estimate.

Jordan elimination has been criticized because it provides no scope for guarding against possible problems of numerical instability, and examples have been constructed which confirm that these problems can occur. For this reason a method due to Bartels and Golub has become increasingly popular. It adapts the method of factorization into upper and lower triangular matrices and uses partial pivoting as a stabilizing device. This procedure is discussed in Section A1.3 (Appendix 1). The operation of modifying U_i by exchanging columns gives

$$\bar{U}_i = \{U_i + (\boldsymbol{\kappa}_{\sigma^c(k)}(W_i) - \boldsymbol{\kappa}_{\sigma(s)}(W_i))\mathbf{e}_s^T\}P \tag{4.31}$$

where P is a permutation matrix specifying a column rearrangement which is favourable for updating the factorization. Here the permutation

$$\begin{pmatrix} s, s+1, \ldots, p \\ p, s, \ldots, p-1 \end{pmatrix}$$

is appropriate. The result is shown diagrammatically in Figure 4.1 in the case $s = 3$, $p = 5$. In general we obtain an upper Hessenberg matrix with nonzero elements in the subdiagonal immediately below the leading diagonal in columns $s, s+1, \ldots, p-1$, and with pth column $\boldsymbol{\kappa}_{\sigma^c(k)}(W_i)$. The updating matrix S_i^{-1} is now defined by the requirement to eliminate the nonzero elements in this subdiagonal to restore upper triangular form and by the requirement to perform interchanges where necessary to ensure that the multipliers in the elimination process have magnitude $\leqslant 1$. In the tableau the elimination procedure is applied to all columns of W_i, and this gives a

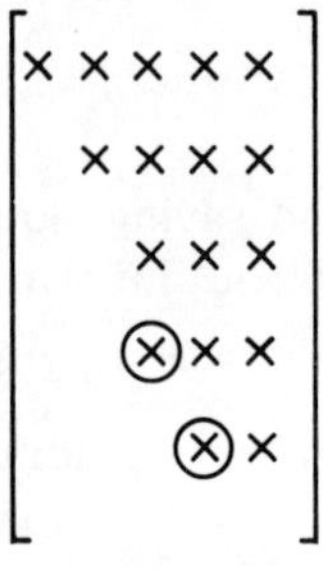

$$\begin{bmatrix} \times & \times & \times & \times & \times \\ & \times & \times & \times & \times \\ & & \times & \times & \times \\ & & \otimes & \times & \times \\ & & & \otimes & \times \end{bmatrix}$$

Figure 4.1 Disposition of nonzero elements in $\bar{U}_i$

complexity estimate of $n(p-s)$ multiplications. Calculation of the α_j, $j=1,2,\ldots,n-p$, leads to a similar estimate as $\mathbf{e}_s^{\mathrm{T}}U_i^{-1}$ is a forward substitution starting with the sth component so that the first $s-1$ elements are zero and do not count in the scalar product (4.28). Again we obtain an asymptotic estimate of np operations on average provided the expected value of s is $p/2$.

(3) Calculation of the NSE test

Given the quantities

$$z_j=\left\|\begin{bmatrix}-\mathbf{v}_j^{\sigma}\\ \mathbf{e}_j\end{bmatrix}\right\|_2^2=1+\|\mathbf{v}_j^{\sigma}\|_2^2, \qquad j=1,2,\ldots,n-p \tag{4.32}$$

the NSE test (4.14) is readily computed. Direct calculation is prohibitive except in the case of the tableau based on Jordan elimination when the $\mathbf{v}_j^{\sigma}$ associated with the $\xi_j>0$ can be read off directly. However, it is possible to establish a recurrence for the z_j which reduces the computation required to a reasonable level.

To set up this recurrence, note that by (4.7) and (4.8), if $j\neq k$,

$$\begin{aligned}\mathbf{v}_j^{\bar{\sigma}}&=\bar{B}^{-1}\boldsymbol{\kappa}_j(\bar{N})=V_{k,s}^{\sigma}B^{-1}\boldsymbol{\kappa}_j(N)\\ &=\mathbf{v}_j^{\sigma}-\alpha_j\{\mathbf{v}_k^{\sigma}-\mathbf{e}_s\}\end{aligned} \tag{4.33}$$

where α_j is given by (4.23) and is required also for computing the ξ_j. Taking the scalar product of each side of (4.33) with itself gives

$$\begin{aligned}\|\mathbf{v}_j^{\bar{\sigma}}\|_2^2&=\|\mathbf{v}_j^{\sigma}\|_2^2-2\alpha_j(\mathbf{v}_j^{\sigma\mathrm{T}}\mathbf{v}_k^{\sigma}-(\mathbf{v}_j^{\sigma})_s)+\alpha_j^2(\|\mathbf{v}_k^{\sigma}\|_2^2-2(\mathbf{v}_k^{\sigma})_s+1)\\ &=\|\mathbf{v}_j^{\sigma}\|_2^2-2\alpha_j\mathbf{v}_j^{\sigma\mathrm{T}}\mathbf{v}_k^{\sigma}+\alpha_j^2(\|\mathbf{v}_k^{\sigma}\|_2^2+1)\end{aligned}$$

so that

$$\bar{z}_j=z_j-2\alpha_j\mathbf{v}_j^{\sigma\mathrm{T}}\mathbf{v}_k^{\sigma}+\alpha_j^2z_k, \qquad j\neq k \tag{4.34}$$

while

$$\bar{z}_k=z_k/(\mathbf{v}_k^{\sigma})_s^2, \tag{4.35}$$

and z_k can be computed from (4.32) as $\mathbf{v}_k^{\sigma}$ is available. It will be seen that the new element is the calculation of the scalar products $\mathbf{v}_j^{\sigma\mathrm{T}}\mathbf{v}_k^{\sigma}$, $j=1,2,\ldots,n-p$. As $\mathbf{v}_j^{\sigma}$ is not available explicitly this can be computed either as

$$\begin{aligned}\mathbf{v}_j^{\sigma\mathrm{T}}\mathbf{v}_k^{\sigma}&=\boldsymbol{\kappa}_j(N)^{\mathrm{T}}B^{-\mathrm{T}}B^{-1}\boldsymbol{\kappa}_k(N)\\ &=\boldsymbol{\kappa}_j(N)^{\mathrm{T}}\mathbf{f}\end{aligned} \tag{4.36}$$

where

$$\mathbf{f}=B^{-\mathrm{T}}B^{-1}\boldsymbol{\kappa}_k(N), \tag{4.37}$$

is independent of j, or

$$\mathbf{v}_j^{\sigma\mathrm{T}}\mathbf{v}_k^{\sigma}=\boldsymbol{\kappa}_{\sigma^c(j)}(W_i)^{\mathrm{T}}R_i^{\mathrm{T}}\mathbf{f} \tag{4.38}$$

where

$$R_i^{\mathrm{T}}\mathbf{f}=U_i^{-\mathrm{T}}U_i^{-1}\boldsymbol{\kappa}_{\sigma^c(k)}(W_i), \tag{4.39}$$

depending on the particular organization (product form of the inverse or tableau) being used. However, the scalar products still involve $p(n-p)$ multiplications, which roughly doubles the asymptotic complexity estimate. It is interesting that this drawback does not occur in the corresponding calculations for the reduced and projected gradient methods applied to LP1 given in the next two sections.

(4) Treatment of degeneracy

If $\mathbf{u}_\sigma \not> 0$ at any stage then the problem is called degenerate. Degeneracy occurs as a consequence of a tie in the selection of the variable to leave the basis in the previous stage so that more than one basic variable is reduced to zero. In this case progress can be made only if

$$(\mathbf{v}_k^\sigma)_j \leq 0, \qquad \forall\, j \ni (\mathbf{u}_\sigma)_j = 0. \tag{4.40}$$

If (4.40) does not hold then $0 \in \Lambda_k$ so that $\mathbf{u}'_\sigma = \mathbf{u}_\sigma$ even though σ is changed. It follows that there is the possibility of cycling repetitively through the possible index sets without making progress, and this possibility must be ruled out if finite termination of the simplex algorithm is to be proved. Here a resolution of the problem of degeneracy is given which starts from the observation that (4.40) possesses an interesting (and useful) geometric interpretation. To develop this we first write LP2 in a form which separates out the basic variables which are at zero level to obtain the problem

$$\max \hat{\mathbf{b}}^{\mathrm{T}}\mathbf{y}$$

subject to

$$\begin{bmatrix}\mathbf{u}_1\\ \mathbf{u}_2\end{bmatrix} + B^{-1}N\mathbf{y} = B^{-1}\mathbf{c} = \begin{bmatrix}\mathbf{d}_1\\ 0\end{bmatrix}$$

and

$$\mathbf{u}_\sigma = \begin{bmatrix}\mathbf{u}_1\\ \mathbf{u}_2\end{bmatrix} \geq 0, \qquad \mathbf{u}_{\sigma^c} = \mathbf{y} \geq 0,$$

where

$$\hat{\mathbf{b}}^{\mathrm{T}} = [\mathbf{b}_{\sigma^c}^{\mathrm{T}} - \mathbf{b}_\sigma^{\mathrm{T}} B^{-1} N]$$

expresses the objective function (apart from a constant term which is equal to $\mathbf{b}_\sigma^{\mathrm{T}}\begin{bmatrix}\mathbf{d}_1\\ 0\end{bmatrix}$) in terms of the nonbasic variables, and degeneracy is implied by $\mathbf{d}_1 > 0$ and $\mathbf{u}_2 = 0$. Let

$$B^{-1}N = \begin{bmatrix}N_1\\ N_2\end{bmatrix}$$

in conformity with the partitioning of $\mathbf{u}_\sigma$. Then, as a consequence of Remark 4.1, (4.40) states that progress can be made in LP2 if and only if the reduced problem

$$\max \hat{\mathbf{b}}^{\mathrm{T}}\mathbf{y} \tag{4.41}$$

subject to

$$\mathbf{u}_2 + N_2\mathbf{y} = 0,$$
$$\mathbf{u}_2 \geq 0, \qquad \mathbf{y} \geq 0,$$

possesses a direction of recession in which it is unbounded. Thus *the resolution of the degeneracy problem is equivalent to computing an unbounded direction of recession for* (4.41). But directions of recession depend only on the constraint matrix so it suffices to compute an unbounded direction of recession for the perturbed and nondegenerate problem

$$\max \hat{\mathbf{b}}^{\mathrm{T}}\mathbf{y} \tag{4.42}$$

subject to

$$\mathbf{u}_2 + N_2\mathbf{y} = \boldsymbol{\varepsilon},$$
$$\mathbf{u}_2 \geq 0, \qquad \mathbf{y} \geq 0$$

where the choice $\boldsymbol{\varepsilon} > 0$ ensures that $[\mathbf{u}_2^{\mathrm{T}}, \mathbf{y}^{\mathrm{T}}] = [\boldsymbol{\varepsilon}^{\mathrm{T}}, 0]$ is a nondegenerate basic feasible solution but is otherwise arbitrary (although ties in the initial Λ_k are undesirable). This can be done using the simplex algorithm, the direction of recession being characterized by $\Lambda_k = \varnothing$ being encountered. Of course the simplex algorithm applied to (4.42) can also encounter degeneracy, but the set of equality constraints now involved in the associated subproblem is reduced by at least one as the successive Jordan elimination steps associated with the pivoting procedure being nonsingular cannot reduce $\boldsymbol{\varepsilon}$ to zero. Thus the degeneracy-removing procedure can be applied recursively, and the above observation shows that the depth of recursion cannot exceed $\dim(\mathbf{u}_2) - 1$. At each level no basis (including degenerate configurations) can repeat because the recursive procedure always must produce either a configuration satisfying (4.40), so that the current objective function is increased, or indicate that (4.40) cannot be satisfied implying that the current point is optimal.

Theorem 4.1 *The simplex algorithm modified to include the recursive treatment of degeneracy terminates at an optimum in a finite number of steps.*

Proof This depends on the observation that the number of possible bases is finite so that the algorithm must terminate if no basis is ever repeated. In the nondegenerate case this is guaranteed as each basis is associated with a particular value of the objective function and this is increased at each step until optimality is attained. If degeneracy occurs, the recursive procedure provides a direction of increase in the objective function in a finite number of steps so that cycling cannot occur and the modified algorithm is also finite.

Remark 4.5 Implementation does not need to depart from that being used in the original problem, except for the additional storage required for the perturbed right-hand sides, as the update operations for the simplex algorithm applied to (4.42) are a subset of the corresponding operations for the simplex algorithm applied to LP2. Thus applying these operations to the full tableau ensures that when the direction of recession is found for (4.42) it can be used immediately in LP2 as it defines a basis satisfying (4.40) and thus a column exchange which increases the objective function.

The above method is local in the sense that only the current basis is relevant, and it has the attraction of being readily generalized to other cases (important applications are given in Sections 2.5, 2.6, and 2.8, and in Chapters 3 and 4). An alternative method of historical importance and considerable elegance, but which depends on the past history of the process, is now described.

Definition 4.2 A vector $\mathbf{g}$ is *lexicographically positive* (written $\mathbf{g} > 0$) if

(i) $\mathbf{g} \neq 0$, and
(ii) the first nonzero component of $\mathbf{g}$ is positive.

This determines an order relation with $\mathbf{g} > \mathbf{h}$ if and only if $\mathbf{g} - \mathbf{h} > 0$ for it is readily verified that

(a) $\mathbf{f} > \mathbf{g}$ and $\mathbf{g} > \mathbf{h} \Rightarrow \mathbf{f} > \mathbf{h}$,
(b) $\mathbf{f} > \mathbf{g} \Leftrightarrow \mathbf{f} + \mathbf{w} > \mathbf{g} + \mathbf{w}$, $\forall\, \mathbf{w}$, and
(c) $\mathbf{f} > \mathbf{g}$ and $\gamma > 0 \Rightarrow \gamma \mathbf{f} > \gamma \mathbf{g}$.

The Lexicographic test is based on the following result.

Theorem 4.2 *Let*

$$F = [\mathbf{u}_\sigma \mid B^{-1}],$$
$$\mathbf{g}^{\mathrm{T}} = \mathbf{b}_\sigma^{\mathrm{T}} F,$$

and assume that $\boldsymbol{\rho}_i(F)$, $i = 1, 2, \ldots, p$ are lexicographically positive. Then the lexicographically smallest element of the set $\{\boldsymbol{\rho}_i(F)/(\mathbf{v}_k^\sigma)_i,\ (\mathbf{v}_k^\sigma)_i > 0\}$ is uniquely determined. If this index is chosen as the index of the column to leave the basis then

(i) *$\{\boldsymbol{\rho}_i(\bar{F}),\ i = 1, 2, \ldots, p\}$ are lexicographically positive, and*
(ii) *$\bar{\mathbf{g}} > \mathbf{g}$.*

Proof If the lexicographically smallest row is not uniquely determined then two rows of F are proportional, and this contradicts the nonsingularity of the basis matrix B. Let the index of the minimizing row be s. Then, by (4.8),

$$\begin{aligned} \boldsymbol{\rho}_i(\bar{F}) &= \mathbf{e}_i^{\mathrm{T}} V_{k,s}^\sigma F \\ &= \boldsymbol{\rho}_i(F) - \frac{(\mathbf{v}_k^\sigma)_i}{(\mathbf{v}_k^\sigma)_s} \boldsymbol{\rho}_s(F), \qquad i \neq s \\ &= \frac{1}{(\mathbf{v}_k^\sigma)_s} \boldsymbol{\rho}_s(F), \qquad i = s. \end{aligned}$$

Thus the rows are clearly lexicographically positive if $i=s$ or $(\mathbf{v}_k^\sigma)_i \leq 0$. If $(\mathbf{v}_k^\sigma)_i > 0$ then

$$\boldsymbol{\rho}_i(\bar{F}) = (\mathbf{v}_k^\sigma)_i\left(\frac{1}{(\mathbf{v}_k^\sigma)_i}\boldsymbol{\rho}_i(F) - \frac{1}{(\mathbf{v}_k^\sigma)_s}\boldsymbol{\rho}_s(F)\right) > 0$$

as a consequence of the definition of s. To verify (ii) we have

$$\begin{aligned}\bar{\mathbf{g}}^{\mathrm{T}} &= \mathbf{b}_{\sigma'}^{\mathrm{T}} F' \\ &= (\mathbf{b}_\sigma + (b_{\sigma^c(k)} - b_{\sigma(s)})\mathbf{e}_s)^{\mathrm{T}} V_{k,s}^{\sigma} F \\ &= \mathbf{g}^{\mathrm{T}} + \frac{\xi_k}{(\mathbf{v}_k^\sigma)_s}\boldsymbol{\rho}_s(F) \\ &> \mathbf{g}^{\mathrm{T}}.\end{aligned}$$

as $\xi_k/(\mathbf{v}_k^\sigma)_s > 0$ unless the optimum has been obtained or (LP2) is unbounded, and $\boldsymbol{\rho}_s(F)$ is lexicographically positive by assumption.

Remark 4.6 The lexicographic test is easy to implement provided B^{-1} is available and the initial conditions are met. The first step is to test for the smallest element in Λ_k. If there are ties then the minimum of $B_{i1}^{-1}/(\mathbf{v}_k^\sigma)_i$ for i running over the ties is computed. If ties remain then the minimum of $B_{i2}^{-1}/(\mathbf{v}_k^\sigma)_i$ is computed for those values of i, and the process is repeated for successive column numbers until the unique minimum is found.

Remark 4.7 The lexicographic test is well adapted for use with both versions of the simplex method which use artificial variables as part of the starting procedure. For example, in (4.17) take $B=I$; then $F=[\mathbf{c}\,|\,I]$ is lexicographically positive as $\mathbf{c} \geq 0$ is assumed.

(5) Termination

A test for termination is provided by Lemma 4.1 which shows that $\mathbf{u}$ is optimal provided $\Xi^\sigma = \varnothing$ so that $\xi_q \leq 0$, $1 \leq q \leq n-p$. In practice this test must be implemented against a small tolerance so it is important for the quantities involved to be consistently scaled to make the comparisons meaningful. Scaling options available that are implemented readily include:

(i) Rescaling of the columns of the constraint matrix by a diagonal matrix:

$$A^{\mathrm{T}}\mathbf{u} \rightarrow A^{\mathrm{T}} D_1 D_1^{-1}\mathbf{u} = A^{\mathrm{T}} D_1 \tilde{\mathbf{u}}.$$

Note that this scaling implies rescaling of $\mathbf{u}$ as well in order to preserve equality constraints.

(ii) Rescaling the rows of the constraint matrix by a diagonal matrix:

$$A^{\mathrm{T}}\mathbf{u} \rightarrow D_2 A^{\mathrm{T}}\mathbf{u} = D_2\mathbf{c}.$$

(iii) Rescaling the objective function:

$$\mathbf{b}^{\mathrm{T}}\mathbf{u} \to \beta \mathbf{b}^{\mathrm{T}}\mathbf{u}.$$

The most important among these scaling options, at least in the sense of directly affecting the ξ_i, is the rescaling of the columns of A^{T}. Consider the rescaling implied by $u_i \to d_i u_i$, $i = 1, 2, \ldots, n$. The consequent changes which preserve equality constraints and objective function value are

$$\boldsymbol{\kappa}_i(A^{\mathrm{T}}) \to d_i^{-1}\boldsymbol{\kappa}_i(A^{\mathrm{T}}),$$

and

$$b_i \to d_i^{-1} b_i$$

so that

$$\xi_i = b_i - \mathbf{b}_\sigma^{\mathrm{T}} B^{-1}\boldsymbol{\kappa}_i(A^{\mathrm{T}}) \to d_i^{-1}\xi_i$$

as

$$\mathbf{b}_\sigma^{\mathrm{T}} B^{-1} \to \mathbf{b}_\sigma^{\mathrm{T}}(D_1^\sigma)^{-1}(B(D_1^\sigma)^{-1})^{-1},$$

where D_1^σ is the restriction of D_1 to the current basis, and is unchanged. It follows that $\xi_i/\|\boldsymbol{\kappa}_i(A)\|_2$ is invariant under column rescaling, and this suggests that a suitable initial scaling is obtained by arranging that

$$\|\boldsymbol{\kappa}_i(A^{\mathrm{T}})\|_2 = 1, \qquad i = 1, 2, \ldots, n.$$

It is probably worth while also to scale the rows of A^{T} so that $\max_i |c_i| = 1$, and the objective function so that $\max_i |b_i| = 1$ with the general intention of keeping quantities commensurate.

It is interesting that the NSE test (4.14) is not independent of column scaling. By (4.6),

$$\begin{bmatrix} \mathbf{v}_k^\sigma \\ -\mathbf{e}_k \end{bmatrix}$$

is chosen so that

$$[B \mid N]\begin{bmatrix} \mathbf{v}_k^\sigma \\ -\mathbf{e}_k \end{bmatrix} = B\mathbf{v}_k^\sigma - \boldsymbol{\kappa}_k(N) = 0$$

so that the equality constraints are preserved implying that the weighting of the unit vector is unchanged by scaling so that

$$\begin{bmatrix} \mathbf{v}_k^\sigma \\ -\mathbf{e}_k \end{bmatrix}$$

does not transform correctly. This problem comes about because the transformation has been restricted to the equality constraints and does not touch the inequalities $\mathbf{u} \geqslant 0$. In contrast, the corresponding test for the reduced gradient algorithm for LP1 given in the next section is independent of the choice of scale for the rows of A. The analogous result here is that the NSE test is independent of the scaling of the rows of A^{T}.

Simplex-like methods for the complementarity problem

The development and exploitation of the simplex algorithm has been accompanied by extensive consideration of variant procedures, and by attempts to incorporate information obtained from the duality structure. One possibility is to consider the complementarity problem associated with the symmetric dual (Exercise 2.2), and this approach proves to be of particular importance. For example, it will be used again in considering quadratic programming problems in Section 5.3. To introduce this topic Lemke's algorithm for the linear complementarity problem is considered. The problem is written in the form:

$$\begin{aligned}&\text{find } (\mathbf{w}, \mathbf{z}) \text{ satisfying}\\ &\mathbf{w} = Z\mathbf{z} + \mathbf{q},\\ &\mathbf{w}, \mathbf{z} \geqslant 0, \qquad \mathbf{w}^{\mathrm{T}}\mathbf{z} = 0.\end{aligned} \tag{4.43}$$

where $Z: R^p \rightarrow R^p$. It is customary to refer to $\mathbf{w}$ as the basic and $\mathbf{z}$ as the nonbasic variables, and this accords with the current usage; for if $\mathbf{q} \geqslant 0$ then $(\mathbf{w}, \mathbf{z}) = (\mathbf{q}, 0)$ is a solution to (4.43). If $\mathbf{q} \not\geqslant 0$ then consider the system

$$\begin{aligned}&\mathbf{w} = Z\mathbf{z} + z_0\mathbf{e} + \mathbf{q}\\ &\mathbf{w}, \mathbf{z}, z_0 \geqslant 0, \qquad \mathbf{w}^{\mathrm{T}}\mathbf{z} = 0.\end{aligned} \tag{4.44}$$

Here $(\mathbf{w}, \mathbf{z}) = (\mathbf{q} + z_0\mathbf{e}, 0)$ is feasible in the sense of satisfying the positivity and complementarity constraints, provided z_0 is chosen large enough. As z_0 is reduced there is a first component w_k which becomes zero (w_k is called a *blocking variable*). Now a pivot step is performed which exchanges w_k and z_0. This corresponds to a basic step of the simplex method and can be implemented using any of the procedures considered in this section. As w_k is nonbasic, feasibility is preserved if z_k is increased, and this can be done until blocked by another basic variable becoming zero. Then a pivot step exchanging z_k and the blocking variable is performed. This frees the complementary variable which is then increased and this procedure is repeated until either:

(i) z_0 becomes a blocking variable, in which case a solution to the complementarity problem has been obtained; or
(ii) the column associated with the complementary variable after the pivoting step is nonnegative so that the complementary variable can be increased without limit (the associated trajectory $(\mathbf{w}, \mathbf{z}, z_0)$ is called a *limiting ray*).

In particular, the procedure cannot cycle under the nondegeneracy assumption that exactly one blocking variable occurs to terminate each step. This is most easily seen by considering Figure 4.2. This details a cycle in which the blocking variables are k_1, k_2, k_3. Note that z_{k_1} increases on the arc joining k_1, k_2, and z_{k_2} on the arc joining k_2, k_3. Also the basic variables $z_{k_1}, z_{k_2} > 0$

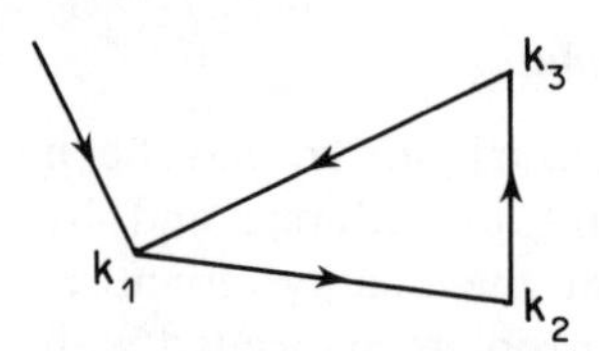

Figure 4.2 Trajectory in which a configuration repeats after three steps

at k_3 which is blocked by w_{k_3}, so both cannot be reduced to zero on k_3, k_1 by the nondegeneracy assumption. This contradicts the assumption that the configuration indicated by k_1 repeats. Clearly the argument generalizes to cycles of any length.

The key step is to investigate the case where the Lemke algorithm terminates in a ray. This requires some assumptions regarding the form of Z, and the following suffice for present purposes.

Definition 4.3 The matrix Z is *copositive plus* if

(i)
$$\mathbf{u}^{\mathrm{T}} Z \mathbf{u} \geqslant 0, \qquad \forall\, \mathbf{u} \geqslant 0, \tag{4.45}$$

and

(ii)
$$(\mathbf{u}^{\mathrm{T}} Z \mathbf{u} = 0, \qquad \mathbf{u} \geqslant 0) \Rightarrow (Z^{\mathrm{T}} + Z)\mathbf{u} = 0. \tag{4.46}$$

Theorem 4.3 *If the Lemke algorithm terminates in a ray, and Z is copositive plus then* (4.43) *has no solution.*

Proof If the procedure terminates in a ray then there exists a nontrivial solution to the homogeneous system

$$\begin{aligned} &\mathbf{s} = Z\mathbf{t} + t_0 \mathbf{e}, \\ &\mathbf{s}, \mathbf{t}, t_0 \geqslant 0, \qquad \mathbf{s}^{\mathrm{T}}\mathbf{t} = 0. \end{aligned} \tag{4.47}$$

Further, $\mathbf{t} \neq 0$, for the contrary implies that $\mathbf{s} > 0$ and (assuming that the trajectory meets the ray at the point $(\mathbf{w}^*, \mathbf{z}^*, z_0^*)$)

$$\mathbf{w}^* - Z\mathbf{z}^* - z_0^* \mathbf{e} = Z\mathbf{t} + \lambda t_0 \mathbf{e} + \mathbf{q}.$$

This can be identified with the initial trajectory by setting

$$\mathbf{w} = \mathbf{w}^* - Z\mathbf{z}^* - z_0 \mathbf{e} + \lambda \mathbf{s}, \qquad \mathbf{z} = \mathbf{t}, \qquad z_0 = \lambda t_0$$

and this implies a cycle, which is a contradiction. From (4.47),

$$\mathbf{t}^{\mathrm{T}} Z \mathbf{t} = -t_0 \sum_{i=1}^{p} t_i \leqslant 0$$

so that, by (4.45),

$$\mathbf{t}^{\mathrm{T}} Z \mathbf{t} = 0, \qquad t_0 = 0. \tag{4.48}$$

Now feasibility along the terminal ray gives

$$(\mathbf{w}^* + \lambda \mathbf{s})^{\mathrm{T}} (\mathbf{z}^* + \lambda \mathbf{t}) = 0$$

so that nonnegativity gives

$$\mathbf{w}^{*\mathrm{T}}\mathbf{t} = 0, \qquad \mathbf{s}^{\mathrm{T}}\mathbf{z}^* = 0. \tag{4.49}$$

Thus

$$0 = \mathbf{t}^{\mathrm{T}}\mathbf{w}^* = \mathbf{t}^{\mathrm{T}}\mathbf{q} + z_0^*\mathbf{t}^{\mathrm{T}}\mathbf{e} + \mathbf{t}^{\mathrm{T}}Z\mathbf{z}^*$$

and

$$0 = \mathbf{s}^{\mathrm{T}}\mathbf{z}^* = \mathbf{t}^{\mathrm{T}}Z^{\mathrm{T}}\mathbf{z}^* + t_0\mathbf{e}^{\mathrm{T}}\mathbf{z}^* = \mathbf{t}^{\mathrm{T}}Z^{\mathrm{T}}\mathbf{z}^*.$$

Adding and using (4.46) with $\mathbf{u} = \mathbf{t}$ gives

$$0 = \mathbf{t}^{\mathrm{T}}\mathbf{q} + z_0^*\mathbf{t}^{\mathrm{T}}\mathbf{e}$$

so that

$$\mathbf{t}^{\mathrm{T}}\mathbf{q} < 0. \tag{4.50}$$

Also

$$0 \leqslant \mathbf{s} = Z\mathbf{t} \Rightarrow Z^{\mathrm{T}}\mathbf{t} \leqslant 0 \tag{4.51}$$

as a consequence of (4.46). But now (4.50) and (4.51) give

$$\mathbf{t}^{\mathrm{T}}\mathbf{w} = \mathbf{t}^{\mathrm{T}}\mathbf{q} + \mathbf{t}^{\mathrm{T}}Z\mathbf{z} < 0$$

which contradicts the existence of a solution to (4.43).

Remark 4.8 It is trivial to verify that

$$Z = \begin{bmatrix} 0 & A \\ -A^{\mathrm{T}} & 0 \end{bmatrix}$$

occurring in the linear complementarity problem derived from the symmetric dual is copositive plus.

Remark 4.9 In the above development the algorithm is described as producing a trajectory. This terminology is justified by noting that at each step the basic variables are linear functions of the complementary variable being increased. In general, each piecewise linear section can be parameterized also by z_0 which then has the character of a continuation or homotopy parameter. However, it need not be true that z_0 decreases steadily to zero (z_0 must change in such a way that the complementary variable increases and in some steps z_0 may increase). This potentially retrogressive behaviour is in contrast to other instances of continuation considered in Sections 4.4 and 5.6.

Remark 4.10 If Z is strictly copositive so that $\mathbf{u}^{\mathrm{T}}Z\mathbf{u} > 0$, $\mathbf{u} \geqslant 0$, $\mathbf{u} \neq 0$, then (4.48) provides a contradiction to the assumption that the algorithm terminates in a ray. Thus this condition suffices to guarantee that the procedure terminates in a solution to the linear complementarity problem. This condition is trivially satisfied if Z is positive definite.

Exercise 4.1 (i) Consider the problem data

$$A^{\mathrm{T}}=\begin{bmatrix}1 & 1 & 1 & 1 & 1 & 0 & 0\\ \frac{1}{2} & -\frac{1}{2} & -\frac{5}{2} & 9 & 0 & 1 & 0\\ \frac{1}{2} & -\frac{3}{2} & -\frac{1}{2} & 1 & 0 & 0 & 1\end{bmatrix}$$

$$\mathbf{b}^{\mathrm{T}}=[-1 \quad 7 \quad 1 \quad 2 \quad 0 \quad 0 \quad 0]$$

$$\mathbf{c}^{\mathrm{T}}=[1 \quad 0 \quad 0].$$

Starting with the basic feasible solution determined by $\sigma=\{5,6,7\}$ verify that this solution is degenerate and that the USE test in conjunction with the simplex algorithm gives as a valid sequence of basic sets $\{5,1,7\}$, $\{5,1,2\}$, $\{5,3,2\}$, $\{5,3,4\}$, $\{5,6,4\}$, $\{5,6,7\}$.

(ii) Set up the reduced problem (4.41) associated with the initial degenerate basic feasible solution and use the recursive procedure to resolve the degeneracy.

(iii) Use the lexicographic procedure to resolve the degeneracy. Can this procedure be interpreted as a pivoting strategy in determining an unbounded solution to (4.42)?

Exercise 4.2 Let $A^{\mathrm{T}}=[B \mid N]$ where B is nonsingular. If

$$B\mathbf{z}=\mathbf{c}$$

$$\mathbf{w}=B\mathbf{e}^{(p)}$$

$$k=\arg\min_{i} z_i$$

and $\tau=\max(-z_k, 0)$, then

$$\begin{bmatrix}\mathbf{u}\\ u_{n+1}\end{bmatrix}=\begin{bmatrix}\mathbf{z}+\tau\mathbf{e}^{(p)}\\ 0\\ \tau\end{bmatrix}$$

is a basic feasible solution to the problem

$$\max \mathbf{b}^{\mathrm{T}}\mathbf{u}-\gamma u_{n+1}$$

subject to

$$[B\,|N|-\mathbf{w}]\begin{bmatrix}\mathbf{u}\\ u_{n+1}\end{bmatrix}=\mathbf{c}, \qquad \begin{bmatrix}\mathbf{u}\\ u_{n+1}\end{bmatrix}\geqslant 0.$$

Show that if $\gamma\geqslant 0$ is large enough then any optimal solution gives $\mathbf{u}$ solving LP2. How large must γ be chosen?

This exercise provides an alternative method of generating an initial basic feasible solution, this time at the cost of only one additional variable. Is it robust against rank deficiency?

Exercise 4.3 Solve the linear complementarity problems specified by the data:

(i)
$$M=\begin{bmatrix}-1 & 1 & 1 & 1\\ 1 & -1 & 1 & 1\\ -1 & -1 & -2 & 0\\ -1 & -1 & 0 & -2\end{bmatrix},\quad \mathbf{q}=\begin{bmatrix}3\\ 5\\ -9\\ -5\end{bmatrix};$$

(ii)
$$M=\begin{bmatrix}1 & 0 & 3\\ -1 & 2 & 5\\ 2 & 1 & 2\end{bmatrix},\quad \mathbf{q}=\begin{bmatrix}-3\\ -2\\ -1\end{bmatrix}.$$

For the first problem the Lemke algorithm terminates in a solution, while for the second it terminates in a ray.

2.5 THE REDUCED GRADIENT ALGORITHM FOR LP1

Because LP1 and LP2 are dual problems there is no loss of generality in considering LP1. This formulation has the attraction that it relates directly to the geometry of extreme points of convex sets discussed in Section 1.2, and it is at least arguable that the algorithm presented here is more readily accessible than the simplex algorithm for LP2. The basic idea is to start with a feasible point, to develop in sequence feasible points lying in the intersections of sets of active constraints of increasing order until an extreme point is produced, and then to move from extreme point to extreme point until the optimality conditions are satisfied. At each point a step is made in the direction determined by a vector $\mathbf{t}$ having the properties that

(i) the direction is a direction of descent so that $\mathbf{c}^T\mathbf{t}<0$; and
(ii) feasibility is preserved so that active constraints either remain active or relax to strict inequalities.

These properties are common both to the algorithm given here and to the projected gradient algorithm described in the next section. These algorithms thus share a common strategy and differ only in the manner in which the descent vector $\mathbf{t}$ is generated at each stage. Here the key idea is based on the following result which also shows how the requirement of feasibility narrows down the search for an appropriate descent direction.

Lemma 5.1 *Let*

$$A_k^T=[\mathbf{a}_1, \mathbf{a}_2, \dots, \mathbf{a}_k],$$
$$H=\text{span}\,(\mathbf{a}_i,\ i=1,2,\dots,k)\subset R^p,$$

and

$$\dim(H)=k.$$

Let E_k be any submatrix formed from columns of I_p so that the matrix B

defined by

$$B = [A_k^T \mid E_k]$$

has rank p, and define

$$\mathbf{t}_s = \boldsymbol{\kappa}_{k+s}(B^{-T}), \qquad s = 1, 2, \ldots, p-k.$$

Then

$$H^\perp = \text{span}\,(\mathbf{t}_s, s = 1, 2, \ldots, p-k).$$

Proof Perhaps the only point that is not immediate is the setting up of B, and it is important to realize that this can be carried out constructively. Let the LU factorization with partial pivoting of A_k^T into upper triangular form (Section A1.3) be

$$A_k^T = P\left[\begin{array}{c|c} L_1 & 0 \\ \hline L_2 & I_{p-k} \end{array}\right]\begin{bmatrix} U \\ 0 \end{bmatrix} \tag{5.1}$$

where P is the permutation matrix of row interchanges, L_1 is $k \times k$ unit lower triangular, and U is $k \times k$ upper triangular. Then a suitable choice for E_k is

$$E_k = P\begin{bmatrix} 0 \\ I_{p-k} \end{bmatrix}. \tag{5.2}$$

Remark 5.1 To see the significance of this construction assume that $\mathbf{a}_1, \ldots, \mathbf{a}_k$ correspond to rows of A associated with constraints which are active at the current point $\mathbf{x}$. Then these constraints remain active if the point is changed to $\mathbf{x} + \gamma\mathbf{t}$ if and only if $\mathbf{t} \in H^\perp$. Thus it is appropriate to seek descent directions in $H^\perp$. Of course E_k could be any matrix which makes B nonsingular. The above choice is an obvious one, is easy to implement, and has proved effective in practice.

For convenience in presenting the main features of the algorithm we assume that

(a) a feasible initial point is available; and
(b) the first constraint to become active in each descent direction, if one exists, is uniquely determined (this is a nondegeneracy assumption).

Both these restrictions are relaxed subsequently. It is not necessary to assume that rank $(A) = p$ as the less restrictive condition

$$\text{rank}\,(A) = K \leqslant p \tag{5.3}$$

can be treated without extra complication (at least in theory).

Remark 5.2 If LP1 has a bounded solution then $\mathbf{c}$ is in the span of any maximally independent subset of the rows of A as $\mathbf{c}$ is in the span of the rows of A by the Kuhn–Tucker conditions.

The algorithm has the characteristic form of a descent method in that its two major components are:

(i) the determination of a descent direction $\mathbf{t}$ in $H^{\perp}$; and
(ii) the calculation of a step in this direction to reduce the objective function subject to the constraint of preserving feasibility.

We consider first the calculation of the vector $\mathbf{t}$ determining the descent direction at the current point $\mathbf{x}$. We assume that the active constraints are pointed to by an index set ν,

$$\nu = \{i; \mathbf{a}_i^{\mathrm{T}}\mathbf{x} = b_i\} \tag{5.4}$$

where $|\nu| = k \leqslant K$, $\mathbf{a}_i^{\mathrm{T}} = \boldsymbol{\rho}_i(A)$, $i = 1, 2, \ldots, n$, and

$$\operatorname{rank}(A_\nu^{\mathrm{T}}) = k \tag{5.5}$$

where

$$A_\nu^{\mathrm{T}} = [\mathbf{a}_{\nu(1)}, \ldots, \mathbf{a}_{\nu(k)}]. \tag{5.6}$$

Let E_ν be the matrix obtained by applying the construction (5.1), (5.2) to A_ν^{T}. Then the matrix B_ν,

$$B_\nu = [A_\nu^{\mathrm{T}} \mid E_\nu], \tag{5.7}$$

has rank p. We define $\mathbf{u}$ to be the solution of the system of equations

$$B_\nu \mathbf{u} = \mathbf{c}. \tag{5.8}$$

There are two cases to consider.

(i) $\mathbf{c} \notin H = \operatorname{span}(\mathbf{a}_i, i \in \nu)$. Let

$$s = \arg\max_i \{|u_i|, i = k+1, \ldots, p\}. \tag{5.9}$$

By assumption $u_s \neq 0$ so that

$$\mathbf{t} = -\operatorname{sgn}(u_s)\boldsymbol{\kappa}_s(B_\nu^{-\mathrm{T}}) \tag{5.10}$$

is an appropriate descent vector as

$$\begin{aligned} \mathbf{t}^{\mathrm{T}}\mathbf{c} &= -\operatorname{sgn}(u_s)\mathbf{e}_s^{\mathrm{T}} B_\nu^{-1}\mathbf{c} \\ &= -\operatorname{sgn}(u_s)\mathbf{e}_s^{\mathrm{T}}\mathbf{u} \\ &= -|u_s| \\ &< 0. \end{aligned}$$

(ii) $\mathbf{c} \in H$. In general it would be expected that this would imply $k = K$. However, this is not necessary and it is not forced by the algorithm. This case is characterized (in exact arithmetic) by

$$u_i = 0, \qquad i = k+1, \ldots, p. \tag{5.11}$$

There are now two possibilities:

(a) $u_i \geq 0$, $i = 1, 2, \ldots, k$, but this implies by (5.7), (5.8), (5.11) that the Kuhn–Tucker conditions hold at $\mathbf{x}$ which is therefore optimal; and
(b) $u_i < 0$ for at least one i, $1 \leq i \leq k$.

In case (ii)(b) choose s as

$$s = \arg\min \{u_i, i = 1, 2, \ldots, k\} \tag{5.12}$$

and set

$$\mathbf{t} = \boldsymbol{\kappa}_s(B^{-\mathrm{T}}) = -\mathrm{sgn}\,(u_s)\boldsymbol{\kappa}_s(B^{-\mathrm{T}}) \tag{5.13}$$

giving

$$\mathbf{t}^{\mathrm{T}}\mathbf{c} = u_s = -|u_s| < 0$$

so that $\mathbf{t}$ defines a direction of descent. However, a new feature emerges here, for

$$\mathbf{t}^{\mathrm{T}}\mathbf{a}_{\nu(s)} = \mathbf{e}_s^{\mathrm{T}} B_\nu^{-1} B_\nu \mathbf{e}_s = 1 > 0$$

so the sth constraint becomes inactive in a step from the current point in the direction defined by $\mathbf{t}$. It follows that

$$\nu := \nu \backslash \{\nu(s)\}. \tag{5.14}$$

Remark 5.3 The tests (5.9), (5.12) are USE tests as they minimize the unnormalized directional derivative $\mathbf{t}^{\mathrm{T}}\mathbf{c}$ over all vectors of the form (5.10), (5.13). Computation of NSE tests is considered subsequently in this Section (5.22).

The second component of the algorithm is the determination of the step λ in the direction determined by $\mathbf{t}$ to give the maximum reduction in the objective function $\mathbf{c}^{\mathrm{T}}(\mathbf{x} + \lambda\mathbf{t})$ while maintaining feasibility. Either the problem is unbounded, so that

$$\mathbf{a}_i^{\mathrm{T}}\mathbf{t} \geq 0, \qquad i \in \nu^{\mathrm{c}}$$

showing that $\mathbf{t}$ is a direction of recession for X, or there is a first constraint (say the qth) which becomes active (exactly one constraint becomes active as a result of the nondegeneracy hypothesis), and the index pointing to this constraint must be added to ν:

$$\nu := \nu \cup \{q\}. \tag{5.15}$$

Corresponding to the deletion (5.14) and the addition (5.15) the matrix B_ν must be updated. The next result guarantees that it maintains full rank.

Lemma 5.2 *The matrix $\bar{B}_\nu$ corresponding to the updated index set ν is nonsingular provided that, when $\mathbf{c} \notin H$, $\mathbf{a}_q$* replaces *$\boldsymbol{\kappa}_{s-k}(E_\nu)$ which can then be interchanged with $\boldsymbol{\kappa}_1(E_\nu)$ to preserve the partitioned form* (5.7).

Proof Let Q be the permutation matrix associated with the column interchange (so that $Q = I_p$ if there is no interchange). Then

$$\begin{aligned}\bar{B}_\nu Q &= B_\nu + (\mathbf{a}_q - \boldsymbol{\kappa}_s(B_\nu))\mathbf{e}_s^{\mathrm{T}} \\ &= B_\nu(I + (B_\nu^{-1}\mathbf{a}_q - \mathbf{e}_s)\mathbf{e}_s^{\mathrm{T}}).\end{aligned} \tag{5.16}$$

The result follows as

$$\begin{aligned}\det([I + (B_\nu^{-1}\mathbf{a}_q - \mathbf{e}_s)\mathbf{e}_s^{\mathrm{T}}]) &= \mathbf{e}_s^{\mathrm{T}} B_\nu^{-1}\mathbf{a}_q \\ &= -\operatorname{sgn}(u_s)\mathbf{t}^{\mathrm{T}}\mathbf{a}_q \\ &\neq 0\end{aligned}$$

as $\mathbf{t}^{\mathrm{T}}\mathbf{a}_q < 0$ if the qth constraint becomes active in the current step.

Remark 5.4 Note that no assumption on rank (A) is required in this development. The pivoting procedure first builds up A_ν^{T} until $\mathbf{c} \in H$, and then exchanges columns until optimality is reached. However, the test $\mathbf{c} \in H$ requires (5.11) to be satisfied and cannot be expected to be robust in the presence of rounding errors which must serve to transform exact zeros into (hopefully) very small numbers. A possible alternative is to combine (5.9), (5.12) in the form

$$s = \arg\max_i\,(-u_i,\ 1 \leqslant i \leqslant k,\ \tau\,|u_i|,\ k+1 \leqslant i \leqslant p) \tag{5.17}$$

where τ is a weight which could be used to ensure that columns of E_ν are favoured for deletion until at some stage the $|u_i|$, $i = k+1, \ldots, p$ are all small. However, (5.17) does possess the interesting feature that it permits relaxing off active constraints associated with large values of $-u_i$ before a (relative) extreme point has been established. Thus what we are doing, in a sense, is deferring a decision on $\mathbf{c} \in H$ and mixing it up with the question of how to terminate the algorithm when no downhill direction exists. Clearly it is important that the quantities occurring in (5.17) should be comparable so that the choice of appropriate problem scalings must be considered carefully. Further information on the choice of τ is given in Section 7.5.

Remark 5.5 If $\mathbf{x}^*$ is the optimum, $\mathbf{x}_0$ the initial feasible point, and E_ν the adjoined matrix which ensures that the final B_ν is nonsingular, then

$$(\mathbf{x}^* - \mathbf{x}_0)^{\mathrm{T}} E_\nu = 0 \tag{5.18}$$

as columns are deleted from E_ν but never added, and $\mathbf{t}$ is orthogonal to the columns of B_ν which are not replaced at each stage. Thus the extra degrees of freedom in the rank-deficient case are taken up by holding certain components of the solution fixed at their initial values.

This remark also permits an immediate proof of finiteness in the nondegenerate case. The key observation is that each point produced is characterized uniquely by the conditions

$$\mathbf{a}_i^{\mathrm{T}}\mathbf{x} = b_i, \qquad i \in \nu$$

and

$$E_\nu^{\mathrm{T}}(\mathbf{x} - \mathbf{x}_0) = 0.$$

Thus there are at most a finite number of reachable points corresponding to a subset of the possible sets (A_ν, E_ν) in which all selections of $|\nu|$ rows of A and $p-|\nu|$ columns of I_p are made for $|\nu| = 0, 1, \ldots, p$. The algorithm sorts through these by asking that the objective function decrease at each step. This ensures that repetition cannot occur and finiteness is an immediate consequence. Finiteness does not depend on the precise form of the steepest-edge test. Finiteness in general requires a method for treating degeneracy and is considered subsequently.

Theorem 5.1 *Provided the nondegeneracy assumption is not violated the reduced gradient algorithm finds the minimum of LP1 in a finite number of steps or shows that it is unbounded.*

In outline the reduced gradient algorithm is as follows:

Algorithm

(i) *Determine an initial feasible point for LP1. Set up the initial B_ν incorporating the active constraints.*
(ii) *Solve* (5.8) *for* $\mathbf{u}$. *Test for termination.*
(iii) *Determine the direction of descent* $\mathbf{t}$. *Adjust* ν *if a constraint becomes inactive in the descent step (case* (ii)(b)).
(iv) *If* $\mathbf{a}_i^{\mathrm{T}}\mathbf{t} \geq 0$, $i = 1, 2, \ldots, n$, *then LP1 is unbounded. Otherwise compute*

$$q = \arg\min_i \left((b_i - \mathbf{a}_i^{\mathrm{T}}\mathbf{x})/\mathbf{a}_i^{\mathrm{T}}\mathbf{t},\ i \in \nu^c,\ \mathbf{a}_i^{\mathrm{T}}\mathbf{t} < 0\right),$$

$$\lambda = (b_q - \mathbf{a}_q^{\mathrm{T}}\mathbf{x})/\mathbf{a}_q^{\mathrm{T}}\mathbf{t},$$

$$\mathbf{x} := \mathbf{x} + \lambda\mathbf{t}.$$

(v) *Update* ν *to include* q, B_ν. *Go to* (ii).

Example 5.1 Consider the problem with the data defined by

$$\mathbf{c}^{\mathrm{T}} = [1, \quad 0, \quad 0]$$

$$A^{\mathrm{T}} = \begin{bmatrix} 1 & 1 & 1 & 1 & 1 & 1 \\ 1 & -1 & 1 & -1 & 1 & -1 \\ 0 & 0 & .5 & -.5 & 1 & -1 \end{bmatrix}$$

$$\mathbf{b}^{\mathrm{T}} = [1, \quad -1, \quad 1.649, \quad -1.649, \quad 2.718, \quad -2.718].$$

It is readily verified that a feasible point is given by

$$\mathbf{x}^{\mathrm{T}} = [.859, \quad 1.859, \quad 0]$$

and

$$\nu = \{2, 5\}.$$

Setting

$$B_\nu = \begin{bmatrix} 1 & 1 & 0 \\ -1 & 1 & 0 \\ 0 & 1 & 1 \end{bmatrix}$$

gives $\mathbf{u}^{\mathrm{T}} = \frac{1}{2}[1, 1, -1]$ so that $\mathbf{c} \notin H$ and an appropriate descent vector is

$$\mathbf{t}^{\mathrm{T}} = \tfrac{1}{2}[-1, \quad -1, \quad 2].$$

The index of the first constraint to be violated in the direction $\mathbf{t}$ is $q = 4$, and the step to this constraint is $\lambda = 1.298$, so

$$\mathbf{x}^{\mathrm{T}} := [.859, \quad 1.859, \quad 0] + 1.298[-.5, \quad -.5, \quad 1] = [.21, \quad 1.21, \quad 1.298],$$

and

$$\nu := \{2, 5, 4\}.$$

In the next iteration we have $\mathbf{c} \in H$,

$$B_\nu = \begin{bmatrix} 1 & 1 & 1 \\ -1 & 1 & -1 \\ 0 & 1 & -.5 \end{bmatrix}$$

$$\mathbf{u}^{\mathrm{T}} = [-.5, \quad .5, \quad 1]$$

and

$$\mathbf{t}^{\mathrm{T}} = \tfrac{1}{2}[-1, \quad -3, \quad 4]$$

corresponding to relaxing off the constraint pointed to by $\nu(1) = 2$. The index of the first constraint violated in the direction $\mathbf{t}$ is $q = 1$. This gives $\lambda = .21$, so that

$$\mathbf{x}^{\mathrm{T}} := [.21, \quad 1.21, \quad 1.298] + .21[-.5, \quad -1.5, \quad 2] = [.105, \quad .895, \quad 1.718]$$

and

$$\nu = \{1, 5, 4\}.$$

Now

$$B_\nu = \begin{bmatrix} 1 & 1 & 1 \\ 1 & 1 & -1 \\ 0 & 1 & -.5 \end{bmatrix}$$

and

$$\mathbf{u}^{\mathrm{T}} = [.25, \quad .25, \quad .5] > 0$$

so that the current point is optimal.

Key points in the development of the algorithm are now considered in greater detail. This involves some repetition where similar points have been made in considering the simplex algorithm. However, this should facilitate comparison between the two methods.

Selection of an initial feasible point

An initial feasible point is required, and this can be provided by adding a single artificial variable to the problem. The argument is similar to the corresponding case for the simplex algorithm. We consider the extended problem

$$\min_{\mathbf{y}\in\bar{X}} [\mathbf{c}^{\mathrm{T}}, \eta]\mathbf{y} : \bar{X} = \left\{\begin{bmatrix}\mathbf{x}\\ x_{p+1}\end{bmatrix}; A\mathbf{x} + x_{p+1}\mathbf{e}^{(n)} \geqslant \mathbf{b},\ x_{p+1} \geqslant 0\right\}. \tag{5.19}$$

This problem has the feasible solution

$$\begin{bmatrix}\mathbf{x}\\ x_{p+1}\end{bmatrix} = \begin{bmatrix}0\\ \max\,(\max_i\,(b_i), 0)\end{bmatrix} \tag{5.20}$$

which can be used for starting the computation. The key result is that $\eta > 0$ can be chosen sufficiently large (but finite) so that $x_{p+1} = 0$ in any optimum solution. It then follows that

$$\begin{bmatrix}\mathbf{x}\\ 0\end{bmatrix}$$

minimizes (5.19) if and only if $\mathbf{x}$ minimizes LP1, so that it suffices to consider (5.19). To show this, consider the Kuhn–Tucker conditions for (5.19). These give

$$\begin{bmatrix}\mathbf{c}\\ \eta\end{bmatrix} = \begin{bmatrix}A^{\mathrm{T}} & 0\\ \mathbf{e}^{(n)\mathrm{T}} & 1\end{bmatrix}\begin{bmatrix}\mathbf{u}\\ u_{n+1}\end{bmatrix}. \tag{5.21}$$

A sufficient condition for $x_{p+1} = 0$ is $u_{n+1} > 0$, in which case it is forced by complementarity. The last equation of (5.21) shows that $u_{n+1} > 0$ provided

$$\eta > \sum_{i=1}^{n} u_i.$$

If η is not chosen large enough, either $x_{p+1} > 0$ at the optimum or (5.19) is unbounded below. To see that the latter case is possible consider the problem

$$\min_{\mathbf{t}\in T} \mathbf{c}^{\mathrm{T}}\mathbf{t} : T = \{\mathbf{t};\ A\mathbf{t} + \mathbf{e}^{(n)} \geqslant 0\}.$$

If this problem has a bounded solution (and this will often be implied by the requirement that LP1 is bounded) then this minimum value is negative

$(=-\zeta, \zeta>0$, say). It follows that

$$\begin{bmatrix} \mathbf{t} \\ 1 \end{bmatrix}$$

is a direction of recession for $\bar{X}$ where $\mathbf{t}$ is a problem minimizer. If

$$\mathbf{c}^{\mathrm{T}}\mathbf{t}+\eta=-\zeta+\eta<0$$

then (5.19) is unbounded below. It follows that ζ is a lower bound for an acceptable η in (5.19). In either exceptional case η must be increased and the computation restarted. If this process generates an unbounded sequence of values of η then LP2 is unbounded. To see this, note that it follows from (5.21) that $\mathbf{u}$ is feasible for LP2. Also

$$\begin{bmatrix} \mathbf{u} \\ u_{n+1} \end{bmatrix}$$

is optimal for the dual of (5.19) by Theorem 2.2 so that

$$\mathbf{c}^{\mathrm{T}}\mathbf{x}+\eta x_{p+1}=\mathbf{u}^{\mathrm{T}}\mathbf{b}.$$

But $x_{p+1}>0$ for all η by assumption so that $\mathbf{u}^{\mathrm{T}}\mathbf{b}$ can be made arbitrarily large by choosing η large enough as Lemma 5.2 ensures that

$$\begin{bmatrix} \mathbf{x} \\ x_{p+1} \end{bmatrix}$$

is bounded independent of η (note that the set of possible values is finite). It now follows from Exercise 2.4 that LP1 has no feasible solution.

Calculation of the NSE test

The tests (5.9), (5.12), and the combined test (5.17) compute a maximum over unnormalized directional derivatives. If these are normalized we obtain the corresponding NSE test. For example, the test corresponding to (5.17) is

$$s=\arg\max_i\,(-u_i/\|\boldsymbol{\kappa}_i(B_\nu^{-\mathrm{T}})\|_2,\ 1\leqslant i\leqslant k,\ \tau\,|u_i|/\|\boldsymbol{\kappa}_i(B_\nu^{-\mathrm{T}})\|_2,\ k+1\leqslant i\leqslant p). \quad (5.22)$$

As in the consideration of the simplex algorithm this test can be computed economically enough when the norm is the Euclidean norm by setting up a recursion for the quantities $z_i=\|\boldsymbol{\kappa}_i(B_\nu^{-\mathrm{T}})\|_2^2$. However, *it is interesting that relatively less work is required here.* From (5.16) the updated inverse matrix is

$$Q\bar{B}_\nu^{-1}=\left\{I-\frac{(B_\nu^{-1}\mathbf{a}_q-\mathbf{e}_s)\mathbf{e}_s^{\mathrm{T}}}{\mathbf{e}_s^{\mathrm{T}}B_\nu^{-1}\mathbf{a}_q}\right\}B_\nu^{-1}$$

so that

$$\begin{aligned}\mathbf{e}_j^{\mathrm{T}}(Q\bar{B}_\nu^{-1})&=\mathbf{e}_j^{\mathrm{T}}B^{-1}-\frac{\mathbf{e}_j^{\mathrm{T}}B_\nu^{-1}\mathbf{a}_q-\delta_{js}}{\mathbf{e}_s^{\mathrm{T}}B_\nu^{-1}\mathbf{a}_q}\,\mathbf{e}_s^{\mathrm{T}}B_\nu^{-1}\\&=\mathbf{e}_j^{\mathrm{T}}B_\nu^{-1}-\zeta_{js}\mathbf{e}_s^{\mathrm{T}}B_\nu^{-1}\end{aligned}$$

and the new column norm becomes

$$\bar{z}_j = \|\mathbf{e}_j^{\mathrm{T}}(Q\bar{B}_\nu^{-1})\|_2^2 = z_j - 2\zeta_{js}\mathbf{e}_j^{\mathrm{T}}B_\nu^{-1}B_\nu^{-\mathrm{T}}\mathbf{e}_s + \zeta_{js}^2 z_s$$
$$= z_j - 2\zeta_{js}y_j + \zeta_{js}^2 z_s \tag{5.23}$$

where the computation of the intermediate vectors

$$\mathbf{y} = B_\nu^{-1}(B_\nu^{-\mathrm{T}}\mathbf{e}_s)$$

and

$$\mathbf{v} = B_\nu^{-1}\mathbf{a}_q$$

(so that $\zeta_{js} = (v_j - \delta_{js})/v_s$) reduces significantly the amount of work necessary, and $z_s = \|B_\nu^{-\mathrm{T}}\mathbf{e}_s\|_2^2$ is available independent of the recurrence at each stage (an important stabilizing element in the calculation). In contrast to the simplex algorithm, use of NSE tests with the reduced gradient algorithm does not change the asymptotic work estimate per iteration. Essentially this is because only p quantities are updated here as opposed to the $n-p$ required in the simplex algorithm.

Remark 5.6 Some cancellation is possible in the computation of the z_j from (5.23). For this reason the recurrence should be monitored. The independent calculation of z_s at each stage permits one important comparison and also serves to stabilize the recurrence. Detection of a $z_j < 0$ indicates catastrophic failure of the recurrence.

Treatment of degeneracy

The assumption that no more than one constraint becomes active in the direction determined by $\mathbf{t}$ means that $|\nu|$ is incremented by at most one, and only one column has to be pivoted into B_ν. If there are ties with more than one constraint becoming active simultaneously then this is no longer true. It is necessary to distinguish two cases depending on whether $\mathbf{c}$ is not or is in the span of the active constraints. The first case is treated by pivoting the columns associated with the constraints into B_ν sequentially, generating $\mathbf{t}$ at each step and then selecting the constraint to be added to minimize $\mathbf{t}^{\mathrm{T}}\mathbf{a}_q$ over the ties not yet dealt with, until progress can be made (this requires that $\mathbf{t}^{\mathrm{T}}\mathbf{a}_j \geqslant 0$ for all remaining ties) or $\mathbf{c} \in H$. If $\mathbf{c} \in H$ then a descent direction must be constructed to satisfy

$$\mathbf{c}^{\mathrm{T}}\mathbf{t} < 0,$$

and

$$A_\nu \mathbf{t} \geqslant 0$$

where now $|\nu| > \dim H$ ($= k + l$ where $\dim H = k$). Certainly such a direction exists if and only if the problem

$$\min_{\mathbf{t} \in T_\nu} \mathbf{c}^{\mathrm{T}}\mathbf{t} : T_\nu = \{\mathbf{t};\, A_\nu \mathbf{t} \geqslant 0\} \tag{5.24}$$

is unbounded below (equivalently there exists $\mathbf{t} \in T_\nu$ such that $\mathbf{c}^T\mathbf{t} < 0$). It follows that the problem of finding a descent direction is equivalent to finding a direction of recession for T_ν satisfying $\mathbf{c}^T\mathbf{t} < 0$. This problem is equivalent to finding an unbounded solution to the *nondegenerate* problem

$$\min_{\mathbf{t} \in \bar{T}_\nu} \mathbf{c}^T\mathbf{t} : \bar{T}_\nu = \{\mathbf{t};\, \mathbf{a}_{\nu(i)}^T\mathbf{t} \geqslant 0,\ 1 \leqslant i \leqslant k,\ \mathbf{a}_{\nu(i)}^T\mathbf{t} \geqslant -\varepsilon_i,\ i > k\} \tag{5.25}$$

as directions of recession are not affected by changes to the right-hand side of the constraint set. The reduced gradient method can be applied to solve this problem using 0 as initial point and B_ν updated by pivoting in an appropriate member of the tie (say the one minimizing $\mathbf{t}^T\mathbf{a}_q$). The method either gives a direction of recession or again encounters degeneracy, but in a smaller problem as each step moves off a constraint. Thus the procedure can be applied recursively (the above argument shows that the depth of recursion is at most $l-1$), and must eventually terminate with a downhill direction (or demonstrate that the current point is optimal). This follows by induction backwards on recursive level because the bottom level is in principle nondegenerate and so terminates by the standard counting argument (there are at most a finite number of allowable A_ν and the objective function is decreased at each step, so repetition of the A_ν is not possible). But this argument now applies for each subproblem at each fixed level because a descent direction is returned in a finite number of operations by each lower-level call, so A_ν cannot repeat in the current subproblem. In particular we have the following result.

Theorem 5.2 *The recursive implementation of the reduced gradient algorithm started from a feasible initial point either finds a minimum of LP1 in a finite number of steps or shows that it is unbounded below.*

Organization

Just as in the case of the simplex algorithm two different types of organization are available depending on problem requirements. If sparsity structure is important then a product form of the inverse approach which keeps factors explicitly will usually be preferable. For smaller, dense problems a tableau approach is convenient and avoids the requirement to maintain the transformation history. Only the approach using a tableau will be indicated here.

We start initially with

$$W = [A^T | I_p | \mathbf{c}] \tag{5.26}$$

and

$$\mathbf{w}^T = [(A\mathbf{x}_0 - \mathbf{b})^T | \mathbf{x}_0^T | \mathbf{c}^T\mathbf{x}_0] \tag{5.27}$$

where $\mathbf{x}_0$ is a suitable feasible point. At the ith step of the algorithm let B_ν

be given in factored form (as in the previous section),

$$B_\nu = R_i U_i \tag{5.28}$$

where the presumption is that both R_i and U_i are readily invertible.

Let

$$W_i = R_i^{-1} W, \qquad \mathbf{w}_i^{\mathrm{T}} = [(A\mathbf{x}_i - \mathbf{b})^{\mathrm{T}} \,|\mathbf{x}_i^{\mathrm{T}}|\, \mathbf{c}^{\mathrm{T}}\mathbf{x}_i], \tag{5.29}$$

and compute

$$\mathbf{u} = U_i^{-1} \boldsymbol{\kappa}_{n+p+1}(W_i) \tag{5.30}$$

noting that the columns of U_i can be read from among the columns of W_i. The edge selection rule can now be evaluated to determine s. The next step computes

$$\hat{\mathbf{t}}^{\mathrm{T}} = -\operatorname{sgn}(u_s)\mathbf{e}_s^{\mathrm{T}} U_i^{-1} = \mathbf{t}^{\mathrm{T}} R_i \tag{5.31}$$

where $\mathbf{t}$ is the descent direction determined by (5.10) or (5.13), and the rates of change of the problem variables (written $\mathbf{Dw}$) satisfy

$$\mathbf{Dw}^{\mathrm{T}} = \hat{\mathbf{t}}^{\mathrm{T}} W_i = \mathbf{t}^{\mathrm{T}}[A^{\mathrm{T}} \,|I|\, \mathbf{c}]. \tag{5.32}$$

The column to be pivoted into B_ν is given by $\mathbf{a}_q$ where (step (iv) of the algorithm)

$$q = \arg\min_j \{\lambda_j, \lambda_j = -(\mathbf{w}_i)_j/(\mathbf{Dw})_j < 0, j \in \nu^c\}. \tag{5.33}$$

Now set

$$\mathbf{w}_{i+1} = \mathbf{w}_i + \lambda_q \mathbf{Dw}, \tag{5.34}$$

update the factors of B_ν to take account of the pivot operation giving

$$B_\nu = R_i S_i U_{i+1}, \tag{5.35}$$

and set

$$W_{i+1} = S_i^{-1} W_i. \tag{5.36}$$

The key point is that the explicit representations (5.28), (5.35) are not required as all the information needed to carry out the algorithm in a compact, convenient, and readily updateable form is given by (5.34), (5.36).

We note briefly two factorizations appropriate to (5.28).

Method of Bartels and Golub

Here the computation is organized to permit the use of an LU decomposition with row interchanges. We order the columns of $\bar{U}_i$ in the form

$$\bar{U}_i = [\boldsymbol{\kappa}_1(U_i), \ldots, \boldsymbol{\kappa}_{s-1}(U_i), \boldsymbol{\kappa}_{s+1}(U_i), \ldots, \boldsymbol{\kappa}_p(U_i), R_i^{-1}\mathbf{a}_q].$$

The nonzero elements in this matrix in the case $p = 5$, $s = 3$ have the pattern shown in Figure 4.1. The situation corresponds exactly to that of the simplex algorithm at this point. Reduction of this matrix to upper triangular form requires eliminating only the single band of elements below the main diagonal in columns $s, s+1, \ldots, p-1$ (the circled elements in Figure 4.1) and pivoting for size where necessary to ensure that the multipliers in the elimination step are $\leqslant 1$ in magnitude. These interchanges plus elimination steps determine S_i^{-1}, and the key to the use of the tableau is the application of these steps to W_i rather than to an explicit form for B_ν. Both U_{i+1} and R_{i+1}^{-1} can be read off from the columns of W_{i+1} so the auxiliary quantities required for the NSE calculation can also be computed easily. For example, the quantities required to evaluate (5.23) are

$$\begin{aligned}\mathbf{y} &= B_\nu^{-1}B_\nu^{-\mathrm{T}}\mathbf{e}_s = -\operatorname{sgn}(u_s)B_\nu^{-1}R_i^{-\mathrm{T}}\hat{\mathbf{t}} \\ &= -\operatorname{sgn}(u_s)U_i^{-1}R_i^{-1}R_i^{-\mathrm{T}}\hat{\mathbf{t}},\end{aligned}$$

and

$$\mathbf{v} = B_\nu^{-1}\mathbf{a}_q = U_i^{-1}\boldsymbol{\kappa}_q(W_i).$$

To estimate the amount of work for large n and fixed p note that the major contributions come from the steps (5.32) and (5.36). In both cases we take the work estimate as $n(p-s)$ multiplications (note that $\hat{\mathbf{t}}$ has zero in positions $1, 2, \ldots, s-1$), giving a cost of $2n(p-s)$ with an expected value of np provided the mean s is $p/2$. This condition is satisfied if s has a symmetric distribution on $\{1, 2, \ldots, p\}$.

Remark 5.7 In the rank-deficient case it is important to order the columns of B_ν so that those of E_ν precede those of A_ν to ensure that the estimate of the mean value of s is as large as possible.

Jordan elimination

This is the procedure traditionally associated with the simplex tableau (Section 2.4). It is somewhat suspect as it does not include provisions to ensure numerical stability. It is characterized by

$$U_i = I_p, \quad R_i = B_\nu.$$

A consequence is that

$$\hat{\mathbf{t}} = \operatorname{sgn}(u_s)\mathbf{e}_s, \quad \mathbf{D}\mathbf{w}^{\mathrm{T}} = -\operatorname{sgn}(u_s)\boldsymbol{\rho}_s(W_i)$$

so that no computation is necessary to obtain the rate of change of the constraint values. However, now the cost of updating (5.36) is np multiplications so that the total is effectively the same as for the method of Bartels and Golub.

Remark 5.8 The Jordan elimination method has the interesting feature that a form of the NSE test is easy to implement. Consider

$$\sum_{k=1}^{n} |(W_i)_{jk}| = \sum_{k=1}^{n} |\mathbf{e}_j^{\mathrm{T}}B_\nu^{-1}\mathbf{a}_k| = \sum_{k=1}^{n} |\mathbf{t}_j^{\mathrm{T}}\mathbf{a}_k|. \tag{5.37}$$

This defines a norm on the orthogonal complement of the final E_ν which contains all possible descent directions (compare (5.18)). Thus an NSE test is available. It has the form

$$s = \arg\min \left(u_j \Big/ \sum_{k=1}^{n} |(W_i)_{jk}|, \; j = 1, 2, \ldots, k \right). \tag{5.38}$$

The test is cheap on the assumption that additions cost less than multiplications. Note that it need only be considered if $u_j < 0$, and that no recursion is required.

Treatment of equality constraints

Two approaches may be used for the treatment of equality constraints. The options are either to use these constraints to eliminate variables or to generate an initial point satisfying them and then force the corresponding constraint rows to be always columns of B_ν so that these constraints are satisfied at every subsequent step. The elimination of variables involves a calculation similar to that used in Remark 5.7 of Chapter 1. Let the equality constraint adjoined to LP1 be

$$C\mathbf{x} = \mathbf{d} \tag{5.39}$$

where $C: R^p \to R^s$, $s < p$, and rank $(C) = s$. To eliminate variables consider an LU factorization of C^{T} using partial pivoting to give

$$P\begin{bmatrix} L_1 & 0 \\ L_2 & I_{p-s} \end{bmatrix}\begin{bmatrix} U \\ 0 \end{bmatrix} = C^{\mathrm{T}} \tag{5.40}$$

where P is the permutation matrix of row interchanges, L_1 is $s \times s$ unit lower triangular, and U is $s \times s$ upper triangular. It follows that

$$U^{\mathrm{T}}[L_1^{\mathrm{T}} \mid L_2^{\mathrm{T}}]\begin{bmatrix} \mathbf{x}_1 \\ \mathbf{x}_2 \end{bmatrix} = \mathbf{d} \tag{5.41}$$

where

$$P\mathbf{x} = \begin{bmatrix} \mathbf{x}_1 \\ \mathbf{x}_2 \end{bmatrix} \tag{5.42}$$

so that

$$\mathbf{x}_1 = L_1^{-\mathrm{T}} U^{-\mathrm{T}} \mathbf{d} - L_1^{-\mathrm{T}} L_2^{\mathrm{T}} \mathbf{x}_2 \tag{5.43}$$

expressing s of the variables $(\mathbf{x}_1)$ in terms of the remaining variables $(\mathbf{x}_2)$. In these free variables the problem becomes

$$\min_{\mathbf{x}_2 \in X_2} (\mathbf{c}_2^{\mathrm{T}}\mathbf{x}_2 - \mathbf{c}_1^{\mathrm{T}} L_1^{-\mathrm{T}} L_2 \mathbf{x}_2) : X_2 = \{\mathbf{x}_2; (A_2 - A_1 L_1^{-\mathrm{T}} L_2^{\mathrm{T}})\mathbf{x}_2 \geq \mathbf{b} - A_1 L_1^{-\mathrm{T}} U^{-\mathrm{T}} \mathbf{d}\} \tag{5.44}$$

where

$$AP = [A_1 \mid A_2],$$

and

$$\mathbf{c}^{\mathrm{T}}P = [\mathbf{c}_1^{\mathrm{T}} \mid \mathbf{c}_2^{\mathrm{T}}].$$

The cost involved in setting up (5.44), estimated by counting multiplications, will be dominated by the cost of

(a) formation of $A_1 L_1^{-\mathrm{T}} L_2^{\mathrm{T}} \approx nps - ns^2/2$; and
(b) factorization of $C^{\mathrm{T}} \approx ps^2/2 - s^3/6$.

This is not large compared with the work necessary in the full algorithm for which an estimate of the form $\gamma(n+p+s)p$ is appropriate, where γ is the number of iterations required. It may be expected to be a slowly increasing function of n and p with typical values several times p (Chapter 7). As reduction in the number of variables appears to reduce problem complexity, it would appear that elimination of variables is unlikely to be a bad strategy, and that it should become increasingly attractive as s becomes comparable with p.

The alternative strategy requires a point satisfying (5.39) to be provided. It need not be feasible for the inequality constraints as this can be taken care of by introducing an artificial variable as in (5.19). There is probably not much to choose between the options when s/p is small, but otherwise elimination would be favoured in general.

Remark 5.9 It may be necessary to change these conclusions when the problem has a particular structure. An amusing example is provided by LP2. In this case the second strategy started from a feasible initial point is exactly the simplex algorithm. Thus the simplex algorithm can be thought of in this context as a way of specializing the organization of the reduced gradient algorithm to take account of problem structure (in this case in the inequality constraints).

Termination

In exact arithmetic, the Kuhn–Tucker conditions are satisfied if

(i) $u_i \geqslant 0,\ i = 1, 2, \ldots, k$, and
(ii) $u_i = 0,\ i = k+1, \ldots, p$.

As these are quantities computed at each step, they provide a natural test to terminate the algorithm. As the second set of quantities are just those that indicate $\mathbf{c} \in H$, the termination question is the alternative to the steepest-edge question. That is, the algorithm is terminated if a satisfactory descent direction cannot be found. The key to the decision is the interpretation of 'satisfactory', and difficulties necessarily will arise if the problem is inherently sensitive to perturbations such as rounding errors. To attempt to reduce these problems it is important to eliminate as much as possible the effects of poor problem scaling. The u_i are directly affected by the scaling of the rows of A. For let D be a diagonal matrix of scale factors, then a row rescaling is achieved by

$$A \to DA.$$

This implies that

$$B_\nu \to B_\nu D_\nu$$

where D_ν rescales the column of A_ν^{T} but leaves E_ν unchanged, so that

$$\mathbf{u} \to D_\nu^{-1}\mathbf{u} \Rightarrow u_i \to u_i/(D_\nu^{-1})_i \tag{5.45}$$

and

$$B_\nu^{-\mathrm{T}} \to B_\nu^{-\mathrm{T}} D_\nu^{-1} \Rightarrow \|\boldsymbol{\kappa}_i(B_\nu^{-\mathrm{T}})\|_2 \to \|\boldsymbol{\kappa}_i(B_\nu^{-\mathrm{T}})\|_2/(D_\nu^{-1})_i. \tag{5.46}$$

It follows that the NSE test (5.22) is invariant under row scaling, but that this is not true of the corresponding USE test (5.17). However, a suitably invariant quantity is

$$\tilde{u}_i = u_i \,\|\mathbf{a}_{\nu(i)}\|_2, \qquad i = 1, 2, \ldots, k, \tag{5.47}$$

suggesting that it is appropriate to scale the rows of A to have unit norm initially if USE tests are to be used. To make tests on $\mathbf{u}$ independent of the scale of $\mathbf{c}$ it suffices to rescale $\mathbf{c}$ to have norm 1.

To terminate the algorithm it is necessary to infer from a computed multiplier estimate $\bar{\mathbf{u}}$ that the exact vector $\mathbf{u}$ would satisfy

$$u_i \geqslant 0, \qquad i = 1, 2, \ldots, k, \qquad u_i = 0, \qquad i = k+1, \ldots, p. \tag{5.48}$$

The above discussion suggests that a test of the following form is appropriate:

$$\begin{aligned} \bar{u}_i &\geqslant -\mathrm{tol}\,\|\bar{\mathbf{u}}\|_2, \qquad i = 1, 2, \ldots, k \\ |\bar{u}_i| &\leqslant \mathrm{tol}\,\|\bar{\mathbf{u}}\|_2, \qquad i = k+1, \ldots, p \end{aligned} \tag{5.49}$$

where tol is an accuracy parameter to be chosen. The major problem with this test is that it takes no account of the sensitivity of the solution of equation (5.8) to the effect of rounding errors. This information is available if the NSE test is used.

Lemma 5.3 *Let the rows of A be scaled to have norm unity. Then*

$$\mathrm{cond}_{\mathrm{F}}(B_\nu) = \left\{ p \sum_{j=1}^{p} z_j \right\}^{1/2} \tag{5.50}$$

where cond $(\cdot)$ is the condition number, and F *indicates that the norm is the Frobenius matrix norm* $\left(\|A\|_{\mathrm{F}}^2 = \sum_{i=1}^{p} \sum_{j=1}^{p} A_{ij}^2 \right)$.

Proof This is by direct calculation. By the definition of z_j,

$$\|B_\nu^{-1}\|_{\mathrm{F}} = \left\{ \sum_{j=1}^{p} z_j \right\}^{1/2}$$

while the assumptions on the row scaling of A implies that each of the columns of B_ν has length unity, so that

$$\|B_\nu\|_F = p^{1/2}.$$

As the Frobenius matrix norm is compatible with the Euclidean vector norm so that

$$\|B\mathbf{u}\|_2 \leq \|B\|_F \|\mathbf{u}\|_2, \forall \mathbf{u},$$

it follows that the standard perturbation results for systems of linear equations apply. Thus, if

$$(B+\delta B)\bar{\mathbf{u}} = \mathbf{c} + \delta\mathbf{c} \tag{5.51}$$

then

$$\|\bar{\mathbf{u}} - \mathbf{u}\|_2/\|\mathbf{u}\|_2 \leq \frac{\text{cond}_F(B)}{1-\|B^{-1}\delta B\|_F}\left\{\frac{\|\delta B\|_F}{\|B\|_F} + \frac{\|\delta \mathbf{c}\|_2}{\|\mathbf{c}\|_2}\right\}. \tag{5.52}$$

Now the problem of termination become one of deciding given a computed $\bar{\mathbf{u}}$, an estimate of $\text{cond}_F(B_\nu)$, and a guess at the magnitudes of $\|\delta B\|_F/\|B\|_F$ and $\|\delta\mathbf{c}\|_2/\|\mathbf{c}\|_2$, if the exact computation would have yielded $\mathbf{u}$ satisfying (5.48). The immediately suspicious elements are those such that

$$\bar{u}_i < 0, \qquad i = 1, 2, \ldots, k, \quad \text{or} \quad \bar{u}_i \neq 0, \qquad i = k+1, \ldots, p.$$

Setting

$$\gamma = \left\{\sum_{i=1}^{k} \min(\bar{u}_i, 0)^2 + \sum_{i=k+1}^{p} \bar{u}_i^2\right\}^{1/2},$$

then an optimal $\mathbf{u}$ can exist only if

$$\gamma \leq \text{cond}_F(B_\nu)\ \text{eps}\ \|\bar{\mathbf{u}}\|_2. \tag{5.53}$$

Here eps is an error tolerance covering the effect of the relative perturbations caused by rounding errors in the numerical computations. A small value of eps implies confidence in the stability of the numerical procedures involved. It is here that the superior behaviour of the Bartels–Golub algorithm in this sense should show itself in comparison with Jordan elimination.

Exercise 5.1 Consider the problem data

$$A^T = \begin{bmatrix} 1 & 0 & 0 & -\frac{1}{3} & -\frac{1}{6} & -\frac{2}{3} \\ 0 & 1 & 0 & \frac{32}{9} & \frac{13}{3} & \frac{16}{9} \\ 0 & 0 & 1 & -\frac{20}{9} & -\frac{5}{18} & -\frac{1}{9} \end{bmatrix}$$

$$\mathbf{b}^T = [1 \quad 1 \quad 1 \quad 1 \quad 1 \quad 1]$$

$$\mathbf{c}^T = [-\tfrac{1}{3} \quad \tfrac{8}{9} \quad \tfrac{4}{9}].$$

The point $\mathbf{x} = \mathbf{e}^{(3)}$ is feasible but degenerate.

(i) Let $\nu=\{1,2,3\}$. Use the recursive procedure to determine if $\mathbf{x}$ is optimal or to resolve the degeneracy.

(ii) Let $\mathbf{c}:=-\mathbf{c}$. Does the degeneracy now restrict the application of the reduced gradient algorithm?

2.6 THE PROJECTED GRADIENT ALGORITHM FOR LP1

The basic strategy employed in this algorithm is identical to that of the reduced gradient algorithm discussed in Section 2.5. It differs only in the manner in which the direction of descent is generated at the current point, and this has the consequence that procedures for generating an initial feasible point and for resolving degeneracy can be taken over without change. This permits attention to be concentrated on such questions as the construction of the descent direction and the implementation of steepest-edge tests in the context of the use of orthogonal transformations as the main computational procedure, and the organization of the computation into a basic tableau form.

The notation follows that of the previous section. Let ν be the index set pointing to the active constraints defined in (5.4),

$$A_\nu^{\mathrm{T}}=[\mathbf{a}_{\nu(1)},\ldots,\mathbf{a}_{\nu(k)}],\qquad |\nu|=k\leqslant \operatorname{rank}(A)=K\leqslant p, \tag{6.1}$$

and

$$H=\operatorname{span}(\mathbf{a}_i,\ i\in\nu)\subseteq R^p. \tag{6.2}$$

It is assumed that A_ν has rank k. The new feature is the way in which the direction of descent is computed in $H^\perp$. It depends on the specification of orthogonal bases for H and $H^\perp$ as a byproduct of an orthogonal factorization of A_ν^{T}. This is summarized in the following result, which is then applied to generate a descent direction.

Lemma 6.1 *Let A_ν^{T} be factorized into the form*

$$A_\nu^{\mathrm{T}}=[Q_1\mid Q_2]\begin{bmatrix}U\\0\end{bmatrix} \tag{6.3}$$

where $Q=[Q_1\mid Q_2]$ is $p\times p$ orthogonal, U is $k\times k$ upper triangular, and the partitions are defined to make multiplication conformable. Then

$$H=\operatorname{span}(\boldsymbol{\kappa}_i(Q_1),\ i=1,2,\ldots,k), \tag{6.4}$$

and

$$H^\perp=\operatorname{span}(\boldsymbol{\kappa}_i(Q_2),\ i=1,2,\ldots,p-k). \tag{6.5}$$

Lemma 6.2 *Let $\mathbf{c}\notin H$, then*

$$\mathbf{t}=-Q_2Q_2^{\mathrm{T}}\mathbf{c} \tag{6.6}$$

is downhill for minimizing LP1.

Proof By assumption $Q_2^T\mathbf{c} \neq 0$. Thus

$$\mathbf{c}^T\mathbf{t} = -\mathbf{c}^T Q_2 Q_2^T \mathbf{c} = -\|Q_2^T\mathbf{c}\|_2^2 < 0.$$

Remark 6.1 The name of the algorithm is suggested by (6.6). Note that $\mathbf{c}$ is the gradient of the objective function in LP1 and $\mathbf{t}$ is obtained by projecting $-\mathbf{c}$ onto $H^\perp$ – hence 'projected gradient'.

The next result gives an alternative characterization of $\mathbf{t}$ in terms of A_ν^T rather than in terms of a specific factorization. This form is important in the subsequent development.

Lemma 6.3 *An equivalent specification of* $\mathbf{t}$ *given in* (6.6) *is that it satisfy the system of equations*

$$\mathbf{t} = -\mathbf{c} + A_\nu^T\mathbf{u}, \tag{6.7}$$

and

$$A_\nu\mathbf{t} = 0. \tag{6.8}$$

This system of equations for $\begin{bmatrix}\mathbf{t}\\ \mathbf{u}\end{bmatrix}$ *has a unique solution if and only if* A_ν *has rank* k.

Proof Because $Q_1Q_1^T$ and $Q_2Q_2^T$ are orthogonal projectors onto H, $H^\perp$ respectively, (6.6) gives

$$\begin{aligned}\mathbf{t} &= -(I - Q_1Q_1^T)\mathbf{c}\\ &= -\mathbf{c} + Q_1Q_1^T\mathbf{c}\\ &= -\mathbf{c} + A_\nu^T\mathbf{u}\end{aligned}$$

using (6.1), (6.4). Thus $\mathbf{t}$ is included in the set of vectors having the general form given by the right-hand side of (6.7). It is specified by the condition that $\mathbf{t} \in H^\perp$, and this is expressed by (6.8). The second part of the lemma follows from the observation that if $\mathbf{c} = 0$ then $A_\nu A_\nu^T\mathbf{u} = 0$ which implies that $\mathbf{u} = 0$ if and only if A_ν^T has full column rank, showing that this is exactly the condition that the homogeneous system has only the zero solution.

Remark 6.2 To compute $\mathbf{u}$ note that by (6.3), $A_\nu^T = Q_1U$, so that

$$U\mathbf{u} = Q_1^T\mathbf{c}. \tag{6.9}$$

Equations (6.7), (6.8) are important for indicating termination. By Lemma 6.3, if $\mathbf{c} \in H$ then $\mathbf{t} = 0$. But then (6.7) is equivalent to the Kuhn–Tucker conditions and gives optimality provided $\mathbf{u} \geq 0$. If $\mathbf{c} \in H$ and $\mathbf{u} \ngeq 0$ then there is at least one component $u_i < 0$ and the current point is not optimal. This information can be used to determine a direction of descent; for if $\mathbf{a}_{\nu(i)}$ is deleted from A_ν^T ($\nu := \nu\backslash\{\nu(i)\}$) then (6.7), (6.8) determine a nontrivial downhill direction in which the $\nu(i)$th constraint becomes inactive. This calculation will now be made. There is no restriction in assuming $i=k$ as the columns of A_ν^T can be rearranged to permit this (it would imply the need to revise the factorization (6.3) in an actual computation). It will be convenient to denote quantities computed after the deletion of $\nu(k)$ from ν by bars.

Lemma 6.4 *Let $u_k<0$. If $\bar{\nu}=\nu\backslash\{\nu(k)\}$ then*

$$\mathbf{a}_{\nu(k)}^T\bar{\mathbf{t}}_k=-U_{kk}^2u_k\geqslant 0 \tag{6.10}$$

so that the $\nu(k)$th constraint becomes inactive in the direction determined by $\bar{\nu}$.

Proof The point about deleting the kth column of A_ν^T is that the revised factorization is obtained just by regrouping the columns of Q. In particular,

$$\bar{Q}_2=[\boldsymbol{\kappa}_k(Q_1)\mid Q_2] \tag{6.11}$$

so that

$$\begin{aligned}\mathbf{a}_{\nu(k)}^T\bar{\mathbf{t}}&=-\mathbf{a}_{\nu(k)}^T\bar{Q}_2\bar{Q}_2^T\mathbf{c}\\&=-U_{kk}\mathbf{e}_1^T\bar{Q}_2^T\mathbf{c}\\&=-U_{kk}\boldsymbol{\kappa}_k(Q_1)^T\mathbf{c}\\&=-U_{kk}^2u_k\\&>0\end{aligned}$$

where $\boldsymbol{\kappa}_k(Q_1)^T\mathbf{c}$ is evaluated using (6.7).

Remark 6.3 This result did not use $\mathbf{c}\in H$. It follows that if at any stage of the algorithm a component of $\mathbf{u}$ is negative then the corresponding column can be deleted from A_ν^T. By (6.8), the resulting descent direction is less constrained and so should reduce the objective function more rapidly. However, it does not follow that the resulting algorithm is more efficient as it may not be possible to go as far in the steeper direction. Thus there are two possible strategies: either to relax off an active constraint if $u_i<0$ at any stage, or to build up ν until $\mathbf{c}\in H$ before using this option. This point is considered further in Section 7.5. The evidence presented favours the second strategy.

The most straightforward application of Lemma 6.4 to determining a direction of descent corresponds to choosing i by

$$i=\arg\min_j\,(u_j,\ j=1,2,\ldots,k) \tag{6.12}$$

to select the column of A_ν^T to be deleted. This is essentially a form of USE test (compare (4.12) or (5.17), for example). To compute the NSE test it is necessary to seek the maximum of the quantities ξ_i corresponding to the magnitude of the normalized directional derivatives. This gives

$$i = \arg\max_{j} \left(\xi_j,\ \xi_j = \frac{|\mathbf{c}^{\mathrm{T}}\bar{\mathbf{t}}_j|}{\|\bar{\mathbf{t}}_j\|_2},\ \forall\, j \ni u_j < 0 \right) \tag{6.13}$$

where $\bar{\mathbf{t}}_j$ is the descent vector computed on the updated index set $\bar{\nu}_j = \nu \backslash \{\nu(j)\}$.

The descent step is completed by finding the first constraint violated in the direction $\mathbf{t} = \bar{\mathbf{t}}_i$. If

$$\mathbf{a}_i^{\mathrm{T}}\mathbf{t} \geq 0, \qquad i = 1, 2, \ldots, n,$$

then LP1 is unbounded. Otherwise there is a first index q such that

$$\mathbf{a}_q^{\mathrm{T}}(\mathbf{x} + \lambda\mathbf{t}) = b_q, \qquad q \notin \nu, \tag{6.14}$$

and

$$\mathbf{a}_q^{\mathrm{T}}\mathbf{t} < 0. \tag{6.15}$$

The index set $\bar{\nu}$ must now be updated to include $q (\nu := \bar{\nu} \cup \{q\})$. Two points should be noted here. The assumption that there is a unique first q corresponds to a nondegeneracy assumption. If degeneracy occurs then it can be resolved by the procedure described in the previous section. Also, there is a result directly corresponding to Lemma 5.2 which ensures that rank $(A_\nu) = |\nu|$.

Lemma 6.5 *If* (6.15) *holds then*

$$\textit{rank}\ ([A_{\bar{\nu}}^{\mathrm{T}} \mid \mathbf{a}_q]) = |\bar{\nu}| + 1. \tag{6.16}$$

Proof By (6.3), and omitting bars except on $\bar{\nu}$,

$$[A_{\bar{\nu}}^{\mathrm{T}} \mid \mathbf{a}_q] = Q\left[\begin{array}{c|c} U & Q_1^{\mathrm{T}}\mathbf{a}_q \\ \hline 0 & Q_2^{\mathrm{T}}\mathbf{a}_q \end{array}\right]$$

and thus has the same rank as $A_{\bar{\nu}}$ only if $Q_2^{\mathrm{T}}\mathbf{a}_q = 0$. But (6.15) gives

$$0 > \mathbf{a}_q^{\mathrm{T}}\mathbf{t} = -\mathbf{a}_q^{\mathrm{T}}Q_2Q_2^{\mathrm{T}}\mathbf{c}$$

showing that $\|Q_2^{\mathrm{T}}\mathbf{a}_q\|_2 \neq 0$.

Remark 6.4 If rank $(A) < p$ then there is a maximal nontrivial subspace L having the property that

$$AL = 0.$$

If LP1 has a bounded solution then necessarily

$$\mathbf{c}^{\mathrm{T}}L = 0$$

also. It then follows from (6.7) that

$$\mathbf{t}^{\mathrm{T}}L = 0$$

for every ν. Thus, if $\mathbf{x}^*$ is the optimum produced by the algorithm,

$$(\mathbf{x}^* - \mathbf{x}_0)^{\mathrm{T}} L = 0. \tag{6.17}$$

This condition corresponds to the condition (5.18) in the reduced gradient algorithm. However, it serves to specify the solution precisely in the rank deficient case only if the conditions

$$p - \dim(L) = \operatorname{rank}(A_{\bar{\nu}}), \qquad \bar{H}^{\perp} = L$$

are satisfied at the optimum, where $\bar{\nu}$ is the index set pointing to the active constraints and $\bar{H}^{\perp}$ is the null space of $A_{\bar{\nu}}$. The qualification is needed as it is not implied by $\mathbf{c} \in H$. If it is not satisfied then the iteration sees only part of the data for the problem, and it is tempting to ask if $\bar{H}^{\perp}$ can replace L in specifying the optimum. This is the case only if

$$\bar{H}^{\perp} \subseteq H^{\perp}$$

for each H encountered. The problem is that updating the rows of A_ν can cause H to change even if its dimension remains fixed, and if the inclusion is not satisfied then the current descent vector need not satisfy $\mathbf{t}^{\mathrm{T}} \bar{H}^{\perp} = 0$ as a consequence of (6.17).

Somewhat similar problems occur in the discussion of the finiteness properties of the algorithm, and to obtain a simple result it is necessary to impose conditions which exclude certain of the possibilities considered. Fortunately, the excluded cases do not appear to be of practical interest (Section 7.5).

Theorem 6.1 *If (a) the nondegeneracy assumption is not violated, and (b) rows are not deleted from A_ν until $\mathbf{c} \in H$, then the projected gradient algorithm finds the minimum of LP1 in a finite number of steps.*

Proof The first assumption ensures that the successive function values are strictly decreasing, and the second ensures that $\mathbf{c} \in H$ is satisfied in at most p steps. The simplest case occurs when the extra condition that $\mathbf{c} \in H$ implies $\dim(H) = \operatorname{rank}(A)$ is satisfied; for then the equations

$$A_\nu \mathbf{x} = \mathbf{b}_\nu,$$

and

$$(\mathbf{x} - \mathbf{x}_0)^{\mathrm{T}} L = 0$$

determine $\mathbf{x}$ uniquely for each ν such that $\mathbf{c} \in H$. The result then follows from the standard counting argument as the index set ν cannot repeat because the objective function is reduced at each step. If this condition does not hold then it is necessary to show that cycling cannot occur. To do this assume that the index set $\hat{\nu}$ with $|\hat{\nu}| < p\text{-}\dim(L)$ repeats and let $\hat{H}$ be the corresponding subspace. Then the descent property of the algorithm ensures that the join of the two $\mathbf{x}$ values is a descent direction in $\hat{H}^{\perp}$. But this implies that the steepest descent direction projected into $\hat{H}^{\perp}$ is nontrivial, contradicting $\mathbf{c} \in \hat{H}$.

Degeneracy can be resolved in precisely the same manner as used in Section 2.5 because the *conditions determining if a descent direction exists are not algorithm-specific*. Thus it is only necessary to insert the projected rather than the reduced gradient method. This does not change the finiteness argument which follows as before.

In outline the form of the algorithm that emerges from the above discussion is as follows.

Algorithm

(i) *Set an initial feasible point, determine the initial* ν, *and compute the orthogonal factorization of* A_{ν}^{T} *(the device of adding an artificial variable to obtain an initial feasible solution to LP1 described in Section* 2.5 *is available here also).*

(ii) *Solve* (6.9) *for* $\mathbf{u}$. *If* $\mathbf{c} \in H$, $\mathbf{u} \geq 0$ *then stop. Option* (a) (*build up* ν *quickly*). *If* $\mathbf{c} \notin H$ *go to* (iii). *Option* (b) (*relax the constraints* (6.8) *on* $\mathbf{t}$ *if possible*). *If* $\mathbf{c} \notin H$ *and* $\mathbf{u} \geq 0$ *then go to* (iii). *Else choose* $u_i < 0$ *to minimize* ξ_i, $\nu := \nu \backslash \{\nu(i)\}$, *update factors of* A_{ν}^{T} *corresponding to column deletion.*

(iii) *Determine the projected gradient vector* $\mathbf{t}$ *for the current* ν.

(iv) *Find first constraint violated in the direction of* $\mathbf{t}$ *to determine new* $\mathbf{x}$. *If no constraint is violated then the problem is unbounded. Let the constraint index be* q.

(v) *If exactly one constraint is violated then* $\nu := \nu \cup \{q\}$, *update factors of* A_{ν}^{T} *to take account of variable addition, go to* (ii). *Else the problem is degenerate, resolve as in Section* 2.5.

Points relating to the implementation of the algorithm are now discussed in greater detail.

Implementation of the NSE test

In (6.13) the NSE test appears to assume computation of the tentative descent directions $\bar{\mathbf{t}}_j$. Computation of each of these requires reordering A_{ν}^{T} and then updating the orthogonal factorization. If the number of comparisons in (6.13) is commensurate with p then the computation involved in calculating the ξ_j will be $O(p^3)$ which could become significant in comparison with the $O(np)$ operations on average per step in the reduced gradient algorithm, and this suggests that it would be worth while to search for an alternative form for the NSE test.

Let $A_{\nu}^{(i)}$ be a rearrangement of the columns of A_{ν}^{T} which interchanges

$\boldsymbol{\kappa}_i(A_\nu^{\mathrm{T}})$ and $\boldsymbol{\kappa}_k(A_\nu^{\mathrm{T}})$. Then the orthogonal factorization of $A_\nu^{(i)}$ can be written

$$A_\nu^{(i)} = [Q_1^{(i)} \mid Q_2]\begin{bmatrix} U^{(i)} \\ 0 \end{bmatrix}$$

where Q_2 depends only on A_ν^{T} and can be chosen independent of the ordering of the columns (for example $Q_1^{(i)}$, $U^{(i)}$ can be determined by Gram–Schmidt orthonormalization without reference to Q_2, which can then be chosen as any orthonormal basis for $H^{\perp}$).

Then

$$\bar{Q}_2^{(i)} = [\boldsymbol{\kappa}_k(Q_1^{(i)}) \mid Q_2],$$

and

$$\xi_i = \frac{|\mathbf{c}^{\mathrm{T}}\bar{\mathbf{t}}_i|}{\|\bar{\mathbf{t}}_i\|_2} = \frac{\|\bar{Q}_2^{(i)\mathrm{T}}\mathbf{c}\|_2^2}{\|\bar{Q}_2^{(i)\mathrm{T}}\mathbf{c}\|_2} = \|\bar{Q}_2^{(i)\mathrm{T}}\mathbf{c}\|_2. \tag{6.18}$$

Direct computation using (6.6) and the dodge employed in the proof of Lemma 6.4 gives

$$\begin{aligned}\|\bar{Q}_2^{(i)\mathrm{T}}\mathbf{c}\|_2^2 &= (\boldsymbol{\kappa}_k(Q_1^{(i)})^{\mathrm{T}}\mathbf{c})^2 + \|\mathbf{t}\|_2^2 \\ &= U_{kk}^{(i)2}u_i^2 + \|\mathbf{t}\|_2^2,\end{aligned} \tag{6.19}$$

where $\mathbf{t}$ is the descent vector associated with ν. This expression still depends on the rearrangement, but this can be removed by noting that

$$U_{kk}^{(i)2} = \frac{1}{(U^{(i)-1}U^{(i)-\mathrm{T}})_{kk}}, \tag{6.20}$$

and that $U^{(i)}$ can be written in factored form as

$$U^{(i)} = Q^{(i)}UP_{ik}, \tag{6.21}$$

where P_{ik} interchanges columns i and k of U, and $Q^{(i)}$ is an orthogonal matrix which restores upper triangular form. Thus

$$\begin{aligned} U_{kk}^{(i)2} &= \frac{1}{\mathbf{e}_k^{\mathrm{T}}(U^{(i)\mathrm{T}}U^{(i)})^{-1}\mathbf{e}_k} \\ &= \frac{1}{\mathbf{e}_k^{\mathrm{T}}(P_{ik}U^{\mathrm{T}}Q^{(i)\mathrm{T}}Q^{(i)}UP_{ik})^{-1}\mathbf{e}_k} \\ &= \frac{1}{\mathbf{e}_i^{\mathrm{T}}(U^{\mathrm{T}}U)^{-1}\mathbf{e}_i} = \frac{1}{(U^{\mathrm{T}}U)_{ii}^{-1}}, \end{aligned} \tag{6.22}$$

and the dependence on the rearrangement has been removed. We have

$$\xi_i^2 = u_i^2/(U^{\mathrm{T}}U)_{ii}^{-1} + \|\mathbf{t}\|_2^2, \tag{6.23}$$

and this shows that ξ_i can be evaluated provided the diagonal elements of the inverse of $U^{\mathrm{T}}U$ are known. These quantities can be updated economically whenever a column is added to or deleted from A_ν^{T} (again note that at no stage is it required that $\mathbf{c} \in H$).

Case (i) $\bar{\nu}=\nu\cup\{q\}$ (addition of a column to A_ν^{T}). Let

$$A_{\bar{\nu}}^{\mathrm{T}}=[A_\nu^{\mathrm{T}}\,|\,\mathbf{a}_q]$$

then

$$Q^{\mathrm{T}}A_{\bar{\nu}}^{\mathrm{T}}=\left[\begin{array}{c|c} U & \bar{\mathbf{a}}_1 \\ \hline 0 & \bar{\mathbf{a}}_2 \end{array}\right]$$

where

$$\bar{\mathbf{a}}=\begin{bmatrix}\bar{\mathbf{a}}_1\\ \bar{\mathbf{a}}_2\end{bmatrix}=\begin{bmatrix}Q_1^{\mathrm{T}}\\ Q_2^{\mathrm{T}}\end{bmatrix}\mathbf{a}_q.$$

Thus

$$\bar{U}^{\mathrm{T}}\bar{U}=\left[\begin{array}{c|c} U^{\mathrm{T}}U & U^{\mathrm{T}}\bar{\mathbf{a}}_1 \\ \hline \bar{\mathbf{a}}_1^{\mathrm{T}}U & \|\bar{\mathbf{a}}_1\|_2^2+\|\bar{\mathbf{a}}_2\|_2^2 \end{array}\right]$$

and

$$(\bar{U}^{\mathrm{T}}\bar{U})^{-1}=\left[\begin{array}{c|c} (U^{\mathrm{T}}U)^{-1}+\dfrac{1}{\|\bar{\mathbf{a}}_2\|_2^2}U^{-1}\bar{\mathbf{a}}_1\bar{\mathbf{a}}_1^{\mathrm{T}}U^{-\mathrm{T}} & \dfrac{-1}{\|\bar{\mathbf{a}}_2\|_2^2}U^{-1}\bar{\mathbf{a}}_1 \\ \hline -\dfrac{1}{\|\bar{\mathbf{a}}_2\|_2^2}\bar{\mathbf{a}}_1^{\mathrm{T}}U^{-\mathrm{T}} & \dfrac{1}{\|\bar{\mathbf{a}}_2\|_2^2} \end{array}\right]$$

so that

$$\begin{aligned}(\bar{U}^{\mathrm{T}}\bar{U})_{jj}^{-1}&=(U^{\mathrm{T}}U)_{jj}^{-1}+\frac{1}{\|\bar{\mathbf{a}}_2\|_2^2}(U^{-1}\bar{\mathbf{a}}_1)_j^2, \qquad j=1,2,\ldots,|\nu|\\ &=\frac{1}{\|\bar{\mathbf{a}}_2\|_2^2}, \qquad j=|\nu|+1.\end{aligned}\tag{6.24}$$

Case (ii) $\bar{\nu}=\nu\backslash\{\nu(i)\}$ (deletion of a column from A_ν^{T}). In this case

$$A_\nu^{\mathrm{T}}=[A_{\bar{\nu}}^{\mathrm{T}}\,|\,\mathbf{a}_{\nu(i)}]P$$

where P is a permutation matrix specifying the rearrangement of the columns of A_ν^{T}. Premultiplying by $\bar{Q}$ and developing $U^{\mathrm{T}}U$ as before gives

$$U^{\mathrm{T}}U=P^{\mathrm{T}}\left[\begin{array}{c|c} \bar{U} & \bar{\bar{a}}_1 \\ \hline 0 & \bar{\bar{a}}_2 \end{array}\right]^{\mathrm{T}}\left[\begin{array}{c|c} \bar{U} & \bar{\bar{\mathbf{a}}}_1 \\ \hline 0 & \bar{\bar{\mathbf{a}}}_2 \end{array}\right]P$$

where

$$\bar{\bar{\mathbf{a}}}=\begin{bmatrix}\bar{\bar{\mathbf{a}}}_1\\ \bar{\bar{\mathbf{a}}}_2\end{bmatrix}=\begin{bmatrix}\bar{Q}_1^{\mathrm{T}}\\ \bar{Q}_2^{\mathrm{T}}\end{bmatrix}\mathbf{a}_{\nu(i)}.$$

Thus

$$(U^{\mathrm{T}}U)^{-1}=P^{\mathrm{T}}\left[\begin{array}{c|c} (\bar{U}^{\mathrm{T}}\bar{U})^{-1}+\dfrac{1}{\|\bar{\bar{\mathbf{a}}}_2\|_2^2}\bar{U}^{-1}\bar{\bar{\mathbf{a}}}_1\bar{\bar{\mathbf{a}}}_1^{\mathrm{T}}\bar{U}^{-\mathrm{T}} & -\dfrac{1}{\|\bar{\bar{\mathbf{a}}}_2\|_2^2}\bar{U}^{-1}\bar{\bar{\mathbf{a}}}_1 \\ \hline -\dfrac{1}{\|\bar{\bar{\mathbf{a}}}_2\|_2^2}\bar{\bar{\mathbf{a}}}_1^{\mathrm{T}}\bar{U}^{-\mathrm{T}} & \dfrac{1}{\|\bar{\bar{\mathbf{a}}}_2\|_2^2} \end{array}\right]P$$

giving

$$(\bar{U}^{\mathrm{T}}\bar{U})^{-1}_{jj} = (U^{\mathrm{T}}U)^{-1}_{P(j)P(j)} - \frac{1}{\|\bar{\bar{\mathbf{a}}}_2\|_2^2}(\bar{U}^{-1}\bar{\bar{\mathbf{a}}}_1)^2_j, \qquad j = 1, 2, \ldots, |\nu|-1. \tag{6.25}$$

where $P(j)$ is the index of the column of A_ν^{T} which is mapped into $\boldsymbol{\kappa}_j(A_{\bar{\nu}}^{\mathrm{T}})$ for $j = 1, 2, \ldots, |\nu|-1$.

It follows that the cost of updating the diagonal elements of $(U^{\mathrm{T}}U)^{-1}$ is dominated by the back-substitution involved as we will see that $\bar{\mathbf{a}}$, $\bar{\bar{\mathbf{a}}}$ are already available if the tableau form of organization is used. This is in fact the least work involved in implementing an NSE test of all the algorithms considered so far.

In this case one diagonal element is computed independently of the recurrence in each addition step. However, an independent check on the recurrence is not available without extra work and in this respect compares unfavourably with the reduced gradient algorithm recurrence. Also (6.25) suggests the possibility of cancellation, in particular if small values of $\|\bar{\bar{\mathbf{a}}}_2\|$ should occur. Experience with the method has proved fairly satisfactory, and it should be noted that it has applications in other contexts – one instance is outlined in Exercise 2.4 of Chapter 5. However, problems have occurred, especially in single precision calculations, and it is recommended that the element produced by (6.25) with the largest relative change be checked for loss of significance by also calculating it directly (the test does not have to be severe as only leading digits matter). This independent value can then be substituted in the same way as in the reduced gradient algorithm. This procedure will cause reinversion if any element goes negative and seems a reasonable compromise between control and additional computation.

Organization in tableau form

The projected gradient algorithm can also be organized in tableau form. Initially set

$$W = [A^{\mathrm{T}} \,|I_p|\, \mathbf{c}] \tag{6.26}$$

and

$$\mathbf{w}^{\mathrm{T}} = [\mathbf{r}_0^{\mathrm{T}} \,|\mathbf{x}_0^{\mathrm{T}}|\, \mathbf{c}^{\mathrm{T}}\mathbf{x}_0] \tag{6.27}$$

where $\mathbf{x}_0$ is the initial feasible point and

$$\mathbf{r}_0 = A\mathbf{x}_0 - \mathbf{b}.$$

If the current stage corresponds to the ith step of the iteration let

$$A_\nu^{\mathrm{T}} = Q_i\begin{bmatrix} U_i \\ 0 \end{bmatrix} \tag{6.28}$$

and

$$W_i = Q_i^{\mathrm{T}} W_j,\ \mathbf{w}_i^{\mathrm{T}} = \{(A\mathbf{x}_i - \mathbf{b})^{\mathrm{T}} \,|\mathbf{x}_i^{\mathrm{T}}|\, \mathbf{c}^{\mathrm{T}}\mathbf{x}_i\}. \tag{6.29}$$

In the deletion phase, assuming that $\boldsymbol{\kappa}_j(A_\nu^{\mathrm{T}})$ is dropped,

$$U_i \to [\boldsymbol{\kappa}_1(U_i)\ldots\boldsymbol{\kappa}_{j-1}(U_i)\boldsymbol{\kappa}_{j+1}(U_i)\ldots\boldsymbol{\kappa}_k(U_i)] = \tilde{U}_i$$

where $\tilde{U}_i$ has the characteristic upper Hessenberg form illustrated in Figure 4.1. In this case it is appropriate to use plane rotations to zero the elements in the subdiagonal positions in order to restore upper triangular form. If $\hat{U}_i$ is the resulting upper triangular matrix then

$$\hat{U}_i = \prod_{l=k-1}^{j} R(l, l+1, (l+1, l))\tilde{U}_i \tag{6.30}$$

and

$$\hat{Q}_i^{\mathrm{T}} = \prod_{l=k-1}^{j} R(l, l+1, (l+1, l))Q_i^{\mathrm{T}} \tag{6.31}$$

where $R(l, l+1, \cdot)$ is the plane rotation which mixes rows l and $l+1$ of the target matrix and makes zero the indicated element. Note that the columns of $\tilde{U}_i$ are included among the columns of W_i so that W_i is correctly updated to $\hat{W}_i$ by applying the rotations to it directly. Now $\hat{Q}_2^{\mathrm{T}}\mathbf{c}$ is held in positions $k, k+1, \ldots, p$ in $\boldsymbol{\kappa}_{n+p+1}(\hat{W}_i)$ so that the rate of change of the constraint values is given by

$$\begin{aligned}(\mathbf{Dw})_s &= \mathbf{a}_s^{\mathrm{T}}\mathbf{t} = -(\hat{Q}_2^{\mathrm{T}}\mathbf{a}_s)^{\mathrm{T}}(\hat{Q}_2^{\mathrm{T}}\mathbf{c})\\ &= -\sum_{j=k}^{p} (\hat{W}_i)_{js}(\hat{W}_i)_{j(n+p+1)}, \qquad s = 1, 2, \ldots, n,\end{aligned}$$

while the rate of change of the components of $\mathbf{x}$ is

$$\begin{aligned}(\mathbf{Dw})_s &= \mathbf{e}_s^{\mathrm{T}}\mathbf{t} = -(\hat{Q}_2^{\mathrm{T}}\mathbf{e}_s)^{\mathrm{T}}(\hat{Q}_2^{\mathrm{T}}\mathbf{c})\\ &= -\sum_{j=k}^{p} (\hat{W}_i)_{js}(\hat{W}_i)_{j(n+p+1)}, \qquad s = n+1, \ldots, n+p,\end{aligned}$$

and a similar calculation shows that $(\mathbf{Dw})_{n+p+1} = \sum_{j=k}^{p} (\hat{W}_i)_{j(n+p+1)}^2$. In summary

$$\mathbf{Dw}^{\mathrm{T}} = -(\hat{Q}_2^{\mathrm{T}}\mathbf{c})^{\mathrm{T}}\hat{Q}_2 W. \tag{6.32}$$

The step λ_i permitted in the direction defined by $\mathbf{t}$ is limited by the first constraint to become active. This gives

$$q = \arg\min_j (-w_j/(\mathbf{Dw})_j, (\mathbf{Dw})_j < 0, j \in \nu^c), \tag{6.33}$$

and $\mathbf{w}_i^{\mathrm{T}}$ is updated by

$$\mathbf{w}_{i+1}^{\mathrm{T}} = \mathbf{w}_i^{\mathrm{T}} + \lambda_i \mathbf{Dw}^{\mathrm{T}} \tag{6.34}$$

where

$$\lambda_i = -w_q/(\mathbf{Dw})_q.$$

The final step is the updating of the factorization of $\hat{W}_i$ to take account of the addition of $\mathbf{a}_q$ to $\hat{A}_\nu^{\mathrm{T}}$. If

$$A_\nu^{\mathrm{T}} := [\hat{A}_\nu^{\mathrm{T}} \mid \mathbf{a}_q] \tag{6.35}$$

then

$$\begin{aligned} U_{i+1} &= H_i \hat{Q}_i^{\mathrm{T}} [\hat{A}_\nu^{\mathrm{T}} \mid \mathbf{a}_q] \\ &= H_i [\hat{U}_i \mid \hat{Q}_i^{\mathrm{T}} \mathbf{a}_q] \\ &= H_i [\hat{U}_i \mid \boldsymbol{\kappa}_q(\hat{W}_i)] \end{aligned} \tag{6.36}$$

where H_i is the orthogonal transformation chosen to reduce the right-hand side to upper triangular form. It is convenient to use a Householder transformation so that, in the notation of Appendix 1,

$$H_i = H(i, \hat{Q}_i^{\mathrm{T}} \mathbf{a}_q). \tag{6.37}$$

Of course U_{i+1} is not generated explicitly, and the transformation is applied to $\hat{W}_i$ to give

$$W_{i+1} = H_i \hat{W}_i. \tag{6.38}$$

Remark 6.5 The work involved in this method is dominated by the operations on the full tableau. These are:

(i) the calculation of **Dw** costing of order $(n+p)(p-k)$ multiplications;
(ii) the transformation of the tableau due to a new constraint becoming active – $2(n+p)(p-k)$ multiplications; and
(iii) the transformation of the tableau due to the deletion of the jth active constraint – $3(n+p)(k-j)$ multiplications.

Here $2(p-k)$ is taken as the figure for applying a Householder transformation to a column, and $3(k-j)$ as the figure for each column due to the plane rotations applied to restore upper traingular form. If there are no deletiôn steps in building up an extreme point then this phase costs $3(n+p)\sum_{k=1}^{p}(p-k) \approx \frac{3}{2}(n+p)p^2$. In the second phase each iterations costs $3(n+p)(p-j)$, giving an expected value of $\approx \frac{3}{2}(n+p)p$ if j has a symmetric discrete distribution. In comparison with the reduced gradient algorithm this implies a 50 per cent increase in cost per iteration. There may be some advantage in stability for the orthogonal transformations.

Remark 6.6 If in the second phase $|\nu| = p-1$, corresponding to the expected situation when rank $(A) = p$, then the reduced and projected gradient methods generate identical iterates from the same extreme point. This follows because $H^\perp$ is one-dimensional, forcing all descent directions to be equivalent. This identity does not extend to the first phase or to the case rank $(A) < p$.

Termination

As in the previous section the decision to terminate is made by deducing from the computed multiplier estimates $\bar{\mathbf{u}}$ that $\mathbf{u} \geqslant 0$, and the same considerations apply. Thus, as far as possible, scale-free quantities should be tested, and an appropriate measure of problem sensitivity taken into account in setting tolerances.

To study the effects of scaling consider

$$A \rightarrow DA$$

where D is a diagonal scaling matrix. The orthogonal matrix taking A_ν^{T} to upper triangular form is unchanged by scaling the rows of A (the directions of the columns of A_ν^{T} are unchanged) so that

$$A_\nu^{\mathrm{T}} \rightarrow Q_1(UD_\nu) \tag{6.39}$$

and

$$\mathbf{u} = U^{-1}Q_1^{\mathrm{T}}\mathbf{c} \rightarrow D_\nu^{-1}\mathbf{u}. \tag{6.40}$$

Thus the USE test transforms in the same manner as in Section 2.5 so that an initial scaling of A to have row norms of unity is advisable. It follows from (6.39) that

$$(U^{\mathrm{T}}U)^{-1} \rightarrow D_\nu^{-1}(U^{\mathrm{T}}U)^{-1}D_\nu^{-1}$$

so that $u_i^2/(U^{\mathrm{T}}U)_{ii}^{-1}$ is invariant under row scaling of A. This shows that the NSE test is independent of row scaling.

Before checking that $\mathbf{u} \geqslant 0$ it is necessary to decide that $\mathbf{t} = 0$ ($\mathbf{c} \in H$) in (6.7). This follows immediately from the nondegeneracy assumption that rank $(A_\nu) = k$ when $k = p$. Otherwise the decision must be made by considering the size of the elements $\bar{Q}_2^{\mathrm{T}}(\mathbf{c} + \delta\mathbf{c})$ recorded in the tableau where $\bar{Q}$ is the computed estimate of Q. The key question is: how big can these elements be and still be capable of being caused by the cumulative effects of rounding error? At this point it is helpful to introduce the orthogonal matrix $\hat{Q}$ which is the matrix which would result if at each stage exact arithmetic were applied to the quantities actually recorded. The results required from the rounding error analysis of orthogonal transformations are:

(i) $$\left\| \hat{Q}^{\mathrm{T}}A_\nu^{\mathrm{T}} - \begin{bmatrix} \bar{U} \\ 0 \end{bmatrix} \right\|_{\mathrm{F}} = \left\| \begin{bmatrix} E_1 \\ E_2 \end{bmatrix} \right\|_{\mathrm{F}},$$

and

(ii) $$\|\hat{Q} - \bar{Q}\|_{\mathrm{F}} = \|F\|_{\mathrm{F}}$$

where estimates of $\|E\|_{\mathrm{F}}$, $\|F\|_{\mathrm{F}}$ have the form

$$\|E\|_{\mathrm{F}} \leqslant \mathrm{eps}\, \|A_\nu\|_{\mathrm{F}},$$
$$\|F\|_{\mathrm{F}} \leqslant \mathrm{eps},$$

and where eps is independent of the particular problem (except possibly for the number of iterations). They say that the computed transformation is nearly orthogonal (this is not the same as saying it is close to Q), and that this neighbouring exact orthogonal transformation gives the computed upper triangular matrix plus a small error.

Now if $\mathbf{c} \in H$ then it can be written

$$\mathbf{c} = Q_1 \tilde{\mathbf{c}}$$

where $\|\mathbf{c}\|_2 = \|\tilde{\mathbf{c}}\|_2$. Thus

$$\begin{aligned} \hat{Q}_2^{\mathrm{T}} \mathbf{c} &= \hat{Q}_2^{\mathrm{T}} Q_1 \tilde{\mathbf{c}} \\ &= \hat{Q}_2^{\mathrm{T}} Q_1 U U^{-1} \tilde{\mathbf{c}} \\ &= E_2 U^{-1} \tilde{\mathbf{c}}. \end{aligned} \tag{6.41}$$

It follows that if A is normalized to have unit row norms then

$$\begin{aligned} \|\hat{Q}_2^{\mathrm{T}} \mathbf{c}\|_2 &\leq \mathrm{eps}\, \|A_\nu\|_{\mathrm{F}}\, \|U^{-1}\|_{\mathrm{F}}\, \|\mathbf{c}\|_2 \\ &= \mathrm{eps} \left\{ p \sum_{i=1}^{k} (U^{\mathrm{T}} U)_{ii}^{-1} \right\}^{1/2} \|\mathbf{c}\|_2 \end{aligned} \tag{6.42}$$

and this bound is available provided the NSE test is used. The interesting feature is the appearance of the condition number of the rectangular matrix

$$\mathrm{cond}_{\mathrm{F}}\,(A_\nu^{\mathrm{T}}) = \left\{ p \sum_{i=1}^{k} (U^{\mathrm{T}} U)_{ii}^{-1} \right\}^{1/2}. \tag{6.43}$$

This result shows that it is potentially important that the tolerance in testing for $\mathbf{t} = 0$ should allow for the sensitivity of the problem.

Assume now that $\mathbf{t} = 0$ is inferred so that the current point is a candidate for optimality and it remains to verify that $\mathbf{u} \geq 0$. Equations (6.7), (6.8) give $\mathbf{u}$ as the solution of a least squares problem with $\mathbf{t}$ playing the role of residual vector. The sensitivity analysis appropriate for the solution of least squares problems by orthogonal factorization methods gives

$$\|\bar{\mathbf{u}} - \mathbf{u}\|_2 \leq \varepsilon\ \mathrm{cond}\,(A_\nu^{\mathrm{T}})(\|\mathbf{u}\|_2 + 1) + \varepsilon\ \mathrm{cond}\,(A_\nu^{\mathrm{T}})^2\, \|\mathbf{t}\|_2 + O(\varepsilon^2) \tag{6.44}$$

where ε is the parameter governing the size of the perturbations and the condition number is the spectral condition number (necessarily $\leq \mathrm{cond}_{\mathrm{F}}\,(A_\nu^{\mathrm{T}})$). It is the negative components of $\bar{\mathbf{u}}$ which attract suspicion, and (6.44) suggests a test of the form

$$\gamma \leq \mathrm{eps}\ \mathrm{cond}_{\mathrm{F}}\,(A_\nu^{\mathrm{T}})(1 + \|\mathbf{u}\|_2) \tag{6.45}$$

where

$$\gamma = \left\{ \sum_{i=1}^{k} (\min\,(\bar{u}_i, 0))^2 \right\}^{1/2}.$$

Even if the deduction that $\|\mathbf{t}\|_2 = 0$ from (6.42) is incorrect, it is straightforward to show that $\|\mathbf{t}\|_2$ is at most the same order of magnitude as $\|\bar{\mathbf{t}}\|_2$ (in the sense of having the same dependence on $\mathrm{cond}\,(A_\nu^{\mathrm{T}})$) so that ignoring the contribution of this term in (6.44) involves ignoring a term proportional to $\varepsilon^2 \operatorname{cond}(A_\nu^T)^3$ by (6.42).

Exercise 6.1 Use the projected gradient algorithm to solve the problem in Example 5.1.

2.7 INTERVAL PROGRAMMING

The interval programming or bounded variables problem (IP) considers a special form of LP1:

$$\min_{\mathbf{x}\in X} \mathbf{c}^T\mathbf{x} : X = \{\mathbf{x};\, \mathbf{b}^- \leqslant A\mathbf{x} \leqslant \mathbf{b}^+\} \qquad \text{IP}$$

where $A: R^p \to R^n$, and $\mathbf{b}^- \leqslant \mathbf{b}^+$ is a necessary condition for the constraints to be consistent. The special form of the constraint inequalities provides structure which can be exploited in solving IP, and in this section we sketch appropriate modifications of both the reduced gradient and simplex algorithms.

To apply the reduced gradient algorithm to IP note that X is defined equivalently by

$$X = \{\mathbf{x};\, A\mathbf{x} \geqslant \mathbf{b}^-,\, -A\mathbf{x} \geqslant \mathbf{b}^+\} \tag{7.1}$$

which is in the form of LP1 without the need of further modification. However, if $b_i^- < b_i^+$ then both sides of the interval constraint cannot be active so that B_ν in (5.7) is specified when we know the index set ν pointing to the rows of A for which equality holds, so that

$$\nu = \{i;\, b_i^- = \mathbf{a}_i^T\mathbf{x} \mid b_i^+ = \mathbf{a}_i^T\mathbf{x}\}, \qquad |\nu| = k, \tag{7.2}$$

and the diagonal matrix S

$$S = \operatorname{diag}\{S_i = 1,\, b_i^- = \mathbf{a}_i^T\mathbf{x},\, S_i = -1,\, b_i^+ = \mathbf{a}_i^T\mathbf{x}\}. \tag{7.3}$$

We have (compare 5.7))

$$\tilde{B}_\nu = [A_\nu^T \mid E_\nu]\left[\begin{array}{c|c} S & 0 \\ \hline 0 & I_{p-k}\end{array}\right] = B_\nu\left[\begin{array}{c|c} S & 0 \\ \hline 0 & I_{p-k}\end{array}\right]. \tag{7.4}$$

The reduced gradient algorithm for interval programming does not differ from the description given in Section 2.5 until a column of A_ν^T has to be dropped to generate a descent direction. Let u_s be selected by an appropriate steepest-edge test. Then if $\mathbf{t}$ is the corresponding direction of descent it follows from (7.4) that

$$\begin{aligned}\mathbf{a}_{\nu(s)}^T\mathbf{t} &= S_s(S_s\mathbf{a}_{\nu(s)}^T)\tilde{B}_\nu^{-T}\mathbf{e}_s,\\ &= S_s.\end{aligned}$$

It follows that a step in the direction determined by $\mathbf{t}$ relaxes off the constraint indicated by $(\nu(s), S_s)$, and it may be that the next constraint encountered is defined by $(\nu(s), -S_s)$. This will be the case if

$$b_s^+ - b_s^- < \min_{q \in \nu^c} \left(\max \left(\frac{b_q^+ - \mathbf{a}_q^T \mathbf{x}}{\mathbf{a}_q^T \mathbf{t}}, \frac{b_q^- - \mathbf{a}_q^T \mathbf{x}}{\mathbf{a}_q^T \mathbf{t}} \right) \right) \tag{7.5}$$

and updating $\tilde{B}_\nu$ *involves changing the sign of* S_s *only*. This results in changing the sign of u_s (removing it from competition in the steepest-edge test) and leaving the other components of $\mathbf{u}$ unchanged. It follows that the runner-up in the competition which defined s now gives the new descent direction. This procedure is continued until (7.5) no longer holds when $\tilde{B}_\nu$ is updated in the manner described in Section 2.5 and the next iteration begins. The new feature here is that a sequence of steps may be possible without the need for other than trivial modifications to the factorization of $\tilde{B}_\nu$ and/or any tableau representations employed. Such steps are therefore cheaper to carry out than the general steps of the reduced gradient algorithm. This feature which depends on the particular form of the problem constraints is called a *multiple pivot sequence.*

Remark 7.1 It is not necessary for all the constraints to be interval constraints in order to exploit the possibility of multiple pivot sequences. It is sufficient to note that the constraint being relaxed is an interval constraint. Interval constraints can be introduced in two (degenerate) cases, but neither of these offer the advantage of multiple pivot sequences. Equality constraints correspond to interval constraints with $b^- = b^+$ so that (7.5) can never hold. Also, if LP1 has a bounded solution then it can be turned into an IP by setting $\mathbf{b}^+ = \gamma \mathbf{e}^{(n)}$ with γ chosen large enough. But there is no information in the adjoined constraint so that 'large enough' implies that the adjoined constraints may never become active. Thus it is again not possible to satisfy (7.5).

An alternative approach to the interval programming problem is provided by the *subopt* algorithm. This is a dual algorithm not only in the sense of not requiring feasibility of the successive iterations (it will be shown that feasibility is equivalent to optimality in this approach), but also because the successive multiplier vectors provide basic feasible solutions to the dual of IP and can be interpreted as steps of a modified simplex algorithm (where 'modified' means modified to allow multiple pivot sequences).

The subopt algorithm takes its origin in the observations that the restricted interval program (RIP):

$$\min_{\mathbf{x} \in X} \mathbf{c}^T \mathbf{a} : X = \{\mathbf{x}; \mathbf{b}_\nu \leqslant A_\nu \mathbf{x} \leqslant \mathbf{b}_\nu\} \qquad \text{RIP}$$

where ν is an index set pointing to a set of $|\nu| = k$ of the rows of A having the properties that:

(a) $\mathbf{c} \in \text{span}\,(\boldsymbol{\rho}_i(A),\ i \in \nu) = H$, and
(b) $\text{rank}\,(A_\nu) = k$

can be solved explicitly, and that this solution can be modified readily to give the solution of the problem expanded by the addition of a further interval constraint.

To solve RIP it is convenient to introduce the new variables

$$\mathbf{y} = B_\nu^{\mathrm{T}} \mathbf{x} \tag{7.6}$$

where B_ν is the matrix defined in (5.7). Then RIP becomes

$$\min_{\mathbf{y} \in Y_\nu} \mathbf{z}_\nu^{\mathrm{T}} \mathbf{y} : Y_\nu = \{\mathbf{y}; \mathbf{b}_\nu^- \leqslant \mathbf{y} \leqslant \mathbf{b}_\nu^+\} \tag{7.7}$$

where

$$z_j = \mathbf{e}_j^{\mathrm{T}} B_\nu^{-1} \mathbf{c}, \qquad j = 1, 2, \ldots, k. \tag{7.8}$$

The solution of (7.7) is

$$y_i = \begin{cases} b_{\nu(i)}^-, & z_i > 0 \\ b_{\nu(i)}^+, & z_i < 0 \\ \theta b_{\nu(i)}^- + (1-\theta) b_{\nu(i)}^+, & 0 \leqslant \theta \leqslant 1, \quad z_i = 0 \end{cases} \tag{7.9}$$

and it will be convenient to specialize $\mathbf{y}$ to be this optimal solution. The corresponding multiplier vector is

$$\mathbf{u}_\nu = S\mathbf{z} \tag{7.10}$$

where S is defined in (7.3). The solution of RIP solves IP provided $\mathbf{x}$ is feasible for IP. This result is given in Lemma 7.1.

Lemma 7.1 *Let* $\mathbf{x}$ *solve RIP and be feasible for IP. Then* $\mathbf{x}$ *is optimal for IP.*

Proof If this were not so then a minimum for a subset of the constraints would be larger than the minimum subject to all constraints.

Consider now RIP augmented by a further interval constraint. If $\mathbf{x}$ is feasible for this constraint then Lemma 7.1 shows that nothing need be done. Accordingly, we assume that $\mathbf{x}$ is not feasible and, to be specific, that the violated constraint is

$$\mathbf{a}_m^{\mathrm{T}} \mathbf{x} - b_m^+ = \Delta > 0. \tag{7.11}$$

Let

$$\mathbf{v} = B_\nu^{-1} \mathbf{a}_m. \tag{7.12}$$

Then a variation

$$\mathbf{y} := \mathbf{y} + \delta \mathbf{e}_s \tag{7.13}$$

moves towards feasibility in (7.11) provided

$$\mathbf{a}_m^{\mathrm{T}} B_\nu^{-\mathrm{T}} \operatorname{sgn}(\delta) \mathbf{e}_s = \operatorname{sgn}(\delta) v_s < 0, \tag{7.14}$$

and maintains feasibility in RIP provided

$$\operatorname{sgn}(\delta) z_s > 0 \tag{7.15}$$

as a consequence of (7.9). Thus any allowable variation of the form (7.13) requires

$$z_s/v_s<0 \tag{7.16}$$

if it is to move towards feasibility in the augmented problem. The ratios satisfying (7.16) summarize the possible moves, and it is convenient to introduce an index set μ pointing to these ratios sorted into *decreasing* order so that the elements of μ rank the set Π where

$$\Pi=\{\pi_i;\ \pi_i=z_{\mu(i)}/v_{\mu(i)},\ 0>\pi_1>\pi_2>\ldots\}. \tag{7.17}$$

To solve the augmented problem, steps of the form (7.13) are made with $s=\mu(j)$, $j=1,2,\ldots,l$, where each constraint $\nu(\mu(j))$ is relaxed to its opposite bound for $j=1,2,\ldots,l-1$, and the constraint (7.11) becomes feasible in the process of relaxing off the $\nu(\mu(l))$th constraint. Let δ_i be the increment in (7.13) in the ith step of this process. The case $i=1$ is typical, and δ_i is found from

$$\mathbf{a}^{\mathrm{T}}_{\nu(\mu(1))}B_\nu^{-\mathrm{T}}(\mathbf{y}+\delta_1\mathbf{e}_{\mu(1)})=\begin{cases}b^+_{\nu(\mu(1))}, & z_{\mu(1)}>0\\ b^-_{\nu(\mu(1))}, & z_{\mu(1)}<0\end{cases}$$

where

$$\mathbf{a}^{\mathrm{T}}_{\nu(\mu(1))}B_\nu^{-\mathrm{T}}\mathbf{y}=\begin{cases}b^+_{\nu(\mu(1))}, & z_{\mu(1)}>0\\ b^-_{\nu(\mu(1))}, & z_{\mu(1)}<0\end{cases}$$

so that

$$\delta_1=\operatorname{sgn}(z_{\mu(1)})(b^+_{\nu(\mu(1))}-b^-_{\nu(\mu(1))}) \tag{7.18}$$

substituting

$$\mathbf{x}_1=\mathbf{x}+\delta_1B_\nu^{-\mathrm{T}}\mathbf{e}_{\mu(1)}$$

into (7.11) and using (7.16) gives

$$\begin{aligned}\mathbf{a}_m^{\mathrm{T}}\mathbf{x}_1-b_m^+&=\Delta+\delta_1\mathbf{a}_m^{\mathrm{T}}B_\nu^{-\mathrm{T}}\mathbf{e}_{\mu(1)}\\ &=\Delta-|v_{\mu(1)}|\,(b^+_{\nu(\mu(1))}-b^-_{\nu(\mu(1))}).\end{aligned}$$

Repeating the calculation j times gives

$$\mathbf{a}_m^{\mathrm{T}}\mathbf{x}_j-b_m^+=\Delta-\sum_{i=1}^{j}\{|v_s|\,(b^+_{\nu(s)}-b^-_{\nu(s)})\}_{s=\mu(i)},\qquad j=1,2,\ldots,l-1, \tag{7.19}$$

while when $j=l$ the increment δ_l in the step in which feasibility is achieved in (7.11) is given by

$$|\delta_l|=\frac{1}{|v_{\mu(l)}|}\left\{\Delta-\sum_{i=1}^{l-1}\{|v_s|\,(b^+_{\nu(s)}-b^-_{\nu(s)})\}_{s=\mu(i)}\right\}. \tag{7.20}$$

The key result is that, as a result of the ordering (7.17), *optimality in the augmented RIP is attained at the same time as feasibility.*

Lemma 7.2 *Let $\bar{\nu}$ be the updated index set so that*

$$\bar{\nu}=\{\nu\backslash\{\nu(q)\}\}\cup\{m\} \tag{7.21}$$

where $q=\mu(l)$. Then $\bar{\mathbf{z}}$ corresponding to the revised solution of (7.8) *is given by*

$$\begin{aligned}\bar{z}_j &= z_j-\left(\frac{v_j}{v_q}\right)z_q, \qquad j\neq q,\\ \bar{z}_q &= z_q/v_q,\end{aligned}\tag{7.22}$$

and

$$\begin{aligned}\operatorname{sgn}(\bar{z}_j) &= -\operatorname{sgn}(z_j), \qquad j=\mu(1),\mu(2),\ldots,\mu(l-1),\\ &= \begin{cases}-1, & j=\mu(l)\\ \operatorname{sgn}(z_j) & \textit{otherwise.}\end{cases}\end{aligned}\tag{7.23}$$

Proof Corresponding to $\bar{\nu}$ we have

$$\begin{aligned}B_{\bar{\nu}} &= B_\nu+(\mathbf{a}_m-\mathbf{a}_{\nu(q)})\mathbf{e}_q^{\mathrm{T}}\\ &= B_\nu[I+(\mathbf{v}-\mathbf{e}_q)\mathbf{e}_q^{\mathrm{T}}]\end{aligned}$$

so that

$$\bar{z}_j=\mathbf{e}_j^{\mathrm{T}}[I+(\mathbf{v}-\mathbf{e}_q)\mathbf{e}_q^{\mathrm{T}}]^{-1}B_\nu^{-1}\mathbf{c}, \qquad j=1,2,\ldots,k,$$

which gives (7.22). Writing this as

$$\bar{z}_j=\left(\frac{z_j}{v_j}-\frac{z_q}{v_q}\right)v_j$$

and using the ordering (7.17) gives (7.23).

Theorem 7.1 *Let* $\mathbf{c}\in\operatorname{span}(\mathbf{a}_j, j\in\bar{\nu})$. *Then*

$$\mathbf{x}_l=\mathbf{x}+\sum_{j=1}^{l}\delta_j B_\nu^{-\mathrm{T}}\mathbf{e}_{\mu(j)}\tag{7.24}$$

solves RIP.

Proof This is a consequence of Lemma 7.2 which shows that $\mathbf{u}_{\bar{\nu}}=\bar{S}\bar{\mathbf{z}}$ provides appropriate multipliers in the Kuhn–Tucker conditions.

Remark 7.2 The condition on $\mathbf{c}$ in Theorem 7.1 is not trivial unless ν points to a maximal linearly independent set of the rows of A. Otherwise it is necessary that $z_j=0$, $j=k+1,\ldots,p$. This requires that $v_j=0$, $j=k+1,\ldots,p$, so that $\mathbf{a}_m$ is required to be in H (which amounts to much the same thing). The argument also assumes that $|z_j|>0$, and that it is possible to move to feasibility in the mth constraint. The first condition corresponds to the nondegeneracy assumption made in developing the simplex algorithm, and the second assumes that the constraints are not contradictory.

Corollary 7.1 *The minimum of the augmented RIP is*

$$\begin{aligned}\mathbf{c}^{\mathrm{T}}\mathbf{x}_l &= \mathbf{c}^{\mathrm{T}}\mathbf{x} + \sum_{j=1}^{l} \delta_j \mathbf{c}^{\mathrm{T}} B_\nu^{-\mathrm{T}} \mathbf{e}_{\mu(j)} \\ &= \mathbf{c}^{\mathrm{T}}\mathbf{x} + \sum_{j=1}^{l-1} \{|z_s|\,(b^+_{\nu(s)} - b^-_{\nu(s)})\}_{s=\mu(j)} + |z_q|\,|\delta_l| \\ &> \mathbf{c}^{\mathrm{T}}\mathbf{x}.\end{aligned} \tag{7.25}$$

An outline of the subopt algorithm can now be given. It assumes that the condition $\mathbf{c} \in H$ is satisfied at each stage and this can be achieved by selecting a maximal linearly independent set of the rows of A to specify the initial RIP (what happens if RIP is degenerate?).

Algorithm (*subopt*)

(i) *Solve RIP to determine* $\mathbf{x}$.
(ii) *If* $\mathbf{x}$ *is feasible then stop.*
(iii) *Find the most violated constraint in IP to determine the constraint to be added to RIP (this is a USE test).*
(iv) *Solve the augmented RIP for* $\mathbf{x}$. *Update* ν.
(v) *Go to* (ii).

Theorem 7.2 *The subopt algorithm solves IP in a finite number of steps.*

Proof This follows by the standard counting argument. By Corollary 7.1 the objective function is increased at each stage. This means that the index sets ν cannot repeat and there exist only a finite number of these sets.

Subopt can be put into context by considering the simplex algorithm applied to the dual problem to IP. It will turn out that the two approaches are closely related, and one result is information on the structure required in LP2 to make possible multiple pivot sequences in a modified simplex algorithm. The dual problem to IP (DIP) is

$$\max_{\mathbf{u} \in U} [\mathbf{b}^{-\mathrm{T}} \mid -\mathbf{b}^{+\mathrm{T}}]\mathbf{u} : U = \{\mathbf{u};\, [A^{\mathrm{T}} \mid -A^{\mathrm{T}}]\mathbf{u} = \mathbf{c},\, \mathbf{u} \geqslant 0\}. \qquad \text{DIP}$$

The first step is to relate the multiplier vector $\mathbf{u}_\nu$ in (7.10) to a basic feasible solution of DIP.

Lemma 7.3 *Let* $\mathbf{u}_\nu$ *be given by* (7.10) *and define* $\mathbf{u} \in R^{2n}$ *by*

$$\begin{aligned} u_{\nu(j)} &= (\mathbf{u}_\nu)_j, \qquad S_j > 0, \\ u_{n+\nu(j)} &= (\mathbf{u}_\nu)_j, \qquad S_j < 0, \quad j = 1, 2, \ldots, k, \\ u_i &= 0 \quad \textit{otherwise}. \end{aligned} \tag{7.26}$$

Then $\mathbf{u}$ *is a basic feasible solution to DIP.*

The next result interprets step (iii) of the subopt algorithm as a USE test in the simplex algorithm applied to DIP.

Lemma 7.4 *The USE test in the simplex algorithm applied to DIP is equivalent to choosing m in* (7.11) *corresponding to the most violated constraint in IP when* $\mathbf{x}$ *is specified by* (7.6), (7.9).

Proof The computation of the steepest edge by (4.11) has to consider two cases corresponding to $\mathbf{a}_j$, $-\mathbf{a}_j$ respectively, with $j \in \nu^c$:

(i)
$$\xi_j = b_j^- - [\mathbf{b}_\nu^T S \mid 0]\tilde{B}_\nu^{-1}\mathbf{a}_j,$$
$$= b_j^- - \mathbf{a}_j^T\mathbf{x};$$

(ii)
$$\xi_{n+j} = -b_j^+ + [\mathbf{b}_\nu^T S \mid 0]\tilde{B}_\nu^{-1}\mathbf{a}_j,$$
$$= -b_j^+ + \mathbf{a}_j^T\mathbf{x},$$

where
$$(\mathbf{b}_\nu)_i = b_{\nu(i)}^-, \qquad S_i > 0,$$
$$= b_{\nu(i)}^+, \qquad S_i < 0, \quad i = 1, 2, \ldots, k.$$

It is the second case which corresponds to (7.11).

Remark 7.3 These calculations use $\tilde{B}_\nu$ defined in (7.4) as basis matrix. This shows how to apply the simplex algorithm when a maximally independent subset of the rows of A has rank less than p. The basis matrix will be produced in this form by the two-phase method using artificial variables for example.

Now consider a step of the simplex algorithm from the basic feasible solution given by (7.26). This gives, using (4.6),

$$\mathbf{u}_\nu \rightarrow \mathbf{u}_1 = \mathbf{u}_\nu - \lambda[\tilde{B}_\nu^{-1}(-\mathbf{a}_m)]_k$$

where $[\cdot]_k$ indicates that only the first k components are taken (the remainder are zero by the maximal independent subset assumption). Thus

$$\mathbf{u}_1 = S(\mathbf{z} + \lambda[\mathbf{v}]_k)$$

so that the smallest positive λ introducting a zero component into $\mathbf{u}_1$ is

$$\lambda = -\pi_1 = -z_{\mu(1)}/v_{\mu(1)}.$$

Increasing λ beyond $-\pi_1$ would make $(\mathbf{u}_1)_{\mu(1)}$ negative ($\mathbf{u}_1$ infeasible). However, *feasibility can be restored by changing the sign of* $S_{\mu(1)}$. It is worth while if the objective function is still increasing because of the trivial nature of the update. More generally, let

$$S^{(i)} = \operatorname{diag}\{S_{\mu(j)}^{(i)} = -S_{\mu(j)}, j = 1, 2, \ldots, i, S_j^{(i)} = S_i \quad \text{otherwise}\} \quad (7.27)$$

then

$$\mathbf{u}_{i+1} = S^{(i)}(\mathbf{z} + \lambda[\mathbf{v}]_k) \tag{7.28}$$

is nonnegative for $-\pi_i \leq \lambda \leq -\pi_{i+1}$. Let $\mathbf{u}(\lambda)$ be defined for $-\pi_i \leq \lambda \leq \pi_{i+1}$, $i = 0, 1, 2, \ldots,$ by

$$\begin{aligned} u_{\nu(j)} &= (\mathbf{u}_{i+1})_j, \qquad S_j^{(i)} > 0 \\ u_{n+\nu(j)} &= (\mathbf{u}_{i+1})_j, \qquad S_j^{(i)} < 0 \\ u_{n+m} &= \lambda \\ u_j &= 0 \quad \text{otherwise.} \end{aligned} \tag{7.29}$$

Then $\mathbf{u}(\lambda)$ is a feasible solution for DIP if $\lambda \geq 0$, which agrees with (7.26) when $\lambda = 0$. Let the corresponding objective function value be $\psi(\lambda)$. Then

$$\psi(\lambda) = [\mathbf{b}^{-\mathrm{T}} \mid \mathbf{b}^{+\mathrm{T}}]\mathbf{u}(\lambda) = \mathbf{b}_\nu^{(i)\mathrm{T}} S^{(i)} \mathbf{u}_{i+1} - \lambda b_m^+ \tag{7.30}$$

where

$$\mathbf{b}_\nu^{(i)} = \mathbf{b}_\nu + \sum_{j=1}^{i} \{\operatorname{sgn}(z_s)(b_{\nu(s)}^+ - b_{\nu(s)}^-)\mathbf{e}_s\}_{s=\mu(j)}. \tag{7.31}$$

To evaluate $\psi(\lambda)$ note that

$$\begin{aligned} \mathbf{b}_\nu^{\mathrm{T}}[\mathbf{v}]_k &= [\mathbf{b}_\nu^{\mathrm{T}} \mid 0]\left[\begin{array}{c|c} S & 0 \\ \hline 0 & I_{p-k} \end{array}\right]\tilde{B}_\nu^{-1}\mathbf{a}_m \\ &= \mathbf{x}^{\mathrm{T}}\mathbf{a}_m \\ &= b_m^+ + \Delta \end{aligned}$$

where use has been made of the specific form of the violated constraint given in (7.11). Substituting for $\mathbf{u}_{i+1}$ from (7.28) in (7.30) gives

$$\begin{aligned} \psi(\lambda) = \psi(0) &+ \sum_{j=1}^{i} \{|z_s|\,(b_{\nu(s)}^+ - b_{\nu(s)}^-)\}_{s=\mu(j)} \\ &+ \lambda\left[\Delta - \sum_{j=1}^{i} \{|v_s|\,(b_{\nu(s)}^+ - b_{\nu(s)}^-)\}_{s=\mu(j)}\right]. \end{aligned} \tag{7.32}$$

This shows that $\psi(\lambda)$ is a piecewise linear function with jumps in derivative at the points $\lambda_i = -\pi_i$, $i = 1, 2, \ldots$. It increases with λ provided

$$\frac{\mathrm{d}\psi}{\mathrm{d}\lambda}(\lambda_i+) = \Delta - \sum_{j=1}^{i} \{|v_s|\,(b_{\nu(s)}^+ - b_{\nu(s)}^-)\}_{s=\mu(j)} > 0, \tag{7.33}$$

that is provided $i \leq l-1$ by (7.19). Thus the optimum value is given by

$$\psi(\lambda_l) = \psi(0) + \sum_{j=1}^{l-1} \{|z_s|\,(b_{\nu(s)}^+ - b_{\nu(s)}^-)\}_{s=\mu(j)} + |\delta_l|\,|z_q|$$

in agreement with (7.25).

Remark 7.4 The above argument shows that determining $q \in \nu$ to be exchanged with m in updating B_ν is equivalent to finding the maximum of a piecewise linear, unimodal (from (7.33)) function of a single variable. Efficient methods for solving the problem are clearly of interest. This point is discussed further in Chapter 7.

The equivalence of subopt with the form of modified simplex algorithm sketched here should now be clear. It is assumed that both start from an initial basis determined by the initial RIP, and this ensures that the elements of π determine the multiple pivot sequences in both algorithms, that the objective function in the simplex algorithm increases in step with the moves to feasibility of the violated constraint in subopt, and that the final basic feasible solution in the simplex step corresponds exactly to the multiplier vector for the new reference in subopt. In summary:

Theorem 7.3 *The subopt algorithm applied to IP is exactly equivalent to the modified simplex algorithm applied to DIP starting from an initial basis determined by the optimal solution to the initial RIP.*

Exercise 7.1 Solve the interval programming problem

$$\min\,(1 \quad 1.284 \quad 1.649 \quad 2.117 \quad 2.718)\mathbf{x}$$

subject to

$$\begin{pmatrix} 1 & 1 & 1 & 1 & 1 \\ 0 & .25 & .5 & .75 & 1 \end{pmatrix}\mathbf{x} = 0,$$

and

$$-\begin{bmatrix} 1 \\ 1 \\ 1 \\ 1 \\ 1 \end{bmatrix} \leqslant \mathbf{x} \leqslant \begin{bmatrix} 1 \\ 1 \\ 1 \\ 1 \\ 1 \end{bmatrix}$$

Also use the (modified) simplex algorithm to solve the dual problem. This example corresponds to the l_1 fitting problem in which e^t is approximated by a linear function on 0(.25)1. Such problems are considered further in Section 3.3.

2.8 A PENALTY FUNCTION METHOD

In a penalty function method the objective function is modified to an unconstrained function which is large when constraints are violated. As a consequence the minimum of the modified objective function is likely to be at least approximately feasible. This idea can be used in solving linear programming problems, and the resulting method has the advantage that a feasible initial point is not required. An appropriate penalty function for LP1 is

$$P(\mathbf{x}, \mu) = \mu \mathbf{c}^{\mathrm{T}}\mathbf{x} + \sum_{i=1}^{n} \psi(r_i(\mathbf{x})) \tag{8.1}$$

where the *penalty parameter* μ is >0,

$$\psi(t) = \max(-t, 0) \tag{8.2}$$

is convex, and

$$r_i(\mathbf{x}) = \mathbf{a}_i^{\mathrm{T}}\mathbf{x} - b_i, \qquad i = 1, 2, \ldots, n. \tag{8.3}$$

A justification for considering penalty function methods is contained in the following results.

Theorem 8.1 *Assume that LP1 has a bounded solution, and let* $\mathbf{x}_i$ *minimize* $P(\mathbf{x}, \mu_i)$, $i = 1, 2, \ldots$ *for a sequence* $\{\mu_i\}\downarrow 0$. *Then if* $i > j$,

(i) $\mathbf{c}^{\mathrm{T}}\mathbf{x}_i \geq \mathbf{c}^{\mathrm{T}}\mathbf{x}_j$;

(ii) $\sum_{k=1}^{n} \psi(r_k(\mathbf{x}_i)) \leq \sum_{k=1}^{n} \psi(r_k(\mathbf{x}_j))$; *and*

(iii) *limit points of* $\{\mathbf{x}_i\} \in X$ *and solve LP1.*

Proof From the definition of $\mathbf{x}_i$, $\mathbf{x}_j$,

$$P(\mathbf{x}_j, \mu_i) \geq P(\mathbf{x}_i, \mu_i)$$

and

$$P(\mathbf{x}_i, \mu_j) \geq P(\mathbf{x}_j, \mu_j).$$

It follows that

$$(\mu_j - \mu_i)\mathbf{c}^{\mathrm{T}}\mathbf{x}_i \geq (\mu_j - \mu_i)\mathbf{c}^{\mathrm{T}}\mathbf{x}_j$$

and

$$\sum_{k=1}^{n} \psi(r_k(\mathbf{x}_j)) - \sum_{k=1}^{n} \psi(r_k(\mathbf{x}_i)) \geq \mu_i(\mathbf{c}^{\mathrm{T}}\mathbf{x}_i - \mathbf{c}^{\mathrm{T}}\mathbf{x}_j) \geq 0,$$

and this establishes (i) and (ii). Also,

$$\mu_i \min_{\mathbf{x}\in X} \mathbf{c}^{\mathrm{T}}\mathbf{x} = \min_{\mathbf{x}\in X} P(\mathbf{x}, \mu_i) \geq P(\mathbf{x}_i, \mu_i) \tag{8.4}$$

so that, as the contribution of the penalty terms is nonnegative,

$$\min_{\mathbf{x}\in X} \mathbf{c}^{\mathrm{T}}\mathbf{x} \geq \mathbf{c}^{\mathrm{T}}\mathbf{x}_i \tag{8.5}$$

showing that $\{\mathbf{c}^{\mathrm{T}}\mathbf{x}_i\}$ converges as a consequence of (i). Now, as $P(\mathbf{x}, 0) \geq 0$, it follows from (8.4) that

$$P(\mathbf{x}_i, \mu_i) \to 0, \qquad i \to \infty,$$

as a consequence of the boundedness of LP1. But this last relation implies

$$\sum_{k=1}^{n} \psi(r_k(\mathbf{x}_i)) \to 0, \qquad i \to \infty,$$

showing that the limit points of $\{\mathbf{x}_i\}$ are feasible. It then follows from (8.5) that they solve LP1.

This result suggests that a sequence $\{\mu_i\}\downarrow 0$ is needed. However, a much better result is possible for the penalty function (7.1), (7.2) applied to the linear programming problem. The solution of LP1 can be found by taking μ *small enough but strictly positive.* We need the following preliminary lemma (see Remark 4.4 of Chapter 1).

Lemma 8.1 *The penalty term $\psi(t)$ is convex and*

$$\partial\psi(t) = \begin{cases} -1, & t<0 \\ [-1,0], & t=0 \\ 0, & t>0. \end{cases}$$

Theorem 8.2 *Let $\mathbf{x}^*$ solve LP1 and $\mathbf{u}$ be the corresponding multiplier vector in the Kuhn–Tucker conditions. If*

$$\mu \, \|\mathbf{u}\|_\infty \leq 1 \tag{8.6}$$

where $\|\cdot\|_\infty$ is the maximum norm, then $\mathbf{x}^*$ minimizes (8.1).

Proof Let the Kuhn–Tucker conditions be

$$\mathbf{c} = \sum_{i\in\sigma} u_i \mathbf{a}_i, \qquad u_i \geq 0,$$

where σ is the index set pointing to the constraints active at $\mathbf{x}^*$. It follows from Lemma 8.1 that $P(\mathbf{x}, \mu)$ is convex in $\mathbf{x}$ for fixed μ and, as $\mathbf{x}^*$ is feasible,

$$\begin{aligned} \partial P(\mathbf{x}^*, \mu) &= \mu \mathbf{c}^{\mathrm{T}} + \sum_{i\in\sigma} \partial\psi(0)\mathbf{a}_i^{\mathrm{T}} \\ &= \mu \sum_{i\in\sigma} \left\{ u_i + \frac{1}{\mu}\, \partial\psi(0) \right\} \mathbf{a}_i^{\mathrm{T}} \end{aligned}$$

so that $0 \in \partial P(\mathbf{x}^*, \mu)$ if (8.6) holds.

Corollary 8.1 *If*

$$\mu \min \|\mathbf{u}\|_\infty > 1$$

where the minimum is taken over all nonnegative multiplier vectors associated with the set of optimum points, then no minimizer of $P(\mathbf{x}, \mu)$ can be feasible for LP1.

Proof It follows from (8.5) that any minimizer of $P(\mathbf{x}, \mu)$ which is feasible for LP1 solves LP1. But this implies the existence of a multiplier vector satisfying (8.6).

The general form of a penalty function algorithm is as follows. It is necessary to reduce the penalty parameter if the minimizer of $P(\mathbf{x}, \mu)$ is not feasible. Here this is done by dividing by 10 but in principle any number >1 would serve. If μ gets too small ($\mu <$ tol) then presumably the feasible region is empty.

Algorithm

(i) *Set initial values for* $\mu = \mu_1$, $\mathbf{x} = \mathbf{x}_0$, tol.
(ii) *Minimize* $P(\mathbf{x}, \mu_i)$ *to compute* $\mathbf{x}_i$.
(iii) *If* $\mathbf{x}_i$ *feasible then stop.*
(iv) $\mu_{i+1} = \mu_i/10$, $i := i+1$.
(v) *If* $\mu_i <$ tol *then report feasible* $\mathbf{x}$ *not found and stop.*
(vi) *Repeat* (ii).

The key step is the computation of the minimum of $P(\mathbf{x}, \mu)$. This can be done efficiently by suitably modified versions of the reduced and projected gradient algorithms considered in Sections 2.5 and 2.6 respectively.

Calculation of descent directions is aided by the following results.

Lemma 8.2 *For any* $\mathbf{t}$, *and* $\lambda > 0$ *small enough,*

$$P(\mathbf{x}+\lambda \mathbf{t}, \mu) = P(\mathbf{x}, \mu) + \lambda \mathbf{g}^{\mathrm{T}}\mathbf{t} + \lambda \sum_{i \in \nu} \psi(\mathbf{a}_i^{\mathrm{T}}\mathbf{t}), \tag{8.7}$$

where

$$\mathbf{g} = \mu \mathbf{c} - \sum_{i \in \alpha} \mathbf{a}_i, \tag{8.8}$$

$$\nu = \{i;\ r_i(\mathbf{x}) = 0\}, \tag{8.9}$$

and

$$\alpha = \{i;\ r_i(\mathbf{x}) < 0\}. \tag{8.10}$$

Corollary 8.2 *The directional derivative of P in the direction determined by* $\mathbf{t}$ *is*

$$P'(\mathbf{x}, \mu : \mathbf{t}) = \mathbf{g}^{\mathrm{T}}\mathbf{t} + \sum_{i \in \nu} \psi(\mathbf{a}_i^{\mathrm{T}}\mathbf{t}). \tag{8.11}$$

Equation (8.11) shows that in order for $\mathbf{t}$ to define a direction of descent it is sufficient that

$$\mathbf{g}^{\mathrm{T}}\mathbf{t} < 0 \tag{8.12}$$

provided

$$\mathbf{a}_i^{\mathrm{T}}\mathbf{t} \geqslant 0, \qquad i \in \nu, \tag{8.13}$$

because then the terms $\psi(\mathbf{a}_i^{\mathrm{T}}\mathbf{t}), i \in \nu$, vanish. Calculating $\mathbf{t}$ to satisfy these inequalities corresponds closely to the strategy employed in Sections 2.5 and 2.6. However, there is another possibility for generating a downhill direction which comes about because $\mathbf{x}$ is not required to be feasible for LP1. This new feature is reflected in the structure of the subdifferential.

Lemma 8.3 *The subdifferential of $P(\mathbf{x}, \mu)$ is given by*

$$\partial P(\mathbf{x}, \mu) = \mu \mathbf{c}^{\mathrm{T}} - \sum_{i \in \alpha} \mathbf{a}_i^{\mathrm{T}} + \sum_{i \in \nu} [-1, 0]\mathbf{a}_i^{\mathrm{T}}. \tag{8.14}$$

The form of ∂P given in Lemma 8.3 shows that here we have *two-sided bounds* for the quantities corresponding to the multipliers in the Kuhn–Tucker conditions for LP1 (the coefficients of the $\mathbf{a}_i$, $i \in \nu$ in the condition $0 \in \partial P(\mathbf{x}, \mu)$ which characterizes the minimizers of P). The implications of this feature are explored in the next result.

Lemma 8.4 *Let*

$$\mathbf{g} = \sum_{i=1}^{k} u_i \mathbf{a}_{\nu(i)} + \hat{\mathbf{g}} \tag{8.15}$$

and

$$H = \operatorname{span}(\mathbf{a}_{\nu(i)}, i \in \nu).$$

If $\mathbf{t}$ *satisfies*

$$A_\nu \mathbf{t} = \theta_j \mathbf{e}_j, \qquad \mathbf{t}^{\mathrm{T}}\hat{\mathbf{g}} = 0 \tag{8.16}$$

where $\theta_j = \pm 1$, then $\mathbf{t}$ is downhill for minimizing $P(\mathbf{x}, \mu)$ provided

(i)
$$\theta_j = 1, \qquad u_j < 0, \tag{8.17a}$$

or

(ii)
$$\theta_j = -1, \qquad u_j > 1. \tag{8.17b}$$

Proof From (8.11)

$$\begin{aligned} P'(\mathbf{x}, \mu : \mathbf{t}) &= \mathbf{g}^{\mathrm{T}}\mathbf{t} + \sum_{i=1}^{k} \psi(\mathbf{a}_{\nu(i)}^{\mathrm{T}}\mathbf{t}) \\ &= \mathbf{u}^{\mathrm{T}} A_\nu \mathbf{t} + \sum_{i=1}^{k} \psi(\theta_i \delta_{ij}) \\ &= \theta_j u_j + \psi(\theta_j) \\ &= \begin{cases} u_j, & \theta_j = 1 \\ 1 - \mu_j, & \theta_j = -1. \end{cases} \end{aligned}$$

Corollary 8.3 *If $\mathbf{g} \in H$ and $u_i \in [0, 1], i = 1, 2, \ldots, k$, then $\mathbf{x}$ minimizes $P(\mathbf{x}, \mu)$.*

Proof This implies $0 \in \partial P(\mathbf{x}, \mu)$ by (8.14), (8.15).

Remark 8.1 Lemma 8.4 shows that if any u_i in (8.15) does not satisfy $u_i \in [0, 1]$ corresponding to the bounds determined by $\partial\psi(0)$ in (8.14), then there is the possibility of generating a downhill direction. If $u_i < 0$ then (8.16) shows that the corresponding constraint becomes inactive for LP1 and no longer contributes to $P(\mathbf{x}, \mu)$. However, if $u_i > 1$ then (8.16) shows that the corresponding constraint becomes infeasible for LP1.

Lemma 8.4 can be used to justify the application of both the reduced and projected gradient algorithms in the minimization of P. In developing these results the simplifying assumption that $|\nu| = k = \text{rank}\,(A_\nu) \leq K = \text{rank}\,(A)$ is made, corresponding to the assumption of nondegeneracy.

Reduced gradient algorithm

Here Lemma 8.4 applies directly. Corresponding to equation (5.7), (5.8) define $\mathbf{u}$ by

$$\mathbf{g} = B_\nu \mathbf{u} = [A_\nu^{\mathrm{T}} \mid E_\nu]\mathbf{u} \tag{8.18}$$

and set

$$\mathbf{t} = -\text{sgn}\,(u_j)\boldsymbol{\kappa}_j(B_\nu^{-\mathrm{T}}) \tag{8.19}$$

where $j \leq k$. Then $\mathbf{t}$ defines a direction of descent provided $u_j \notin [0, 1]$.

Projected gradient algorithm

A (possibly null) descent vector $\mathbf{t}$ is defined by the system of equations

$$-\mathbf{t} = \mathbf{g} - A_\nu^{\mathrm{T}}\mathbf{u}$$

$$A_\nu \mathbf{t} = 0$$

which provide the direct analogue of (6.7), (6.8) in this case. Consider the modified system with solution $(\bar{\mathbf{t}}, \bar{\mathbf{u}})$ defined by

$$-\bar{\mathbf{t}} = \mathbf{g} - A_\nu^{\mathrm{T}}\bar{\mathbf{u}} \tag{8.20}$$

$$A_\nu \bar{\mathbf{t}} = \theta_j \gamma_j \mathbf{e}_j \tag{8.21}$$

where γ_j is a scale factor to be determined. Multiplying (8.20) by A_ν gives

$$-\theta_j\gamma_j\mathbf{e}_j = A_\nu \mathbf{g} - A_\nu A_\nu^{\mathrm{T}}\bar{\mathbf{u}} \tag{8.22}$$

so that

$$\bar{\mathbf{u}} = \mathbf{u} + \theta_j\gamma_j\boldsymbol{\kappa}_j((A_\nu A_\nu^{\mathrm{T}})^{-1}), \tag{8.23}$$

and

$$\bar{\mathbf{t}} = \mathbf{t} + \theta_j\gamma_j A_\nu^{\mathrm{T}}\boldsymbol{\kappa}_j((A_\nu A_\nu^{\mathrm{T}})^{-1}). \tag{8.24}$$

Lemma 8.4 can now be applied to $\mathbf{z}_j = \dfrac{1}{\gamma_j}(\bar{\mathbf{t}} - \mathbf{t})$. As $\mathbf{t}$ is either a descent vector or null, it follows that $\bar{\mathbf{t}} = \mathbf{t} + \gamma_j \mathbf{z}_j$ is necessarily a descent vector provided $u_j \notin [0, 1]$ and $\gamma_j > 0$. To determine γ_j note that the system (8.20), (8.21) can be made to correspond to the updated projected gradient system in which the $\nu(j)$th constraint becomes feasible ($\theta_j = 1$) or infeasible ($\theta_j = -1$). This requires the updating steps $\bar{\nu} = \nu \backslash \{\nu(j)\}$, corresponding to $\bar{\alpha} = \alpha$, $\theta_j = 1$, and $\bar{u}_j = 0$ or $\bar{\alpha} = \alpha \cup \{\nu(j)\}$, $\theta_j = -1$, and $\bar{u}_j = 1$. This determines γ_j by substituting in (8.23) which gives

$$\bar{u}_j = u_j - \theta_j \gamma_j (A_\nu A_\nu^T)_{jj}^{-1}. \tag{8.25}$$

Thus $\gamma_j > 0$ is a consequence of $u_j \notin [0, 1]$.

Remark 8.2 If $u_j > 1$ then $\theta_j = -1$ in order to give a downhill direction. This shows that $\mathbf{a}_{\nu(j)}^T \mathbf{t} < 0$, corresponding to a move away from the feasible region. As the penalty procedure aims to force a feasible point as part of the solution procedure, this seems intuitively unsatisfactory. However, a move away from the feasible region must occur if μ is too large and forces $P(\mathbf{x}_j, \mu)$ to have a minimum at an infeasible point. This suggests that if a constraint must become infeasible in order to generate a downhill direction then this could indicate that μ is too large. Thus the appropriate action could be to reduce μ and this is attractive if the extra computation required to minimize P for the current μ is significant. The next result reinforces this idea by showing that if $X \neq \emptyset$ then a downhill direction can be found without making an active constraint infeasible by reducing μ.

Lemma 8.5 *If $X \neq \emptyset$ and $\mathbf{x} \notin X$ then there exists a downhill direction for minimizing $P(\mathbf{x}, 0) = \sum_{i=1}^{n} \psi(r_i(\mathbf{x}))$ which does not require any constraint feasible at $\mathbf{x}$ to become infeasible.*

Proof Let $\mathbf{y} \in X$. Then the subgradient inequality gives

$$0 = P(\mathbf{y}, 0) \geq P(\mathbf{x}, 0) + \mathbf{v}^T(\mathbf{y} - \mathbf{x}), \qquad \mathbf{v} \in \partial P(\mathbf{x}, 0)$$
$$\Rightarrow \mathbf{v}^T(\mathbf{y} - \mathbf{x}) < 0, \qquad \forall\, \mathbf{v} \in \partial P(\mathbf{x}, 0)$$
$$\Rightarrow \mathbf{y} - \mathbf{x} \text{ downhill for minimizing } P \text{ at } \mathbf{x}$$

as $P(\mathbf{x}, 0) > 0$ by assumption. No constraint feasible at $\mathbf{x}$ can become infeasible on the line joining $\mathbf{x}$ to $\mathbf{y}$ as $\mathbf{y} \in X$ and a straight line can properly intersect a hyperplane in at most one point.

The preceding discussion gives a procedure for finding a descent direction at the current $\mathbf{x}$. Either reduced or projected gradient methods can be used or μ must be reduced and the vector of tentative multipliers $\mathbf{u}$ recomputed.

The next step is to compute the minimum of P in the direction thus defined. This is the subject of the next lemma. In practice a fast algorithm for computing this minimum is important. This point is discussed further in Section 7.5.

Lemma 8.6 *Let*

$$\Lambda = \{\lambda_i;\ \lambda_i = -r_i/\mathbf{a}_i^{\mathrm{T}}\mathbf{t},\ \lambda_i \geqslant 0,\ i \in \nu^c\} \tag{8.26}$$

and π be an index set pointing to the elements of Λ sorted in increasing order. Then the minimum of $P(\mathbf{x}+\lambda\mathbf{t}, \mu)$ is attained at

$$\mathbf{x}' = \mathbf{x} + \lambda_{\pi(q)}\mathbf{t} \tag{8.27}$$

where q is the first index such that

$$\mathbf{g}^{\mathrm{T}}\mathbf{t} + \sum_{j=1}^{q} |\mathbf{a}_{\pi(j)}^{\mathrm{T}}\mathbf{t}| \geqslant 0. \tag{8.28}$$

Proof The directional derivative of P in the direction defined by $\mathbf{t}$ changes only when a constraint changes from feasible to infeasible or infeasible to feasible. The directional derivative at

$$\mathbf{y}_j + 0 = \mathbf{x} + (\lambda_{\pi(j)} + 0)\mathbf{t}$$

is given by

$$P'(\mathbf{y}_j + 0, \mu : \mathbf{t}) = P'(\mathbf{y}_j - 0, \mu : \mathbf{t}) \begin{cases} +\mathbf{a}_{\pi(j)}^{\mathrm{T}}\mathbf{t} : \text{infeasible} \rightarrow \text{feasible} \\ -\mathbf{a}_{\pi(j)}^{\mathrm{T}}\mathbf{t} : \text{feasible} \rightarrow \text{infeasible} \end{cases}$$

and $\mathbf{a}_j^{\mathrm{T}}\mathbf{t} > 0$ if the constraint becomes feasible, <0 if it becomes infeasible. The inequality (8.28) just expresses the condition that the directional derivative has become positive so that no further reduction in P is possible.

The index determined by (8.28) points to the row of A to be added to A_ν, while $\boldsymbol{\rho}_{\nu(s)}(A)$ must be deleted if the corresponding constraint has become feasible. As in Lemmas 5.2 and 6.5, the successive A_ν are guaranteed to have full rank as $|\mathbf{a}_{\pi(q)}^{T}\mathbf{t}| > 0$ so that organization of the penalty algorithm can use exactly the same tableau structure already described in Sections 2.5 and 2.6 (Exercise 8.1(iii)). However, although the idea behind the basic approach is the same, there are some differences in detail in the treatment of degeneracy which occurs when more than one constraint becomes active in the current descent step. If this happens then it is necessary to find $\mathbf{t}$ satisfying

$$\phi(\mathbf{t}) = \mathbf{g}^{\mathrm{T}}\mathbf{t} + \sum_{i=1}^{k} \psi(\mathbf{a}_{\nu(i)}^{\mathrm{T}}\mathbf{t}) + \sum_{i=k+1}^{k+l} \psi(\mathbf{a}_{\nu(i)}^{\mathrm{T}}\mathbf{t}) < 0 \tag{8.29}$$

where $l > 0$ in the case of degeneracy. If there exists $\mathbf{t}$ satisfying (8.29) then it is satisfied for all $\lambda\mathbf{t}, \lambda > 0$, showing that $\phi(\mathbf{t})$ is unbounded below. It is convenient to call $\mathbf{t}$ a *direction of recession* for ϕ.

Lemma 8.7 *If and only if* $\mathbf{t}$ *is a direction of recession for* ϕ *then it is a* *direction of recession for ϕ_ε given by*

$$\phi_\varepsilon(\mathbf{t}) = \mathbf{g}^{\mathrm{T}}\mathbf{t} + \sum_{i=1}^{k} \psi(\mathbf{a}_{\nu(i)}^{\mathrm{T}}\mathbf{t}) + \sum_{i=k+1}^{k+l} \psi(\mathbf{a}_{\nu(i)}^{\mathrm{T}}\mathbf{t} + \varepsilon_i) \tag{8.30}$$

for arbitrary perturbations ε_i.

Proof A direct calculation gives for λ large enough

$$\phi_\varepsilon(\lambda\mathbf{t}) = \lambda\phi(\mathbf{t}) + \sum_{i=k+1}^{k+l} \begin{cases} 0, & \mathbf{a}_{\nu(i)}^{\mathrm{T}}\mathbf{t} > 0 \\ \psi(\varepsilon_i), & \mathbf{a}_{\nu(i)}^{\mathrm{T}}\mathbf{t} = 0 \\ -\varepsilon_i, & \mathbf{a}_{\nu(i)}^{\mathrm{T}}\mathbf{t} < 0 \end{cases}$$

showing that $\phi_\varepsilon(\lambda\mathbf{t}) \to -\infty$, $\lambda \to \infty \Leftrightarrow \phi(\lambda\mathbf{t}) \to -\infty$, $\lambda \to \infty$.

This result shows that the problem of finding a descent direction in the case of degeneracy can be reduced to the problem of minimizing $\phi_\varepsilon(\mathbf{t})$ given by (8.29). This is in the form of a *nondegenerate* penalty problem (compare 8.1)), and either the reduced gradient or projected gradient forms of the penalty algorithm can be applied to determine a direction of recession by minimizing ϕ_ε. Of course the resulting problem can also encounter degeneracy. But this problem can be treated by introducing a new modified subproblem corresponding to (8.30), and it is important to notice that this can involve at most $k+l-1$ rows of A as each step moves off one constraint so that it cannot be involved in the tie. Thus this procedure can be applied recursively and the recursion can have a maximum depth of $l-1$. It follows that a descent direction satisfying (8.29) is returned in a finite number of steps (finiteness follows by induction because no reference can be repeated at any one node of the search tree because progress is always made as a downhill direction is always returned). Thus we have the following result

Theorem 8.3 *The penalty function procedure solves LP1 in a finite number of steps.*

Exercise 8.1 (i) Discuss the case $\mathbf{x}$ feasible and $u_s > 1$ in (8.11).

(ii) What tests correspond to (5.9), (5.12), (5.22) in this case?

(iii) What changes are necessary to adapt the organization of the reduced gradient algorithm? How about the test (5.38)? Equality constraints? What changes are necessary to adopt the organization of the projected gradient algorithm.

(iv) Consider the development of a penalty method for LP2 in which the equality constraints are satisfied explicitly at each step and the positivity constraints are forced by penalty terms. Consider, in particular, implementation of NSE tests, and the treatment of degeneracy.

(v) Use penalty methods to solve the problem in Example 5.1.

Exercise 8.2 The second procedure discussed in Section 2.4 and the procedure given in Section 2.5 for modifying the problem so that an initial feasible solution can be written down are penalty methods. Discuss.

NOTES ON CHAPTER 2

In contrast to the work on determining conditions under which linear inequalities can be solved, linear programming is very much a modern subject. For example, Dantzig (1963), in attempting a historical survey of antecedents, reports that Fourier in 1826 and de la Vallée Poussin in 1911 consider descent algorithms for the solution of overdetermined systems of linear equations in the maximum norm (the problem considered in Section 3.4) before moving forward to 1939 to discuss the contributions of Kantorovich who gave linear programming formulations of certain production problems. Hadley (1962) sums up this early period:

> 'Programming problems first arose in economics, where the optimal allocation of resources has long been of interest to economists. More specifically, however, programming problems seem to be a direct outgrowth of the work done by a number of individuals in the 1930s. One outstanding theoretical model developed then was Von Neumann's linear model of an expanding economy, which was part of the efforts of a number of Austrian and German economists and mathematicians who were studying generalizations of Walrasian equilibrium models of an economy. A more practical approach was made by Leontief, who developed input–output models of the economy. His work was concerned with determining how much various industries would have to produce to meet a specified bill of consumer demands. Input–output models did not actually involve any optimization; instead they required the solution of a system of simultaneous linear equations.
>
> During World War II, a group under the direction of Marshall K. Wood worked on allocation problems for the United States Air Force. Generalizations of Leontief-type models were developed to allocate resources in such a way as to maximize or minimize some linear objective function. George B. Dantzig was a member of the Air Force group; he formulated the general linear programming problem and devised the simplex method of solution in 1947. His work was not generally available until 1951, when the Cowles Commission Monograph No. 13 was published (Koopmans (1951).'

The Cowles report edited by Koopmans provides the first open publication of the development of linear programming. Koopmans introduces the volume as follows:

> 'The immediate occasion for it is to report on a conference on "linear programming", held in Chicago at the Cowles Commission for Research in Economics on June 20–24, 1949. In this conference, scientists classifiable as economists, mathematicians, statisticians, administrators, or combinations thereof, pooled their knowledge, experience, and points of view to discuss the theory and practice of efficient utilization of resources. The mathematicians brought new tools of analysis essential to the progress of economics. The administrators introduced an element of closeness to actual operations and

decisions not otherwise attainable. Those speaking as statisticians adduced data and discussed their limitations. The economists contributed an awareness of the variety of institutional arrangements that may be utilized to achieve efficient allocation.'

He then goes on to explain the name 'linear programming'.

'The name of the conference topic, "linear programming", requires explanation. In earlier phases of work reported on in this volume, contacts and exchanges of ideas among its authors were stimulated by a common interest in the formal problem of maximization of a linear function of variables subject to linear inequalities. The term "linear programming" became a convenient designation for the class of allocation or programming problems which give rise to that maximization problem. The different title of this volume is intended to convey that the work has in part already outgrown the designation and may be expected to outgrow it further.'

Dantzig (1963) is more specific in crediting the name to Koopmans. Dantzig does not use the word 'simplex' in the title of the paper which provides the first account of the algorithm but in a footnote gives the attributions:

'The author wishes to acknowledge that his work on this subject stemmed from discussions in the spring of 1947 with Marshall K. Wood, in connection with Air Force programming methods. The general nature of the "simplex" approach (as the method discussed here is known) was stimulated by discussions with Leonid Hurwicz.

The author is indebted to T. C. Koopmans, whose constructive observations regarding properties of the simplex led directly to a proof of the method in the early fall of 1947.'

However, the simplex algorithm was already well established as the titles of several papers testify. Recently Dantzig (1982) has again reviewed the early history.

A selection from the many available accounts of linear programming and its applications could include Vajda (1958), Hadley (1962), Dantzig (1963), Luenberger (1973), Murty (1976), McLewin (1980), and Chvatal (1983). These provide a wide enough spectrum of treatments to permit the claims that the development of the fundamental theorems given here can be regarded as standard. However, it is more usual to develop duality (generally attributed to Von Neumann) without recourse to the convex theory presented in Chapter 1. On the other hand, such a treatment has didactic value as it permits linear programming to be seen as a simple example of a more general problem class.

The treatment of the basic solution methods concentrates intentionally on algorithmic details. Here the major development since about 1970 has been a move away from schemes based on Jordan elimination to product form procedures based on stable factorization and updating techniques. In particular, the Bartels–Golub method (Bartels (1971), Bartels, Golub, and Saunders

(1970)) has been adopted widely – for example, the highly regarded implementation for large-scale problems of Reid (1976). Recent developments are discussed in Reid (1982). The description of the implementation of NSE tests follows Goldfarb and Reid (1977) who improve on proposals of Harris (1973). Numerical experiments, for example by Quandt and Kuhn (1964), had suggested advantages for NSE tests but lack of an economical method for computing them had proved a disincentive to their use. The treatment of degeneracy follows Perold (1980) but the key idea of finding a direction of recession is in Wolfe (1963). The other advantages of this approach are that it seeks the direction of recession by a problem-solving technique completely in accord with the overall solution strategy, and that it generalizes immediately to other problems (examples are given in Sections 2.5, 2.8, 3.3, and 4.4). These advantages appear to be conclusive. Other methods include the perturbation method of Charnes (1952), the lexicographic method of Dantzig, Orden and Wolfe (1955), and the systematic pivoting rule of Bland (1977). Examples demonstrating cycling go back to Hoffman (1953) and Beale (1955). The exercises here are based on examples in Murty (1976). The treatment of complementarity problems follows Cottle and Dantzig (1968).

The terminology 'active set method' is used in Fletcher (1980) who gives a valuable survey of a general class of algorithms containing the reduced and projected gradient algorithms for LP1 treated here as special cases. The solution of LP1 by these methods goes back at least to Rosen (1960) who gave a projected gradient method which he tested against the simplex algorithm. He found his method incurred a cost disadvantage from the need to evaluate the projection matrices explicitly (the technology of economical stable updating of matrix factors not being then available). The reduced gradient algorithm was introduced by Wolfe (1967), and both he and Rosen were primarily interested in nonlinear problems. The increase in interest in stable methods led Gill and Murray (1973) and Saunders (1972) to look again at these algorithms. An important aspect of their work had to be the development and implementation of fast, stable updating techniques for matrix factorizations, and a direct result is the important summary Gill, Golub, Murray, and Saunders (1974). With their interest in large problems both papers seek to avoid recording orthogonal transformations by computing

$$LL^{\mathrm{T}} = B^{\mathrm{T}}B,$$

$$\mathbf{d} = B^{\mathrm{T}}\mathbf{c}, \quad LL^{\mathrm{T}}\mathbf{u} = \mathbf{d},$$

and

$$LL^{\mathrm{T}}\mathbf{y} = \mathbf{e}_s, \quad \mathbf{t} = B\mathbf{y}.$$

Paige (1973) shows that in this case the (implicit) formation of $B^{\mathrm{T}}B$ does not have deleterious effects. The standard reference on rounding errors in linear systems is Wilkinson (1965), the error analysis for linear least squares problems is due to Golub and Wilkinson (1966). A comprehensive error analysis for linear programming which takes account of the solution trajectory does not seem to be well developed; but one reference is Müller–

Merbach (1970). A thorough treatment of the behaviour of linear inequalities under perturbation is due to Robinson (1975). The implementation of NSE tests adapts Goldfarb and Reid's approach (1977) to this case. The method of Remark 5.8 is due to Bloomfield and Steiger (1981). The dual simplex algorithm of Lemke (1954) belongs with the class of descent algorithms for LP1, but its usual application is in post-optimality studies to determine the behaviour of a solution to LP2 determined by the simplex algorithm as a function of systematic variations in the objective function. In this case a feasible initial vector is given by the optimum multiplier vector. In other circumstances the need for a feasible starting vector is often given mistakenly as a difficulty with this method.

The treatment of interval constraints follows Robers and Ben Israel (1969, 1970). The importance of this work, in particular as it serves to consolidate several approaches to the l_1 approximation problem, is discussed further in Section 3.3. However, the savings possible by taking advantage of the special form of interval constraints is widely appreciated (reflecting their frequent occurrence) and is treated in the textbooks cited.

A general treatment of penalty function methods can be found in Fiacco and McCormick (1968). The application to linear programming is due to Conn (1976) who advocates a projected gradient algorithm. Bartels (1980) treats positivity constraints in LP2 by penalty methods (Exercise 8.1(iv)). Because the line-search phase increases the complexity of these algorithms it is necessary for them to show better convergence behaviour if they are to be preferred. This has not been demonstrated.

Not all methods for solving linear programming problems use the geometry of the constraint set explicitly. One such method is given in Chapter 7. A second associated with the name of Karmarkar considers the problem

$$\min_{\mathbf{x}\in X} \mathbf{c}^{\mathrm{T}}\mathbf{x};\ X=\{\mathbf{x};\ A^{\mathrm{T}}\mathbf{x}=0,\ \mathbf{e}^{\mathrm{T}}\mathbf{x}=1,\ \mathbf{x}\geqslant 0\}.$$

If the current point is $\mathbf{x}_i>0$ then a descent direction is given by $-\mathbf{t}_i$ where

$$\mathbf{t}_i=\arg\min_{\mathbf{t}\in T} \mathbf{t}^{\mathrm{T}}W_i\mathbf{t};\qquad T=\{\mathbf{c}-[A\mid \mathbf{e}]\mathbf{u},\ \mathbf{u}\in R^{p+1}\}$$

and

$$W_i=\operatorname{diag}\{(\mathbf{x}_i)_j,\ j=1,2,\ldots,n\}.$$

If $\mathbf{x}_i$ is optimal then $\mathbf{t}, \mathbf{u}$ provide the Kuhn–Tucker multipliers. Otherwise $\mathbf{x}_i$ is used to estimate the optimal W very much in the manner of the IRLS methods considered in Section 4 of Chapter 5. The novelty of Karmarkar's approach lies in the use of a transformation applied at each stage to take $\mathbf{x}\rightarrow(1/n)\mathbf{e}$ thus ensuring the positivity condition is satisfied. Also, the descent step is made only to the surface of the largest sphere that can be inscribed in X, a conservative choice that tends to avoid introducing very small x components too soon. There appear to be conflicting claims made for this method; but the connection with IRLS suggests that in some cases it could be extremely slowly convergent.

CHAPTER 3

Applications of linear programming in discrete approximation

3.1 INTRODUCTION

An important application of the kinds of procedures described in the previous chapter has been to the development of algorithms for discrete approximation problems. Frequently these problems can be cast in the form of a linear programming problem to which the simplex algorithm can be applied directly. However, it is often the case that special structures can be exploited to provide more efficient procedures.

Consider the residual vector $\mathbf{r}$ defined by the systems of equations

$$\mathbf{r}(\mathbf{x}) = M\mathbf{x} - \mathbf{f} \tag{1.1}$$

where $M: R^p \to R^n$ can be thought of as defining a linear model (for example of some physical process), $\mathbf{f}$ is an associated vector which could be measured responses, and typically $n > p$ (frequently n/p is large although there is some interest when this is not so). In most cases it will not be necessary to indicate the dependence of $\mathbf{r}$ on $\mathbf{x}$. The requirement is now to find $\mathbf{x}$ in the situation in which it cannot be guaranteed that (1.1) has a solution in which $\mathbf{r} = 0$. If $\mathbf{x}$ cannot be found to reduce $\mathbf{r}$ to zero then it is reasonable to ask that $\mathbf{x}$ be chosen so that an appropriate measure of $\mathbf{r}$ is small. This problem is called the *discrete approximation problem*. The resulting minimization problem is convex if the measure chosen is convex. In particular, this will be so when the measure is a norm.

The application of linear programming to the discrete approximation problem has been studied extensively in the two cases:

(i) l_1 approximation,

$$\|\mathbf{r}\|_1 = \sum_{i=1}^{n} |r_i|; \tag{1.2a}$$

and

(ii) maximum norm (l_∞) approximation,

$$\|\mathbf{r}\|_\infty = \max_{1 \leq i \leq n} |r_i| \tag{1.2b}$$

Before considering the algorithmic developments, we give some of the basic properties of the problems (1.2a), (1.2b). It is convenient to denote the norm on the range space of M by a subscript R and that on the domain space by the subscript D. While the properties of the range space norm are clearly fundamental in characterizing the problem under consideration, the norm on the domain space is hardly ever critical.

Theorem 1.1 *Let M have rank p. Then all solutions to the problem*

$$\min \|\mathbf{r}\|_R : \mathbf{r} = M\mathbf{x} - \mathbf{f} \tag{1.3}$$

are bounded.

Proof It follows from the boundedness of M that

$$\max_{\|\mathbf{t}\|_D = 1} \|M\mathbf{t}\|_R = \chi < \infty \tag{1.4}$$

and from the assumed rank of M that

$$\min_{\|\mathbf{t}\|_D = 1} \|M\mathbf{t}\|_R = \tau > 0. \tag{1.5}$$

It follows from (1.4) that

$$\big| \|\mathbf{r}(\mathbf{x})\|_R - \|\mathbf{r}(\mathbf{y})\|_R \big| \leqslant \|\mathbf{r}(\mathbf{x}) - \mathbf{r}(\mathbf{y})\|_R \leqslant \chi \, \|\mathbf{x} - \mathbf{y}\|_D$$

so that $\|\mathbf{r}(\mathbf{x})\|_R$ is continuous on compact sets. From (1.5) we obtain

$$\|\mathbf{r}(\mathbf{x}) - \mathbf{r}(\mathbf{y})\|_R \geqslant \tau \, \|\mathbf{x} - \mathbf{y}\|_D$$

In particular, setting $\mathbf{y} = 0$,

$$\|\mathbf{r}(\mathbf{x}) + \mathbf{f}\|_R \geqslant \tau \, \|\mathbf{x}\|_D$$

so that choosing $\mathbf{x}$ such that

$$\|\mathbf{x}\|_D > \frac{2}{\tau} \|\mathbf{f}\|_R$$

gives

$$\|\mathbf{r}(\mathbf{x})\|_R > \|\mathbf{f}\|_R = \|\mathbf{r}(0)\|_R$$

It follows that such an $\mathbf{x}$ cannot be a competitor in the minimization process. Thus consideration can be restricted to minimizing $\|\mathbf{r}\|_R$ on the compact set

$$S = \{\mathbf{x}; \|\mathbf{x}\|_D \leqslant 2 \, \|\mathbf{f}\|_R / \tau\}.$$

This reduces the problem to that of minimizing a continuous function on a compact set and guarantees that the solution is bounded.

Remark 1.1. To obtain a bounded solution in the rank-deficient case it is necessary to adjoin extra conditions to the problem. In constructing algorithms no assumptions are made on the rank of M and the subsidiary conditions developed as part of the overall procedure.

Theorem 1.1 answers the question concerning the existence of solutions of (1.2). However, it is easy to show that M having full column rank is not sufficient for uniqueness in either the l_1 or the l_∞ norm.

Example 1.1 (i) Nonuniqueness in the l_1 norm: Let

$$\mathbf{r} = \begin{pmatrix} 1 \\ -1 \end{pmatrix} x - \begin{pmatrix} 1 \\ 1 \end{pmatrix}.$$

The rank $(M) = 1 = p$, and

$$|r_1| + |r_2| = |1 - x| + |1 + x|$$

has a minimum value of 2 which is attained for $-1 \leqslant x \leqslant 1$.

(ii) Nonuniqueness in the l_∞ norm: Let

$$\mathbf{r} = -\begin{pmatrix} 1 \\ 0 \end{pmatrix} x + \begin{pmatrix} 1 \\ 1 \end{pmatrix}.$$

Then rank $(M) = 1 = p$, and

$$\max_{1 \leqslant i \leqslant 2} |r_i| = \max \{1, |1 - x|\}$$

has a minimum of 1 which is attained for $0 \leqslant x \leqslant 2$.

The conditions for uniqueness can be explored further by considering the conditions for a minimum of (1.3),

$$0 \in \partial \|\mathbf{r}(\mathbf{x})\|_{\mathrm{R}}. \tag{1.6}$$

The subdifferential is readily calculated in the l_1 norm by noting that

$$\partial |t| = \begin{cases} -1, & t < 0 \\ [-1, 1], & t = 0 \\ 1, & t > 0 \end{cases}$$

whence

$$\partial \|\mathbf{r}(\mathbf{x})\|_1 = \sum_{i \in \mu^c} \theta_i \mathbf{m}_i^{\mathrm{T}} + \sum_{i \in \mu} \lambda_i \mathbf{m}_i^{\mathrm{T}} \tag{1.7}$$

where

$$\theta_i = \operatorname{sgn}(r_i),$$
$$\mu = \{i; r_i(\mathbf{x}) = 0\},$$

and

$$-1 \leqslant \lambda_i \leqslant 1, \qquad i \in \mu.$$

Theorem 1.2 *Let* $\mathbf{x}^*$ *minimize* (1.3), *and* dim $(\mathbf{m}_i, i \in \mu) = p$. *Then the* l_1 *problem has a unique minimum provided*

$$-1 < \lambda_i^* < 1, \qquad i \in \mu \tag{1.8}$$

where the λ_i^* *are the coefficients in* (1.7) *corresponding to the condition* (1.6). *If* $|\mu|=p$ *then* (1.8) *is also necessary.*

Proof A direct calculation making use of (1.7) gives, for $\delta>0$ small enough,

$$\|\mathbf{r}(\mathbf{x}+\delta\mathbf{t})\|_1=\|\mathbf{r}(\mathbf{x})\|_1+\delta\left\{\sum_{i\in\mu^c}\theta_i\mathbf{m}_i^{\mathrm{T}}\mathbf{t}+\sum_{i\in\mu}|\mathbf{m}_i^{\mathrm{T}}\mathbf{t}|\right\} \tag{1.9}$$

$$\geqslant\|\mathbf{r}(\mathbf{x})\|_1+\delta\min_{i\in\mu}(1-|\lambda_i^*|)\sum_{i\in\mu}|\mathbf{m}_i^{\mathrm{T}}\mathbf{t}| \tag{1.10}$$

from which uniqueness follows provided $\min_{i\in\mu}(1-|\lambda_i^*|)>0$.

On the other hand, if $|\lambda_k|=1$ and $|\mu|=p$ then a unique $\mathbf{t}$ can be found satisfying

$$\mathbf{m}_k^{\mathrm{T}}\mathbf{t}=\operatorname{sgn}(\lambda_k),$$
$$\mathbf{m}_i^{\mathrm{T}}\mathbf{t}=0,\qquad i\in\mu\backslash\{k\},$$

and this corresponds to a direction of non-uniqueness as then it gives equality in (1.9) for all $\delta>0$ small enough.

Remark 1.2 The condition that $\dim(\mathbf{m}_i, i\in\mu)=p$ is clearly necessary to make the above argument work. The case $|\mu|>p$ is analogous to the condition of degeneracy in linear programming. In particular, the multipliers λ_i^* are not uniquely determined by (1.6), (1.7). However, the condition (1.8) is too strong – see Exercise 1.2.

In the l_∞ norm the subdifferential is calculated using the results (5.14) and (5.20) of Chapter 1. These give

$$\partial\|\mathbf{r}(\mathbf{x})\|_\infty=\operatorname{conv}\{\theta_i\mathbf{m}_i^{\mathrm{T}}, i\in\mu\} \tag{1.11}$$

where

$$\mu=\{i;|r_i|=\|\mathbf{r}\|_\infty\}.$$

The explicit condition corresponding to (1.6) is that there exist multipliers λ_i^* such that

$$0=\sum_{i\in\mu}\lambda_i^*\theta_i\mathbf{m}_i,\qquad \lambda_i^*\geqslant0,\qquad \sum_{i\in\mu}\lambda_i^*=1. \tag{1.12}$$

The result corresponding to Theorem 1.2 is as follows.

Theorem 1.3 *Let* $\mathbf{x}^*$ *minimize* (1.3), *and* $\dim(\mathbf{m}_i, i\in\mu)=p$. *Then the* l_∞ *problem has a unique solution provided*

$$\lambda_i^*>0,\qquad i\in\mu \tag{1.13}$$

in (1.12). *If* $|\mu|=p+1$ *then this condition is also necessary.*

Proof A direct calculation gives, for $\delta > 0$ small enough,

$$\|\mathbf{r}(\mathbf{x}+\delta\mathbf{t})\|_\infty = \|\mathbf{r}(\mathbf{x})\|_\infty + \delta \max_{i\in\mu} \theta_i \mathbf{m}_i^T \mathbf{t} \tag{1.14}$$

By the rank condition, $\mathbf{m}_i^T\mathbf{t} \neq 0$ for at least one i. From (1.12), as all terms cannot vanish for any $\mathbf{t}$,

$$\min_{\|\mathbf{t}\|_D=1} \max_{i\in\mu} \theta_i \mathbf{m}_i^T t = \omega > 0$$

and uniqueness follows from this inequality. To estimate how small ω can be let

$$\min_{\|\mathbf{t}\|_D} \max_{i\in\mu} |\mathbf{m}_i^T t| = \tau > 0.$$

Note that $\omega < \tau$ is possible as $\max_{i\in\mu} \theta_i \mathbf{m}_i^T \mathbf{t} \leq \max_{i\in\mu} |\mathbf{m}_i^T \mathbf{t}|$ and that for all $\mathbf{t}$ such that strict inequality holds (from (1.12))

$$\max_{i\in\mu} \theta_i \mathbf{m}_i^T \mathbf{t} > \min_{i\in\mu} \lambda_i^* |\min_i \theta_i \mathbf{m}_i^T \mathbf{t}| > \min_{i\in\mu} \lambda_i^* \tau$$

As the right hand side is independent of $\mathbf{t}$

$$\omega > \min_{i\in\mu} \lambda_i^* \tau \tag{1.15}$$

However, if $\lambda_k^* = 0$ and $|\mu| = p+1$, then from (1.12)

$$\dim(\mathbf{m}_i, i \in \mu\backslash\{k\}) < p,$$

and there exists $\mathbf{t}$ satisfying

$$\mathbf{m}_i^T\mathbf{t} = 0, \qquad i \in \mu\backslash\{k\},$$
$$\mathbf{m}_k^T\mathbf{t} = -\theta_k.$$

It follows from (1.14) that this defines a direction of nonuniqueness for small enough $\delta > 0$ as it gives equality in (1.14) ($\|\mathbf{r}(\mathbf{x}+\delta\mathbf{t})\|_\infty = |r_i(\mathbf{x}+\delta\mathbf{t})| = |r_i(\mathbf{x})| = \|\mathbf{r}(\mathbf{x})\|_\infty$, $i \in \mu\backslash\{k\}$, $|r_k(\mathbf{x}+\delta\mathbf{t})| < \|\mathbf{r}(\mathbf{x})\|_\infty$).

Remark 1.3 If $\dim(\mathbf{m}_i, i\in\mu) = p$ and $|\mu| = p+1$, then the λ_i^*, $i \in \mu$, are uniquely determined by (1.6). If $|\mu| > p+1$ then this provides the situation analogous to degeneracy in linear programming.

Remark 1.4 If $|\mu| = p+1$ then the condition (1.13) must be satisfied if every $p \times p$ submatrix of M has rank p. This condition is known as the *Haar condition.* It is frequently assumed in studying l_∞ problems.

The uniqueness results obtained above have the form

$$\|\mathbf{r}(\mathbf{x}+\delta\mathbf{t})\|_R \geq \|\mathbf{r}\|_R + \gamma\delta\,\|\mathbf{t}\|_D$$

where

$$\gamma = \begin{cases} \min_{i\in\mu}(1-\lambda_i^*) \min_{\|\mathbf{t}\|_D=1} \sum_{i\in\mu} |\mathbf{m}_i^T\mathbf{t}|, & l_1 \text{ problem} \\ \min_{i\in\mu} \lambda_i^* \left\{ \min_{\|\mathbf{t}\|_D=1} \max_{i\in\mu} |\mathbf{m}_i^T\mathbf{t}| \right\}, & l_\infty \text{ problem.} \end{cases}$$

These are particular cases of the property known as *strong uniqueness.*

Definition 1.1 The convex function $C(\mathbf{x})$ has a strong unique minimum at $\mathbf{x}^*$ provided there exists $\gamma > 0$ such that

$$C(\mathbf{x}) \geq C(\mathbf{x}^*) + \gamma \|\mathbf{x} - \mathbf{x}^*\|_D, \qquad \forall\, \mathbf{x} \in R^p. \tag{1.16}$$

Theorem 1.4 *The following three propositions are equivalent:*

(i) *the definition of strong uniqueness*

$$C(\mathbf{x}) \geq C(\mathbf{x}^*) + \gamma \|\mathbf{x} - \mathbf{x}^*\|_D, \qquad \forall\, \mathbf{x} \in R^p;$$

(ii) *a condition that the directional derivative is strictly positive in all directions from* $\mathbf{x}^*$;

$$\forall\, \mathbf{t} \in R^p\ \exists\, \mathbf{v} \in \partial C(\mathbf{x}^*) \ni \mathbf{v}^T\mathbf{t} \geq \gamma \|\mathbf{t}\|_D;$$

and

(iii) *a strict 'zero in the convex hull' condition,*

$$\exists\, \varepsilon > 0 \ni \|\mathbf{t}\|_D < \varepsilon \Rightarrow \mathbf{t} \in \partial C(\mathbf{x}^*).$$

Remark 1.5 Proposition (iii) can be restated as $0 \in \operatorname{int} \partial C(\mathbf{x}^*)$.

Proof (i) $\Leftrightarrow$ (ii). This result follows by noting that both (i) and (ii) imply and are implied by

$$C'(\mathbf{x}^* : \mathbf{t}) \geq \gamma \|\mathbf{t}\|$$

as a consequence of Lemma 4.6 of Chapter 1 and the observation that the difference quotients in (1.4.9) are nondecreasing functions of λ.

(ii) $\Leftrightarrow$ (iii). If (ii) does not hold for any $\gamma > 0$ then there exists $\mathbf{t}$ such that

$$(\mathbf{v}^T\mathbf{t} \leq 0,\ \forall\, \mathbf{v}^T \in \partial C(\mathbf{x}^*)) \Rightarrow \partial C(\mathbf{x}^*) \subseteq \{\mathbf{u}^T;\ \mathbf{t}^T\mathbf{u} \leq 0\}$$

showing that $\partial C(\mathbf{x}^*)$ is contained in $H^-(\mathbf{t}, 0)$ so that $0 \notin \operatorname{int} \partial C(\mathbf{x}^*)$. On the other hand, if $0 \notin \operatorname{int} \partial C(\mathbf{x}^*)$ then there exists a hyperplane supporting the closed convex set $\partial C(\mathbf{x}^*)$ at zero. If $\mathbf{t}$ is the normal to this hyperplane $\mathbf{t}^T\mathbf{v} \leq 0$, all $\mathbf{v}^T \in \partial C(\mathbf{x}^*)$.

Remark 1.6 The conditions given in Theorems 1.2 and 1.3 are readily interpreted in terms of the third proposition of Theorem 1.4 as they require the multipliers λ_i^*, $i \in \mu$, to lie properly in the interior of the range stated in the specification of the subdifferential.

Strong uniqueness turns out to be exactly the property required to characterize uniqueness in both the l_1 and l_∞ problems. The above results require tightening only in the case of degeneracy, and it is convenient to do this by considering first strong uniqueness for LP1. This requires interpreting LP1 as a problem of minimizing a convex function. Let

$$C(\mathbf{x}) = \mathbf{c}^T\mathbf{x} + \sum_{i=1}^{n} \delta(b_i - \mathbf{a}_i^T\mathbf{x}) \tag{1.17}$$

where $\delta(t)$ is the one-dimensional indicator function defined by

$$\delta(t)=\begin{cases}0, & t\leqslant 0\\ +\infty & t>0.\end{cases} \tag{1.18}$$

Then $C(\mathbf{x})=\mathbf{c}^{\mathrm{T}}\mathbf{x}$ for all $\mathbf{x}\in X=\{\mathbf{x};\ A\mathbf{x}\geqslant\mathbf{b}\}$, and is $+\infty$ otherwise. Thus the problem of finding $\mathbf{x}^*$ minimizing $C(\mathbf{x})$ is equivalent to solving LP1.

To compute the subdifferential of $C(\mathbf{x})$ we use

$$\partial\delta(t)=\begin{cases}0, & t<0\\ [0,\infty), & t=0\\ \varnothing, & t>0\end{cases} \tag{1.19}$$

so that

$$\partial C(\mathbf{x}^*)=\mathbf{c}^{\mathrm{T}}-\sum_{i\in\sigma}u_i\mathbf{a}_i^{\mathrm{T}}$$

where

$$\sigma=\{i;\ \mathbf{a}_i^{\mathrm{T}}\mathbf{x}^*-b_i=0\},$$

and

$$0\leqslant u_i<\infty.$$

Theorem 1.5 *Uniqueness is equivalent to strong uniqueness for LP1.*

Proof It is convenient to use here the second proposition of Theorem 1.4. Let $\|\mathbf{t}\|_{\mathrm{D}}=1$. We note two cases:

(i) $\mathbf{a}_i^{\mathrm{T}}\mathbf{t}\geqslant 0,\ i\in\sigma$. Uniqueness implies that $\mathbf{c}^{\mathrm{T}}\mathbf{t}>0$ for all $\mathbf{t}$ satisfying the inequalities. Therefore

$$\min \mathbf{c}^{\mathrm{T}}\mathbf{t}=\tau>0$$

But $\mathbf{c}\in\partial C(\mathbf{x}^*)$.

(ii) $\mathbf{a}_k^{\mathrm{T}}\mathbf{t}<0$ for at least one $k\in\sigma$. In this case

$$\mathbf{c}^{\mathrm{T}}\mathbf{t}-\sum_{i\in\sigma\backslash\{k\}}u_i\mathbf{a}_i^{\mathrm{T}}\mathbf{t}+u_k|\mathbf{a}_k^{\mathrm{T}}\mathbf{t}|$$

can be made arbitrarily large by fixing u_i, $i\in\sigma\backslash\{k\}$, and choosing u_k arbitrarily large. This shows that the directional derivative is unbounded for directions out of the feasible region.

To apply this result it is necessary to write the l_1 and l_∞ problems as linear programming problems. Several methods for doing this will be discussed, and here the procedure followed is to introduce the concept of a polyhedral norm which can then be specialized as required. For our purposes it suffices to consider the consistent set of linear inequalities.

$$P\mathbf{x}\leqslant\mathbf{e}^{(N)} \tag{1.20}$$

where $P: R^n \to R^N$, having the properties:

(i) the set $S_p = \{\mathbf{x}; P\mathbf{x} \leqslant \mathbf{e}^{(N)}\}$ is bounded and has a proper interior in R^n, and

(ii) $P\mathbf{x} \leqslant \mathbf{e}^{(N)} \Leftrightarrow P|\mathbf{x}| \leqslant \mathbf{e}^{(N)}$ where $|\mathbf{x}|_i = |x_i|$, $\quad i = 1, 2, \ldots, n$.

Definition 1.2 Let

$$\|\mathbf{x}\|_p = \min \lambda\,;\, P\mathbf{x} \leqslant \lambda\mathbf{e}^{(N)}, \tag{1.21}$$

then $\|\cdot\|_p$ is a *polyhedral norm*. It is readily verified that $\|\cdot\|_p$ is the norm corresponding to S_p as unit ball. Condition (ii) is sufficient to ensure that $\|\gamma\mathbf{x}\|_p = |\gamma|\,\|\mathbf{x}\|_p$ but is stronger than necessary.

Example 1.2 (i) The l_1 norm: define P to have the $N = 2^n$ rows corresponding to the number of different ways of putting ± 1 into n locations.

(ii) The l_∞ norm.

$$P = \begin{bmatrix} I \\ -I \end{bmatrix}$$

In this case $N = 2n$.

The discrete approximation problem to find $\mathbf{x}$ to minimize $\|\mathbf{r}(\mathbf{x})\|_p$ is equivalent to the linear programming problem

$$\min h$$

subject to

$$P\{M\mathbf{x} - \mathbf{f}\} \leqslant h\mathbf{e}^{(N)}. \tag{1.22}$$

Theorem 1.6 *Uniqueness is equivalent to strong uniqueness for polyhedral norm approximation problems.*

Proof Values of $\|\mathbf{r}\|_p$ correspond to feasible points $(\mathbf{x}, h)$ of (1.22) for which at least one inequality holds with equality. The allowable variations are thus more constrained, so that the result is a consequence of Theorem 1.5.

Strong uniqueness can be used to generate the results which characterize the minimum of $\|\mathbf{r}\|_p$. Recall (Example 4.4 of Chapter 1) that

$$\mathbf{v}^{\mathrm{T}} \in \partial_r \|\mathbf{r}\|_p \Leftrightarrow \mathbf{v}^{\mathrm{T}}\mathbf{r} = \|\mathbf{r}\|_p, \qquad \|\mathbf{v}\|^* = 1$$

so the the subdifferential of $\|\mathbf{r}\|_p$ is just the convex hull of the rows of P aligned with $\mathbf{r}$, and these rows have unit norm in the dual norm.

Lemma 1.1 *If* $\mathbf{x}^*$ *is the strong unique minimizer of* $\|\mathbf{r}(\mathbf{x})\|_p$ *then*

$$\dim\,(\partial_r \|\mathbf{r}(\mathbf{x}^*)\|_p) \geqslant p + 1 \tag{1.23}$$

Proof From Lemma 5.5 of Chapter 1 it follows that

$$\partial_x \|\mathbf{r}\|_p = \partial_r \|\mathbf{r}\|_p M$$

so that if $\dim(\partial_r \|r(\mathbf{x}^*)\|_p) < p$ then $\partial_x \|r(\mathbf{x}^*)\|_p$ is contained in a hyperplane through the origin; while if this is not true when $\dim = p$ then it is not possible to satisfy $0 \in \partial_x \|\mathbf{r}\|_p$. The result is thus a consequence of Theorem 1.4, Proposition (iii).

Example 1.3 (i) l_1 approximation. Consider the set S of 2^k vectors determined by the index set μ in (1.7) with $|\mu| = k$ by the requirements $i \in \mu^c \Rightarrow v_i$ has the same value θ_i where θ_i is either $+1$ or -1 for all $\mathbf{v} \in S$; $i \in \mu \Rightarrow v_i$ can have either value $+1$ or -1. Then $\dim(S) = k+1$. In fact all $\mathbf{v} \in S$ are linear combinations of the vectors

$$\mathbf{w}_i = \mathbf{e}_{\mu(i)}, \qquad i = 1, 2, \ldots, k,$$

with weights ± 1, and

$$\mathbf{w}_{k+1} = \sum_{i \in \mu^c} \theta_i \mathbf{e}_i.$$

The application to l_1 approximation is immediate, for if S specifies the rows of P aligned with $\mathbf{r}(\mathbf{x}^*)$ then Lemma 1.1 gives $k \geqslant p$ if the minimum is unique.

(ii) l_∞ approximation. Here (compare (1.11))

$$\theta_i \mathbf{e}_i^{\mathrm{T}} \mathbf{r} = \|\mathbf{r}\|_p \Rightarrow \theta_i \mathbf{e}_i^{\mathrm{T}} \in \partial \|\mathbf{r}\|_p.$$

Thus it follows from Lemma 1.1 that there must be at least $p+1$ components of $\mathbf{r}(\mathbf{x}^*)$ equal in magnitude to $\|\mathbf{r}\|_p$ if $\mathbf{x}^*$ is a strong unique minimizer.

Exercise 1.1 Two norms to which the above results can be applied are as follows. Consider index sets μ_i, $i = 1, 2, \ldots, m$, such that:

(a) $$\mu_i \cap \mu_j = 0, \; i \neq j, \quad \text{and}$$

(b) $$\bigcup_{i=1}^{m} \mu_i = \{1, 2, \ldots, n\}.$$

Verify that the following are norms:

(i) $$\max_{1 \leqslant k \leqslant m} \sum_{i \in \mu_k} |r_i|,$$

(ii) $$\sum_{k=1}^{m} \max_{i \in \mu_k} |r_i|.$$

Derive the characterization results.

Exercise 1.2 Show that $\mathbf{x}$ is a unique solution to the l_1 problem if
(i) $0 \in \partial \|\mathbf{r}\|_1$, and
(ii) there exist multipliers λ_i, $|\lambda_i| < 1$, $i \in \nu$, such that $\operatorname{rank}(\mathbf{m}_{\nu(i)}$, $i = 1, 2, \ldots, k$, $|\lambda_{\nu(i)}| < 1) = p$,

where it is assumed that the degenerate situation $k > p$ is possible. Compare the result with Lemma 2.3 of Chapter 2, and use it to derive Proposition (iii) of Theorem 1.4.

What is the corresponding result for the l_∞ problem?

Exercise 1.3 If $\boldsymbol{\kappa}_j(M) = \mathbf{e}^{(n)}$ then the model expressed by (1.1) is said to contain an intercept term, and there is no restriction in assuming $j = 1$. For the l_1 problem show that at an optimum, the presence of an intercept term forces the inequality

$$||\mu_+^c| - |\mu_-^c|| \leq |\mu|$$

where μ_+^c, μ_-^c point to the positive and negative residuals respectively. In the l_∞ problem show that the intercept term forces the equation

$$\sum_{i \in \mu} \lambda_i \theta_i = 0.$$

3.2 DESCENT METHODS FOR l_1 APPROXIMATION

In this section the reduced and projected gradient algorithms are adapted to solve the l_1 approximation problem. Recall that for LP1 the basic idea is to construct a descent vector having the property that certain active constraints remain active in the descent step. This permits a strategy in which first a (relative) extreme point is located, followed by a sequence of moves from extreme point to extreme point until the minimum is determined. To adapt this strategy, note that extreme points are suitable points at which to test if the Kuhn–Tucker conditions characterizing a minimum are satisfied. The characterization theorem for l_1 approximation (Theorem 1.2) gives a special role to the set of zero residuals which correspond here to the set of active constraints in LP1; and this suggests a strategy of first building up an appropriate set of zero residuals (it is only assumed that rank $(M) = K \leq p$) and then moving from one such set to another until the characterization theorem is satisfied.

There is also a very close connection between the l_1 problem and the penalty function problem considered in Section 2.8, as both are special cases of piecewise linear functions.

Definition 2.1 Let $C(\mathbf{x})$ be defined by

$$C(\mathbf{x}) = \tau \mathbf{c}^T \mathbf{x} + \sum_{i=1}^{n} \Psi(r_i, \alpha_i, \beta_i) \tag{2.1}$$

where

$$\Psi(t, \alpha, \beta) = \max(-\alpha t, \beta t) \tag{2.2}$$

and $\alpha, \beta \geq 0$. Then $C(\mathbf{x})$ is a *piecewise linear function*. It is clear from (2.2) that $C(\mathbf{x})$ is convex. Also the penalty objective function corresponds to

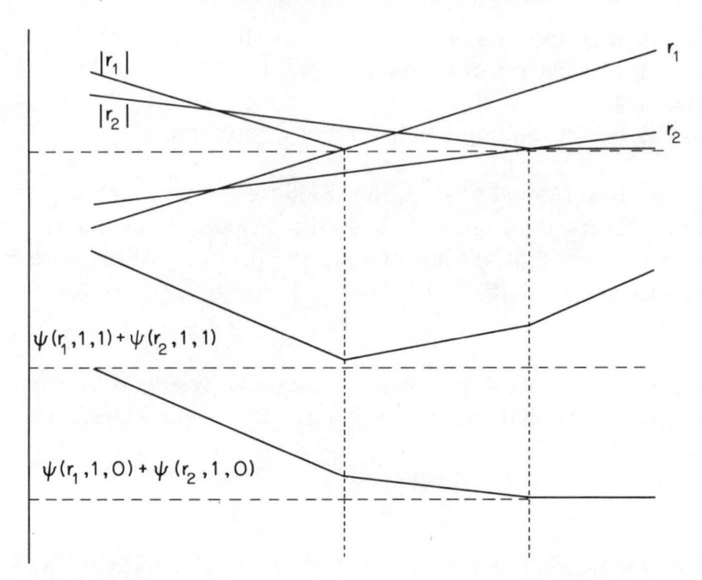

Figure 2.1 Simple cases of piecewise linear functions

choosing $\tau>0$, $\alpha=1$, $\beta=0$, and the l_1 objective function to $\tau=0$, $\alpha=\beta=1$. These are illustrated in Figure 2.1 in the case $n=2$, $p=1$. The point to note is that it is at the zero components of $\mathbf{r}$ that the graph of $C(\mathbf{x})$ changes.

Remark 2.1 If $\tau=0$ and $\alpha>0$, $\beta>0$ then $C(\mathbf{x})$ defines a guage function on R^n for

(i) $\Psi(t, \alpha, \beta)\geqslant 0, =0$ only if $t=0$; and
(ii) $\Psi(t_1+t_2, \alpha, \beta)\leqslant\Psi(t_1, \alpha, \beta)+\Psi(t_2, \alpha, \beta)$; while
(iii) $\Psi(\gamma t, \alpha, \beta)=|\gamma|\,\Psi(t, \alpha, \beta)$

holds only if $\gamma\geqslant 0$ or $\alpha=\beta$. In the latter case $C(\mathbf{x})$ defines a norm.

The descent algorithms are worked out for $C(\mathbf{x})$ piecewise linear. The approach follows closely that of Section 2.8 which is almost a special case (the difference relates to the sign of the multiplier vector which in Section 2.8 follows the convention of the Kuhn–Tucker conditions, but here adopts the opposite sign which occurs naturally in forming the subdifferential of its objective function). The basic results on which the calculation of the descent vector at the current point depend, are now summarized.

Lemma 2.1 (*calculation of* $\partial\Psi$)

$$\partial\Psi(t, \alpha, \beta)=\begin{cases}-\alpha, & t<0\\ [-\alpha, \beta], & t=0,\\ \beta, & t>0\end{cases} \tag{2.3}$$

Lemma 2.2 *Let* μ *be the index set*

$$\mu = \{i; r_i = 0\}. \tag{2.4}$$

Then

$$\partial C(\mathbf{x}) = \mathbf{g}^{\mathrm{T}} + \sum_{i \in \mu} [-\alpha_i, \beta_i]\mathbf{m}_i^{\mathrm{T}} \tag{2.5}$$

where

$$\mathbf{g} = \tau \mathbf{c} + \sum_{i \in \mu^c} \frac{\Psi(r_i, \alpha_i, \beta_i)}{r_i} \mathbf{m}_i. \tag{2.6}$$

Corollary 2.1 *The directional derivative of C is given by*

$$C'(\mathbf{x} : \mathbf{t}) = \mathbf{g}^{\mathrm{T}}\mathbf{t} + \sum_{i \in \mu} \Psi(\mathbf{m}_i^{\mathrm{T}}\mathbf{t}, \alpha_i, \beta_i) \tag{2.7}$$

Corollary 2.2 *Let*

$$H = \text{span}\,(\mathbf{m}_i, i \in \mu). \tag{2.8}$$

Then $C(\mathbf{x})$ *is minimized at* $\mathbf{x}^*$ *if and only if* $\mathbf{g} \in H$ *and there exist multipliers* u_i, $i = 1, 2, \ldots, k = |\mu|$, *such that*

$$0 = \mathbf{g} + \sum_{i=1}^{k} u_i \mathbf{m}_{\mu(i)} \tag{2.9a}$$

and

$$-\alpha_{\mu(i)} \leqslant u_i \leqslant \beta_{\mu(i)}, \qquad i = 1, 2, \ldots, k. \tag{2.9b}$$

Proof By Lemma 2.2 this is just the statement that $0 \in \partial C(\mathbf{x}^*)$.

The main result is similar to Lemma 8.4 of Chapter 2 (but note the sign differences).

Lemma 2.3 *Let*

$$\mathbf{g} = -\sum_{i=1}^{k} u_i \mathbf{m}_{\mu(i)} + \hat{\mathbf{g}}. \tag{2.10}$$

If $\mathbf{t}$ *satisfies*

$$[M_\mu \mathbf{t} = \theta_j \mathbf{e}_j, \qquad \mathbf{t}^{\mathrm{T}}\hat{\mathbf{g}} = 0] \tag{2.11}$$

where $\boldsymbol{\rho}_i(M_\mu) = \mathbf{m}_{\mu(i)}$, $i = 1, 2, \ldots, k$, *and* $\theta_j = \pm 1$, *then* $\mathbf{t}$ *is downhill for minimizing* $C(\mathbf{x})$ *provided*

(i) $$\theta_j = -1, \qquad u_j < -\alpha_{\mu(j)}, \quad \text{or} \tag{2.12a}$$

(ii) $$\theta_j = 1, \qquad u_j > \beta_{\mu(j)}. \tag{2.12b}$$

Proof From (2.7),

$$C'(\mathbf{x}:\mathbf{t}) = \mathbf{g}^{\mathrm{T}}\mathbf{t} + \sum_{i=1}^{k} \Psi(\mathbf{m}_{\mu(i)}^{\mathrm{T}}\mathbf{t}, \alpha_{\mu(i)}, \beta_{\mu(i)})$$

$$= -\mathbf{u}^{\mathrm{T}}M_{\mu}\mathbf{t} + \sum_{i=1}^{k} \Psi(\theta_j\delta_{ij}, \alpha_{\mu(i)}, \beta_{\mu(i)})$$

$$= -\theta_j u_j + \Psi(\theta_j, \alpha_{\mu(j)}, \beta_{\mu(j)})$$

$$= -u_j + \beta_{\mu(j)}, \qquad \theta_j = 1,$$

$$u_j + \alpha_{\mu(j)}, \qquad \theta_j = -1.$$

Remark 2.2 In minimizing the penalty objective function it was possible to distinguish between the two cases in the above lemma corresponding to active constraints becoming either inactive or infeasible (violated). This distinction does not hold in general and residuals of either sign have essentially the same importance. Thus both alternatives are used to generate descent directions.

To apply Lemma 2.3 to generate descent vectors it is convenient to make the usual nondegeneracy assumptions that $k = \mathrm{rank}\,(M_{\mu}) \leq K = \mathrm{rank}\,(M)$ and that exactly one new zero residual occurs to terminate each descent step. Application to the two main cases are considered separately.

Reduced gradient algorithm

Let

$$B_{\mu} = [M_{\mu}^{\mathrm{T}} \mid E_{\mu}]$$

where as usual, E_{μ} is made up of columns of I_p chosen so that B_{μ} has full rank. Corresponding to equations (5.7), (5.8) of Chapter 2, define $\mathbf{u}$ by

$$\mathbf{g}_{\mu} = -B_{\mu}\mathbf{u} \tag{2.13}$$

and, for $j = 1, 2, \ldots, p$, set

$$\mathbf{t}_j = \mathrm{sgn}\,(u_j)\boldsymbol{\kappa}_j(B_{\mu}^{-\mathrm{T}}). \tag{2.14}$$

If $j \leq k$ then the conditions of Lemma 2.3 are satisfied so that $\mathbf{t}_j$ defines a descent vector provided $u_j \notin [\alpha_{\mu(i)}, \beta_{\mu(i)}]$ and $\mathrm{sgn}\,(r_j(\mathbf{x} + \delta\mathbf{t}_j)) = \theta_j$ if $\delta > 0$. Also, if $\mathbf{g} \notin H$ then $\max\,(|u_i|, i = k+1, \ldots, p) > 0$. Let

$$j = \arg\max\,(|u_i|, i = k+1, \ldots, p),$$

then (2.14) again defines a descent vector. Thus the strategies available for determining a direction of descent are essentially the same as in the case of LP1.

Projected gradient algorithm

The basic equations of this method have the form

$$-\mathbf{t} = \mathbf{g} + M_\mu^T \mathbf{u} \tag{2.15}$$

and

$$M_\mu \mathbf{t} = 0. \tag{2.16}$$

They prove the direct analogues of (6.7), (6.8) of Chapter 2 in this case. If $\mathbf{g} \notin H$ then $\mathbf{t}$ given by (2.15), (2.16) is downhill for minimizing C as, using (2.7),

$$C'(\mathbf{x} : \mathbf{t}) = \mathbf{t}^T \mathbf{g} = -\|\mathbf{t}\|_2^2 \tag{2.17}$$

If $\mathbf{g} \in H$ then $\mathbf{t}$ is null and an alternative must be sought. An appropriate form is proved by the system

$$-\bar{\mathbf{t}} = \mathbf{g} + M_\mu^T \bar{\mathbf{u}} \tag{2.18}$$

and

$$M_\mu \bar{\mathbf{t}} = \theta_j \gamma_j \mathbf{e}_j \tag{2.19}$$

defining $(\bar{\mathbf{t}}, \bar{\mathbf{u}})$ where γ_j is a scale factor to be determined. Eliminating $\bar{\mathbf{t}}$ from (2.18) using (2.19) gives

$$\begin{aligned}\bar{\mathbf{u}} &= -(M_\mu M_\mu^T)^{-1} M_\mu \mathbf{g} - \theta_j \gamma_j (M_\mu M_\mu^T)^{-1} \mathbf{e}_j \\ &= \mathbf{u} - \theta_j \gamma_j (M_\mu M_\mu^T)^{-1} \mathbf{e}_j\end{aligned} \tag{2.20}$$

while substituting for $\bar{\mathbf{u}}$ in (2.18) gives

$$\bar{\mathbf{t}} = \mathbf{t} + \theta_j \gamma_j M_\mu^T (M_\mu M_\mu^T)^{-1} \mathbf{e}_j. \tag{2.21}$$

Lemma 2.3 can now be applied to

$$\mathbf{z}_j = \frac{1}{\gamma_j} (\bar{\mathbf{t}} - \mathbf{t})$$

and shows that this is a descent vector provided $u_j \notin [-\alpha_{\mu(j)}, \beta_{\mu(j)}]$ and $\theta_j u_j = |u_j|$. As $\mathbf{t}$ is either a descent vector or null it follows that

$$\bar{\mathbf{t}} = \mathbf{t} + \gamma_j \mathbf{z}_j$$

is a descent vector provided $\gamma_j > 0$ and $u_j \notin [-\alpha_{\mu(j)}, \beta_{\mu(j)}]$. To determine γ_j, interpret $\bar{\mathbf{t}}$ as the projected gradient vector determined by $\bar{\mu} = \mu \backslash \{\mu(j)\}$. By (2.6), (2.19), this requires

$$\bar{u}_j = \frac{\Psi(\theta_j \gamma_j, \alpha_{\mu(j)}, \beta_{\mu(j)})}{\theta_j \gamma_j}$$

in order that $\mathbf{g}$ be updated correctly. Substituting in (2.20) gives

$$\gamma_j (M_\mu M_\mu^T)_{jj}^{-1} = |u_j| - \frac{\Psi(\theta_j \gamma_j, \alpha_{\mu(j)}, \beta_{\mu(j)})}{\gamma_j} \tag{2.22}$$

showing that $\gamma_j > 0$ is a consequence of $u_j \notin [-\alpha_{\mu(j)}, \beta_{\mu(j)}]$ as $M_\mu M_\mu^T$ is positive definite by the nondegeneracy assumptions.

Remark 2.3 This development of the projected gradient algorithm shows clearly the options available for the generation of a descent vector. There are two cases.

(i) $\mathbf{g} \notin H$. Here a descent vector can always be found from (2.15), (2.16). However, there is a further option, for if any component of $\mathbf{u}$ violates the constraints (2.9b) then the corresponding residual can be allowed to relax away from zero and thus reduce the constraints on the descent direction expressed by (2.16).
(ii) $\mathbf{g} \in H$. In this case the current point is optimal unless u_j violates (2.9b) for at least one j. It is necessary to relax off $r_{\mu(j)}$ to develop a descent vector, which turns out to be $\gamma_j \mathbf{z}_j$.

Line search

The next stage in the implementation of a descent algorithm is the calculation of the length of the step in the descent direction which has been generated. In the algorithms that treat LP1 directly this is usually just a step to the point at which the first new constraint becomes active, and a larger step is possible only when another device is available to restore feasibility as in the generation of multiple pivot sequences. Here the situation is just a somewhat more general restatement of that for the penalty algorithm, and feasibility does not impose a constraint. The points on the line $\mathbf{x}+\lambda\mathbf{t}$ at which it could be necessary to revise $\mathbf{t}, \mathbf{u}$ are those at which an additional residual vanishes, and these are the only points which have to be considered in seeking the point at which $C(\mathbf{x}+\lambda\mathbf{t})$ stops decreasing (consider Figure 2.1). This minimum is characterized by a value λ^* such that

$$C'(\mathbf{x}+(\lambda^*-\delta)\mathbf{t}:\mathbf{t})<0$$

and

$$C'(\mathbf{x}+(\lambda^*+\delta)\mathbf{t}:\mathbf{t})\geq 0 \tag{2.23}$$

where $\delta > 0$. The change in C as a zero residual is passed in the direction $\mathbf{t}$ is readily computed from (2.7). Assume r_j vanishes at $\mathbf{x}+\lambda\mathbf{t}$; then

$$\begin{aligned} C'(\mathbf{x}+(\lambda+\delta)\mathbf{t}:\mathbf{t}) &= C'(\mathbf{x}+(\lambda-\delta)\mathbf{t}:\mathbf{t})+\left(\frac{\Psi(-\theta_j, \alpha_j, \beta_j)}{\theta_j}+\frac{\Psi(\theta_j, \alpha_j, \beta_j)}{\theta_j}\right)\mathbf{m}_j^T\mathbf{t} \\ &= C'(\mathbf{x}+(\lambda-\delta)\mathbf{t}:\mathbf{t})+(\alpha_j+\beta_j)\,|\mathbf{m}_j^T\mathbf{t}| \end{aligned} \tag{2.24}$$

where δ is chosen sufficiently small so that only r_j changes sign in the interval $(\lambda-\delta, \lambda+\delta)$ and $\theta_j = \text{sgn}\,(\mathbf{m}_j^T\mathbf{t})$. Combining (2.23) and (2.24) gives the following result.

Lemma 2.4 *Let*

$$\Lambda = \{\lambda_i;\ \lambda_i = -r_i/\mathbf{m}_i^{\mathrm{T}}\mathbf{t},\ \lambda_i > 0\}$$

and let π be an index set pointing to the elements of Λ sorted into increasing order. Then $\lambda^* = \lambda_{\pi(m)}$ where m points to the first index in π such that

$$C'(\mathbf{x}:\mathbf{t}) + \sum_{i=1}^{m} (\alpha_{\pi(i)} + \beta_{\pi(i)})\,|\mathbf{m}_{\pi(i)}^{\mathrm{T}}\mathbf{t}| \geq 0. \tag{2.25}$$

There is an alternative approach to the line search which is completely equivalent to the above result for the reduced gradient algorithm, but only partially equivalent in the case of the projected gradient algorithm. We consider first the reduced gradient algorithm.

Assume that the descent vector at the current point is

$$\mathbf{t} = \operatorname{sgn}(u_q)\boldsymbol{\kappa}_q(B_\mu^{-\mathrm{T}})$$

and that the first new zero residual corresponds to r_s. If the line search is halted at this point then B_μ is updated to give

$$\begin{aligned}\bar{B}_\mu &= B_\mu + (\mathbf{m}_s - \boldsymbol{\kappa}_q(B_\mu))\mathbf{e}_q^{\mathrm{T}} \\ &= B_\mu\{I + (B_\mu^{-1}\mathbf{m}_s - \mathbf{e}_q)\mathbf{e}_q^{\mathrm{T}}\}\end{aligned} \tag{2.26}$$

while

$$\bar{\mathbf{g}} = \tilde{\mathbf{g}} + \frac{\Psi(-\theta_s, \alpha_s, \beta_s)}{\theta_s}\mathbf{m}_s \tag{2.27}$$

where $\tilde{\mathbf{g}}$ is evaluated at $\mathbf{x} + \delta\mathbf{t}$ for $\delta > 0$ small enough, and $\theta_s = \operatorname{sgn}(\mathbf{m}_s^{\mathrm{T}}\mathbf{t})$. In particular,

$$\begin{aligned}\bar{u}_q &= -\mathbf{e}_q^{\mathrm{T}}\bar{B}_\mu^{-1}\bar{\mathbf{g}} \\ &= -\mathbf{e}_q^{\mathrm{T}}\left(I - \frac{(B_\mu^{-1}\mathbf{m}_s - \mathbf{e}_q)\mathbf{e}_q^{\mathrm{T}}}{\mathbf{e}_q^{\mathrm{T}}B_\mu^{-1}\mathbf{m}_s}\right)B_\mu^{-1}\bar{\mathbf{g}} \\ &= \frac{-\mathbf{e}_q^{\mathrm{T}}B_\mu^{-1}\bar{\mathbf{g}}}{\mathbf{e}_q^{\mathrm{T}}B_\mu^{-1}\mathbf{m}_s} \\ &= -\frac{\mathbf{t}^{\mathrm{T}}\bar{\mathbf{g}}}{\mathbf{t}^{\mathrm{T}}\mathbf{m}_s} \\ &= -\left(\frac{C'(\mathbf{x}:\mathbf{t})}{\mathbf{t}^{\mathrm{T}}\mathbf{m}_s} + \frac{\Psi(-\theta_s, \alpha_s, \beta_s)}{\theta_s}\right)\end{aligned} \tag{2.28}$$

Remark 2.4 This result shows that $\bar{u}_q$ can be computed using quantities currently available.

Now

$$\theta_s\bar{u}_q = -\frac{C'(\mathbf{x}:\mathbf{t})}{|\mathbf{t}^{\mathrm{T}}\mathbf{m}_s|} - \Psi(-\theta_s, \alpha_s, \beta_s) > 0$$

provided

$$C'(\mathbf{x}:\mathbf{t}) + \Psi(-\theta_s, \alpha_s, \beta_s)\,|\mathbf{t}^{\mathrm{T}}\mathbf{m}_s| < 0.$$

If this holds,

$$|\bar{u}_q| - \Psi(\theta_s, \alpha_s, \beta_s) = -(C'(\mathbf{x}:\mathbf{t}) + (\alpha_s + \beta_s)|\mathbf{t}^T\mathbf{m}_s|)\frac{1}{|\mathbf{t}^T\mathbf{m}_s|} \tag{2.29}$$

showing that $\bar{u}_q$ *lies outside its bounds if and only if* $\mathbf{t}$ *is still downhill for minimizing* C by (2.25). If this is the case then a descent vector $\bar{\mathbf{t}}$ can be obtained by relaxing off $r_s = 0$. This gives

$$\begin{aligned}\bar{\mathbf{t}}^T &= \text{sgn}\,(\bar{u}_q)\boldsymbol{\kappa}_q(\bar{B}_\mu^{-T})^T \\ &= \theta_s\mathbf{e}_q^T\left(I - \frac{(B_\mu^{-1}\mathbf{m}_s - \mathbf{e}_q)\mathbf{e}_q^T}{\mathbf{e}_q^T B_\mu^{-1}\mathbf{m}_s}\right)B_\mu^{-1} \\ &= \theta_s\frac{\mathbf{e}_q^T B_\mu^{-1}}{\mathbf{e}_q^T B_\mu^{-1}\mathbf{m}_s} \\ &= \frac{\mathbf{t}^T}{|\mathbf{t}^T\mathbf{m}_s|}\end{aligned} \tag{2.30}$$

so that $\bar{\mathbf{t}}$ *is proportional to* $\mathbf{t}$.

Lemma 2.5 *For the reduced gradient algorithm there is an exact equivalence between the steps in the line search to find the minimum of* $C(\mathbf{x}+\lambda\mathbf{t})$ *and the steps in the multiple pivot sequence in which a sequence of multiplier estimates are generated by* (2.28) *and violate the bound test* (2.29) *so that the corresponding residual can be relaxed immediately, thus avoiding the need to update the factored form of* B_μ.

The equivalence is not so complete for the projected gradient algorithm. However, it is still the case that the multiplier corresponding to the first new zero residual is available. Let μ point to the residuals held at zero level in the direction defined by $\mathbf{t}$ and assume that r_s is the new residual. If $|\mu| = k-1$,

$$M_\mu^T = [Q_1 \mid Q_2]\begin{bmatrix} U \\ 0 \end{bmatrix},$$

and

$$\mathbf{y} = P\mathbf{m}_s / \|P\mathbf{m}_s\|_2$$

where $P = Q_2 Q_2^T$ projects onto the columns of Q_2, then

$$\bar{Q} = [(Q_1 \mid \mathbf{y}) \mid \bar{Q}_2],$$

$$\bar{U} = \begin{bmatrix} Q_1^T \\ \mathbf{y}^T \end{bmatrix}[M_\mu^T \mid \mathbf{m}_s] = \left[\begin{array}{c|c} U & Q_1^T\mathbf{m}_s \\ \hline 0 & \mathbf{y}^T\mathbf{m}_s \end{array}\right]$$

and

$$\bar{u}_k = -\frac{\mathbf{y}^T\bar{\mathbf{g}}}{\mathbf{y}^T\mathbf{m}_s}$$
$$= -\frac{1}{\|P\mathbf{m}_s\|_2}\left(\frac{\mathbf{m}_s^T Q_2 Q_2^T \mathbf{g}}{\|P\mathbf{m}_s\|_2} + \frac{\Psi(-\theta_s, \alpha_s, \beta_s)}{\theta_s}\mathbf{y}^T\mathbf{m}_s\right)$$
$$= \frac{\mathbf{m}_s^T\mathbf{t}}{\|P\mathbf{m}_s\|_2^2} - \frac{\Psi(-\theta_s, \alpha_s, \beta_s)}{\theta_s}. \tag{2.31}$$

If

$$\theta_s\bar{u}_k = \frac{|\mathbf{m}_s^T\mathbf{t}|}{\|P\mathbf{m}_s\|_2^2} - \Psi(-\theta_s, \alpha_s, \beta_s) > 0$$

then progress can be made by relaxing off $r_s = 0$ provided

$$|\bar{u}_k| - \Psi(\theta_s, \alpha_s, \beta_s) = \frac{|\mathbf{m}_s^T\mathbf{t}|}{\|P\mathbf{m}_s\|_2^2} - \Psi(-\theta_s, \alpha_s, \beta_s) - \Psi(\theta_s, \alpha_s, \beta_s) > 0 \tag{2.32}$$

Remark 2.5 This inequality differs from (2.29) except under the particular circumstances discussed below. These give conditions under which $\mathbf{t}$ and $\bar{\mathbf{t}}$ are parallel.

As $\mathbf{g}$ must be updated on passing through a zero residual, it follows from (2.15) that these conditions cannot always be satisfied.

Lemma 2.6 *Let*

$$H = \text{span}\,(\mathbf{m}_{\mu(i)}, i = 1, 2, \ldots, k-1, \mathbf{m}_j)$$

where $\mathbf{m}_j$ is the row of M associated with the residual relaxed to product $\mathbf{t}$. If

$$\mathbf{m}_s, \mathbf{g}, \bar{\mathbf{g}} \in H$$

then

$$\mathbf{t}, \bar{\mathbf{t}}, P\mathbf{m}_j, P\mathbf{m}_s, \mathbf{y}$$

are parallel vectors where $\bar{\mathbf{t}}$ is the descent vector obtained by relaxing $r_s = 0$.

Proof Only the case of $\mathbf{t}$ is considered, as the others are similar. By assumption $\mathbf{g}$ can be written

$$\mathbf{g} = [M^T \mid \mathbf{m}_j]\begin{bmatrix}\mathbf{z}\\ \zeta\end{bmatrix}$$

so that (with $\theta_j = \text{sgn}\,(r_j)$)

$$\mathbf{t} = -P\tilde{\mathbf{g}} = -P\left\{[M_\mu^T \mid \mathbf{m}_j]\begin{bmatrix}\mathbf{z}\\ \zeta\end{bmatrix} + \frac{\Psi(\theta_j, \alpha_j, \beta_j)}{\theta_j}\mathbf{m}_j\right\}$$
$$= -\left(\zeta + \frac{\Psi(\theta_j, \alpha_j, \beta_j)}{\theta_j}\right)P\mathbf{m}_j$$

If Lemma 2.6 holds then

$$\begin{aligned} C'(\mathbf{x}:\mathbf{t}) &= \tilde{\mathbf{g}}^{\mathrm{T}}\mathbf{t} \\ &= -\tilde{\mathbf{g}}^{\mathrm{T}}\mathbf{y}\mathbf{y}^{\mathrm{T}}\tilde{\mathbf{g}} \\ &= -\frac{(\mathbf{m}_s^{\mathrm{T}}\mathbf{t})^2}{\|P\mathbf{m}_s\|_2^2} \end{aligned}$$

so that (2.32) can be written

$$|\bar{u}_k| - \Psi(\theta_s, \alpha_s, \beta_s) = \frac{|C'(\mathbf{x}:\mathbf{t})|}{|\mathbf{m}_s^{\mathrm{T}}\mathbf{t}|} - (\alpha_s + \beta_s)$$

which is identical with the expression (2.29) obtained for the reduced gradient algorithm.

Implementation

Implementation of both algorithms need not differ significantly from that described for the corresponding applications of the algorithms to LP1. In particular, the organization and updating of the tableau is unchanged. New features come from the form of the objective function – for example, the two-sided tests on the multipliers u_i. This has some implication for scaling. Consider row scaling of M so that

$$M \to DM$$

or, equivalently,

$$r_i \to d_i r_i = s_i, \qquad i = 1, 2, \ldots, n.$$

The objective function must remain unaltered or the problem changes. This requires

$$\Psi(r_i, \alpha_i, \beta_i) \to \Psi\left(s_i, \frac{\alpha_i}{d_i}, \frac{\beta_i}{d_i}\right) \tag{2.33}$$

and this is compatible with the transformation rule for the multipliers. To see this, note that

$$\frac{\Psi(r_i, \alpha_i, \beta_i)}{r_i}\mathbf{m}_i = \frac{\Psi(s_i, \alpha_i/d_i, \beta_i/d_i)}{s_i} d_i \mathbf{m}_i$$

so that $\mathbf{g}$ is unchanged. Now from (2.10) (or equivalently (2.15) with $\hat{\mathbf{g}}$ replacing $\mathbf{t}$)

$$\mathbf{g} - \hat{\mathbf{g}} = -\sum_{i=1}^{|\mu|} u_i \mathbf{m}_{\mu(i)} = -\sum_{i=1}^{|\mu|} \frac{u_i}{d_i} d_i \mathbf{m}_{\mu(i)}$$

so that

$$u_i \to u_i/d_i. \tag{2.34}$$

The appropriate NSE test for the reduced gradient algorithm can be determined using Lemma 2.3. It takes the form (compare (5.22) of Chapter 2)

$$j = \arg\max \left(\frac{|u_j| - \Psi(\theta_j, \alpha_{\mu(j)}, \beta_{\mu(j)})}{\|\kappa_j(B_\mu^{-T})\|_2}, \quad j = 1, 2, \ldots, |\mu|, \right.$$

$$\left. \frac{\tau\,|u_j|}{\|\kappa_j(B_\mu^{-T})\|_2}, \quad j = |\mu| + 1, \ldots, p \right) \tag{2.35}$$

Here $\Psi(\theta, \alpha, \beta)$ is defined in (2.2) and provides a shorthand for β if $\theta > 0$ else α, while τ is the weighting factor discussed in Section 7.5. The recurrence relation (5.23) of Chapter 2 is available to compute the column norms. It is clear that the NSE test is independent of row scaling.

Termination

Termination occurs when it is inferred from consideration of quantities computed at the current stage of the algorithm that the conditions

$$|u_j| - \Psi(\theta_j, \alpha_{\mu(j)}, \beta_{\mu(j)}) \leq 0, \; j = 1, 2, \ldots, |\mu|, \qquad |\mu_j| = 0, \; j = |\mu| + 1, \ldots, p$$

would be satisfied if the calculations had been performed exactly. Again, the situation is very close to that discussed in Section 2.5. In particular, the recurrence (5.23) of Chapter 2 used in conjunction with the NSE test provides the information necessary to estimate $\operatorname{cond}_F(B_\mu)$. The direct analogue of the test (5.50) is

$$\left\{ \sum_{j=1}^{|\mu|} \max\,(|u_j| - \Psi(\theta_j, \alpha_{\mu(j)}, \beta_{\mu(j)}), 0)^2 + \sum_{j=|\mu|+1}^{p} u_j^2 \right\}^{1/2} \leq \operatorname{cond}_F(B_\mu)\, \text{eps}\, \|\mathbf{u}\|_2$$

For the projected gradient algorithm the appropriate form of NSE test can be obtained from Lemma 2.3. With $\bar{\mathbf{t}}$ given by (2.21) it follows that

$$\begin{aligned} \|\bar{\mathbf{t}}\|_2^2 = -C'(\mathbf{x} : \bar{\mathbf{t}}) &= \|\mathbf{t}\|_2^2 + \gamma_j(|u_j| - \Psi(\theta_j, \alpha_{\mu(j)}, \beta_{\mu(j)})) \\ &= \|\mathbf{t}\|_2^2 + \frac{(|u_j| - \Psi(\theta_j, \alpha_{\mu(j)}, \beta_{\mu(j)}))^2}{(M_\mu M_\mu^T)_{jj}^{-1}} \\ &= \|\mathbf{t}\|_2^2 + (|u_j| - \Psi(\theta_j, \alpha_{\mu(j)}, \beta_{\mu(j)}))^2 / (U^T U)_{jj}^{-1} \end{aligned} \tag{2.37}$$

where U is the upper triangular matrix obtained in the orthogonal factorization of M_μ^T. This result is exactly equivalent to (5.23) of Chapter 2. Moreover, the argument which shows that this test is independent of row scaling extends trivially as a consequence of (2.33), (2.34).

The discussion of termination in Section 2.6 also applies. Thus (6.42) of Chapter 2, which tests for $\mathbf{t} = 0$, becomes (where the quantities referred to are the computed estimates)

$$\|Q_2^T \mathbf{g}\|_2 \leq \text{eps}\, \operatorname{cond}_F(M_\mu)\, \|\mathbf{g}\|_2 \tag{2.38}$$

while the test (6.45) for the multipliers being in range becomes

$$\left\{\sum_{i=1}^{|\mu|} \max\left(|u_j| - \Psi(\theta_j, \alpha_{\mu(j)}, \beta_{\mu(j)}), 0\right)^2\right\}^{1/2} \leq \text{eps}\ \text{cond}_F(M_\mu)(1+\|\mathbf{u}\|_2) \tag{2.39}$$

Treatment of degeneracy

The foregoing discussion is predicated on the assumption that degeneracy, which here takes the form of more than one new zero residual being introduced at the new point $\mathbf{x}+\gamma^*\mathbf{t}$, does not occur. This is a problem only if $\mathbf{g}$ is in the span of a proper subset of the columns of M_μ^T, and in this case we follow the general approach introduced in Section 2.5 and seek a descent vector $\mathbf{t}$ to satisfy

$$\phi(\mathbf{t}) = \mathbf{g}^T\mathbf{t} + \sum_{i\in\mu} \Psi(\mathbf{m}_i^T\mathbf{t}, \alpha_i, \beta_i) < 0 \tag{2.40}$$

Note that if there exists $\mathbf{t}$ satisfying (2.40) then the problem of minimizing $\phi(\mathbf{x})$ is unbounded below as $\phi(\lambda\mathbf{t}) = \lambda\phi(\mathbf{t})$, $\lambda > 0$ so that it is compatible with previous usage to call $\mathbf{t}$ *a direction of recession for* ϕ. Note also that $\phi(\mathbf{x})$ is a piecewise linear function according to Definition 2.1, and this suggests using the algorithms developed above to minimize ϕ until a direction of recession appears. This has the advantage that a suitably factored form for $M_{\mu_1}^T$, where $\mu = \mu_1 \cup \mu_2$ and $\mathbf{g} \in \text{span}(\boldsymbol{\kappa}_i(M_{\mu_1}^T))$, is in general available after an additional updating step, and this suggests starting the minimizing of ϕ from $\mathbf{x}_0 = 0$. However, this point is degenerate for the new problem, but this difficulty can be overcome by applying the minimization algorithm instead to the modified objective function ϕ_ε defined by

$$\phi_\varepsilon(\mathbf{x}) = \mathbf{g}^T\mathbf{x} + \sum_{i\in\mu_1} \Psi(\mathbf{m}_i^T\mathbf{x}, \alpha_i, \beta_i) + \sum_{i\in\mu_2} \Psi(\mathbf{m}_i^T\mathbf{x}+\varepsilon_i, \alpha_i, \beta_i) \tag{2.41}$$

where the $\varepsilon_i \neq 0$ but are otherwise arbitrary. The key result needed is that $\mathbf{t}$ is a direction of recession for ϕ if and only if it is a direction of recession for ϕ_ε. This follows directly from

$$\phi_\varepsilon(\lambda\mathbf{t}) = \phi(\lambda\mathbf{t}) + \sum_{i\in\mu_2} \begin{cases} \beta_i\varepsilon_i, & \mathbf{m}_i^T\mathbf{t} > 0 \\ \Psi(\varepsilon_i, \alpha_i, \beta_i), & \mathbf{m}_i^T\mathbf{t} = 0 \\ -\alpha_i\varepsilon_i, & \mathbf{m}_i^T\mathbf{t} < 0 \end{cases} \tag{2.42}$$

which holds for all $\lambda > 0$ sufficiently large. Thus it suffices to minimize $\phi_\varepsilon(\mathbf{x})$ to determine $\mathbf{t}$, and the recursive procedure described in Section 2.5 is available to resolve any further degeneracies.

Remark 2.6 Given a treatment of degeneracy, it is possible to prove finiteness of the algorithms. It is first necessary to know that the nondegenerate cases are finite. In the reduced gradient algorithm it suffices to note that $\mathbf{x}$ is uniquely determined by μ, E_μ, and that these configurations cannot repeat because the objective function always decreases. Thus finiteness follows because the number of configurations is finite. In the projected gradient algorithm the proof can be modelled on the proof of Theorem 6.1 of Chapter 2. Again the experimental evidence suggests that zero residuals should not be relaxed until $\mathbf{g} \in H$. Degeneracy can now be dealt with by the recursive argument given in Section 2.5 which applies to give the finiteness of both algorithms.

Example 2.1 Consider the reduced gradient algorithm for the l_1 problem for the model

$$M = \begin{pmatrix} 1 & 0 \\ 1 & .5 \\ 1 & 1 \end{pmatrix}, \qquad \mathbf{f} = \begin{pmatrix} 1 \\ 1.649 \\ 2.718 \end{pmatrix}.$$

The reduced model with $x_2 = 0$ can be solved readily to give $\mathbf{x}^T = [1.649, 0]$ and provides a convenient starting point. Here $\mu = \{2\}$,

$$\mathbf{g} = \begin{pmatrix} 1 \\ 0 \end{pmatrix} - \begin{pmatrix} 1 \\ 1 \end{pmatrix} = -\begin{pmatrix} 0 \\ 1 \end{pmatrix}, \qquad B_\mu = \begin{pmatrix} 1 & 0 \\ .5 & 1 \end{pmatrix}, \qquad \mathbf{u} = \begin{pmatrix} 0 \\ 1 \end{pmatrix}.$$

Then a descent direction is

$$\mathbf{t}^T = (-.5, 1),$$

and

$$\Lambda = \{1.298, 2.138\}, \qquad \Pi = \{1, 3\}.$$

The first step in the line search gives

$$\mathbf{g}^T\mathbf{t} + 2\left|\mathbf{t}^T\begin{pmatrix} 1 \\ 0 \end{pmatrix}\right| = 0$$

so the minimum is achieved for $\lambda = \lambda_1$,

$$\mathbf{x}^T := [1.649, 0] + 1.298[-.5, 1] = [1, 1.298],$$

and

$$\mu := \{2, 1\}.$$

Now

$$\mathbf{g} = -\begin{pmatrix} 1 \\ 1 \end{pmatrix}, \qquad B_\mu = \begin{pmatrix} 1 & 1 \\ .5 & 0 \end{pmatrix}, \qquad \mathbf{u} = \begin{pmatrix} 2 \\ -1 \end{pmatrix}.$$

As $2 \notin [-1, 1]$ the descent direction is obtained by relaxing off the equation pointed to by $\mu(1)$. This gives

$$\mathbf{t}^T = [0, 2],$$

and

$$\Lambda = \{.21\}, \qquad \Pi = \{3\}.$$

In this case the line search is trivial. Thus

$$\mathbf{x}^{\mathrm{T}} := [1, 1.298] + .21[0, 2] = [1, 1.718]$$

and

$$\mu := \{1, 3\}.$$

Now

$$\mathbf{g} = \begin{pmatrix} 1 \\ .5 \end{pmatrix}, \quad B_{\mu} = \begin{bmatrix} 1 & 1 \\ 1 & 0 \end{bmatrix}, \quad \mathbf{u} = \begin{bmatrix} -.5 \\ -.5 \end{bmatrix}$$

Here $-1 \leqslant u_i \leqslant 1$, $i = 1, 2$, so the current point is optimal.

To check the multiplier estimate from (2.25) at the end of the first step we have

$$\bar{u}_2 = -\left\{\frac{-1}{-.5} + \frac{1}{-1}\right\} = -1$$

agreeing with the explicit computation given above.

Exercise 2.1 It is an important point that the tableau organization is available to implement both the projected and reduced gradient algorithms. Let

$$W = [I \,|M^{\mathrm{T}}|\, \mathbf{g}],$$
$$\mathbf{w}^{\mathrm{T}} = [\mathbf{x}^{\mathrm{T}} \,|\mathbf{r}^{\mathrm{T}}|\, C].$$

(i) Discuss the transformation of this tableau organization and show that this follows essentially the discussion in Chapter 2. In particular, verify that the quantities required to compute **u**, **t** and update **x**, **r** are readily available.

(ii) Note that here **g** is not fixed but must be modified whenever a residual either becomes zero or is allowed to relax off zero. Derive the formulae for updating **g** and C in each case (note that there are two cases to consider in the line search depending on whether the line search terminates at r_j or continues).

It should be noted that when relaxing off $r_j = 0$ in order to generate a descent direction $\mathbf{t}_j$ it may not be true that $\operatorname{sgn}(r_j + \lambda \mathbf{m}^{\mathrm{T}}\mathbf{t}) = \operatorname{sgn}(u_j)$ for $\lambda > 0$ small enough as a consequence of the effects of rounding error. This may lead to inconsistencies in the line search if j is included in the set of indices allowed in the competition in Lemma 2.4 (of course indices in μ need not be considered).

3.3 LINEAR PROGRAMMING ALGORITHMS IN l_1 APPROXIMATION

Perhaps the most important step towards making l_1 approximation readily accessible as a basic tool in data analysis was its formulation as a tractable linear programming problem, as the simplex algorithm could then be used to compute numerical solutions on a routine basis. To make this formulation it is convenient to start by writing the defining equations in terms of nonnegative variables. This gives

$$\mathbf{x}=\mathbf{u}-\mathbf{v},$$

and

$$r_i = y_i - z_i = \mathbf{m}_i^{\mathrm{T}}(\mathbf{u}-\mathbf{v}) - f_i, \qquad i=1,2,\ldots,n \tag{3.1}$$

where

$$\mathbf{u}\geqslant 0, \qquad \mathbf{v}\geqslant 0, \qquad \mathbf{y}\geqslant 0, \qquad \mathbf{z}\geqslant 0.$$

also

$$y_i z_i = 0 \Rightarrow |r_i| = y_i + z_i \tag{3.2}$$

so that the objective function becomes a sum of nonnegative variables provided the complementarity condition is satisfied. This suggests considering the linear program (in the form LP2)

$$\max_{\mathbf{u},\mathbf{v},\mathbf{y},\mathbf{z}} -[0\,|0\,|\mathbf{e}^{(n)\mathrm{T}}|\,\mathbf{e}^{(n)\mathrm{T}}]\begin{bmatrix}\mathbf{u}\\ \mathbf{v}\\ \mathbf{y}\\ \mathbf{z}\end{bmatrix} \tag{3.3}$$

subject to the constraints

$$[M\,|-M\,|-I_n|\,I_n]\begin{bmatrix}\mathbf{u}\\ \mathbf{v}\\ \mathbf{y}\\ \mathbf{z}\end{bmatrix}=\mathbf{f}$$

and

$$\mathbf{u}\geqslant 0, \qquad \mathbf{v}\geqslant 0, \qquad \mathbf{y}\geqslant 0, \qquad \mathbf{z}\geqslant 0.$$

Lemma 3.1 *The linear program* (3.3) *is bounded and any basic feasible optimizing solution of* (3.3) *solves the* l_1 *problem.*

Proof The constraints imply that the objective function in (3.3) is nonpositive so that boundedness is immediate. The matrix of equality constraints has full row rank so this is true also of the optimal basis matrix. It follows that no pair u_i, v_i or y_i, z_i can be basic variables simultaneously as this would lead to two columns of the basis matrix being proportional. Thus the complementarity condition (3.2) is satisfied automatically by any basic solution. An immediate consequence is that any basic solution to (3.3) gives a valid $(\mathbf{x}, \mathbf{r})$ for the l_1 problem and hence an upper bound to its minimum. But to any $(\mathbf{x}, \mathbf{r})$ minimizing the l_1 problem there corresponds a feasible solution to (3.3) constructed by satisfying (3.1), (3.2), and hence a lower bound to the maximum of (3.3). This is possible only if the extremal values agree in magnitude.

Remark 3.1 In proving Lemma 3.1 no assumption is needed on rank (M). It follows that *the simplex algorithm can be used directly to solve the* l_1 *problem without the need to make any assumptions on* rank (M). In the rank-deficient case it is clear from the form of the equality constraints that the extra degrees of freedom are taken up by fixing certain components of $\mathbf{x}$ at zero level.

Thus it only remains to ask if it is possible to exploit the special structure in (3.3) in order to economize in the computation. This question possesses a clean, if possibly unexpected, answer. *There is an exact equivalence between the possible options (USE/NSE tests, multiple pivot sequences) in the simplex algorithm applied to* (3.3) *and the possible options in the reduced gradient algorithm applied to the* l_1 *problem directly.* Thus the algorithms are equivalent in a very complete sense. It would seem to be advisable to concentrate on the reduced gradient form which provides about as compact a formulation of the solution process as can be expected.

The equivalence requires showing that the actual calculations in the two methods are identical and the only problems involved are in reconciling the organizational differences (for example, the simplex method builds up a basis for (3.3) by adding a column of M at a time, while B_μ in (2.13) is built up by adding a row of M at a time). It would appear to be sufficient to consider a typical calculation and the one chosen here is the rate of change of the residuals in the two cases.

(i) Reduced gradient algorithm: in this case the quantities of interest are

$$\mathrm{d}r_i = \mathbf{m}_i^{\mathrm{T}}[\mathrm{M}_\mu^{\mathrm{T}} \mid E_\mu]^{-\mathrm{T}}\mathbf{e}_s, \qquad i = 1, 2, \ldots, n \tag{3.4}$$

where s specifies the current descent direction. After suitable reordering of rows and columns if necessary, this can be specialized to

$$\begin{aligned}\mathrm{d}r_i &= \mathbf{m}_i^{\mathrm{T}}\left[\begin{array}{c|c} M_{\mu,\nu}^{\mathrm{T}} & 0 \\ \hline \bar{M}_{\mu,\nu} & I_{p-k}\end{array}\right]^{-\mathrm{T}}\mathbf{e}_s \\ &= \mathbf{m}_i^{\mathrm{T}}\left[\begin{array}{c|c} M_{\mu,\nu}^{-\mathrm{T}} & 0 \\ \hline -\bar{M}_{\mu,\nu}M_{\mu,\nu}^{-\mathrm{T}} & I_{p-k}\end{array}\right]^{\mathrm{T}}\mathbf{e}_s\end{aligned} \tag{3.5}$$

where ν is an index set pointing to the zero rows of E_μ. There are two cases to consider:

(a) $\quad 1 \leq s \leq k = |\mu|$;

$$\mathrm{d}r_i = \mathbf{m}_i^{\mathrm{T}}\boldsymbol{\kappa}_s\left(\begin{bmatrix} M_{\mu,\nu}^{-1} \\ 0 \end{bmatrix}\right) \tag{3.6}$$

(b) $\quad s > k$;

$$\mathrm{d}r_i = \mathbf{m}_i^{\mathrm{T}}\boldsymbol{\kappa}_{s-k}\left(\left[\begin{array}{c} -M_{\mu,\nu}^{-1}\bar{M}_{\mu,\nu}^{\mathrm{T}} \\ \hline I_{p-k}\end{array}\right]\right) \tag{3.7}$$

(ii) Simplex algorithm: the relevant quantities in the comparison are the $\mathbf{v}_q^\sigma = B^{-1}\boldsymbol{\kappa}_q(N)$ (compare with (4.6) of Chapter 2, for example). The basis matrix can be written

$$B = \left[\begin{array}{c|c} M_{\mu,\nu}S_1 & 0 \\ \hline M_{\mu^c,\nu}S_1 & S_2 \end{array}\right] \tag{3.8}$$

where

$$\mu = \{i;\, r_i = 0 \Rightarrow y_i,\, z_i \text{ nonbasic}\},$$
$$\nu = \{i;\, \boldsymbol{\kappa}_i(B) = \pm\boldsymbol{\kappa}_{\nu(i)}(M),\, i = 1, 2, \ldots, k = |\nu| = |\mu|\},$$
$$S_1 = \text{diag}\,\{\text{sgn}\,(x_{\nu(j)}),\, j = 1, 2, \ldots, k\},$$
$$S_2 = \text{diag}\,\{-\text{sgn}\,(r_{\mu^c(j)}),\, j = 1, 2, \ldots, n-k\},$$

and the equivalence of the definitions of μ, ν in the two approaches should be noted. Again there are two cases to consider in calculating $\mathbf{v}_q^\sigma$:

(a) $\boldsymbol{\kappa}_q(N) = \pm\boldsymbol{\kappa}_l(M), \qquad l \notin \nu.$

$$B^{-1}M\mathbf{e}_l = \left[\begin{array}{c|c} S_1 M_{\mu,\nu}^{-1} & 0 \\ \hline -S_2 M_{\mu^c,\nu} M_{\mu,\nu}^{-1} & S_2 \end{array}\right]\boldsymbol{\kappa}_l(M). \tag{3.9}$$

Now $-\mathbf{v}_q^\sigma$ partitions to give the change in the basic $(\mathbf{u}\,|\,\mathbf{v})$ defining the nonzero components of $\mathbf{x}$ in the first k components, and the change in the basic $(\mathbf{y}\,|\,\mathbf{z})$ defining the nonzero components of $\mathbf{r}$ in the final $n-k$ components. Seeking out $\mathbf{m}_i^{\mathrm{T}}$ among the rows of $M_{\mu^c,\nu}$ and noting that the first k elements of $\boldsymbol{\kappa}_l(M)$ correspond to a row of $\bar{M}_{\mu,\nu}$ permits the identity of this case with (3.7) to be deduced immediately.

(b) $\boldsymbol{\kappa}_q(N) = \pm\mathbf{e}_l, \qquad l \in \mu.$

$$B^{-1}\mathbf{e}_l = \begin{bmatrix} S_1\boldsymbol{\kappa}_l(M_{\mu,\nu}^{-1}) \\ -S_2 M_{\mu^c,\nu}\boldsymbol{\kappa}_l(M_{\mu,\nu}^{-1}) \end{bmatrix} \tag{3.10}$$

Arguing as in the previous case permits the identification to be made with (3.6) above.

An alternative approach considers the dual of (3.3). This is

$$\min \mathbf{f}^{\mathrm{T}}\mathbf{z} \tag{3.11}$$

subject to

$$\begin{bmatrix} M^{\mathrm{T}} \\ -M^{\mathrm{T}} \\ -I_n \\ I_n \end{bmatrix}\mathbf{z} \geqslant \begin{bmatrix} 0 \\ 0 \\ -\mathbf{e}^{(n)} \\ -\mathbf{e}^{(n)} \end{bmatrix},$$

and these constraint inequalities simplify to

$$M^{\mathrm{T}}\mathbf{z} = 0$$

and

$$-\mathbf{e}^{(n)} \leqslant \mathbf{z} \leqslant \mathbf{e}^{(n)}. \tag{3.12}$$

Thus *the dual is an interval programming problem.* This fact permits the identity of the modified simplex algorithm (modified to permit multiple pivot sequences) and subopt applied to (3.4), (3.5) to be deduced from Theorem 7.3 of Chapter 2.

One other advantage of the relation between the l_1 problem and its formulation as a tractable linear programming problem is that it shows that consideration of the l_1 problem subject to linear constraints is straightforward. The additional constraints make only one significant change in that if one of these constraints becomes active in the reduced gradient line search (modified simplex algorithm multiple pivot sequence), then it will block further progress in the current direction in general.

Exercise 3.1 (a) Deduce the optimality conditions for the l_1 problem by applying the Kuhn–Tucker conditions to (3.3).

(b) Deduce the optimality conditions for the constrained l_1 problem both directly and by considering the equivalent linear programming formulation. Extend Lemma 2.3 to cover this case.

(c) Complete the demonstration of the equivalence of the reduced gradient and modified simplex algorithms by considering
(i) the USE tests for selecting the current descent direction, and
(ii) the computation of the minimum in this descent direction.

Exercise 3.2 Show that any linear programming problem in the form LP2 which has a bounded solution has an equivalent formulation as (3.12) and hence as an l_1 problem.

Exercise 3.3 How does the linear program (3.5) transform under row scaling of the data matrix M. In particular, deduce the transformation rule (2.35) for the multiplier.

Exercise 3.4 Derive linear programming formulations for discrete approximation problems based on the norms of Exercise 1.1.

3.4 DESCENT ALGORITHMS FOR THE l_∞ PROBLEM

The l_∞ problem can be posed directly as a linear programming problem in $p+1$ variables.

Lemma 4.1 *If and only if $h, \mathbf{x}$ solve the linear programming problem*

$$\min_{h,\mathbf{x}} h$$

subject to the constraints

$$-h \leqslant \mathbf{m}_i^{\mathrm{T}}\mathbf{x} - f_i \leqslant h, \qquad i = 1, 2, \ldots, n,$$

then $\mathbf{x}$ solves the l_∞ problem.

Remark 4.1 The constraints on (4.1) can be written in the equivalent form

$$\begin{aligned} h + \mathbf{m}_i^{\mathrm{T}}\mathbf{x} &\geq f_i, \\ h - \mathbf{m}_i^{\mathrm{T}}\mathbf{x} &\geq -f_i, \qquad i = 1, 2, \ldots, n. \end{aligned} \tag{4.2}$$

Thus the problem (4.1) is exactly in the form LP1. Note that $h \geq 0$ is a consequence of the inequalities (4.2), and that this is not an interval program in the sense used in Section 2.7. Note also that the characterization theorem (Theorem 1.3) can be interpreted as specifying an extreme point of the feasible region in the linear programming formulation.

In fact little more need be said about the development of descent algorithms as these amount essentially to a straightforward application of the basic descent algorithms for LP1. However, some account can be taken of the special structure of (4.2) to reduce the amount of tableau processing needed, as the basic tableau would require storage of the full constraint matrix

$$\begin{bmatrix} \mathbf{e}^{(n)\mathrm{T}} & \mathbf{e}^{(n)\mathrm{T}} \\ M^{\mathrm{T}} & -M^{\mathrm{T}} \end{bmatrix}$$

We consider the reduced gradient algorithm and note that either member of the pair of constraints (4.2) can be referenced given i and the sign s_i associated with $\mathbf{m}_i$. If the constraints are referenced by an index set μ then B_μ in (5.7) of Chapter 2 can be written

$$B_\mu = \left[\begin{array}{c|c|c} 1 & \mathbf{e}^{(k)\mathrm{T}} & 0 \\ \hline s_{\mu(1)}\mathbf{m}_{\mu(1)} & M_\mu^{\mathrm{T}} S_\mu & E_\mu \end{array}\right] \tag{4.3}$$

where $|\mu| = k+1$, $\kappa_j(M_\mu^{\mathrm{T}}) = \mathbf{m}_{\mu(j+1)}$, and $S_\mu = \operatorname{diag}\{s_{\mu(i)};\ i = 1, 2, \ldots, k+1\}$. Equivalently,

$$\begin{aligned} B_\mu &= \left[\begin{array}{c|c} 1 & 0 \\ \hline s_{\mu(1)}\mathbf{m}_{\mu(1)} & I_p \end{array}\right] \left[\begin{array}{c|c} 1 & [\mathbf{e}^{(k)\mathrm{T}} \mid 0] \\ \hline 0 & [M_\mu^{\mathrm{T}} S_\mu \mid E_\mu] - s_{\mu(1)}\mathbf{m}_{\mu(1)}[\mathbf{e}^{(k)\mathrm{T}} \mid 0] \end{array}\right] \\ &= \left[\begin{array}{c|c} 1 & 0 \\ \hline s_{\mu(1)}\mathbf{m}_{\mu(1)} & I_p \end{array}\right] \left[\begin{array}{c|c} 1 & [\mathbf{e}^{(k)\mathrm{T}} \mid 0] \\ \hline 0 & \hat{B}_\mu \end{array}\right] \end{aligned} \tag{4.4}$$

where

$$\hat{B}_\mu = [M_\mu S_\mu - s_{\mu(1)}\mathbf{m}_{\mu(1)}\mathbf{e}^{(k)\mathrm{T}} \mid E_\mu] \tag{4.5}$$

so that

$$\boldsymbol{\kappa}_j(\hat{B}_\mu) = s_{\mu(j+1)}\mathbf{m}_{\mu(j+1)} - s_{\mu(1)}\mathbf{m}_{\mu(1)}, \qquad j = 1, 2, \ldots, k. \tag{4.6}$$

Now, writing (5.8) of Chapter 2 as

$$B_\mu \begin{bmatrix} u_1 \\ \hat{\mathbf{u}} \end{bmatrix} = \mathbf{e}_1$$

and making use of the factorization (4.4) gives

$$\left[\begin{array}{c|c} 1 & [\mathbf{e}^{(k)\mathrm{T}} \mid 0] \\ \hline 0 & \hat{B}_\mu \end{array}\right] \begin{bmatrix} u_1 \\ \hat{\mathbf{u}} \end{bmatrix} = \begin{bmatrix} 1 \\ -s_{\mu(1)}\mathbf{m}_{\mu(1)} \end{bmatrix}.$$

Thus the multiplier vector $\mathbf{u}$ can be found by solving

$$\hat{B}_\mu \hat{\mathbf{u}} = -s_{\mu(1)}\mathbf{m}_{\mu(1)} \tag{4.7}$$

as

$$u_1 = 1 - [\mathbf{e}^{(k)\mathrm{T}} \mid 0]\hat{\mathbf{u}}. \tag{4.8}$$

Calculation of the descent direction requires evaluating

$$\begin{aligned}
\mathbf{e}_j^{\mathrm{T}} B_\mu^{-1} &= \mathbf{e}_j^{\mathrm{T}} \left[\begin{array}{c|c} 1 & -[\mathbf{e}^{(k)\mathrm{T}} \mid 0]\hat{B}_\mu^{-1} \\ \hline 0 & \hat{B}_\mu^{-1} \end{array}\right] \left[\begin{array}{c|c} 1 & 0 \\ \hline -s_{\mu(1)}\mathbf{m}_{\mu(1)} & I_p \end{array}\right] \\
&= \mathbf{e}_j^{\mathrm{T}} \left[\begin{array}{c|c} 1 - [\mathbf{e}^{(k)\mathrm{T}} \mid 0]\hat{\mathbf{u}} & -[\mathbf{e}^{(k)\mathrm{T}} \mid 0]\hat{B}_\mu^{-1} \\ \hline \hat{\mathbf{u}} & \hat{B}_\mu^{-1} \end{array}\right] \\
&= \begin{cases} [u_1, -[\mathbf{e}^{(k)\mathrm{T}} \mid 0]\hat{B}_\mu^{-1}], & j = 1, \\ [\hat{u}_{j-1}, \mathbf{e}_{j-1}^{\mathrm{T}}\hat{B}_\mu^{-1}], & j = 2, \ldots, p. \end{cases}
\end{aligned} \tag{4.9}$$

These results show that all the information needed to carryout the reduced gradient algorithm can be obtained by working $\hat{B}$. This is defined on the basic tableau structure $[M^{\mathrm{T}} \mid I_p]$. However, note that the matrix factorization

$$R_i U_i = \hat{B}_\mu$$

is defined on a function of the columns of the tableau èxpressed by (4.6).

Remark 4.2 The reduction discussed above will occur naturally in the general context of minimizing polyhedral convex functions in Chapter 4. There the specialized column $s_{\mu(1)}\mathbf{m}_{\mu(1)}$ is referred to as the origin of the group of current extrema. The derivation shows that the algorithm does not depend on the choice of origin. However, updating the tableau is affected by change of origin corresponding to deleting $\mu(1)$ from μ. The choice of new origin could then be made to reduce to a minimum the work involved in updating the factorization of $\hat{B}_\mu$ and transforming the tableau.

Example 4.1 Example 5.1 of Chapter 2 corresponds to the linear program form (4.2) of the l_∞ problem for the model

$$\mathbf{r} = \begin{bmatrix} 1 & 0 \\ 1 & .5 \\ 1 & 1 \end{bmatrix} \mathrm{x} - \begin{bmatrix} 1 \\ 1.649 \\ 2.718 \end{bmatrix}$$

already treated in the l_1 case in Example 2.1. The first iteration with the reduced system gives

$$\hat{B}_\mu \left[\begin{pmatrix} 1 \\ 1 \end{pmatrix} - \begin{pmatrix} -1 \\ 0 \end{pmatrix}, \begin{matrix} 0 \\ 1 \end{matrix}\right], \qquad \hat{\mathbf{u}} = \begin{bmatrix} \frac{1}{2} \\ -\frac{1}{2} \end{bmatrix}, \qquad u_1 = 1 - [1, 0]\hat{\mathbf{u}} = \tfrac{1}{2},$$

and

$$\mathbf{t}^{\mathrm{T}} = [\hat{u}_2 \mid \mathbf{e}_2^{\mathrm{T}}\hat{B}_\mu^{-1}] = [-\tfrac{1}{2} \mid [-\tfrac{1}{2}, 1]].$$

The second iteration gives

$$\hat{B}_\mu=\left[\begin{pmatrix}1\\1\end{pmatrix}-\begin{pmatrix}-1\\0\end{pmatrix},\begin{pmatrix}-1\\-.5\end{pmatrix}-\begin{pmatrix}-1\\0\end{pmatrix}\right],\qquad \hat{\mathbf{u}}=\begin{pmatrix}\frac{1}{2}\\1\end{pmatrix},\qquad u_1=[1-[1,1]\hat{\mathbf{u}}]=-\tfrac{1}{2}$$

Thus the descent vector in the next step is formed by dropping the origin constraint. Equation (4.9) gives

$$\mathbf{t}^T=[u_1,-[1,1]\hat{B}_\mu^{-1}]=[-\tfrac{1}{2}\,|\,[-\tfrac{3}{2},2]].$$

To verify that the next point is optimal we have

$$\hat{B}_\mu=\left[\begin{pmatrix}1\\0\end{pmatrix}-\begin{pmatrix}1\\1\end{pmatrix},\begin{pmatrix}-1\\-.5\end{pmatrix}-\begin{pmatrix}1\\1\end{pmatrix}\right],\qquad \hat{\mathbf{u}}=\begin{pmatrix}.25\\.5\end{pmatrix},\qquad u_1=[1-[1,1]\hat{\mathbf{u}}]=.25.$$

Exercise 4.1 Verify the independence of the algorithm on change of origin by carrying out the calculations for $s_{\mu(l)}\mathbf{m}_{\mu(l)}$ as origin.

Exercise 4.2 Show that multiple pivot sequences are not possible for this problem by computing the multiplier associated with the constraint to become active in the current descent step and showing that this is positive. Is there any possibility of a line search for this problem?

Exercise 4.3 Consider the projected gradient algorithm for this problem and work out the reduction to the compact tableau. Also consider Exercises 4.1 and 4.2 in this case.

Exercise 4.4 Consider the updating of the method of Bartels and Golub (Section 2.5) under change of origin constraint. If the new origin is $s_{\mu(j+1)}\mathbf{m}_{\mu(j+1)}$ then

$$\boldsymbol{\kappa}_s(\bar{U})=\boldsymbol{\kappa}_s(U)-\boldsymbol{\kappa}_j(U),\qquad s<j,$$
$$\boldsymbol{\kappa}_s(\bar{U})=\boldsymbol{\kappa}_{s+1}(U)-\boldsymbol{\kappa}_j(U),\qquad j<s+1\leq k,$$

and the new column corresponding to the new extremal residual encountered in the current descent step is inserted in the last position. Determine the value of j that minimizes the work involved in restoring $\bar{U}$ to upper triangular form. Is the same j appropriate for the orthogonal factorization appropriate to the projected gradient algorithm?

Exercise 4.5 How does this problem transform under row scaling of the data matrix M?

3.5 ASCENT ALGORITHMS FOR THE DISCRETE APPROXIMATION PROBLEM

Descent algorithms for the l_∞ problem have not proved efficient for problems in which the data varies systematically, including the important problem of computing an approximation to a function by a linear combination of given functions by minimizing the maximum residual on a finite set of points in the case that the functions are smoothly varying.

Example 5.1 Consider the problem of constructing an approximation to e^x by a polynomial of degree $p-1$ on $0 \leqslant x \leqslant 1$ by minimizing the maximum error on a grid of n equispaced points. In Table 5.1 the successive index sets μ pointing to the sets of active constraints generated by the projected gradient algorithm are listed for $p=3$, $n=10$, 20, and initial $\mathbf{x}^{\mathrm{T}}=[1, .5, 0]$. In both cases the results are given for the usual two options to build up to $|\mu|=p+1$ quickly (option (i)), and to relax off the constraint corresponding to the multiplier estimate which most violates its bounds (option (ii)). It will be seen that the number of steps appears to be proportional to n with adjacent points tending to provide the new extrema. What is happening is that the rate of decrease of the deleted extrema is not a local minimum, so that there will be residuals at adjacent points, which have values close to the extremum as a consequence of the systematic nature of the problem data, which also decrease more slowly and so become candidates for an extremum in the current step. This behaviour has not been encountered with randomly generated test problems where the number of steps appears to be a small multiple of p and is typical of the problems where descent methods work well (Section 7.4).

In contrast, the simplex method applied to the dual of (4.1) performs well on just these problems. Also, this approach can be used to solve continuous approximation problems by an iterative process in which a suitably defined discrete problem is solved at each stage, and it generalizes substantially – even to problems with more general objective functions than norms. The resulting methods are called *exchange algorithms*. They are known to be rapidly convergent for the l_∞ problem under reasonable assumptions (this point is discussed further in Appendix 2). It is interesting that the basic algorithms predate the simplex algorithm (see the notes at the end of the chapter).

The problem dual to (4.1) is

$$\max_{\mathbf{u} \in U} [\mathbf{f}^{\mathrm{T}} \mid -\mathbf{f}^{\mathrm{T}}]\mathbf{u} : U = \left\{\mathbf{u};\; \begin{bmatrix} \mathbf{e}^{(n)\mathrm{T}} & \mathbf{e}^{(n)\mathrm{T}} \\ M^{\mathrm{T}} & -M^{\mathrm{T}} \end{bmatrix} \mathbf{u} = \mathbf{e}_1, \mathbf{u} \geqslant 0 \right\}. \tag{5.1}$$

The simplex algorithm applies directly to (5.1) so the main interest lies in taking advantage of special structure in the problem. The two-phase method is available to generate a basic feasible solution which will contain artificial variables at zero level if M has rank $K<p$. Thus for simplicity, it is assumed that a basic feasible solution to (5.1) is known and that M has rank p. If the system for referencing constraints introduced in the previous section is used

Table 5.1 Projected gradient algorithm. Successive index sets pointing to the active constraints in the l_∞ approximation of e^x by a polynomial of degree 2 on n equispaced points in [0, 1].

n	10		20	
i \ option	(i)	(ii)	(i)	(ii)
1	10	10	20	20
2	10, 1	10, 1	20, 1	20, 1
3	10, 1, 2	10, 1, 2	20, 1, 2	20, 1, 2
4	10, 1, 2, 3	10, 2, 3	20, 1, 2, 3	20, 2, 3
5	10, 1, 3, 4	10, 3, 4	20, 1, 3, 4	20, 3, 4
6	10, 1, 4, 5	10, 4, 5	20, 1, 4, 5	20, 4, 5
7	10, 1, 5, 6	10, 5, 6	20, 1, 5, 6	20, 5, 6
8	10, 1, 6, 7	10, 6, 7	20, 1, 6, 7	20, 6, 7
9	10, 1, 7, 8	10, 7, 1	20, 1, 7, 8	20, 7, 8
10	10, 1, 8, 3	10, 7, 1, 2	20, 1, 8, 9	20, 8, 9
11		10, 7, 2, 8	20, 1, 9, 10	20, 9, 10
12		10, 2, 8, 3	20, 1, 10, 11	20, 10, 11
13		10, 8, 3, 1	20, 1, 11, 12	20, 11, 12
14			20, 1, 12, 13	20, 12, 13
15			20, 1, 13, 14	20, 13, 1
16			20, 1, 14, 15	20, 13, 1, 14
17			20, 1, 15, 16	20, 1, 14, 2
18			20, 1, 16, 6	20, 14, 2, 3
19				20, 14, 3, 15
20				20, 3, 15, 4
21				20, 15, 4, 5
22				20, 15, 5, 6
23				20, 15, 6, 16
24				20, 6, 16, 1

then the current basis matrix is

$$B=\begin{bmatrix} 1 & \mathbf{e}^{(p)\mathrm{T}} \\ s_{\mu(1)}\mathbf{m}_{\mu(1)} & M_\mu^{\mathrm{T}} S_\mu \end{bmatrix} \tag{5.2}$$

where μ is the index set pointing to the constraint pairs, and s_i gives the sign of the appropriate member of the pair. There is no loss of generality in assuming that $\operatorname{rank}(M_\mu)=p$.

Lemma 5.1 *Repetition of rows of M in M_μ^{T} in an optimal basis matrix is possible only if there exists $\mathbf{x}$ such that $\mathbf{r}=M\mathbf{x}-\mathbf{f}=0$.*

Proof There can be at most one repetition as otherwise B is singular (the sum of columns corresponding to each repeated pair is $2\mathbf{e}_1$). The result now follows from duality because if a variable is in the dual basis then equality holds in the corresponding primal constraint.

Remark 5.1 It is a slight simplification to ignore the case $\mathbf{r}=0$. The starting

The starting point for the exchange algorithm is the observation that if $n=p+1$ then the solution of the l_∞ problem can be found directly. It can be padded out with zero to give a basic feasible solution to (5.1).

Lemma 5.2 *Let $n=p+1$,*

$$M_\mu^{-\mathrm{T}}\mathbf{m}_{\mu(1)}=\mathbf{z}, \tag{5.3}$$

and

$$\bar{S}=\operatorname{diag}\{-\operatorname{sgn}(z_i),\ i=1,2,\ldots,p\}. \tag{5.4}$$

Then

$$\mathbf{u}_\sigma=\frac{1}{1+\sum_{j=1}^{p}|z_j|}\begin{bmatrix}1\\ -\bar{S}\mathbf{z}\end{bmatrix} \tag{5.5}$$

is an optimal basic feasible solution to (5.1). *Also,*

$$S_\mu=\theta\bar{S} \tag{5.6}$$

where $\theta=\pm1$ is chosen to give a positive value to

$$h=[f_{\mu(1)},\mathbf{f}_\mu^{\mathrm{T}}]S\mathbf{u}_\sigma \tag{5.7}$$

and

$$S=\left[\begin{array}{c|c}\theta & \\ \hline & S_\mu\end{array}\right]. \tag{5.8}$$

Proof The rank assumption on M ensures that there is essentially a unique relation of linear dependence among the $\mathbf{m}_{\mu(i)}$, $i=1,2,\ldots,p+1$, so that finding a nonnegative $\mathbf{u}$ to satisfy the constraints is equivalent to finding the right combination of signs for the $\mathbf{m}_{\mu(i)}$. This leaves just one degree of freedom to be specified by (5.7). The corresponding solution to the dual is given by

$$\begin{bmatrix} 1 & \theta \mathbf{m}_{\mu(1)}^{T} \\ \mathbf{e}^{(p)} & S_\mu M_\mu \end{bmatrix} \begin{bmatrix} h \\ \mathbf{x} \end{bmatrix} = \begin{bmatrix} \theta f_{\mu(1)} \\ S_\mu \mathbf{f}_\mu \end{bmatrix} \tag{5.9}$$

It is feasible (each inequality in (4.2) is satisfied at one or other bound when $n=p+1$), and (5.7) expresses the equality of primal and dual objective functions.

Example 5.2 To illustrate Lemma 5.2 consider the problem with $n=p+1=3$;

$$M=\begin{bmatrix} 1 & 0 \\ 1 & .5 \\ 1 & 1 \end{bmatrix}, \qquad \mathbf{f}=\begin{bmatrix} 1 \\ 1.649 \\ 2.718 \end{bmatrix}.$$

Then, with $\mu=\{3, 1, 2\}$,

$$\mathbf{z}=\begin{bmatrix} 1 & 1 \\ 0 & .5 \end{bmatrix}^{-1}\begin{bmatrix} 1 \\ 1 \end{bmatrix}=\begin{bmatrix} -1 \\ 2 \end{bmatrix}, \qquad \bar{S}=\begin{bmatrix} 1 & 0 \\ 0 & -1 \end{bmatrix}, \qquad \mathbf{u}_\sigma=\frac{1}{4}\begin{bmatrix} 1 \\ 1 \\ 2 \end{bmatrix}.$$

To determine θ, h we have

$$\mathbf{f}^{T}\left[\begin{array}{c|c} 1 & \\ \hline & \bar{S} \end{array}\right]\mathbf{u}_\sigma=\tfrac{1}{4}[2.718 \quad 1 \quad -1.649]\begin{bmatrix} 1 \\ 1 \\ 2 \end{bmatrix}=.105>0$$

giving $\theta=1$, $h=.105$. Equation (5.9) becomes

$$\begin{bmatrix} 1 & 1 & 1 \\ 1 & 1 & 0 \\ 1 & -1 & -.5 \end{bmatrix}\begin{bmatrix} h \\ \mathbf{x} \end{bmatrix}=\begin{bmatrix} 2.718 \\ 1 \\ -1.649 \end{bmatrix}$$

so that

$$\mathbf{x}=\begin{bmatrix} .895 \\ 1.718 \end{bmatrix}, \qquad \mathbf{r}=\begin{bmatrix} -.105 \\ .105 \\ -.105 \end{bmatrix}.$$

Remark 5.2 The proof of Lemma 5.2 shows that feasibility and optimality are the same for basic solutions when $n=p+1$. Thus the simplex method applied to (5.1) generates the optimal solutions to a sequence of subproblems each with $n=p+1$ in such a way that the objective function increases at each step.

It turns out that the rule for modifying the current subproblem is also very easy to interpret.

Lemma 5.3 *The USE test for the variable to enter the basis is equivalent to finding the residual of largest magnitude when* $\begin{bmatrix} h \\ \mathbf{x} \end{bmatrix}$ *is given by* (5.9).

Proof Evaluating equation (4.11) of Chapter 2 using (5.9) gives

$$\begin{aligned}\xi_s &= \theta_t f_t - [f_{\mu(1)} \mid \mathbf{f}_\mu^{\mathrm{T}}]SB^{-1}\begin{bmatrix} 1 \\ \theta_t \mathbf{m}_t \end{bmatrix} \\ &= \theta_t f_t - [h \mid \mathbf{x}^{\mathrm{T}}]\begin{bmatrix} 1 \\ \theta_t \mathbf{m}_t \end{bmatrix} \\ &= -(h + \theta_t r_t) \\ &= |r_t| - h\end{aligned}$$

if $\theta_t = -\operatorname{sgn}(r_t)$. As only values of $\xi_s > 0$ as s runs over the columns of N are relevant in determining a direction of ascent, the result follows.

Lemma 5.3 provides a compact characterization of the column to become basic and also of optimality (if all residuals are $\leqslant h$ in magnitude then the current point is optimal). To determine the column to leave the basis, it is possible to use the special form of the problem by noting that what is required is a relation of linear dependence among the columns left after deleting a single column from the extended array

$$[s_{\mu(1)}\mathbf{m}_{\mu(1)} \mid M_\mu^{\mathrm{T}} S_\mu \mid \theta_t \mathbf{m}_t] \tag{5.10}$$

One relation of linear dependence is determined by $\mathbf{u}_\sigma$. To construct a second, let

$$\mathbf{z}_t = \theta_t M_\mu^{-\mathrm{T}} \mathbf{m}_t, \tag{5.11}$$

and set

$$\mathbf{q}_1 = \begin{bmatrix} 1 \\ -\bar{S}\mathbf{z} \\ 0 \end{bmatrix}, \qquad \mathbf{q}_2 = \begin{bmatrix} 0 \\ -S_\mu \mathbf{z}_t \\ 1 \end{bmatrix},$$

and define

$$\mathbf{y}(\lambda) = \mathbf{q}_1 + \lambda \mathbf{q}_2. \tag{5.12}$$

By construction, $\mathbf{y}(\lambda)$ provides a relation of linear dependence on the columns of (5.10) for all λ. Also, $\mathbf{y}(\lambda) \geqslant 0$ for $\lambda > 0$ small enough, and the particular value of λ of interest is the smallest such that a new component of $\mathbf{y}$ (say y_l) becomes zero. The argument is familiar, and

$$l = \arg\min_j \{\lambda_j = -(\mathbf{q}_1)_j/(\mathbf{q}_2)_j, \forall\, j \ni \lambda_j > 0, j = 1, 2, \ldots, p+1\} \tag{5.13}$$

The new basic feasible solution is

$$u_i' = \begin{cases} \phi_l y_i(\lambda_l), & i \neq l, \quad i = 1, 2, \ldots, p+1 \\ \phi_l, & i = l \end{cases} \tag{5.14}$$

where $\phi_l = 1/\sum_{i=1}^{p+2} y_i(\lambda_l)$ is a scale factor chosen to ensure that the inhomogeneous constraint on (5.1) is satisfied. Note that $\lambda_l = +\infty$ is allowed and corresponds to deleting $\mathbf{m}_{\mu(1)}$ from (5.10). The other updating necessary corresponds to inserting $\mathbf{m}_t$ instead of $\mathbf{m}_{\mu(l)}$ so that

$$\mu' = (\mu \backslash \{\mu(l)\}) \cup \{t\}, \qquad s_{\mu(l)} = \theta_t.$$

The evaluation of the objective function for the new basis is an important calculation. Substituting in (5.7) and using $y_l = 0$ gives

$$\begin{aligned}
h'/\phi_l &= \sum_{i=1}^{p+1} y_i(\lambda_l) s_{\mu(i)} f_{\mu(i)} + \lambda_l \theta_t f_t \\
&= \left(1 + \sum_{i=1}^{p} |z_i|\right) h + \lambda_l [0 \mid -\mathbf{z}_t^{\mathrm{T}} S_\mu] \begin{bmatrix} s_{\mu(1)} f_{\mu(1)} \\ S_\mu \mathbf{f}_\mu \end{bmatrix} + \lambda_l \theta_t f_t \\
&= \left(1 + \sum_{i=1}^{p} |z_i|\right) h + \lambda_l [-\mathbf{z}_t^{\mathrm{T}} S_\mu \mathbf{e}^{(p)} \mid -\theta_t \mathbf{m}_t^{\mathrm{T}}] \begin{bmatrix} h \\ \mathbf{x} \end{bmatrix} + \lambda_l \theta_t f_t \\
&= \left(1 + \lambda_l + \sum_{i=1}^{p} \{|z_i| - \lambda_l s_{\mu(i)} (\mathbf{z}_t)_i\}\right) h + \lambda_l \{-h - \theta_t (\mathbf{m}_t^{\mathrm{T}} \mathbf{x} - f_t)\} \\
&= h/\phi_l + \lambda_l (|r_t| - h)
\end{aligned}$$

where (5.9) is used in the passage from the second to the third of these equations. Thus

$$h' = h + \lambda_l \phi_l (|r_t| - h) \tag{5.15}$$

showing that the objective function is strictly increased, provided:

(i) the current subproblem solution is not optimal for the full problem so that $\max_s |r_s| - h > 0$, $s \in \mu^c$; and

(ii) the current basic feasible solution is nondegenerate so that $u_l > 0$ ($\mathbf{y} > 0$, $\lambda > 0$ small enough).

This step of the simplex algorithm does not solve the l_∞ problem defined on the extended array (5.10) with $n = p+2$ in general. It is in the solution of this problem that the possibility of something analogous to the multiple pivot sequences of interval programming occurs ((4.2) is not covered by our previous discussion of interval constraints because h occurs in the interval bounds). To solve the extended problem it is necessary to sort through the

$$\binom{p+2}{p+1} = p+2$$

possible subproblems defined by the subsets of $(p+1)$ columns of (5.10). To generate the optimal solution for each subproblem requires the unique positive relation of linear dependence (to within a scale factor) to be determined systematically. This can be done by:

(a) choosing $\lambda = \lambda_j$ so that $\mathbf{y}(\lambda)$ has a zero component corresponding to the omitted column ($\mathbf{y}(\lambda)$ will also have negative components corresponding to values $0<\lambda_s<\lambda_j$ in the competition (5.13));
(b) changing the signs in S corresponding to the negative components of $\mathbf{y}(\lambda_j)$ to obtain S^j – this corresponds to moving to the opposite bound in each pair of the primal inequalities (4.2) involved; and
(c) checking that the resulting objective function is positive and replacing S^j by $-S^j$ if it is not.

To specify the quantities involved let

$$E^j = \operatorname{diag}\{-1,\ y_i(\lambda_j)<0;\ 1,\ y_i(\lambda_j)\geq 0\}. \tag{5.16}$$

Then the positive relation of linear dependence is

$$\mathbf{y}^j = E^j\mathbf{y}(\lambda_j), \tag{5.17}$$

and the corresponding matrix of signs identifying the active constraints is

$$S^j = \operatorname{diag}\{E^j_iS_i,\ i\neq j,\ i=1,2,\ldots,p+1;\ \theta_t,\ i=j\} \tag{5.18}$$

The new objective function values is (arguing as in the derivation of (5.15))

$$\begin{aligned}
h^j/\phi_j &= \left|\sum_{i=1}^{p+1} f_{\mu(i)}S^j_iE^j_iy_i(\lambda_j)+\lambda_j\theta_tf_t\right| \\
&= \left|\sum_{i=1}^{p+1} f_{\mu(i)}S_iy_i(\lambda_j)+\lambda_j\theta_tf_t\right| \\
&= \left|([1\,|-\mathbf{z}^{\mathrm{T}}\bar{S}]+\lambda_j[0\,|-\mathbf{z}_t^{\mathrm{T}}S_\mu])\begin{bmatrix} 1 & s_{\mu(i)}\mathbf{m}^{\mathrm{T}}_{\mu(i)} \\ \mathbf{e}^{(p)} & S_\mu M_\mu \end{bmatrix}\begin{bmatrix} h \\ \mathbf{x}\end{bmatrix}+\lambda_j\theta_tf_t\right| \\
&= \left|\left(\sum_{i=1}^{p+1} y_i(\lambda_j)\right)h+\lambda_j\theta_t(-\mathbf{m}_t^{\mathrm{T}}\mathbf{x}+f_t)\right| \\
&= \left|\left(\sum_{i=1}^{p+2} y_i(\lambda_j)\right)h+\lambda_j(|r_t|-h)\right|
\end{aligned}$$

so that, as $\phi_j = \sum_{i=1}^{p+2} E^j_iy_i(\lambda_j) = \sum_{i=1}^{p+2} |y_i(\lambda_j)|$,

$$h(\lambda_j) = h^j = \frac{\left|\left(\sum_{i=1}^{p+2} y_i(\lambda_j)\right)h(0)+\lambda_j(|r_s|-h(0))\right|}{\sum_{i=1}^{p+2} |y_i(\lambda_j)|} \tag{5.19}$$

To solve the l_∞ problem on the extended array (5.10) requires finding j to maximize h^j.

The numerator in (5.19) has the form $\pi(\lambda)(a+b\lambda)$ where

$$a=\left(1+\sum_{i=1}^{p}|z_i|\right)h(0), \qquad b=h(0)\sum_{i=2}^{p+1}(\mathbf{q}_2)_i+|r_s|, \quad \text{and} \quad \pi(\lambda)=\operatorname{sgn}(a+\lambda b).$$

The denominator is a piecewise linear function of λ with breaks at the λ_j. To specify it let τ be an index set pointing to the λ_j sorted in order on the extended interval $(0+, \infty, -\infty, 0-)$. Then on any subinterval $[\lambda_{\tau(j)}, \lambda_{\tau(j+1)}]$ the denominator has the form $c+\lambda d$, where

$$c=-\sum_{i=1}^{j}(\mathbf{q}_1)_{\tau(i)}+\sum_{i=j+1}^{p+2}(\mathbf{q}_1)_{\tau(i)}, \tag{5.20}$$

$$d=-\sum_{i=1}^{j}(\mathbf{q}_2)_{\tau(i)}+\sum_{i=j+1}^{p+2}(\mathbf{q}_2)_{\tau(i)}. \tag{5.21}$$

If $\bar{\lambda}=-\dfrac{a}{b}\notin[\lambda_{\tau(j)}, \lambda_{\tau(j+1)}]$ then we can differentiate

$$h(\lambda)=\pi(\lambda)\frac{a+\lambda b}{c+\lambda d} \tag{5.22}$$

to obtain

$$\frac{\mathrm{d}h}{\mathrm{d}\lambda}=\pi(\lambda)\frac{bc-ad}{(c+\lambda d)^2} \tag{5.23}$$

It follows that

$$\operatorname{sgn}\left(\frac{\mathrm{d}h}{\mathrm{d}\lambda}\right)=\pi(\lambda)\operatorname{sgn}(bc-ad) \tag{5.24}$$

is constant on any interval $[\lambda_{\tau(j)}, \lambda_{\tau(j+1)}]$ not containing $\bar{\lambda}$. Now, from (5.20), (5.21), the change in c, d as λ increases past λ_j, $-\infty<\lambda<\infty$, is

$$\Delta c=-\operatorname{sgn}(\lambda_j)2(\mathbf{q}_1)_j, \qquad \Delta d=-\operatorname{sgn}(\lambda_j)2(\mathbf{q}_2)_j$$

as when $\lambda<0$ the number of negative components of $\mathbf{y}$ decreases as λ increases, until there are none when $\lambda=0$. Further increasing λ reverses this process and the number of negative components increases as λ increases to $+\infty$. The change in $bc-ad$ is

$$\begin{aligned}\Delta(bc-ad)&=b\Delta c-a\Delta d\\&=\operatorname{sgn}(\lambda_j)2(\mathbf{q}_2)_j(a-((\mathbf{q}_1)_j/(\mathbf{q}_2)_j)b)\end{aligned}$$

so that

$$\pi(\lambda_j)\,\Delta(bc-ad)=-2\,|(\mathbf{q}_2)_j|\,|a+\lambda_j b| \tag{5.25}$$

as $\operatorname{sgn}(\lambda_j)=-\operatorname{sgn}((\mathbf{q}_2)_j)$ as $(\mathbf{q}_1)_j\geqslant 0$. Thus $\pi(\lambda)(bc-ad)$ is a decreasing step function of $\tau(j)$ as j increases, except in the interval containing $\bar{\lambda}$, where it jumps back to a positive value as a consequence of

$$\frac{\mathrm{d}h}{\mathrm{d}\lambda}(\bar{\lambda}+0)=\left|\frac{\mathrm{d}h}{\mathrm{d}\lambda}(\bar{\lambda}-0)\right|.$$

When $\lambda=0$,

$$\pi(\lambda)(bc-ad)=\left(1+\sum_{i=1}^{p}|z_i|\right)(|r_t|-h(0))>0$$

so that the maximum of h must lie in the interval $(0,\bar{\lambda})$ which is searched in the order $(0,+\infty,-\infty,\bar{\lambda})$ if $\bar{\lambda}<0$ for a change in sign of $\pi(\lambda)(bc-ad)$. In summary, we have the following result.

Lemma 5.4 *The optimum subproblem from the class determined by the subsets of $p+1$ columns of the array (5.10) is found by maximizing*

$$h(\lambda)=\frac{\left|h(0)\sum_{i=1}^{p+2}y_i(\lambda)+\lambda(|r_s|-h(0))\right|}{\sum_{i=1}^{p+2}|y_i(\lambda)|}$$

with respect to λ. The optimum occurs for $\lambda=\lambda_{\tau(j)}=-(\mathbf{q}_1)_{\tau(j)}/(\mathbf{q}_2)_{\tau(j)}$ where

$$\frac{dh}{d\lambda}(\lambda_{\tau(j)}-0)\geqslant 0 \quad \textit{and} \quad \frac{dh}{d\lambda}(\lambda_{\tau(j)}+0)\leqslant 0.$$

It is characterized by

$$\left(1+\sum_{i=1}^{p}|z_i|\right)(|r_t|-h(0))-2\sum_{i=1}^{j-1}|(\mathbf{q}_2)_{\tau(i)}|\,|a+\lambda_{\tau(i)}b|\geqslant 0$$

and

$$\left(1+\sum_{i=1}^{p}|z_i|\right)(|r_t|-h(0))-2\sum_{i=1}^{j}|(\mathbf{q}_2)_{\tau(i)}|\,|a+\lambda_{\tau(i)}b|\leqslant 0$$

where $0<\lambda_{\tau(1)}<\lambda_{\tau(2)}\ldots<\lambda_{\tau(j)}$, the order being defined on the extended interval $(0,+\infty,-\infty,\bar{\lambda})$ if necessary, and the strict inequalities being a consequence of nondegeneracy assumed for basic feasible solutions.

Remark 5.3 The region in which $\pi(\lambda)=-1$ corresponds to the case in which all constraints must be moved to their opposite bounds in order to obtain a positive value of h.

Example 5.3 To illustrate Lemma 5.4, consider the problem with

$$\mathbf{r}=\begin{bmatrix}1\\1\\1\end{bmatrix}x-\begin{bmatrix}1\\1.649\\2.718\end{bmatrix}, \qquad n=3, \qquad p=1.$$

To construct a basic feasible solution let $\mu=\{1,2\}$. Then (5.9) becomes

$$\begin{bmatrix}1 & \theta_1\\1 & \theta_2\end{bmatrix}\begin{bmatrix}h\\x\end{bmatrix}=\begin{bmatrix}\theta_1\\1.649\theta_2\end{bmatrix}$$

and $z=1$, so that $\theta_2=-\theta_1$. It follows that

$$2h=\theta_1+1.649\theta_2$$

giving $h=.3245$, $\theta_1=-1$, $\theta_2=1$, $x=1.3245$, and $r_3=-1.3935$. Thus the extended array is

$$[-1, 1, 1],$$

$$\mathbf{q}_1=\begin{bmatrix}1\\1\\0\end{bmatrix}, \quad \mathbf{q}_2=\begin{bmatrix}0\\-1\\1\end{bmatrix},$$

and

$$\Lambda=\{-\infty, 1\}.$$

A direct calculation now gives

$$h(\lambda)=\frac{|.649+1.069\lambda|}{\sum_{i=1}^{3}|y_i|}$$

where

$$\sum_{i=1}^{3}|y_i|=\begin{cases}2, & 0\leqslant\lambda\leqslant1\\ 2\lambda, & 1\leqslant\lambda<\infty\\ 2+2\lambda, & -\infty<\lambda\leqslant0.\end{cases}$$

The maximum value of h is .859 corresponding to $\lambda=1$, so the second column is omitted in the optimal basis. Solving the resulting equations gives $x=1.859$, $r_2=.21$.

Two extensions of the ascent algorithm are important: to l_∞ problems subject to subsidiary constraints, and to polyhedral norm approximation problems (more generally, polyhedral convex objective functions).

The constrained l_∞ problem

Consider the problem

$$\min h$$

subject to

$$-h\leqslant\mathbf{m}_i^{\mathrm{T}}\mathbf{x}-f_i\leqslant h, \qquad i=1,2,\ldots,n,$$

and

$$A\mathbf{x}\geqslant b, \tag{5.26}$$

where $A:R^p\rightarrow R^m$. The dual problem is

$$\max\,[\mathbf{f}^{\mathrm{T}}|\,-\mathbf{f}^{\mathrm{T}}\,|\mathbf{b}^{\mathrm{T}}]\mathbf{u}$$

subject to

$$\begin{bmatrix} \mathbf{e}^{(n)\mathrm{T}} & \mathbf{e}^{(n)\mathrm{T}} & 0 \\ M^{\mathrm{T}} & -M^{\mathrm{T}} & A^{\mathrm{T}} \end{bmatrix} \mathbf{u} = \mathbf{e}_1, \qquad \mathbf{u} \geqslant 0 \tag{5.27}$$

The main features of an algorithm can be developed very much as described above. We assume that

$$\operatorname{rank}\left(\begin{bmatrix} M \\ A \end{bmatrix}\right) = p \tag{5.28}$$

so that a basis matrix for (5.27) takes the form

$$B = \begin{bmatrix} \mathbf{e}^{(k_1)\mathrm{T}} & 0 \\ M_{\mu_1} S & A_{\mu_2} \end{bmatrix} \tag{5.29}$$

where

$$\mu = \mu_1 \cup \mu_2, \qquad |\mu_1| = k_1, \qquad |\mu_2| = k_2, \qquad k_1 + k_2 = p + 1 \tag{5.30}$$

is the index set defining the columns of M^{T}, A^{T} in the basis, and S is the sign matrix. Selection of the variable to enter the basis parallels the calculation given in Lemma 5.3. We have the following result.

Lemma 5.5 *Define* $\begin{bmatrix} h \\ \mathbf{x} \end{bmatrix}$ *by means of the equations*

$$B^{\mathrm{T}} \begin{bmatrix} h \\ \mathbf{x} \end{bmatrix} = \begin{bmatrix} S\mathbf{f}_{\mu_1} \\ \mathbf{b}_{\mu_2} \end{bmatrix}. \tag{5.31}$$

Then the USE test for the variable to enter the basis is equivalent to selecting the most violated constraint in (5.26).

Proof An application of equation (4.11) of Chapter 2 gives, allowing for the two different classes of constraints indicated by the two possibilities within the bracketed terms,

$$\begin{aligned} \xi_s &= \begin{Bmatrix} \theta_t f_t \\ b_t \end{Bmatrix} - [\mathbf{f}_{\mu_1}^{\mathrm{T}} S \mid \mathbf{b}_{\mu_2}^{\mathrm{T}}] B^{-1} \left\{ \begin{bmatrix} 1 \\ \theta_t \mathbf{m}_t \end{bmatrix}, \begin{bmatrix} 0 \\ \mathbf{a}_t \end{bmatrix} \right\} \\ &= \begin{Bmatrix} \theta_t f_t \\ b_t \end{Bmatrix} - [h \mid \mathbf{x}^{\mathrm{T}}] \left\{ \begin{bmatrix} 1 \\ \theta_t \mathbf{m}_t \end{bmatrix}, \begin{bmatrix} 0 \\ \mathbf{a}_t \end{bmatrix} \right\} \\ &= \begin{Bmatrix} |r_t| - h \\ b_t - \mathbf{a}_t^{\mathrm{T}} \mathbf{x} \end{Bmatrix} \end{aligned} \tag{5.32}$$

where $\theta_t = -\operatorname{sgn}(r_t)$.

Selection of the variable to leave the basis is again made by choosing λ to make a component of $\mathbf{y}(\lambda)$ vanish, where $\mathbf{y}(\lambda)$ expresses a relation of linear dependence on the extended array

$$[M_{\mu_1}^{\mathrm{T}} S \mid A_{\mu_2}^{\mathrm{T}} \mid \{\theta_t \mathbf{m}_t, \mathbf{a}_t\}] \tag{5.33}$$

and $\mathbf{y}(0) = \begin{bmatrix} \mathbf{u}_\sigma \\ 0 \end{bmatrix}$. However, it may not be possible to choose λ to maximize $h(\lambda)$ over all possible subproblems defined by (5.33) because the search must terminate, in general, if the zero component is associated with a column of A_μ^T. Only if the same column with opposite sign is available to absorb the minus introduced in the component of $\mathbf{y}$ to restore feasibility can the process be continued. Otherwise the search is essentially as described before.

The polyhedral norm problem

In this case it follows from (1.21) that the primal problem is

$$\min h$$

subject to

$$P[M\mathbf{x} - \mathbf{f}] \leq h\mathbf{e}^{(N)}. \tag{5.34}$$

The assumption that $\|\mathbf{x}\|_P = \|-\mathbf{x}\|_P$ for all $\mathbf{x} \in R^n$ implies that P can be partitioned into

$$P = \begin{bmatrix} \bar{P} \\ -\bar{P} \end{bmatrix} \tag{5.35}$$

It follows that (5.34) can be written in the equivalent form

$$\min h$$

subject to

$$\begin{bmatrix} \bar{\mathbf{e}} & \bar{P}M \\ \bar{\mathbf{e}} & -\bar{P}M \end{bmatrix} \begin{bmatrix} h \\ \mathbf{x} \end{bmatrix} \geq \begin{bmatrix} \bar{P}\mathbf{f} \\ -\bar{P}\mathbf{f} \end{bmatrix} \tag{5.36}$$

where $\bar{\mathbf{e}}^{(N)T} = [\bar{\mathbf{e}}^T \mid \bar{\mathbf{e}}^T]$. This problem is formally equivalent to (4.2). It follows that the algorithms developed for the l_∞ problem can be applied to this more general problem. However, there are potential difficulties. Some of these are structural, and some insight can be gained on this point by considering the discussion of singular problems in Appendix 2.

The other class of difficulties relate to implementation and come about because $\bar{P}$ can have an enormous number of rows. In particular, it may not make any sense at all to consider an implementation that requires explicit storage of the constraint matrix (as, for example, in the simplex tableau). However, the actual basis matrix is reasonably compact, so that the main problem is computing ξ_s in the USE test (for example). One possibility, given that

$$\mathbf{r} = M\mathbf{x} - \mathbf{f}$$

is known, is to generate $\boldsymbol{\rho}_s(P)$ satisfying

$$\boldsymbol{\rho}_s(P)\mathbf{r} = \|\mathbf{r}\|_P.$$

This automatically maximizes $\boldsymbol{\rho}_s(P)\mathbf{r}$ for s running over the rows of P and so defines the column to be added to the basis by the analogue of Lemma 5.3. Thus column generation can be used to avoid storing P. For example, if

$$\|\mathbf{r}\|_p = \sum_{i=1}^{n} |r_i|$$

then

$$\boldsymbol{\rho}_s(P) = [\theta_1, \theta_2, \ldots, \theta_n]$$

where $\theta_i = \operatorname{sgn}(r_i)$ and is available when $\mathbf{r}$ is known provided that no $r_i = 0$. As zero residuals are not preserved in general in this application of the exchange algorithm the possible ambiguity from this cause has not proved a problem. *However, the algorithm proves inefficient in just those l_1 problems which correspond to the problems described in Example* 5.1 *in which the descent algorithm does not do well.*

Exercise 5.1 (i) Display explicitly the primal linear programs corresponding to the polyhedral norm approximation problems which are obtained by fitting to e^t by $x_1 + x_2 t$ on $t_i = .25(i-1)$, $i = 1, 2, \ldots, 5$, in the l_1 and l_∞ norms subject to the constraint that the approximations never lie below the curve. (ii) Use the exchange algorithm to solve both problems.

3.6 BARELY OVERDETERMINED PROBLEMS

If $(n-p)/p$ is small then computational saving may be made by transforming the discrete approximation problem. The idea is very simple. Let Z be the matrix defined by

(i) $Z: R^k \to R^n$, rank $(Z) = k$,
(ii) $Z^T M = 0$
(iii) span $(\boldsymbol{\kappa}_i(Z), i = 1, 2, \ldots, k) =$ span $(\boldsymbol{\kappa}_i(M), i = 1, 2, \ldots, p)^\perp$.

Then

$$\operatorname{rank}(Z) = k = n - K$$

where $K = \operatorname{rank}(M)$, and

$$\mathbf{r} = M\mathbf{x} - \mathbf{f} \Leftrightarrow Z^T\mathbf{r} = -Z^T\mathbf{f} \tag{6.1}$$

The forward implication follows from (ii) above. The reverse implication follows from

$$Z^T(\mathbf{r}+\mathbf{f}) = 0 \Rightarrow \mathbf{r}+\mathbf{f} \in \text{column space of } M$$
$$\Rightarrow \mathbf{r}+\mathbf{f} = M\mathbf{x}$$

It follows that the problem (1.3)

$$\min_{\mathbf{x}} \|\mathbf{r}\|_R : \mathbf{r} = M\mathbf{x} - \mathbf{f}$$

is exactly equivalent to the problem

$$\min_{\mathbf{r}} \|\mathbf{r}\|_R : Z^T\mathbf{r} = -Z^T\mathbf{f}, \tag{6.2}$$

and if $(n-K)/K$ is small then (6.2) has a much smaller constraint set than (1.3). It seems often to be the case that such problems are solved in relatively fewer iterations so that provided the overhead in computing Z is not excessive there is the possibility of savings. To estimate the cost consider the reduction of M to upper triangular form using an elimination scheme with partial pivoting. This gives

$$P\begin{bmatrix} L_1 & 0 \\ L_2 & I_k \end{bmatrix}\begin{bmatrix} U \\ 0 \end{bmatrix} = M \tag{6.3}$$

where P is the permutation matrix of row interchanges, L_1 is $K \times K$ unit lower triangular, and rank $(U) = K$. Note that U is upper triangular if $K = p$, and truncated upper triangular otherwise. An immediate consequence is

$$[-L_2 L_1^{-1} \quad I_k] P^{-1} M = 0 \tag{6.4}$$

so this equation defines an appropriate Z^T. Counting multiplications gives contributions from

(a) the factorization: $npk-(n+p)\dfrac{K^2}{2}+\dfrac{K^3}{3}$

and

(b) formation of $L_2 L_1^{-1}: kK^2/2$

with a total in the case $K = p$ of $np^2 - 2p^3/3$. To estimate the overall cost it is necessary to add to this overhead cost a figure representing the number of iterations by the cost of updating a tableau based on Z^T (say $\gamma_1 nk$). This is to be compared with a figure representing number of iterations times the updating cost of a tableau based on M^T (say $\gamma_2 nK$). It seems that the transformed problem can be competitive even when $(n-p)/p$ is as large as 1.

We now consider algorithmic aspects of the l_1 and l_∞ problems. It is possible to draw substantially on material already developed, and to avoid repetition only the more significant differences will be considered.

In the l_1 problem the argument presented in Section 3.3 leads to the linear programming problem in LP2 form:

$$\max -[\mathbf{e}^{(n)T} \mid \mathbf{e}^{(n)T}]\begin{bmatrix} \mathbf{y} \\ \mathbf{z} \end{bmatrix}$$

subject to

$$[Z^T \mid -Z^T]\begin{bmatrix} \mathbf{y} \\ \mathbf{z} \end{bmatrix} = Z^T\mathbf{f}, \qquad \mathbf{y} \geqslant 0, \qquad \mathbf{z} \geqslant 0 \tag{6.5}$$

This corresponds under duality to the interval program

$$\min -\mathbf{f}^T Z\mathbf{u}$$

subject to

$$-\mathbf{e}^{(n)} \leqslant Z\mathbf{u} \leqslant \mathbf{e}^{(n)}. \tag{6.6}$$

The application of supopt to this problem, and of the modified simplex method to (6.5) have been discussed in Section 2.6.

In the l_∞ norm, (6.2) becomes the linear program

$$\min h$$

subject to

$$-h\mathbf{e}^{(n)} \leqslant \mathbf{r} \leqslant h\mathbf{e}^{(n)}$$

and

$$Z^{\mathrm{T}}\mathbf{r} = -Z^{\mathrm{T}}\mathbf{f} \tag{6.7}$$

Note that this problem has $n+1$ variables and k equality constraints. Thus at a (relative) extreme point, $n-k+1$ of the inequalities must hold with equality. The dual problem is

$$\max\,[-\mathbf{f}^{\mathrm{T}}Z \mid \mathbf{f}^{\mathrm{T}}Z]\mathbf{u}_2$$

subject to

$$\begin{bmatrix} \mathbf{e}^{(n)\mathrm{T}} & \mathbf{e}^{(n)\mathrm{T}} & 0 & 0 \\ I_n & -I_n & Z & -Z \end{bmatrix}\begin{bmatrix} \mathbf{u}_1 \\ \mathbf{u}_2 \end{bmatrix} = \mathbf{e}_1, \qquad \mathbf{u}_1 \geqslant 0, \qquad \mathbf{u}_2 \geqslant 0. \tag{6.8}$$

This can be solved using an exchange (ascent) algorithm which is developed by a similar argument to that used in the previous section. Here the key point is that the solution of the restricted l_∞ problem with $n-k+1$ pairs of inequality constraints can be specified exactly. The equations determining a basic feasible solution can be written in partitioned form

$$B\mathbf{u}^{\sigma} = \begin{bmatrix} \mathbf{e}^{(n-k)T} & 1 & 0 \\ S^{-} & 0 & Z_2T \\ 0 & \theta_s\mathbf{e}_s & Z_1T \end{bmatrix}\begin{bmatrix} \mathbf{u}_1^{\sigma} \\ \mathbf{u}_2^{\sigma} \end{bmatrix} = \begin{bmatrix} 1 \\ 0 \\ 0 \end{bmatrix} \tag{6.9}$$

where

$$S = \begin{bmatrix} S^{-} & \\ & \theta_s \end{bmatrix}$$

and T are diagonal matrices with elements ± 1 corresponding to selection of the columns from (6.8) ($S_i = -1$ if the inequality $r_i \leqslant h$ is basic); Z_1 is a $k \times k$ submatrix of Z and it gives $(k-1)$ equations for the components of $\mathbf{u}_2^{\sigma}$ uncoupled from the rest of the system corresponding to the components of $\mathbf{r}$ not at their bounds in the primal problem, and the remaining row couples the calculation of $\mathbf{u}_1^{\sigma}$ and $\mathbf{u}_2^{\sigma}$ but not in a particularly nasty way. The solution to (6.9) can be found by setting

$$(\mathbf{u}_1^{\sigma})_{n-k+1} = \alpha \tag{6.10}$$

and solving the last k equations to give

$$T\mathbf{u}_2^{\sigma} = -\alpha\theta_s Z_1^{-1}\mathbf{e}_s. \tag{6.11}$$

The next $(n-k)$ equations give

$$\mathbf{u}_1^{\sigma} = \alpha\begin{bmatrix} \theta_s S^{-} Z_2 Z_1^{-1}\mathbf{e}_s \\ 1 \end{bmatrix} \tag{6.12}$$

while the first equation gives

$$\alpha = \frac{1}{1+\theta_s \mathbf{e}^{(n-k)\mathrm{T}} S^- Z_2 Z_1^{-1} \mathbf{e}_s} \tag{6.13}$$

The objective function value is

$$\begin{aligned} h &= -\mathbf{f}^\mathrm{T} Z T \mathbf{u}_2^\sigma \\ &= \alpha \theta_s \mathbf{f}^\mathrm{T} Z Z_1^{-1} \mathbf{e}_s, \end{aligned} \tag{6.14}$$

and it should be noted that it is independent of T. As h must be positive, this fixes θ_s. S^- is now determined by the condition that $\mathbf{u}_1^\sigma \geqslant 0$ in (6.12), and T by the condition the $\mathbf{u}_2^\sigma \geqslant 0$ in (6.11). This completes the specification of the solution to the dual of the restricted l_∞ problem. By the duality theorem the solution of the primal problem satisfies

$$\mathbf{r} = \begin{bmatrix} \mathbf{r}_1 \\ \mathbf{r}_2 \end{bmatrix} = \begin{bmatrix} \mathbf{r}_1 \\ -hS^- \mathbf{e}^{(n-k)} \end{bmatrix} \tag{6.15}$$

where $\mathbf{r}_1$, h are determined by the conditions

$$\mathbf{e}_s^\mathrm{T} \mathbf{r}_1 = -h\theta_s \tag{6.16}$$

and

$$Z_1^\mathrm{T} \mathbf{r}_1 = -Z^\mathrm{T} \mathbf{f} + h Z_2^\mathrm{T} S^- \mathbf{e}^{(n-k)}$$

so that

$$\mathbf{r}_1 = -Z_1^{-\mathrm{T}} Z^\mathrm{T} \mathbf{f} + h Z_1^{-\mathrm{T}} Z_2^\mathrm{T} S^- \mathbf{e}^{(n-k)} \tag{6.17}$$

It is readily verified that eliminating $\mathbf{r}_1$ between (6.16), (6.17) gives (6.14).

To develop the mechanics of the simplex algorithm note that the only candidates to enter the basis are associated with the columns

$$\begin{bmatrix} 1 \\ 0 \\ \theta_t \mathbf{e}_t \end{bmatrix}, \qquad t = 1, 2, \ldots, k, \qquad t \neq s,$$

as the other possibilities must lead to duplicated columns of Z or replication of columns of I_n, and this latter case can be ruled out by the arguments of the previous section (Lemma 5.1). We calculate

$$\mathbf{v}_t = \begin{bmatrix} \mathbf{v}_1 \\ \mathbf{v}_2 \end{bmatrix} = B^{-1} \begin{bmatrix} 1 \\ 0 \\ \theta_t \mathbf{e}_t \end{bmatrix} \tag{6.18}$$

Let

$$(\mathbf{v}_1)_{n-k+1} = \beta \tag{6.19}$$

then

$$T\mathbf{v}_2 = Z_1^{-1}\{\theta_t \mathbf{e}_t - \beta \theta_s \mathbf{e}_s\} \tag{6.20}$$

and

$$\mathbf{v}_1 = \begin{bmatrix} -S^- Z_2 Z_1^{-1}\{\theta_t \mathbf{e}_t - \beta\theta_s \mathbf{e}_s\} \\ \beta \end{bmatrix}$$

$$= \frac{\beta}{\alpha}\mathbf{u}_1 - \begin{bmatrix} \theta_t S^- Z_2 Z_1^{-1}\mathbf{e}_t \\ 0 \end{bmatrix} \tag{6.21}$$

Substituting in the first equation of (6.18) gives

$$\frac{\beta}{\alpha} = 1 + \theta_t \mathbf{e}^{(n-k)\mathrm{T}} S^- \boldsymbol{\kappa}_t(Z_2 Z_1^{-1}). \tag{6.22}$$

We can now show, as in Lemma 5.3, that the USE test is equivalent to selecting the column to enter the basis as that corresponding to the most violated constraint. The USE test considers

$$\begin{aligned} \xi &= \mathbf{f}^{\mathrm{T}} Z T \mathbf{v}_2 \\ &= \mathbf{f}^{\mathrm{T}} Z Z_1^{-1}\{\theta_t \mathbf{e}_t - \beta\theta_s \mathbf{e}_s\} \\ &= \theta_t \mathbf{f}^{\mathrm{T}} Z Z_1^{-1}\mathbf{e}_t - \frac{\beta}{\alpha} h. \end{aligned}$$

Now, by (6.17),

$$\begin{aligned} \mathbf{e}_t^{\mathrm{T}}\mathbf{r} &= -\mathbf{e}_t^{\mathrm{T}} Z_1^{-\mathrm{T}} Z^{\mathrm{T}}\mathbf{f} + h\mathbf{e}_t^{\mathrm{T}} Z_1^{-\mathrm{T}} Z_2^{\mathrm{T}} S^- \mathbf{e}^{(n-k)} \\ &= -\mathbf{e}_t^{\mathrm{T}} Z_1^{-\mathrm{T}} Z^{\mathrm{T}}\mathbf{f} + \theta_t\left(\frac{\beta}{\alpha} - 1\right)h, \end{aligned}$$

so the equation for ξ_t becomes

$$\begin{aligned} \xi_t &= -\theta_t(\mathbf{r}_1)_t - h \\ &= |(\mathbf{r}_1)_t| - h \end{aligned} \tag{6.23}$$

as $\theta_t = -1$ corresponding to $(\mathbf{r}_1)_t > 0$.

To determine the variable to leave the basis, note that the sign matrix T does not affect the calculation of $\mathbf{u}_1^\sigma$ or of the objective function. Thus a component of $\mathbf{u}_2^\sigma$ which becomes infeasible can be restored to feasibility by changing the sign of the associated column of T without impacting the rest of the calculation. Variations in the basic feasible solution for the multiple pivot sequence obtained by changing the signs of the components of S can be studied as in the previous section by considering two linearly independent relations of linear dependence on the augmented array

$$\begin{bmatrix} S^- & 0 & 0 & Z_2 T \\ 0 & \theta_s \mathbf{e}_t & \theta_t \mathbf{e}_t & Z_1 T \end{bmatrix} \tag{6.24}$$

which defines the extended subproblem with $n-k+2$ pairs of inequality constraints. As before, our aim is to solve the extended subproblem. Two suitable relations of linear dependence are given by

$$\mathbf{q}_1 = \begin{bmatrix} \mathbf{q}_{11} \\ \hline \mathbf{q}_{12} \end{bmatrix} = \begin{bmatrix} -\theta_s S^- \boldsymbol{\kappa}_s (Z_2 Z_1^{-1}) \\ 1 \\ 0 \\ \hline -\theta_s T Z_1^{-1} \mathbf{e}_s \end{bmatrix} \tag{6.25}$$

and

$$\mathbf{q}_2 = \begin{bmatrix} \mathbf{q}_{21} \\ \hline \mathbf{q}_{22} \end{bmatrix} = \begin{bmatrix} -\theta_t S^- \boldsymbol{\kappa}_t (Z_2 Z_1^{-1}) \\ 0 \\ 1 \\ \hline -\theta_t T Z_1^{-1} \mathbf{e}_t \end{bmatrix} \tag{6.26}$$

We define

$$\mathbf{y}(\lambda) = \begin{bmatrix} \mathbf{y}_1(\lambda) \\ \mathbf{y}_2(\lambda) \end{bmatrix} = \mathbf{q}_1 + \lambda \mathbf{q}_2, \tag{6.27}$$

and note that the values of λ corresponding to the possible optimizing basic feasible solutions are contained in the set

$$\Lambda = \{\lambda_i = -(\mathbf{q}_{11})_i / (\mathbf{q}_{21})_i, \qquad i = 1, 2, \ldots, n-k+1\}. \tag{6.28}$$

The value of the objective function associated with $\mathbf{y}(\lambda_j)$ is

$$\begin{aligned} h(\lambda_j) &= \left| -\sum_{i=1}^{k} (\mathbf{f}^{\mathrm{T}} Z)_i \operatorname{sgn}((\mathbf{y}_2)_i) \, |(\mathbf{y}_2)_i| \right| \Big/ \sum_{i=1}^{n-k+2} |(\mathbf{y}_1)_i| \\ &= |\theta_s \mathbf{f}^{\mathrm{T}} Z Z_1^{-1} \mathbf{e}_s + \lambda_j \theta_t \mathbf{f}^{\mathrm{T}} Z Z_1^{-1} \mathbf{e}_t| \Big/ \sum_{i=1}^{n-k+2} |(\mathbf{y}_1)_i|. \end{aligned} \tag{6.29}$$

The numerator is thus the modulus of a linear function of λ so that (6.29) has the same form as (5.19), and the development of the multiple pivot sequence follows as before.

Remark 6.1 The economies in this case are obtained because the only inversion that is required is that of the $k \times k$ matrix Z_1. Thus the appropriate estimate of computational cost corresponds to that of a tableau based on Z. The interchanges that occur at each stage are

$$\boldsymbol{\kappa}_j(Z) \to Z_1$$

$$\boldsymbol{\kappa}_s(Z) \to Z_2$$

where j is the index of λ maximizing $h(\lambda)$.

Example 6.1 (a) Consider the problem of Example 5.1. This is

$$\min_{\mathbf{x}} \max_i |r_i|$$

where

$$\mathbf{r}=\begin{bmatrix}1 & 0\\ 1 & \frac{1}{2}\\ 1 & 1\end{bmatrix}\begin{bmatrix}x_1\\ x_2\end{bmatrix}-\begin{bmatrix}1\\ 1.649\\ 2.718\end{bmatrix}$$

Here $n=3$, $p=2$, and $k=1$. An appropriate choice for Z is

$$Z^{\mathrm{T}}=[1 \quad -2 \quad 1]$$

so the equivalent problem is

$$\min \max_i |r_i|$$

subject to

$$r_1-2r_2+r_3=-.42$$

and

$$-h\leqslant r_i\leqslant h, \qquad i=1,2,3.$$

The solution is clearly $r_1=-.105$, $r_2=.105$, $r_3=-.105$.

(b) Consider now the l_∞ problem (compare Example 5.2) with

$$\mathbf{r}=\begin{bmatrix}1\\ 1\\ 1\end{bmatrix}\mathbf{x}-\begin{bmatrix}1\\ 1.649\\ 2.718\end{bmatrix}.$$

Here $n=3$, $p=1$, and $k=2$. An appropriate choice for Z is

$$Z^{\mathrm{T}}=\begin{bmatrix}1 & -2 & 1\\ 1 & 0 & -1\end{bmatrix}$$

Now $n-k+2=n$ so this case gives the extended subproblem corresponding to a multiple pivot sequence. If the first two constraint pairs are chosen to define the initial subproblem then equation (6.9) becomes

$$\begin{bmatrix}1 & 1 & 0 & 0\\ \theta_1 & 0 & 1 & 1\\ 0 & \theta_2 & -2 & 0\\ 0 & 0 & 1 & -1\end{bmatrix}\begin{bmatrix}\mathbf{u}_1\\ T\mathbf{u}_2\end{bmatrix}=\begin{bmatrix}1\\ 0\\ 0\\ 0\end{bmatrix}$$

Solving for $T\mathbf{u}_2$ gives

$$\begin{bmatrix}-2 & 0\\ 1 & -1\end{bmatrix}T\mathbf{u}_2=-\alpha\theta_2\begin{bmatrix}1\\ 0\end{bmatrix}, \qquad \mathbf{u}_2=-\alpha\theta_2\begin{bmatrix}-.5\\ -.5\end{bmatrix}$$

so that

$$\theta_1 u_{11}=\alpha\theta_2[1 \quad 1]\begin{bmatrix}-.5\\ -.5\end{bmatrix}=-\theta_2\alpha$$

and this implies that

$$\alpha=\tfrac{1}{2}, \qquad \theta_2=-\theta_1,$$

while (6.14) gives

$$h = .3245, \qquad \theta_2 = +1.$$

The active inequality constraints give

$$x - 1 = .3245,$$
$$x - 1.649 = -.3245,$$

so that $x = 1.3245$, $r_3 = -1.3835$, $\theta_3 = 1$.
The extended array corresponding to (6.24) is

$$\begin{bmatrix} -1 & 0 & 0 & 1 & 1 \\ 0 & 1 & 0 & -2 & 0 \\ 0 & 0 & 1 & 1 & -1 \end{bmatrix},$$

the linear dependence relations (6.25), (6.26) are

$$\mathbf{q}_1 = \begin{bmatrix} 1 \\ 1 \\ 0 \\ .5 \\ .5 \end{bmatrix}, \qquad \mathbf{q}_2 = \begin{bmatrix} 1 \\ 0 \\ 1 \\ 0 \\ 1 \end{bmatrix},$$

and

$$\Lambda = \{-1, \infty\}.$$

Evaluating (6.29) gives

$$h(\lambda) = \frac{|.42 + 1.718\lambda|}{\sum_{i=1}^{3} |(\mathbf{y}_1)_i|}$$

where

$$\sum_{i=1}^{3} |(\mathbf{y}_1)_i| = \begin{cases} 2 + 2\lambda, & 0 \leqslant \lambda < \infty, \\ -2\lambda, & -\infty < \lambda \leqslant -1, \\ 2, & -1 \leqslant \lambda \leqslant 0. \end{cases}$$

It is verified readily that h is maximized when $\lambda = \infty$, $h = .859$, corresponding to deleting the second pair of constraints. The active inequality constraints give

$$x - 1 = .859$$
$$x - 2.718 = -.859$$

so that $x = 1.859$, $r_2 = .21$.

Exercise 6.1 Consider the case $n = p + 2$ treated in detail in Section 3.5. What is Z in this case? What is the relationship betwen the two algorithms?

NOTES ON CHAPTER 3

For extended treatments of the discrete approximation problem see Cheney (1966) and Watson (1980). The approach to strong uniqueness follows Jittorntrum and Osborne (1980) and Smith (1979). Polyhedral norms as a way to provide a unified approach is considered in Anderson and Osborne (1976). Linear programming in l_1 approximation goes back at least to Wagner (1959). Other references include Barrodale and Young (1966). The modified simplex algorithm is described by Barrodale and Roberts (1973) and Spyropoulous, Kiontousis, and Young (1973). The application of subopt is given by Robers and Ben-Israel (1969), and Abdelmalik (1975) describes the same algorithm from the point of view of the dual simplex method. The reduced gradient algorithm of Section 3.2 would appear to have the advantages of direct derivation and compact implementation. Armstrong and Godfrey (1979) show the equivalence of the algorithm of Abdelmalik and that of Barrodale and Roberts, and their numerical results (1979) would appear to indicate they are aware of the importance of efficient line searches. However, Robers and Ben-Israel had already shown the equivalence of subopt and the modified simplex algorithm in their 1969 paper. The projected gradient method is due to Bartels, Conn, and Sinclair (1978). Barrodale and Roberts (1978) consider constrained l_1 problems. All algorithms indicated use a USE test based on Lemma 2.3. An NSE test is given in Bloomfield and Steiger (1980), and the consistently good performance of this algorithm indicates this as an important consideration. The application of a reduced gradient algorithm to l_∞ problems is considered in Cheney (1966) where earlier references are given. Some difficulties in formulating a consistent starting procedure are experienced, and restrictive rank assumptions made, but the possibility of applications to polyhedral convex functions is noted. Cline (1976) gives an algorithm similar to that described here while a projected gradient algorithm is described by Bartels, Conn, and Charalambous (1978). The ascent methods have an extensive literature going back to Remes (1934) at least. Kelley (1958) used the dual simplex method explicitly, and Stiefel (1959) gave an exchange algorithm and then demonstrated the equivalence with the simplex algorithm (1960). Bittner (1961) and Osborne and Watson (1967) discussed the relaxation of the Haar condition using linear programming. A first-phase procedure was suggested by Barrodale and Philips (1975). The multiple pivot sequence calculation is due to Hopper and Powell (1977) who suggest that it might have value also in avoiding ill-conditioned basis matrices. An equivalent algorithm but starting from the interval programming point of view is given in Armstrong and Kung (1980). The incorporation of constraints is due to Watson (1973, 1974) and Taylor and Winter (1970). The application to polyhedral norms is due to Anderson and Osborne (1976). An extension to constraints and multiple pivot sequences is given in Osborne (1978). All the

Table N.1 A summary of the main algorithms and their properties.

Problem	Algorithm	Extensions	Equivalences	Modifications
l_1	Primal simplex	Multiple pivot sequences	Barrodale and Roberts, Kiontousis, Spiropoulous and Young, Abdelmalik, Robers and Ben-Israel (Reduced gradient).	Bloomfield and Steiger (NSE test), Bartels, Conn and Sinclair (Projected gradient)
		Constraints		
l_∞	Reduced gradient	Constraints	Cline, Cheney, and Goldstein.	Bartels, Conn and Charalambous. (Projected gradient)
l_∞	Dual simplex (exchange algorithm)	Multiple pivot sequences	Hopper and Powell, Armstrong and Kung.	Bartels and Joe (Penalty algorithm)
		Constraints		

l_∞ algorithms reported here use a form of the USE test given in Lemma 5.1. The advantages that can be obtained by transforming the barely overdetermined problem are pointed out in the l_1 case by Senta and Steiger (1984). Generally, degeneracy is assumed away. An exception is Bartels and Joe (1983) who give an adaptation of Bartels penalty linear programming algorithm for LP2 to the l_∞ problem. Hopper and Powell (1977) use Wolfe's form of the recursive scheme.

In summary, the algorithms and their interrelations are given in Table N.1.

CHAPTER 4

Polyhedral convex functions

4.1 INTRODUCTION

Polyhedral convex functions are introduced in this chapter as a means of providing a convenient framework for a unified treatment of the problems considered up till now. In principle this approach does not contribute anything new as the problem of minimizing a polyhedral convex function can always be posed as a linear programming problem (and vice versa). However, polyhedral convex functions can offer considerable advantages in practice, both in compactness of the problem statement and in facilitating the use of inherent problem structure. These advantages show themselves, in particular, in problems related to the discrete approximation problems considered in Chapter 3, and for this reason the representation of these problems in this form is discussed in some detail in order to motivate subsequent developments.

The key to a general treatment of descent algorithms is a compact parametrization of the subdifferential of the polyhedral convex function, and a suitable form is developed here quite generally. It has the advantage that the inherent structure of the function appears naturally, and this facilitates the construction in particular cases. It is then applied to provide a general setting for reduced and projected gradient algorithms. However, no attempt is made here to get involved in fine implementation details. These have been discussed extensively in Chapters 2 and 3 for important special cases, and this material provides suitable guidelines for other applications of the general theory.

Definition 1.1 A *polyhedral convex function* (PCF) assuming bounded values on a polyhedral convex subset X of R^P is the pointwise maximum of a finite number of affine functions. That is, $C: X \subseteq R^P \rightarrow R$ is the closed convex function given by

$$C(\mathbf{x}) = \begin{cases} \max\limits_{1 \leq i \leq N} \{\mathbf{c}_i^{\mathrm{T}}\mathbf{x} - d_i\}, & \mathbf{x} \in X \\ +\infty & \mathbf{x} \in X^{\mathrm{c}} \end{cases} \tag{1.1}$$

where X is the polyhedral convex set

$$X = \{\mathbf{x};\, \mathbf{a}_i^T\mathbf{x} \geq b_i,\ i = 1, 2, \ldots, n\}. \tag{1.2}$$

Remark 1.1 An explicit form for C can be given using the indicator function for X to write

$$C(\mathbf{x}) = \max_{1 \leq i \leq N} \{\mathbf{c}_i^T\mathbf{x} - d_i\} + \delta(\mathbf{x} \mid X). \tag{1.3}$$

Further structure is obtained by introducing the elementary PCF (1.18) of Chapter 3, and writing C as

$$C(\mathbf{x}) = \max_{1 \leq i \leq N} \{\mathbf{c}_i^T\mathbf{x} - d_i\} + \sum_{i=1}^{N} \delta(b_i - \mathbf{a}_i^T\mathbf{x}). \tag{1.4}$$

A consequence is a representation of $\partial C(\mathbf{x})$ which follows from Lemma 4.7 and equation (5.14) of Chapter 1 on recalling the form of $\partial\delta(t)$ given in (1.19) of Chapter 3.

Lemma 1.1

$$\partial C(\mathbf{x}) = \operatorname{conv}(\mathbf{c}_i^T;\, i \in \nu) - \sum_{i \in \sigma} \partial\delta(0)\mathbf{a}_i^T \tag{1.5}$$

where

$$\nu = \{i;\, \mathbf{c}_i^T\mathbf{x} - d_i = C(\mathbf{x})\},$$

and

$$\sigma = \{i;\, \mathbf{a}_i^T\mathbf{x} - b_i = 0\}.$$

This expression turns out to be too cumbersome and does not reflect clearly enough the detailed structure of the PCF to be of much use in many applications, so that an alternative form is sought. To motivate this form, Examples 1.1–1.7 are considered.

Example 1.1 The linear programming problem LP1 is equivalent to minimizing the PCF (3.1.17). Here $N = 1$, $\mathbf{c}_1 = \mathbf{c}$, and the condition for a minimum, $0 \in \partial C(\mathbf{x})$, gives

$$\mathbf{c}^T = \sum_{i \in \sigma} u_i\mathbf{a}_i^T, \qquad 0 \leq u_i < \infty. \tag{1.6}$$

This will be recognized as the Kuhn–Tucker conditions.

Example 1.2 The polyhedral norm (3.1.21) defines a PCF in which

$$\mathbf{c}_i^T = \boldsymbol{\rho}_i(P), \qquad i = 1, 2, \ldots, N. \tag{1.7}$$

Here $X = R^P$, and

$$\partial\, \|\mathbf{x}\|_p = \operatorname{conv}(\boldsymbol{\rho}_i(P),\, i \in \nu) \tag{1.8}$$

so that, by the definition of ν, $\partial \|\mathbf{x}\|_P$ is the convex hull of the rows of P aligned with $\mathbf{x}$.

Example 1.3 Consider the *composite function*

$$C(\mathbf{x}) = h(\mathbf{r}(\mathbf{x})) \tag{1.9}$$

where h is the PCF $h: R^n \to R$ given by

$$h(\mathbf{r}) = \max_{1 \leq i \leq N} \{\mathbf{h}_i^T \mathbf{r} - o_i\} \tag{1.10}$$

and, in conformity with previous usage,

$$\mathbf{r} = M\mathbf{x} - \mathbf{f} \tag{1.11}$$

where $M: R^P \to R^n$. In particular, h can be a polyhedral norm (implying, in particular, that $k_i = 0$, $i = 1, 2, \ldots, N$), and a typical application occurs when $\mathbf{r}$ is the residual due to an estimate $\mathbf{x}$ of the parameter vector in a linear model. The idea is to determine $\mathbf{x}$ by minimizing $h(\mathbf{r})$. Familar examples (compare Example 1.2 of Chapter 3) include:

(a)
$$\mathbf{h}_i^T = \begin{cases} \mathbf{e}_i^T, & i = 1, 2, \ldots, n \\ -\mathbf{e}_i^T, & i = n+1, \ldots, 2n \end{cases}$$

so that $h(\mathbf{r}) = \max_{1 \leq i \leq n} |r_i|$ and $N = 2n$.

(b) $\mathbf{h}_i^T$ given by the possible ways of putting ± 1 into n locations so that $h(\mathbf{r}) = \sum_{i=1}^{n} |r_i|$ and $N = 2^n$.

It follows from (1.10), (1.11) that

$$\mathbf{c}_i^T = \mathbf{h}_i^T M, \qquad d_i = o_i + \mathbf{h}_i^T \mathbf{f}, \qquad i = 1, 2, \ldots, N. \tag{1.12}$$

Applying (1.5) gives

$$\partial C(\mathbf{x}) = \text{conv}\,(\mathbf{h}_i^T M, i \in \nu). \tag{1.13}$$

Note that in the second case $|\nu| = 2^k$ if $\mathbf{r}$ has k zero components. Thus $\partial C(\mathbf{x})$ is potentially the convex hull of a large set of vectors. It is this phenomenon which causes this representation of $\partial C(\mathbf{x})$ to lose its attractiveness.

Example 1.4 The piecewise linear function defined in (2.1) of Chapter 3 provides an important example of a composite PCF. Here

$$C(\mathbf{x}) = \tau \mathbf{c}^T \mathbf{x} + h(\mathbf{r}) \tag{1.14}$$

where

$$h(\mathbf{r}) = \sum_{i=1}^{n} \Psi(r_i, \alpha_i, \beta_i), \tag{1.15}$$

and the $\mathbf{h}_i$ are the set of 2^n vectors obtained by forming all possible combinations in which the sth location is filled by either $-\alpha_s$ or β_s for $s=1,2,\ldots,n$. In this case the subdifferential is

$$\partial C(\mathbf{x})=\tau\mathbf{c}^{\mathrm{T}}+\operatorname{conv}(\mathbf{h}_i^{\mathrm{T}}M,\ i\in\nu) \tag{1.16}$$

However, an alternative form has been given in (2.5) of Chapter 3:

$$\partial C(\mathbf{x})=\mathbf{g}^{\mathrm{T}}+\sum_{i\in\mu}[-\alpha_i,\beta_i]\mathbf{m}_i^{\mathrm{T}}$$

where

$$\mu=\{i;\ r_i=0\},$$

and

$$\mathbf{g}=\tau\mathbf{c}+\sum_{i\in\mu^c}\frac{\Psi(r_i,\alpha_i,\beta_i)}{r_i}\mathbf{m}_i$$

The important feature of this second representation is the explicit use made of the structural information contained in the zero residuals. For future reference it is noted that (2.5) of Chapter 3 can be written in the equivalent form

$$\mathbf{z}^{\mathrm{T}}\in\partial C(\mathbf{x})\Leftrightarrow\exists\ u_i,\qquad -\alpha_i\leqslant u_i\leqslant\beta_i,\qquad i=1,2,\ldots,|\mu|$$

such that

$$\mathbf{z}=\mathbf{g}+V\mathbf{u} \tag{1.17}$$

where

$$\boldsymbol{\kappa}_i(V)=\mathbf{m}_{\mu(i)},\qquad i=1,2,\ldots,|\mu|$$

Equation (1.17) provides an explicit parametrization of ∂C in terms of a number of parameters equal to the number of zero residuals. This should be contrasted with the parametrization implied by (1.16) which involves $2^{|\mu|}-1$ parameters.

Example 1.5 A somewhat similar application of polyhedral convex functions occurs in the case of stochastic linear programming when the resources b_i, $i=1,\ldots,n$ in LP1 are uncertain. Seeking protection against the worst that can happen by choosing $\mathbf{x}$ to minimize worst-case losses under the assumption that b_j is independently distributed in the interval $[\alpha_j,\beta_j]$ with mean $\bar{b}_j$, $j=1,2,\ldots,n$, but that information otherwise available is minimal, then an appropriate PCF is

$$C(\mathbf{x})=\mathbf{c}^{\mathrm{T}}\mathbf{x}+\sum_{j=1}^{n}\xi_j\Psi^{(j)}(r_j) \tag{1.18}$$

where

$$\xi_j\geqslant 0,$$

$$\Psi^{(j)}=\begin{cases}0, & r_j\leqslant\alpha_j-\bar{b}_j\\ \dfrac{\beta_j-\bar{b}_j}{\beta_j-\alpha_j}(r_j+\bar{b}_j-\alpha_j), & \alpha_j-\bar{b}_j\leqslant r_j\leqslant\beta_j-\bar{b}_j\\ r_j, & r_j\geqslant\beta_j-\bar{b}_j\end{cases} \tag{1.19}$$

and

$$r_j = \mathbf{a}_j^{\mathrm{T}}\mathbf{x} - \bar{b}_j, \qquad j = 1, 2, \ldots, n.$$

Now $\Psi^{(j)}$ can also be expressed as the maximum of the indicated functions so that $C(\mathbf{x})$ is a PCF where the $\mathbf{h}_i$ are the set of $N = 3^n$ vectors obtained by forming all possible combinations in which the sth location is filled by 0, $\dfrac{\beta_s - \bar{b}_s}{\beta_s - \alpha_s}$, 1, for $s = 1, 2, \ldots, n$.

To form the subdifferential let the index set μ be given by

$$\mu = \mu_1 \cup \mu_{12} \cup \mu_2 \cup \mu_{23} \cup \mu_3$$

where the subsets are defined by

$$\begin{aligned}
j \in \mu_1 &\Rightarrow r_j < \alpha_j - \bar{b}_j,\\
j \in \mu_{12} &\Rightarrow r_j = \alpha_j - \bar{b}_j,\\
j \in \mu_2 &\Rightarrow \alpha_j - \bar{b}_j < r_j < \beta_j - \bar{b}_j,\\
j \in \mu_{23} &\Rightarrow r_j = \beta_j - \bar{b}_j,\\
j \in \mu_3 &\Rightarrow r_j > \beta_j - \bar{b}_j,
\end{aligned}$$

for $j = 1, 2, \ldots, n$. Then

$$\partial C(\mathbf{x}) = \mathbf{g}^{\mathrm{T}} + \sum_{j \in \mu_{12}} \xi_j \left[0, \frac{\beta_j - \bar{b}_j}{\beta_j - \alpha_j}\right] \mathbf{m}_j^{\mathrm{T}} + \sum_{j \in \mu_{23}} \xi_j \left[\frac{\beta_j - \bar{b}_j}{\beta_j - \alpha_j}, 1\right] \mathbf{m}_j^{\mathrm{T}} \tag{1.20}$$

where

$$\mathbf{g} = \mathbf{c} + \sum_{j \in \mu_2} \xi_j \frac{\beta_j - \bar{b}_j}{\beta_j - \alpha_j} \mathbf{m}_j + \sum_{j \in \mu_3} \xi_j \mathbf{m}_j. \tag{1.21}$$

Clearly (1.20) can also be written in the form (1.17) where the constraints on the u_j follow from the particular form of $\partial\Psi^{(j)}$ in the sums over the index sets μ_{12}, μ_{23}.

Example 1.6 This example provides a PCF which occurs in statistical estimation problems. It also introduces another structural element in which ties in the ranking of the residuals are significant. Let *scores* η_i, $i = 1, 2, \ldots, n$ satisfy:

(i) $\eta_1 \leqslant \eta_2 \leqslant \ldots \leqslant \eta_n$,
(ii) $\eta_i = -\eta_{n-i+1}$, $i = 1, 2, \ldots, n$, and
(iii) $\eta_1 < \eta_n$ (so the η_i cannot vanish identically).

Definition 1.2 The *dispersion* $\Delta(\mathbf{r})$ is given by

$$\Delta(\mathbf{r}) = \sum_{i=1}^{n} \eta_i r_{\pi(i)} \tag{1.22}$$

where π *is an index set pointing to the components of* $\mathbf{r}$ *sorted into increasing order.*

To see that $\Delta(\mathbf{r})$ is a PCF let H be the matrix with rows consisting of all distinct permutations of the η_i, $i=1,2,\ldots,n$. If the number of rows of H is N, and

$$\mathbf{h}_i=\boldsymbol{\rho}_i(H), \qquad i=1,2,\ldots,N,$$

then

$$\Delta(\mathbf{r})=\max_{1\leq i\leq N} \mathbf{h}_i^{\mathrm{T}}\mathbf{r}. \tag{1.23}$$

The procedure in which $\mathbf{x}$ is estimated by minimizing $\Delta(\mathbf{r})$ is called *rank regression*. Certain aspects of the use of this procedure in data analysis are considered in Section 5.5. Note that it follows from the definition of the η_i, $i=1,2,\ldots,n$ that

$$\Delta(\mathbf{e}^{(n)})=0 \tag{1.24}$$

Thus rank regression cannot estimate a constant or intercept term in a model and such a term must be estimated separately.

In this case (1.13) gives

$$\partial\Delta(\mathbf{r})=\partial C(\mathbf{x})=\operatorname{conv}(\mathbf{h}_i^{\mathrm{T}}M,\, i\in\nu). \tag{1.25}$$

The set ν contains more than a single element whenever there are ties in the ranking of the residuals; for then any permutation of the η_j associated with the tied residuals does not change the dispersion. Also N and $|\nu|$ can be very large (like $n!$ and $p!$ respectively if the η_i are distinct) making it important to seek alternative ways to specify $\partial C(\mathbf{x})$. Here the structural information is provided by the tied residuals, and these are specified by an index set μ partitioned such that

$$\mu=\mu_1\cup\mu_2\cup\ldots\cup\mu_{n_g} \tag{1.26}$$

where μ_i points to the ith distinct group of ties, $i=1,2,\ldots,n_g$,

$$|\mu_i|=k_i+1, \qquad i=1,2,\ldots,n_g \tag{1.27}$$

and

$$\sum_{i=1}^{n_g} k_i=k. \tag{1.28}$$

It is convenient to specialize $\mathbf{m}_{\mu_i(1)}$ to be the *origin* of the ith subgroup of the ties, to define the matrix V_i by

$$\boldsymbol{\kappa}_j(V_i)=\mathbf{v}_j^{(i)}=\mathbf{m}_{\mu_i(j+1)}-\mathbf{m}_{\mu_i(1)}, \qquad j=1,2,\ldots,k_i, \tag{1.29}$$

and to set

$$V=[V_1\,|\,V_2\,|\ldots|\,V_{n_g}] \tag{1.30}$$

The compact form of $\partial C(\mathbf{x})$ is now given by the following result.

Lemma 1.2 *Let* $k\leq K=\operatorname{rank}(M)\leq p$, *and* $\operatorname{rank}(V)=k$. *Then*

$$\mathbf{z}^{\mathrm{T}}\in\partial C(x)\Leftrightarrow\mathbf{z}=\mathbf{g}+V\mathbf{u} \tag{1.31}$$

where

(i)
$$\mathbf{g} = \sum_{i \in \mu^c} \eta_{\chi(i)} \mathbf{m}_i + \sum_{i=1}^{n_g} \left\{ \sum_{j=1}^{k_i+1} \eta_{\chi(\mu_i(j))} \right\} \mathbf{m}_{\mu_i(1)} \tag{1.32}$$

and χ is the index set such that

(a) $\eta_{\chi(j)}$ *is the score associated with* r_j, *and*

(*b*) *within each group* $\eta_{\chi(\mu_i(j))} \geq \eta_{\chi(\mu_i(l))}$ *if* $j \geq l$,

and

(ii)
$$\mathbf{u} = \mathbf{u}^{(1)} \times \mathbf{u}^{(2)} \times \ldots \times \mathbf{u}^{(n_g)}$$

where $\mathbf{u}^{(i)}$ *is the vector of multipliers associated with the ith group and*

$$\sum_{j=1}^{l} \eta_{\chi(\mu_i(j))} \leq \sum_{j=1}^{l} u^{(i)}_{\omega(j)} \leq \sum_{j=1}^{l} \eta_{\chi(\mu_i(k_i-j+2))} \tag{1.33}$$

for $l = 1, 2, \ldots, k_i$ *and* ω *any permutation of* $1, 2, \ldots, k_i$.

Remark 1.2 The inequalities (1.33) do not depend on the choice of group origin. To see this let the contribution to $\partial C(\mathbf{x})$ from the ith group with $\mu_i(s)$ pointing to the group origin be $\mathbf{g}_i^{(s)} + V_i^{(s)} \mathbf{u}^{(i,s)}$. Then a change of origin to $\mu_i(1)$ gives

$$\begin{aligned} \mathbf{g}_i^{(s)} + V_i^{(s)} \mathbf{u}^{(i,s)} &= \mathbf{g}_i^{(1)} + \zeta^{(i)} \mathbf{v}_{s-1}^{(i)} + [V_i - \mathbf{v}_{s-1}^{(i)} \mathbf{e}^{(k_i)\mathrm{T}} - \mathbf{v}_{s-1}^{(i)} \mathbf{e}_{s-1}^{\mathrm{T}}] \mathbf{u}^{(i,s)} \\ &= \mathbf{g}_i^{(1)} + V_i \left\{ \mathbf{u}^{(i,s)} + \mathbf{e}_{s-1} \left(\zeta^{(i)} - \sum_{j=1}^{k_i} u_j^{(i,s)} - u_{s-1}^{(i,s)} \right) \right\} \\ &= \mathbf{g}_i^{(1)} + V_i \mathbf{u}^{(i)} \end{aligned}$$

where

$$\zeta^{(i)} = \sum_{j \in \mu_i} \eta_{\chi(j)},$$

and

$$u_j^{(i)} = u_j^{(i,s)}, \qquad j \neq s, \qquad j = 1, 2, \ldots, k_i,$$

$$u_s^{(i)} = \zeta^{(i)} - \sum_{j=1}^{k_i} u_j^{(i,s)}.$$

Direct substitution shows that the inequalities (1.33) hold for $\mathbf{u}^{(i,s)}$ if and only if they hold for $\mathbf{u}^{(i)}$.

Proof It is readily seen that the elements in ∂C must have the declared form by collecting terms in (1.25). The more difficult part is determining the constraints on $\mathbf{u}$. To do this the subgradient inequality is used directly. Let

$$W = [V \mid E_\mu]^{-\mathrm{T}}$$

where (as usual) E_μ is made up of columns of I_p chosen so that the inverse exists. Then any **t** can be written in an obvious notation as

$$\mathbf{t}=\sum_{j=1}^{p}\alpha_j\boldsymbol{\kappa}_j(W)=\sum_{i=1}^{n_g}\sum_{j=1}^{k_i}\alpha_j^{(i)}\mathbf{w}_j^{(i)}+\sum_{j=m+1}^{p}\alpha_j\mathbf{w}_j.$$

Because the ordering of the columns of V_i is disposable, and because the system of inequalities (1.33) is invariant under change of group origin, there is no loss of generality in assuming that the $\alpha_j^{(i)}$ are ordered for each i such that

$$0\leqslant\alpha_1^{(i)}\leqslant\alpha_2^{(i)}\leqslant\ldots\leqslant\alpha_{k_i}^{(i)}.$$

The only point to note is that the group origin can be chosen to ensure that $\alpha_1^{(i)}$ is nonnegative. To see this, consider the change of origin to $\mathbf{m}_{\mu_i(s)}$. Then $\alpha_{s-1}^{(i)}:=-\alpha_{s-1}^{(i)}$, while for $j\neq s-1$,

$$\mathbf{v}_j^{(i)}:=\mathbf{v}_j^{(i)}-\mathbf{v}_{s-1}^{(i)},\qquad \alpha_j^{(i)}:=\alpha_j^{(i)}-\alpha_{s-1}^{(i)}.$$

Thus starting with any ordering it is necessary only to choose the origin corresponding to the most negative of the $\alpha_j^{(i)}$ and then to reorder the columns of the resulting V_i.

With these assumptions, and assuming also that $\|\mathbf{t}\|_2$ is sufficiently small that only rankings within groups are disturbed, then the subgradient inequality gives

$$\begin{aligned}C(\mathbf{x}+\mathbf{t})&=C(\mathbf{x})+\mathbf{g}^{\mathrm{T}}\mathbf{t}+\sum_{i=1}^{n_g}\sum_{j\in\mu_i}\eta_{\chi(j)}(\mathbf{m}_j-\mathbf{m}_{\mu_i(1)})^{\mathrm{T}}\mathbf{t}\\&=C(\mathbf{x})+\mathbf{g}^{\mathrm{T}}\mathbf{t}+\sum_{i=1}^{n_g}\sum_{j=1}^{k_i}\alpha_j^{(i)}\eta_{\chi(\mu_i(1))+j}\\&\geqslant C(\mathbf{x})+\mathbf{g}^{\mathrm{T}}\mathbf{t}+\sum_{i=1}^{n_g}\sum_{j=1}^{k_i}\alpha_j^{(i)}u_j^{(i)}.\end{aligned}$$

Proceeding in similar fashion but replacing **t** by $-\mathbf{t}$ gives

$$\begin{aligned}C(\mathbf{x}-\mathbf{t})&=C(\mathbf{x})-\mathbf{g}^{\mathrm{T}}\mathbf{t}-\sum_{i=1}^{n_g}\sum_{j=1}^{k_i}\alpha_j^{(i)}\eta_{\chi(\mu_i(k_i+1))-j}\\&\geqslant C(\mathbf{x})-\mathbf{g}^{\mathrm{T}}\mathbf{t}-\sum_{i=1}^{n_g}\sum_{j=1}^{k_i}\alpha_j^{(i)}u_j^{(i)}\end{aligned}$$

Thus $\mathbf{u}\in\partial C(\mathbf{x})$ if and only if

$$\sum_{j=1}^{k_i}\alpha_j^{(i)}\eta_{\chi(\mu_i(k_i+1)-j}\leqslant\sum_{j=1}^{k_i}\alpha_j^{(i)}u_{\omega(j)}^{(i,s)}\leqslant\sum_{j=1}^{k_i}\alpha_j^{(i)}\eta_{\chi(\mu_i(1))+j}$$

as each group of $\alpha_j^{(i)}$ can be varied independently for $j=1,2,\ldots,n_g$. Note that the inequalities now take account of the choice of origin and the reordering of the columns of V_i required to order the $\alpha_j^{(i)}$ to permit arbitrary

t. These inequalities can be rearranged to give the equivalent set

$$\beta_1 \sum_{j=1}^{k_i} \eta_{\chi(\mu_i(k_i+1))-j} + \sum_{l=2}^{k_i} \beta_l \sum_{j=l}^{k_i} \eta_{\chi(\mu_i(k_i+1))-j}$$

$$\leq \beta_1 \sum_{j=1}^{k_i} u_{\omega(j)}^{(i,s)} + \sum_{l=2}^{k_i} \beta_l \sum_{j=l}^{k_i} u_{\omega(j)}^{(i,s)}$$

$$\leq \beta_1 \sum_{j=1}^{k_i} \eta_{\chi(\mu_i(1))+j} + \sum_{l=2}^{k_i} \beta_l \sum_{j=l}^{k_i} \eta_{\chi(\mu_i(1))+j}$$

where $\beta_1 = \alpha_1^{(i)}$, $\beta_j = \alpha_j^{(1)} - \alpha_{j-1}^{(1)}$, $j = 2, \ldots, k_i$, are arbitrary nonnegative numbers. The inequalities (1.33) are an immediate consequence.

Remark 1.3 To interpret the bounds in (1.33) note that they correspond to scores associated with the possible ways of splitting the ith group into two subgroups with one ranked below the other. Note that the splitting reduces $\dim(V)$ by one as a new origin has to be found for one of the subgroups. The importance of this observation will become clear in discussing the generation of descent directions.

Example 1.7 A related approach to parameter estimation based on ranking procedures but which permits an intercept term to be determined considers the minimization of

$$h(\mathbf{r}) = \sum_{i=1}^{n} \eta_i \, |r_{\mu(i)}| \tag{1.34}$$

where now it is the $|r_i|$ that are ranked (absolute rank) and the scores satisfy

$$0 \leq \eta_1 \leq \eta_2 \leq \ldots \leq \eta_n.$$

Let the matrix H have rows made up of all distinct permutations of $\pm \eta_i$ (so that $N = 2^n n!$ when the η_i are distinct). Then

$$\mathbf{h}_i^{\mathrm{T}} = \boldsymbol{\rho}_i(H), \qquad i = 1, 2, \ldots, N,$$

and (1.13) provides one representation of $\partial C(\mathbf{x})$. The compact form analogous to (1.31) follows by a similar argument when $|r_i| \neq 0$, $i = 1, 2, \ldots, n$. The only change needed is in the definition of $\mathbf{v}_j^{(i)}$ which becomes

$$\mathbf{v}_j^{(i)} = \boldsymbol{\kappa}_j(V_i) = \theta_{\mu_i(j+1)} \mathbf{m}_{\mu_i(j+1)} - \theta_{\mu_i(1)} \mathbf{m}_{\mu_i(1)}, \qquad j = 1, 2, \ldots, k_i, \tag{1.35}$$

where $\theta_i = \operatorname{sgn}(r_i)$. In particular, the inequality (1.33) still constrains the allowable values of $\mathbf{u}$. However, a group of ties at zero level requires separate treatment. Let this group be pointed to by the index set μ_0. It is appropriate to choose

$$\mathbf{v}_j^{(0)} = \mathbf{m}_{\mu_0(j)}, \qquad j = 1, 2, \ldots, |\mu_0|. \tag{1.36}$$

There is no contribution to $\mathbf{g}$ from this group, and the constraints on the multiplier vector $\mathbf{u}^{(0)}$ are

$$\sum_{l=s}^{|\mu_0|} |u_{\omega(l)}^{(0)}| \leq \sum_{l=s}^{|\mu_0|} \eta_l \tag{1.37}$$

for $s = 1, 2, \ldots, |\mu_0|$ and all permutations ω of $1, 2, \ldots, |\mu_0|$.

Remark 1.4 Actually, all the examples considered permit the subdifferential to be put into the particular form (1.17), (1.31), strongly suggesting that this form is generic in character. The two cases where this has not been indicated explicitly are:

(i) The linear programming problem. Here (1.5) gives

$$\mathbf{z}^{\mathrm{T}} \in \partial C(\mathbf{x}) \Leftrightarrow \mathbf{z} = \mathbf{g} + V\mathbf{u}$$

where

$$\mathbf{g} = \mathbf{c}, \qquad \boldsymbol{\kappa}_i(V) = -\mathbf{a}_{\sigma(i)},$$

and

$$u_i \geq 0, \qquad i = 1, 2, \ldots, |\sigma|.$$

The result is important as it shows that problems involving linear constraints can be considered as a special case of our formalism.

(ii) Discrete approximation in the maximum norm. Here the right hint is provided in Section 3.4, and (1.11) of Chapter 3 is reorganized to give

$$\mathbf{z}^{\mathrm{T}} \in \partial C(\mathbf{x}) \Leftrightarrow \mathbf{z} = \mathbf{g} + V\mathbf{u}$$

where $\theta_{\nu(1)}\mathbf{m}_{\nu(1)}$ is the group origin for the group of ties corresponding to the maximum norm value,

$$\mathbf{g} = \theta_{\nu(1)}\mathbf{m}_{\nu(1)},$$

$$\boldsymbol{\kappa}_i(V) = \theta_{\nu(i+1)}\mathbf{m}_{\nu(i+1)} - \theta_{\nu(1)}\mathbf{m}_{\nu(1)},$$

$$u_i \geq 0, \qquad i = 1, 2, \ldots, |\nu| - 1,$$

and

$$\sum_{i=1}^{|\nu|-1} u_i \leq 1.$$

Exercise 1.1

(i) What is $C^*(\mathbf{u})$, the conjugate convex function to $C(\mathbf{x})$, if:
   (a) $C(\mathbf{x}) = \mathbf{c}^{\mathrm{T}}\mathbf{x}$,
   (b) $C(\mathbf{x}) = |\mathbf{x}|$.

(ii) If $C(\mathbf{x} + \gamma\mathbf{t}) \to \infty$, $\gamma \to \infty$, all $\mathbf{t}$, what can be said about dom $C^*(\mathbf{u})$.

Exercise 1.2 Show that $\Delta(\mathbf{r}) \geq 0$ so that the dispersion $\Delta(\cdot)$ is a semi-norm.

4.2 SUBDIFFERENTIAL STRUCTURE OF POLYHEDRAL CONVEX FUNCTIONS

The examples considered in the previous section suggest a generic form for $\partial C(\mathbf{x})$ possessing an economical parametrization. This is explored further in this section where the main features of this representation are characterized in some detail. The development is carried through for composite functions (the simple case corresponds to $M = I_p$), and for dom $C = R^P$ (Example 1.1 shows that the case dom C polyhedral and of simple enough form does not present additional difficulties). The necessary extensions are sketched in Exercise 4.1 and it is an advantage of our formalism that problems with linear constraints appear as a special case of our general approach.

For any PCF $h(\mathbf{r})$ there must be a rule for generating the defining family of affine functions. Here this is assumed to consist of:

(a) a specified set of real numbers $\eta_1, \eta_2, \ldots, \eta_s$, and
(b) a set of vectors $\mathbf{h}_i$, $i = 1, 2, \ldots, N$, defined on the η_i, $i = 1, 2, \ldots, s$, such that

$$\mathbf{h}_i = \mathbf{h}_i(\eta_1, \eta_2, \ldots, \eta_s), \qquad i = 1, 2, \ldots, N. \tag{2.1}$$

As in (1.10) the affine functions need not be homogeneous but the displacement term o_i is not important for differential considerations.

Example 2.1 It is easy to make these identifications in the examples considered in the previous section. For example:

(i) The l_∞ problem. Here $\eta_1 = 1$, $\eta_2 = -1$, $\mathbf{h}_{2i-1} = \eta_1\mathbf{e}_i$, $\mathbf{h}_{2i} = \eta_2\mathbf{e}_i$, $i = 1, 2, \ldots, n$.

(ii) The piecewise linear functions. Here $\eta_{2i-1} = -\alpha_i$, $\eta_{2i} = \beta_i$, $i = 1, 2, \ldots, n$. The $\mathbf{h}_i$ list all possible arrangements such that $(\mathbf{h}_i)_j$ is either η_{2j-1} or η_{2j}.

(iii) Rank regression. Here scores η_i, $i = 1, 2, \ldots, n$ are given and the $\mathbf{h}_i$ list the vectors consisting of all distinct permutations of the η_i, $i = 1, 2, \ldots, n$.

The subdifferential $\partial_r h(\mathbf{r})$ consists of the convex hull of the $\mathbf{h}_i^T$, $i \in \nu = \{i; \mathbf{h}_i^T\mathbf{r} + o_i = h(\mathbf{r})\}$. If one element $\mathbf{h}^{*T} \in \partial_r h(\mathbf{r})$ is specified then all other elements can be obtained if there is available a minimal linearly independent set of functionals such that

$$\phi_j(\mathbf{r}) = \boldsymbol{\phi}_j^T\mathbf{r} = \alpha_j, \qquad j = 1, 2, \ldots, k \tag{2.2}$$

having the property that there exist numbers $u_j^{(i)}$ such that

$$\begin{aligned}\mathbf{h}_i &= \mathbf{h}^* + \sum_{j=1}^{k} u_j^{(i)}\boldsymbol{\phi}_j, \qquad i \in \nu \\ &= \mathbf{h}^* + \Phi\mathbf{u}^{(i)}.\end{aligned} \tag{2.3}$$

The functionals in this minimal representation are called *structure functionals*. The α_j in (2.2) are determined by the o_i in the definition of h. The comment immediately following (2.1) applies here also. In our examples it is only in Example 1.5 that the α_j differ from 0.

Remark 2.1 Such a set of functionals always exist. For example, starting with

$$\boldsymbol{\phi}_i^* = \mathbf{h}_i - \mathbf{h}^*, \qquad \alpha_i^* = o_i - \mathbf{r}^T\mathbf{h}^*, \qquad i \in \nu, \tag{2.4}$$

which gives the convex hull representation, it is necessary only to construct a minimal basis for the span of the $\boldsymbol{\phi}_i^*$. However, the structure functionals contain the information about the nature of the nondifferentiability of h and frequently can be deduced from the problem statement.

There is an ambiguity inherent in the definition of the quantities in (2.3) for $\mathbf{h}^*$ can be modified by an arbitrary element in the span of the columns of Φ by changing the $\mathbf{u}^{(i)}$, $i \in \nu$, appropriately. But once these are specified it follows from (2.3) that

$$\mathbf{z}^T \in \partial_r h(\mathbf{r}) \Leftrightarrow \mathbf{z} = \mathbf{h}^* + \Phi\mathbf{u} \tag{2.5}$$

where

$$\mathbf{u} \in U = \operatorname{conv}(\mathbf{u}^{(i)}, i \in \nu) \tag{2.6}$$

In addition to $\mathbf{h}^*$ *and the matrix* Φ *of the structure functionals, the set* U *which constrains the allowable (multiplier) vectors* $\mathbf{u}$ *is the third essential feature in the generic representation of the subdifferential.* It follows from (2.6) that U is a bounded polyhedral convex set. Of course, boundedness is linked to the assumption that $\operatorname{dom} C = R^P$. Thus directions of recession in ∂C are linked to constraints on $\operatorname{dom} C$ and to the property of the elementary indicator function that $\partial\delta(0) = [0, \infty)$.

Example 2.2 Continuing the identifications begun in Example 2.1 gives:

(i) The l_∞ problem. As in Remark 1.4 set

$$\mathbf{h}^* = \theta_{\nu(1)}\mathbf{e}_{\nu(1)}, \qquad \phi_s(\mathbf{r}) = \theta_{\nu(s+1)}r_{\nu(s+1)} - \theta_{\nu(1)}r_{\nu(1)}, \qquad s = 1, 2, \ldots, |\nu| - 1$$

$$U = \left\{\mathbf{u};\, u_s \geq 0,\, s = 1, 2, \ldots, |\nu| - 1,\, 1 - \sum_{s=1}^{|\nu|-1} u_s \geq 0\right\}.$$

(ii) Piecewise linear functions. Here the appropriate identifications follow from (2.5) of Chapter 3. The structure functionals are determined by the zero residuals so that

$$\mathbf{h}^* = \sum_{j \in \mu^c} \theta_j \Psi(\theta_j, \alpha_j, \beta_j)\mathbf{e}_j, \qquad \phi_j(\mathbf{r}) = r_{\mu(j)},$$

while the inequalities $-\alpha_{\mu(j)} \leq u_j \leq \beta_{\mu(j)}$, $j = 1, 2, \ldots, |\mu|$, determine U.

(iii) Rank regression. In this example the structure is in the tied residuals so we choose $\phi_{\mu_i(j)}(\mathbf{r}) = r_{\mu_i(j+1)} - r_{\mu_i(1)}$, $j = 1, 2, \ldots, k_i$, $i = 1, 2, \ldots, n_g$. An element in ∂h is readily constructed by taking the multipliers in the convex hull representation to each have the same value. This gives

$$\mathbf{h}^* = \sum_{j\in\mu^c} \eta_{\chi(j)}\mathbf{e}_j + \sum_{i=1}^{n_g}\left(\sum_{j=1}^{k_i+1} \eta_{\chi(\mu_i(j))}\right)\left\{\frac{1}{k_i+1}\sum_{j=1}^{k_i} \boldsymbol{\phi}_{\mu_i(j)} + \mathbf{e}_{\mu_i(1)}\right\}.$$

Note that it differs from the $\mathbf{h}^*$ which determines $\mathbf{g}$ in (1.32) by an element in the span of the columns of Φ (see the remark after (2.4) above). It is clear that $\mathbf{h}^*$ plus the representers $\boldsymbol{\phi}_j$ of the structure functionals form a basis for ∂h. The inequalities determining U corresponding to this choice of $\mathbf{h}^*$ are

$$q\sum_{j=1}^{k_i-q+1} \eta_{\chi(\mu_i(j))} - (k_i-q+1)\sum_{j=k_i-q+2}^{k_i+1} \eta_{\chi(\mu_i(j))} \leqslant (k_i+1)\sum_{j=q}^{k_i} u^{(i)}_{\omega(j)}$$

$$\leqslant -(k_i-q+1)\sum_{j=1}^{q} \eta_{\chi(u_i(j))} + q\sum_{j=q+1}^{k_i+1} \eta_{\chi(\mu_i(j))}$$

for $q = 1, 2, \ldots, k_i$. They are related to the inequalities (1.33) by the correction which gives the $\mathbf{h}^*$ appropriate to (1.32).

The subdifferential formula for composite functions now follows. Let

$$C(\mathbf{x}) = h(M\mathbf{x} - \mathbf{f})$$

then the chain rule for subdifferentials, (5.20) of Chapter 1, gives

$$\mathbf{z}^{\mathrm{T}} \in \partial C(\mathbf{x}) \Leftrightarrow \mathbf{z} = \mathbf{g} + V\mathbf{u} \tag{2.7}$$

where

$$\mathbf{g} = M^{\mathrm{T}}\mathbf{h}^*, \tag{2.8}$$

$$\boldsymbol{\kappa}_j(V) = \mathbf{v}_j = M^{\mathrm{T}}\boldsymbol{\phi}_j, \qquad j = 1, 2, \ldots, k, \tag{2.9}$$

and

$$\mathbf{u} \in U.$$

Remark 2.2 The assumption that the ϕ_j, $j = 1, 2, \ldots, k$ are linearly independent on the subspace generated by the columns of M corresponds to the standard nondegeneracy assumption and ensures that $V = M^{\mathrm{T}}\Phi$ has rank k. If $k > K = \operatorname{rank}(M)$ then the ϕ_j cannot be linearly independent on the column space of M and the problem is called *degenerate.*

The examples confirm that an appropriate set of structure functionals can frequently be deduced from the problem specification – a point that the name is intended to convey. Thus the result which makes the general approach work is that *given the structure functionals it is possible to deduce an explicit form for U in terms of a system of linear inequalities.* The appropriate tool is the directional derivative of a convex function. In fact, let $\mathbf{t}$ be any direction; then it folows from (2.7) that

$$\mathbf{z}^{\mathrm{T}}\mathbf{t} = \mathbf{g}^{\mathrm{T}}\mathbf{t} + \mathbf{t}^{\mathrm{T}}V\mathbf{u} \leqslant C'(\mathbf{x} : \mathbf{t})$$

which is already in the form of a linear constraint on $\mathbf{u}$.

It remains to specialize $\mathbf{t}$, and the manner in which this is done depends on the geometry of the PCF. Note that (2.7) resembles the equation for a flat (Definition 2.7 of Chapter 1) so that it makes sense to work relative to the affine hull of all $\mathbf{y}$ such that $\partial C(\mathbf{y}) = \partial C(\mathbf{x})$. With this convention the current configuration is a point, and the edges E_j leading from it are connected sets on which ∂C is constant and characterized by

(i) rank $(V(E_j)) = \text{rank}\,(V(\mathbf{x})) - 1$, and

(ii) conv $(\partial C(E_j)\ \forall\ E_j) = \partial C(\mathbf{x})$.

Sets containing the given point on which rank $(V) = \text{rank}\,(V(\mathbf{x})) - 2$ are called *facets*. They are bounded by edges, and directional derivatives in facets are linear combinations of directional derivatives in the bounding edges. Higher-dimensional facets can be generated in similar fashion but are of less immediate interest.

The idea of a structure functional makes it easy to specify the directions $\mathbf{t}$ corresponding to the edges. This follows from the assumption that they have a particular canonical form characteristic of a particular problem (for example, expressing tied, zero, or extremal residuals), and this means that the possible forms on an edge can be deduced immediately. This suggests that consideration should be restricted to bases that are 'strictly structural', in the sense that an edge is obtained by the reduction of dimension resulting from deleting any element of the basis, and in all the examples considered it turns out that all edges can be obtained in this way. For example, if at the current point the possible structure is expressed by (say) certain groups of tied residuals and certain zero residuals, then the possibilities that reduce the rank of V by exactly one correspond either to splitting a group of ties into two subgroups which separate on the edge (one degree of freedom is removed by the requirement to find a new origin constraint for one of the new groups which could be the trivial group of one – see Remark 1.3); or allowing one zero residual to become nonzero; or allowing a pair of zeros to relax from zero as a tie. Each of these possibilities has a direct representation in terms of deleting a structure functional from an appropriate basis. This reflects a simple structure for the transformations between the possible bases which are allowable in this sense (Exercise 2.1). It will be convenient to restrict attention to this case.

Consider now the specification of an edge E_q by relaxing one of the structure functionals active at $\mathbf{x}$ but preserving the remaining structure information. If the structure functional selected is ϕ_q then in general it is necessary to seek a new allowable basis $\phi_i^{(q)}$, $i = 1, 2, \ldots, \ldots, k-1$, such that

$$\begin{aligned} &\text{(i)}\quad \phi_i^{(q)}(E_q) = \alpha_i^{(q)}, i = 1, 2, \ldots, k-1, \quad \text{and} \\ &\text{(ii)}\quad (\phi_i^{(q)}(\mathbf{r}) = \alpha_i^{(q)}, i = 1, 2, \ldots, k-1, \quad \text{and} \\ &\qquad \phi_q(\mathbf{r}) = \alpha_q) \Leftrightarrow \phi_j(\mathbf{r}) = \alpha_j,\ j = 1, 2, \ldots, k. \end{aligned} \tag{2.10}$$

Selecting this new basis requires finding a nonsingular matrix $S^{(q)}$ and $\mathbf{s}_q$

such that:

(i)
$$[\Phi^{(q)} \mid \boldsymbol{\phi}_q]\left[\begin{array}{c|c} S^{(q)} & 0 \\ \hline \mathbf{s}_q^{\mathrm{T}} & 1 \end{array}\right] = \Phi P_q \tag{2.11}$$

where P_q is a permutation matrix taking account of any interchanges necessary; and

(ii) determining $\mathbf{t}_q \in R^P$ such that

$$\phi_q(M\mathbf{t}_q) = \mathbf{v}_q^{\mathrm{T}}\mathbf{t}_q \neq 0$$

and

$$\Phi^{(q)\mathrm{T}} M\mathbf{t}_q = V^{(q)\mathrm{T}}\mathbf{t}_q = 0 \tag{2.12}$$

Here (2.11) just expresses the equivalence of the two bases, while (2.12) expresses the way in which the residual structure information is preserved in the displacement in E_q characterized by $\mathbf{t}_q$.

The directional derivative in the direction $\mathbf{t}_q$ is

$$C'(\mathbf{x} : \mathbf{t}_q) = \max_{\mathbf{z}^{\mathrm{T}} \in \partial C(\mathbf{x})} \mathbf{z}^{\mathrm{T}}\mathbf{t}_q, \tag{2.13}$$

and this is attained for

$$\mathbf{z} = \mathbf{g} + \zeta_q \mathbf{v}_q \tag{2.14}$$

where

$$\zeta_q = \begin{cases} \zeta_q^+ = \max\limits_{\mathbf{u} \in U} (\mathbf{s}_q^{\mathrm{T}} \mid 1) P_q^{-1}\mathbf{u}, & \mathbf{v}_q^{\mathrm{T}}\mathbf{t}_q > 0, \\ \zeta_q^- = \min\limits_{\mathbf{u} \in U} (\mathbf{s}_q^{\mathrm{T}} \mid 1) P_q^{-1}\mathbf{u}, & \mathbf{v}_q^{\mathrm{T}}\mathbf{t}_q < 0 \end{cases} \tag{2.15}$$

can be computed from a knowledge of the directional derivative. This gives explicit constraints

$$\zeta_q^- \geqslant (\mathbf{s}_q^{\mathrm{T}} \mid 1) P_q^{-1}\mathbf{u} \quad \text{or} \quad (\mathbf{s}_q^{\mathrm{T}} \mid 1) P_q^{-1}\mathbf{u} \leqslant \zeta_q^+ \tag{2.16}$$

where q ranges over all possible edges (it is not intended to preclude repetitions of the structure functional ϕ_q which means different $\Phi^{(q)}$ associated with the same ϕ_q). Also *these constraints do not depend on how* $\mathbf{t}_q$ *is specified on the orthogonal complement of* V. The inequalities (2.16) will have an interval form if the residual structure is preserved both in the direction $\mathbf{t}_q$ and in the direction $-\mathbf{t}_q$. Typically this will be the case if the structure is expressed in terms of tied or zero residuals but not if the structure is expressed in terms of extreme values.

The inequalities (2.16) specify U precisely. This is a consequence of the geometry of the PCF. For example, let $\mathbf{t}$ be a direction in a facet. Then

$$\mathbf{t} = \delta_1\mathbf{t}_1 + \delta_2\mathbf{t}_2 \tag{2.17}$$

where $\mathbf{t}_1$ and $\mathbf{t}_2$ are directions in the bounding edges. Then linearity gives

$$C'(\mathbf{x} : \mathbf{t}) = \delta_1 C'(\mathbf{x} : \mathbf{t}_1) + \delta_2 C'(\mathbf{x} : \mathbf{t}_2). \tag{2.18}$$

This is possible if and only if the maximizations in (2.13) implied by (2.18) can be carried out simultaneously (so the same $\mathbf{u}$ in (2.7) gives both directional derivatives). This shows:

(a) new inequalities on U are not obtained by considering configurations of higher dimension than edges; and
(b) higher-dimensional configurations are determined by finding just those inequalities (2.15) which can be maximized simultaneously by the same $\mathbf{u}$.

Remark 2.3 Let $\mathbf{t}_q$ be in the direction defining the edge E_q. Then, for $\delta > 0$ small enough,

$$\mathbf{z}^{\mathrm{T}} \in \partial C(\mathbf{x} + \delta \mathbf{t}_q) \Rightarrow \mathbf{z} = \mathbf{g}_q + V^{(q)}\mathbf{u} \tag{2.19}$$

by construction. As $C'(\mathbf{x}:\mathbf{t}_q) = C'(\mathbf{x} + \delta\mathbf{t}_q : \mathbf{t}_q)$ by linearity in the edge, it follows from (2.14) that

$$\mathbf{g}_q^{\mathrm{T}}\mathbf{t}_q = \mathbf{g}^{\mathrm{T}}\mathbf{t}_q + \zeta_q \mathbf{v}_q^{\mathrm{T}}\mathbf{t}_q$$

so that it is appropriate to set

$$\mathbf{g}_q = \mathbf{g} + \zeta_q \mathbf{v}_q \tag{2.20}$$

Example 2.3

(i) The l_∞ problem. The structure functionals for this problem are given in Example 2.2(i). Preserving structure on an edge requires maintaining $k-1$ residuals of the same magnitude and allowing one to grow less rapidly (decrease more rapidly) in the direction $\mathbf{t}$. If this residual is $r_{\nu(q+1)}$ then

$$S^{(q)} = I, \qquad \mathbf{s}_q = 0,$$

and

$$\begin{aligned} \mathbf{z}^{\mathrm{T}}\mathbf{t} &= \theta_{\nu(1)}\mathbf{m}_{\nu(1)}^{\mathrm{T}}\mathbf{t} + u_q \mathbf{v}_q^{\mathrm{T}}\mathbf{t} \\ &\leqslant C'(\mathbf{x}:\mathbf{t}) = \theta_{\nu(1)}\mathbf{m}_{\nu(1)}^{\mathrm{T}}\mathbf{t} \end{aligned}$$

so that $u_q \geqslant 0$ as $\mathbf{v}_q^{\mathrm{T}}\mathbf{t} < 0$. On the other hand, if $r_{\nu(1)}$ decreases in magnitude in the direction $\mathbf{t}$ relative to the other residuals, then it is necessary to modify the structure functionals and a suitable set is $\phi_i - \phi_1$, $i = 2, \ldots, k-1$ as $\boldsymbol{\phi}_s^{(1)} = \theta_{\nu(s+1)}\mathbf{e}_{\nu(s+1)} - \theta_{\nu(2)}\mathbf{e}_{\nu(2)} = \boldsymbol{\phi}_s - \boldsymbol{\phi}_1$. In this case $\mathbf{t}$ is generated by relaxing ϕ_1 and (2.11) gives

$$[\mathbf{v}_2 - \mathbf{v}_1, \ldots, \mathbf{v}_{k-1} - \mathbf{v}_1 \mid \mathbf{v}_1] \left[\begin{array}{ccc|c} 1 & & & \\ & \ddots & & 0 \\ & & 1 & \\ \hline 1 & \ldots & 1 & 1 \end{array}\right] = VP_q$$

so that

$$\begin{aligned} \mathbf{z}^{\mathrm{T}}\mathbf{t} &= \theta_{\nu(1)}\mathbf{m}_{\nu(1)}^{\mathrm{T}}\mathbf{t} + \mathbf{t}^{\mathrm{T}}\mathbf{v}_1 \sum_{i=1}^{k-1} u_i \\ &\leqslant C'(\mathbf{x}:\mathbf{t}) = \theta_{\nu(2)}\mathbf{m}_{\nu(2)}^{\mathrm{T}}\mathbf{t} \end{aligned}$$

giving

$$\sum_{i=1}^{k-1} u_i \leq 1.$$

Thus the convex hull form of ∂C is recovered. Note that the constraints do not have an interval form. Why?

(ii) Piecewise linear functions. Here the structure is contained in the zero residuals so that $\boldsymbol{\phi}_q = \mathbf{e}_{\mu(q)}$, $q = 1, 2, \ldots, k$. Thus $S^{(q)} = I$, $\mathbf{s}_q = 0$, and the inequalities (2.16) become interval constraints

$$\zeta_q^- \leq u_q \leq \zeta_q^+$$

From (2.7) of Chapter 3,

$$C'(\mathbf{x} : \mathbf{t}) = \mathbf{g}^{\mathrm{T}}\mathbf{t} + \Psi(\mathbf{m}_{\mu(q)}^{\mathrm{T}}\mathbf{t}, \alpha_{\mu(q)}, \beta_{\mu(q)})$$

and it follows from the definition of Ψ that

$$\zeta_q^- = -\alpha_{\mu(q)}, \qquad \zeta_q^+ = \beta_{\mu(q)}.$$

(iii) The rank regression problem. In this case the only way to reduce the dimension of V by exactly one is by splitting a group of tied residuals into two subgroups. If the group to split is the ith then, after possible re-indexing, the subgroups of tied residuals on the edge can be taken as $r_{\mu_i(1)}, \ldots, r_{\mu_i(q)}$, and $r_{\mu_i(q+1)}, \ldots, r_{\mu_i(k_i+1)}$. If the representers of the structure functionals associated with the ith group are

$$\boldsymbol{\phi}_j = \mathbf{e}_{\mu_i(j+1)} - \mathbf{e}_{\mu_i(1)}, \qquad j = 1, 2, \ldots, k_i,$$

then the edge is specified by relaxing off $\boldsymbol{\phi}_q$ and (2.11) gives

$$[[\mathbf{v}_1, \ldots, \mathbf{v}_{q-1}] \,|\, [\mathbf{v}_{q+1} - \mathbf{v}_q, \ldots, \mathbf{v}_{k_i} - \mathbf{v}_q] \,|\, \mathbf{v}_q] \left[\begin{array}{c|c|c} \begin{matrix}1 & & \\ & \ddots & \\ & & 1\end{matrix} & & 0 \\ \hline & \begin{matrix}1 & & \\ & \ddots & \\ & & 1\end{matrix} & 0 \\ \hline 0 \ \ldots \ 0 & 1 \ \ldots \ 1 & 1 \end{array}\right] = V_i P_q$$

This gives constraints on $\mathbf{u}^{(i)}$ of the form

$$\zeta_q^- \leq \sum_{j=q}^{k_i} u_j^{(i)} \leq \zeta_q^+$$

To estimate ζ_q note that the conventions adopted in Lemma 1.2 give

$$C(\mathbf{x} + \delta\mathbf{t}) = \begin{cases} C(\mathbf{x}) + \delta\left\{\mathbf{g}^{\mathrm{T}}\mathbf{t} + \mathbf{v}_q^{\mathrm{T}}\mathbf{t} \displaystyle\sum_{j=q+1}^{k_i+1} \eta_{\chi(\mu_i(j))}\right\}, & \mathbf{v}_q^{\mathrm{T}}\mathbf{t} > 0, \\ C(\mathbf{x}) + \delta\left\{\mathbf{g}^{\mathrm{T}}\mathbf{t} + \mathbf{v}_q^{\mathrm{T}}\mathbf{t} \displaystyle\sum_{j=1}^{k_i-q+1} \eta_{\chi(\mu_i(j))}\right\}, & \mathbf{v}_q^{\mathrm{T}}\mathbf{t} < 0, \end{cases}$$

so that

$$\zeta_q^+ = \sum_{j=q+1}^{k_i+1} \eta_{x(\mu_i(j))}, \qquad \zeta_q^- = \sum_{j=1}^{k_i-q+1} \eta_{x(\mu_i(j))}$$

The resulting inequalities are exactly equivalent to (1.33) when all the possible splittings into distinct subgroups are taken into account.

Remark 2.4 The (relative) extreme points of epi C are those points differing by an element in the null space of M and satisfying

$$\{\partial C(\mathbf{x}) = \partial C(\mathbf{y})\} \Leftrightarrow M(\mathbf{x}-\mathbf{y}) = 0.$$

In particular, if rank $(M) = p$ then $\mathbf{x} = \mathbf{y}$.

Lemma 2.1 *If* $\mathbf{x}$ *is a relative extreme point of* epi C *and the* ϕ_j *are linearly independent on the column space of* M *then* $k = |\mu| = K = \text{rank}\,(M)$.

Proof Two points having the same subdifferential must satisfy

$$\Phi^T \mathbf{r}(\mathbf{x}) = \Phi^T \mathbf{r}(\mathbf{y})$$

so that

$$0 = \Phi^T M(\mathbf{x}-\mathbf{y}).$$

Now M can be reduced by pre- and post-multiplying by orthogonal transformations to give

$$Q^T M S = \left[\begin{array}{c|c} R & 0 \\ \hline 0 & 0 \end{array}\right]$$

where R is $K \times K$ upper triangular. It follows that

$$M = Q_1 R S_1^T$$

where the columns of Q_1 provide an orthonormal basis for the column space of M and the rows of S_1^T provide an orthonormal basis for the row space of M. Now $\Phi^T Q_1$ is $k \times K$ and has rank $k \leqslant K$ by the independence assumption. If $k < K$ it is always possible to find a linear combination of the columns of S_1^T in the null space of $\Phi^T Q_1$ as rank $(S_1^T) = K$. Thus $M(\mathbf{x}-\mathbf{y}) = 0$ is forced only if $k = K$.

The next result has a direct application to the construction of descent directions.

Lemma 2.2 *If* $k = K$ *and the* ϕ_j *are linearly independent on the column space of* M *then* $\mathbf{g}$ *is in the span of the columns of* V.

Proof It follows from (2.8) that $\mathbf{g}$ is in the space of the rows of M so that

$$S_2^T \mathbf{g} = 0$$

This implies the consistency of the linear system

$$\mathbf{g} = M^{\mathrm{T}}\Phi\mathbf{u}.$$

Exercise 2.1

(i) What are the structure functionals for Example 1.7? Derive the inequalities determining U in this case.

(ii) Derive the transformations giving the different bases for $\partial C(\mathbf{x})$ in the cases of:
   (a) change of origin in a group of non-zero tied residuals, and
   (b) the different representations of a group of zero residuals.

(iii) Show that the analytic specification of U does not change under these transformations.

Exercise 2.2 Represent the norms in Exercise 1.1 of Chapter 3 as polyhedral convex functions. What are the structure functionals? Determine the compact form for the subdifferential in each case.

Exercise 2.3 The generic form for the subdifferential has value for more general functions. Let $f = \sqrt{(x_1^2 + x_2^2)}$ and determine $\mathbf{g}$, V, U in the representation of $\partial f(0, 0)$.

4.3 UNIQUENESS QUESTIONS FOR THE MINIMA OF POLYHEDRAL CONVEX FUNCTIONS

The form (2.7) for the subdifferential of a PCF lends itself to consideration of the uniqueness question. It should be no surprise that strong uniqueness is the property to consider. This follows from the next result.

Theorem 3.1 *Uniqueness is equivalent to strong uniqueness for PCFs.*

Proof This is by a similar argument to that used in proving Theorem 1.6 of Chapter 3 which is a special case. First note that minimizing C is equivalent to solving the linear program

$$\min x_{p+1} \tag{3.1}$$

subject to

$$\mathbf{c}_i^{\mathrm{T}}\mathbf{x} - d_i \leqslant x_{p+1}, \qquad i = 1, 2, \ldots, N$$

and

$$A^{\mathrm{T}}\mathbf{x} - \mathbf{b} \geqslant 0.$$

Now $C(\mathbf{x}) = x_{p+1}$ provided $\begin{bmatrix} \mathbf{x} \\ x_{p+1} \end{bmatrix}$ is feasible for (3.1) and $\nu \neq \varnothing$. This is a subset of the variations allowed in Theorem 1.5 of Chapter 3, so the result follows.

The main result is a direct generalization of the corresponding results for l_1 and l_∞ approximation (Theorems 1.2 and 1.3 of Chapter 3 respectively). These results become easy corollaries by using the explicit representation of ∂C available in each case.

Theorem 3.2 *Let $\mathbf{x}^*$ minimize $C(\mathbf{x})$ and assume that $k = K = p$ so that C is nondegenerate at $\mathbf{x}^*$. Then $\mathbf{x}^*$ is a unique minimum if and only if $\mathbf{u}^* \in \text{int } U$ where*

$$0 = \mathbf{g} + V\mathbf{u}^* \tag{3.2}$$

is the representation of 0 in $\partial C(\mathbf{x}^*)$.

Proof If C is nondegenerate at $\mathbf{x}^*$ then the columns of V are linearly independent. Thus $\partial C(\mathbf{x}^*)$ contains a sphere about 0 if and only if $\mathbf{u}^* \in \text{int } U$ and $k = p$. To construct a direction of nonuniqueness when $\mathbf{u}^* \notin \text{int } U$, note that U can then be supported at $\mathbf{u}^*$, and this implies the existence of a vector $\mathbf{w}$ such that

$$\mathbf{w}^{\mathrm{T}}(\mathbf{u} - \mathbf{u}^*) \leqslant 0, \qquad \forall\, \mathbf{u} \in U.$$

We require to find $\mathbf{t}$ such that $C'(\mathbf{x} : \mathbf{t}) = 0$. But

$$\begin{aligned} C'(\mathbf{x} : \mathbf{t}) &= \sup_{\mathbf{z}^{\mathrm{T}} \in \partial C(\mathbf{x}^*)} \mathbf{z}^{\mathrm{T}}\mathbf{t} \\ &= \sup_{\mathbf{u} \in U} (\mathbf{g} + V\mathbf{u})^{\mathrm{T}}\mathbf{t} \\ &= \sup_{\mathbf{u} \in U} (V(\mathbf{u} - \mathbf{u}^*))^{\mathrm{T}}\mathbf{t} \end{aligned}$$

using (3.2). As $C'(\mathbf{x}^* : \mathbf{t}) \geqslant 0\ \forall\, \mathbf{t}$ as $\mathbf{x}^*$ is a minimizer, it is necessary only to choose $\mathbf{t}$ such that

$$V^{\mathrm{T}}\mathbf{t} = \mathbf{w}$$

This can be done always by the rank assumption on V.

Exercise 3.1 Carry out the construction of the direction of nonuniqueness in the l_1 and l_∞ cases.

4.4 DESCENT METHODS FOR MINIMIZING POLYHEDRAL CONVEX FUNCTIONS

The compact form for the subdifferential of a PCF given in Section 4.2 is here applied to develop a general treatment of reduced and projected gradient methods. The optimality condition $0 \in \partial C$ shows that we are seeking a point such that $\mathbf{g} \in H = \text{span}\,(\mathbf{v}_i, i = 1, \ldots, k)$, and the two phases of the descent algorithm correspond first to finding a point such that $\mathbf{g} \in H$ and then searching such points until a representation of $\mathbf{g}$ is obtained giving $\mathbf{u} \in U$. The basic result needed for the construction of descent directions is a direct generalization of Lemma 2.3 of Chapter 3.

Lemma 4.1 *Let* span $(\mathbf{v}_i, i=1,2,\ldots,k)=H$,

$$\mathbf{g}=-V\mathbf{u}+\hat{\mathbf{g}}, \tag{4.1}$$

and $\mathbf{t}$ *satisfy*

$$\mathbf{t}^{\mathrm{T}}[V^{(q)} \mid \mathbf{v}_q]=\mathbf{e}_k^{\mathrm{T}}\theta_q, \qquad \mathbf{t}^{\mathrm{T}}\hat{\mathbf{g}}=0 \tag{4.2}$$

Then $\mathbf{t}$ *defines a descent direction if*

(i) $\quad \zeta_q^+ - [\mathbf{s}_q^{\mathrm{T}} \mid 1]P_q^{-1}\mathbf{u} < 0$ *and* $\theta_q=+1$, *or*

(ii) $\quad \zeta_q^- - [\mathbf{s}_q^{\mathrm{T}} \mid 1]P_q^{-1}\mathbf{u} > 0$ *and* $\theta_q=-1$. $\qquad$ (4.3)

Proof It follows from (2.14) that

$$C'(\mathbf{x}:\mathbf{t})=\mathbf{g}^{\mathrm{T}}\mathbf{t}+\zeta_q\theta_q.$$

Now, using (4.1), (4.2),

$$\mathbf{t}^{\mathrm{T}}\mathbf{g}=-\mathbf{t}^{\mathrm{T}}[V^{(q)} \mid \mathbf{v}_q]\left[\begin{array}{c|c} S^{(q)} & 0 \\ \hline \mathbf{s}_q^{\mathrm{T}} & 1 \end{array}\right]P_q^{-1}\mathbf{u}$$

$$=-\theta_q[\mathbf{s}_q^{\mathrm{T}} \mid 1]P_q^{-1}\mathbf{u}$$

so that

$$C'(\mathbf{x}:\mathbf{t})=\theta_q\{\zeta_q-[\mathbf{s}_q^{\mathrm{T}} \mid 1]P_q^{-1}\mathbf{u}\}<0$$

provided (4.3) holds.

To apply this result it is convenient to make the usual nondegeneracy assumptions which here take the form that rank $(V)=k \leqslant K=$ rank (M). This is implied by exactly one independent new structure functional becoming active at the end of each descent step.

Reduced gradient algorithm

Let

$$B_\mu=[V \mid E_\mu] \tag{4.4}$$

where E_μ is chosen, as usual, to ensure that B_μ is invertible and may be constructed from the columns of I_p. Then $\mathbf{u}$ in (4.1) can be found as the first k components in the solution of

$$\mathbf{g}=-B_\mu\hat{\mathbf{u}} \tag{4.5}$$

Given $\mathbf{u}$, the structure functional to be relaxed is found by determining the most violated of the constraints (4.3) (this corresponds to a USE test) in order to specify q and θ_q. Then $\mathbf{t}$ is computed as

$$\begin{aligned}\mathbf{t}^{\mathrm{T}} &= \theta_q \mathbf{e}_q^{\mathrm{T}}[[V^{(q)} \mid \mathbf{v}_q] \mid E_\mu]^{-1} \\ &= \theta_q \mathbf{e}_q^{\mathrm{T}}\left[VP_q\left[\begin{array}{c|c} S^{(q)} & 0 \\ \hline \mathbf{s}_q^{\mathrm{T}} & 1\end{array}\right]^{-1} \Bigg| E_\mu\right]^{-1} \\ &= \theta_q[[\mathbf{s}_q^{\mathrm{T}} \mid 1]P_q^{-1} \mid 0]B_\mu^{-1} \end{aligned} \tag{4.6}$$

The options available are very much as before. Provided $k < K$ then there are two possibilities:

(a) To fix the descent direction by letting

$$j = \arg\max\,(|\hat{u}_i|,\ i = k+1, \ldots, p)$$

and setting

$$\mathbf{t} = -\operatorname{sgn}(u_j)B_\mu^{-\mathrm{T}}\mathbf{e}_j.$$

This choice preserves all the currently active structure functionals and builds up to $k = K$ quickly.

(b) To determine $\mathbf{t}$ by (4.6) if any of the constraints defining U are violated.

Of course, a mix of these strategies as in (5.17) of Chapter 2 is also possible. The second strategy must be used when $k = K$, as it follows from Lemma 2.2 that $\mathbf{g} \in H$ in this case.

Remark 4.1 The NSE test seeks q given by

$$q = \arg\max\left(\frac{|C'(\mathbf{x}:\mathbf{t}_j)|}{\|\mathbf{t}_j\|_2},\ \forall\, j \ni C'(\mathbf{x}:\mathbf{t}_j) < 0\right) \tag{4.7}$$

where the $\mathbf{t}_j$ define the downward-pointing edges at the current point. However, it would seem that establishing a recurrence for calculating the $\|\mathbf{t}_j\|_2$ is potentially a high-cost operation unless the constraints on U have a sufficiently simple form. Consider, for example, the rank regression problem. Here each edge leading from the current point gives a constraint on U and a possible descent direction. The distinct edges correspond to the number of ways each group can be split into two distinct, non-null subgroups. The number of subgroups that can be formed from the ith group is $2^{k_i} - 1$ so the number of edges is

$$n_{\mathrm{E}} = \sum_{i=1}^{n_g} (2^{k_i} - 1)$$

This rapidly becomes an uncomfortably large number of recurrences to be updated. Fortunately, however, this number is not relevant to testing the constraints on U for feasibility. This can be done by sorting the components of $\mathbf{u}^{(i)}$ into order and checking

$$\sum_{j=1}^{l} \eta_{\chi(\mu_i(j))} \leqslant \sum_{j=1}^{l} u^{(i)}_{\pi_i(j)}$$

and

$$\sum_{j=l}^{k_i} u^{(i)}_{\pi_i(j)} \leqslant \sum_{j=l+1}^{k_i+1} \eta_{\chi(\mu_i(j))},$$

for $l = 1, 2, \ldots, k_i$ and $i = 1, 2, \ldots, n_g$, where π_i is the index set which ranks the components of $\mathbf{u}^{(i)}$. The reason for this is that the USE test is concerned only with the largest constraint violations for given $\mathbf{u}$.

Projected gradient algorithm

The compact form for $\partial C(\mathbf{x})$ relates directly to the equations used for developing the projected gradient descent direction. These equations consist of an unconstrained version of (2.7),

$$-\mathbf{t} = \mathbf{g} + V\mathbf{u}, \tag{4.8}$$

and an equation expressing the way in which structure is preserved in the descent step,

$$V^{\mathrm{T}}\mathbf{t} = 0. \tag{4.9}$$

If $\mathbf{g} \notin H$ then a nontrivial $\mathbf{t}$ exists satisfying (4.8), (4.9) and is downhill for minimizing C for

$$C'(\mathbf{x} : \mathbf{t}) = \mathbf{t}^{\mathrm{T}}\mathbf{g} = -\|\mathbf{t}\|_2^2.$$

If $\mathbf{g} \in H$ then $\mathbf{t}$ is null as a consequence of Lemma 2.2 and an alternative must be sought. A suitable system is provided by

$$-\bar{\mathbf{t}} = \mathbf{g} + [V^{(q)} \mid \mathbf{v}_q]\bar{\mathbf{u}} \tag{4.10}$$

and

$$[V^{(q)} \mid \mathbf{v}_q]^{\mathrm{T}}\bar{\mathbf{t}} = \theta_q \gamma_q \mathbf{e}_k. \tag{4.11}$$

As in (2.18), (2.19) of Chapter 3, these equations define $(\bar{\mathbf{t}}, \bar{\mathbf{u}})$, while γ_q is a scale factor to be determined. Solving for $(\bar{\mathbf{t}}, \bar{\mathbf{u}})$ gives

$$\bar{\mathbf{u}} = \left[\begin{array}{c|c} S^{(q)} & 0 \\ \hline \mathbf{s}_q^{\mathrm{T}} & 1 \end{array}\right] P_q^{-1}\left(\mathbf{u} - \theta_q \gamma_q (V^{\mathrm{T}}V)^{-1} P_q^{-\mathrm{T}} \begin{bmatrix} \mathbf{s}_q \\ 1 \end{bmatrix}\right) \tag{4.12}$$

and

$$\bar{\mathbf{t}} = \mathbf{t} + \theta_q \gamma_q V (V^{\mathrm{T}}V)^{-1} P_q^{-\mathrm{T}} \begin{bmatrix} \mathbf{s}_q \\ 1 \end{bmatrix}. \tag{4.13}$$

Lemma 4.1 can be applied to $\mathbf{z}_q = \dfrac{1}{\gamma_q}(\bar{\mathbf{t}} - \mathbf{t})$. It shows that $\mathbf{z}_q$ is a descent vector provided $\mathbf{u}$, θ_q satisfy (4.3). As $\mathbf{t}$ is either a descent vector or null, it follows that

$$\bar{\mathbf{t}} = \mathbf{t} + \gamma_q \mathbf{z}_q \tag{4.14}$$

is a descent vector provided $\gamma_q > 0$. To fix γ_q it is appropriate to choose $\bar{u}_k$ so that (4.10), (4.11) correspond to a projected gradient system on the edge determined by $\bar{\mathbf{t}}$. This gives $\bar{u}_k = \zeta_q$ by (2.20). By (4.12), this requires that

$$\zeta_q = [\mathbf{s}_q^{\mathrm{T}} \mid 1] P_q^{-1} \mathbf{u} - \theta_q \gamma_q [\mathbf{s}_q^{\mathrm{T}} \mid 1] P_q^{-1} (V^{\mathrm{T}} V)^{-1} P_q^{-\mathrm{T}} \begin{bmatrix} \mathbf{s}_q \\ 1 \end{bmatrix}$$

so that

$$\gamma_q = -\frac{\theta_q (\zeta_q - [\mathbf{s}_q^{\mathrm{T}} \mid 1] P_q^{-1} \mathbf{u})}{[\mathbf{s}_q^{\mathrm{T}} \mid 1] P_q^{-1} (V^{\mathrm{T}} V)^{-1} P_q^{-\mathrm{T}} \begin{bmatrix} \mathbf{s}_q \\ 1 \end{bmatrix}}. \tag{4.15}$$

This shows that $\gamma_q > 0$ is a consequence of (4.3).

Thus a descent direction can be found always provided the current point is not optimal. There are two cases:

(a) $\mathbf{g} \notin H$. Here $\mathbf{t} \neq 0$ and provides a descent direction. However, if $\mathbf{u}$ violates any of the constraints (4.3) then the associated structure functional can be relaxed, thus removing one of the constraints on the descent direction.

(b) $\mathbf{g} \in H$. Here $\mathbf{t} = 0$ and Lemma 4.1 must be applied to generate a descent direction. The current point is optimal if $\mathbf{u}$ satisfies the constraints on U. Otherwise, choosing the most violated constraint to determine a descent direction corresponds to a USE test. The resulting descent vector is $\gamma_q \mathbf{z}_q$.

Remark 4.2 As in the reduced gradient method it is not difficult to specify an NSE test. By (6.18) of Chapter 2 it suffices to consider $C'(\mathbf{x} : \mathbf{t})$. By (2.20),

$$\begin{aligned} C'(\mathbf{x} : \mathbf{t}) &= \mathbf{t}^{\mathrm{T}} (\mathbf{g} + \zeta_q \mathbf{v}_q) \\ &= -\|\mathbf{t}\|_2^2 + \theta_q \gamma_q [\mathbf{g} + \zeta_q \mathbf{v}_q]^{\mathrm{T}} V (V^{\mathrm{T}} V)^{-1} P_q^{-\mathrm{T}} \begin{bmatrix} \mathbf{s}_q \\ 1 \end{bmatrix} \\ &= -\|\mathbf{t}\|_2^2 - \theta_q \gamma_q ([\mathbf{s}_q^{\mathrm{T}} \mid 1] P_q^{-1} \mathbf{u} - \zeta_q) \\ &= -\|\mathbf{t}\|_2^2 - \frac{(\zeta_q - [\mathbf{s}_q^{\mathrm{T}} \mid 1] P_q^{-1} \mathbf{u})^2}{[\mathbf{s}_q^{\mathrm{T}} \mid 1] P_q^{-1} (V^{\mathrm{T}} V)^{-1} P_q^{-\mathrm{T}} \begin{bmatrix} \mathbf{s}_q \\ 1 \end{bmatrix}}. \end{aligned} \tag{4.16}$$

This expression generalizes equation (2.37) of Chapter 3. However, its utility is governed by the same considerations as those outlined in Remark 4.1. It would appear to be most useful when the structure of U is sufficiently simple to permit efficient recursive updating of the scaling term.

The line search

The second stage of the descent algorithm fixes the length of step in the direction of the descent vector. Now $C'(\mathbf{x}+\lambda\mathbf{t}:\mathbf{t})$ can change only at a point at which a new structure functional becomes active (say ϕ_s). If the corresponding point is $\mathbf{x}_s$ then a suitable direction in the edge corresponding to relaxing ϕ_s is determined by $\mathbf{t}$ as the remaining structure is preserved in this direction. To compute the change in the directional derivative at $\mathbf{x}_s$, note that for $\delta>0$ sufficiently small the definition of directional derivative gives

$$C'(\mathbf{x}_s-\delta\mathbf{t}:\mathbf{t})=-C'(\mathbf{x}_s:-\mathbf{t})$$

so that

$$\begin{aligned}C'(\mathbf{x}_s+\delta\mathbf{t}:\mathbf{t})-C'(\mathbf{x}_s-\delta\mathbf{t}:\mathbf{t})&=C'(\mathbf{x}_s:\mathbf{t})+C'(\mathbf{x}_s:-\mathbf{t})\\&=\begin{cases}(\zeta_s^+-\zeta_s^-)(\mathbf{v}_s^{\mathrm{T}}\mathbf{t}), & \mathbf{v}_s^{\mathrm{T}}\mathbf{t}>0,\\(\zeta_s^--\zeta_s^+)(\mathbf{v}_s^{\mathrm{T}}\mathbf{t}), & \mathbf{v}_s^{\mathrm{T}}\mathbf{t}<0,\end{cases}\\&=(\zeta_s^+-\zeta_s^-)\,|\mathbf{v}_s^{\mathrm{T}}\mathbf{t}|.\end{aligned}\tag{4.17}$$

This result permits the characterization of the minimum of $C(\mathbf{x}+\lambda\mathbf{t})$ as a function of λ.

Lemma 4.2 *Let s run over all structure functionals,*

$$\Lambda=\left\{\lambda_s;\,\lambda_s=-\frac{\phi_s(\mathbf{r})}{\phi_s(M\mathbf{t})},\,\lambda_s>0\right\},\tag{4.18}$$

and π be an index set pointing to the elements of Λ sorted into increasing order. Let m be the first index in π such that

$$C'(\mathbf{x}:\mathbf{t})+\sum_{i=1}^{m}(\zeta_{\pi(i)}^+-\zeta_{\pi(i)}^-)\,|\mathbf{v}_{\pi(i)}^{\mathrm{T}}\mathbf{t}|\geqslant 0\tag{4.19}$$

then $\lambda_{\pi(m)}$ minimizes $C(\mathbf{x}+\lambda\mathbf{t})$.

Example 4.1 Consider the rank regression problem in the particular case that the structure increases by two groups $\{r_1, r_2\}$ and $\{r_3, r_4\}$ with associated scores (η_1, η_2), (η_3, η_4) coming together, with the first group catching up to the second to form the enlarged group $\{r_1, r_2, r_3, r_4\}$. Then the new structure functional is r_3-r_1, and $\mathbf{v}_3^{\mathrm{T}}\mathbf{t}<0$. Direct calculation gives

$$C(\mathbf{x}+\delta\mathbf{t})=C(\mathbf{x})+\delta\left\{\sum_{j=5}^{n}\eta_{\chi(j)}\mathbf{m}_j^{\mathrm{T}}\mathbf{t}+\eta_3\mathbf{m}_1^{\mathrm{T}}\mathbf{t}+\eta_4\mathbf{m}_2^{\mathrm{T}}\mathbf{t}+\eta_1\mathbf{m}_3^{\mathrm{T}}\mathbf{t}+\eta_2\mathbf{m}_4^{\mathrm{T}}\mathbf{t}\right\}$$

and

$$C(\mathbf{x}-\delta\mathbf{t})=C(\mathbf{x})-\delta\left\{\sum_{j=5}^{n}\eta_{\chi(j)}\mathbf{m}_j^{\mathrm{T}}\mathbf{t}+\eta_1\mathbf{m}_1^{\mathrm{T}}\mathbf{t}+\eta_2\mathbf{m}_2^{\mathrm{T}}\mathbf{t}+\eta_3\mathbf{m}_3^{\mathrm{T}}\mathbf{t}+\eta_4\mathbf{m}_4^{\mathrm{T}}\mathbf{t}\right\}$$

so that

$$\frac{C(\mathbf{x}+\delta\mathbf{t})-C(\mathbf{x})}{\delta}-\frac{C(\mathbf{x})-C(\mathbf{x}-\delta\mathbf{t})}{\delta}=-(\eta_3+\eta_4-\eta_1-\eta_2)\mathbf{v}_3^{\mathrm{T}}\mathbf{t}$$

$$=(\zeta_3^+-\zeta_3^-)\,|\mathbf{v}_3^{\mathrm{T}}\mathbf{t}|.$$

This calculation should be compared with Example 2.3(iii).

Remark 4.3 There is a catch in Lemma 4.2 which has its origin in a problem similar to that which affects the general use of NSE tests. The problem is that there can be substantial numbers of structure functionals which must be inspected in setting up Λ so that sorting Λ to find the optimum λ can be a significant computation. For example, consider the problem of minimizing C given by

$$C=\sum_{i=1}^{n} i\,|r_{\mu(i)}|, \qquad r_i=x-f_i,$$

where the $\mu(i)$ are the absolute ranks. This one-variable case of Example 1.7 is minimized when x is the median of the $n(n+1)/2$ pairwise averages $(f_i+f_j)/2$, $1\leqslant i\leqslant j\leqslant n$. (This result follows from

$$\frac{\mathrm{d}C}{\mathrm{d}x}\simeq\sum_{i=1}^{n} i\,\mathrm{sgn}\,(r_{\mu(i)})\simeq\sum_{i=1}^{n}\sum_{j=1}^{i}\mathrm{sgn}\,(r_{\mu(i)}+r_{\mu(j)})).$$

It is known that to find the median of K objects requires $\mathrm{O}(K)$ comparisons, and this suggests that $\mathrm{O}(n^2)$ comparisons will be needed in computing x. This number is commensurate with the number of structure functionals in the problem ($\pm r_i \pm r_j$, $1\leqslant i<j\leqslant n$, and r_i, $i=1,2,\ldots,n$). However, this argument does not take account of the property that the set of objects whose median is required is of the form $X+X$, where $|X|=n$, and it is known that this additional structure permits the median to be found in $\mathrm{O}(n\log n)$ comparisons. It is an open question if this result extends to the form appropriate to the line search for Example 1.7. Here $r_i=w_ix-f_i$. Ties occur when $|r_i|=|r_j|$, so that

$$\theta_i(w_ix_{ij}-f_i)=\theta_j(w_jx_{ij}-f_j)$$

giving

$$x_{ij}=\frac{\theta_if_i-\theta_jf_j}{\theta_iw_i-\theta_jw_j}.$$

This shows that the set to be partitioned is not a simple cartesian sum except when $w_i=w_j=1$ giving $x_{ij}=(f_i+f_j)/2$, agreeing with the above result. However, it still retains the property that the x_{ij} can be constructed from just $2n$ pieces of information provided by the f_i and w_i. In comparison, the l_1 problem to minimize C given by

$$C=\sum_{i=1}^{n}|r_i|, \qquad r_i=x-f_i,$$

requires the computation of the median of the f_i, $i = 1, 2, \ldots, n$, and costs $O(n)$ comparisons. Again this is commensurate with the number of structure functionals (r_i, $i = 1, 2, \ldots, n$). The implications are that the complexity of the line-search phase of the descent algorithms can vary substantially from problem to problem, and that it is important that this phase be implemented efficiently.

The alternative approach to the line search, in which the feasibility of the multiplier vector associated with the updated structure is tested to see if the new structure functional can be relaxed immediately, can also be developed in general. Here the key point is that the constraint associated with the new structure functional can be evaluated without the need to update the particular matrix factorizations being used.

In the reduced gradient algorithm this approach is completely equivalent to the line search procedure. Let the new structure functional be ϕ_s and the corresponding point $\mathbf{x}_s$. Then (4.6) gives

$$\mathbf{t}^T = (\mathbf{t}^T\mathbf{v}_s)[[\mathbf{s}_s^T \mid 1]P_s^{-1} \mid 0]B_\mu^{-1}$$

(note the slightly more general normalization as here $\mathbf{t}$ is given, not computed) so that, using (4.5)

$$\mathbf{t}^T\mathbf{g} = -(\mathbf{t}^T\mathbf{v}_s)[\mathbf{s}_s^T \mid 1]P_s^{-1}\mathbf{u} \tag{4.20}$$

Thus

$$\begin{aligned}(\mathbf{t}^T\mathbf{v}_s)[\zeta_s - [\mathbf{s}^T \mid 1]P_s^{-1}\mathbf{u}] &= \mathbf{g}^T\mathbf{t} + \zeta_s\mathbf{v}_s^T\mathbf{t} \\ &= C'(\mathbf{x}_s : \mathbf{t})\end{aligned}$$

confirming the equivalence.

If $\hat{\mathbf{t}}$ is the descent vector in the edge leading to $\mathbf{x}_s$ in the projected gradient case then $\hat{\mathbf{t}}$ corresponds to $\bar{\mathbf{t}}$ in (4.10), (4.11), but with the normalization $\bar{\mathbf{t}}^T\mathbf{v}_q = -\hat{\mathbf{t}}^T\mathbf{v}_q$ instead of $\theta_q\gamma_q$ in (4.11). If $\mathbf{z}$ is the projection of $\mathbf{v}_s$ onto the null space of $V^{(s)}$, then

$$\mathbf{z} \in \operatorname{span}(\boldsymbol{\kappa}_i(V^{(s)}), i = 1, 2, \ldots, k-1, \mathbf{v}_s)$$

$$-\bar{\mathbf{t}}^T\mathbf{z} = \mathbf{z}^T\mathbf{g} + \|\mathbf{z}\|_2^2\, \bar{u}_k$$

from equation (4.10), and

$$\bar{\mathbf{t}}^T\mathbf{z} = (\bar{\mathbf{t}}^T\mathbf{v}_q)\, \|\mathbf{z}\|_2^2\, [\mathbf{s}_s^T \mid 1]P_s^{-1}(V^TV)^{-1}P_s^{-T}\begin{bmatrix}\mathbf{s}_s \\ 1\end{bmatrix}$$

from equation (4.13). Equation (4.12) gives

$$\bar{u}_k = [\mathbf{s}_s^T \mid 1]P_s^{-1}\mathbf{u} - (\bar{\mathbf{t}}^T\mathbf{v}_q)[\mathbf{s}_s^T \mid 1]P_s^{-1}(V^TV)^{-1}P_s^{-T}\begin{bmatrix}\mathbf{s}_s \\ 1\end{bmatrix},$$

and combining these results we obtain

$$0 = \mathbf{z}^T\mathbf{g} + \|\mathbf{z}\|_2^2\, [\mathbf{s}_s^T \mid 1]P_s^{-1}\mathbf{u}. \tag{4.21}$$

Equation (4.21) permits evaluation of the constraint on U derived from the new structure functional without the requirement to update the factorization of V. As in Section 3.2 (and see Exercise (4.2) below), relaxing off ϕ_s gives a vector parallel to $\hat{\mathbf{t}}$ only in certain circumstances.

Degeneracy

Degeneracy occurs in these algorithms when more than one structure functional becomes active in a descent step and when $\mathbf{g}$ is contained in the span of a subset of the resulting structure vectors. At the current point the problem of computing a direction of descent is to find $\mathbf{t}$ such that

$$\hat{C}(\mathbf{t}) = \max_{i \in \nu} \mathbf{c}_i^T \mathbf{t} < 0, \tag{4.22}$$

where ν is defined in (1.5). If no such $\mathbf{t}$ exists then the current point is optimal. Otherwise,

$$\min_{\mathbf{t}} \hat{C}(\mathbf{t}) = -\infty, \tag{4.23}$$

and it is compatible with previous usage to call the set of directions in which $\hat{C}$ is unbounded directions of recession for $\hat{C}$. Thus resolution of degeneracy is a matter of computing a direction of recession for $\hat{C}$ (compare, for example, Sections 2.5 and 3.2). Let ν be partitioned into

$$\nu = \nu_1 \cup \nu_2 \tag{4.24}$$

where ν_1 determines a set of independent structure functionals appropriate to a nondegenerate representation of ∂C, and ν_2 points to the additional terms complicating the situation. Now define

$$\hat{C}_\varepsilon(\mathbf{t}) = \max \left\{ \max_{i \in \nu_1} (\mathbf{c}_i^T \mathbf{t}), \max_{i \in \nu_2} (\mathbf{c}_i^T \mathbf{t} - \varepsilon_i) \right\} \tag{4.25}$$

and consider the problem of minimizing $\hat{C}_\varepsilon$. If $\varepsilon_i > 0$, $i \in \nu_2$, then (4.25) is nondegenerate and either of the basic algorithms can be used for minimization. The result that justifies this approach to finding a direction of recession for $\hat{C}$ is given in the next lemma, which follows directly from the definition.

Lemma 4.3 *Let* $\mathbf{t}$ *be a direction of recession for* $\hat{C}$, *then* $\mathbf{t}$ *is a direction of recession for* $\hat{C}_\varepsilon$. *Let* $\mathbf{t}$ *be a direction of recession for* $\hat{C}_\varepsilon$, *then it is a direction of recession for* $\hat{C}$.

Thus it suffices to minimize $\hat{C}_\varepsilon$ which is initially a nondegenerate problem until either a direction in which the problem is unbounded is found (at which point the original degeneracy problem is resolved), or a further occurrence of degeneracy is encountered. On the second occasion a structure functional is being relaxed so the recursive use of the procedure for resolving degeneracy outlined here involves successively smaller problems and so must terminate in a finite number of steps.

Resolution of degeneracy in a finite number of steps is an important component in proving the finiteness of the descent algorithms. The other component is a finiteness proof in the nondegenerate case. The necessary arguments can be found in the proofs of Theorems 5.1 and 6.1 of Chapter 2, which generalize immediately to this somewhat more general situation.

Scaling

The need to preserve the minimum of $C(\mathbf{x})$ (and this is done most easily by leaving function values unchanged) reduces the scope for problem scaling. Assume that C is composite and that $h(\cdot)$ defines a fitting procedure, while the form of the model defining $\mathbf{r}$ changes from application to application. Then it is transformations such as row scaling of the data matrix that correspond to desirable additional manipulations. If $\mathbf{r} \to D\mathbf{r} = \mathbf{s}$, then preservation of the function values implies that

$$h \to \hat{h} = h \circ D^{-1}$$

and this has the immediate consequence for the representation of the subdifferential that

$$\mathbf{h}^* \to \hat{\mathbf{h}}^* = D^{-1}\mathbf{h}^*$$

and

$$\boldsymbol{\phi}_j \to \hat{\boldsymbol{\phi}}_j = D^{-1}\boldsymbol{\phi}_j, \qquad j = 1, 2, \ldots, k.$$

It follows that the constraint set U is unchanged $(\mathbf{h}_i = \mathbf{h}_i^* + \sum u_j^{(i)}\mathbf{h}_j \Leftrightarrow \hat{\mathbf{h}}_i = \hat{\mathbf{h}}_i^* + \sum u_j^{(i)}\hat{\mathbf{h}}_j)$ and only the structure functionals transformed. Also

$$V = M^{\mathrm{T}}\Phi = M^{\mathrm{T}}DD^{-1}\Phi = \hat{V}$$

so that scaling does not affect either NSE or USE tests. This appears to contradict the considerations of Section 3.2 – especially (2.33) of Chapter 3 and the associated discussion. However, here $\mathbf{h}_i = \mathbf{e}_{\nu(i)}$ so that

$$\begin{aligned} VU &= M^{\mathrm{T}}DD^{-1}\Phi U \\ &= M^{\mathrm{T}}D\Phi D_{\nu}^{-1}U \end{aligned}$$

showing that in this particular case the structure functionals can be held constant and the effect of the transformation transferred to the constraint set.

Exercise 4.1 Consider the problem

$$\min_{\mathbf{x} \in X} C(\mathbf{x}) : X = \{\mathbf{x} : A\mathbf{x} \geq \mathbf{b}\}$$

corresponding to minimizing a PCF subject to linear inequality constraints.

(i) Show that the Kuhn–Tucker conditions characterizing an optimum $\mathbf{x}$ are that there exist $\mathbf{u} \in U$, $\mathbf{w} \geq 0$ such that

$$\mathbf{g} + V\mathbf{u} = \sum_{i \in \sigma} w_i \mathbf{a}_i$$

where $\mathbf{g}^{\mathrm{T}} + \mathbf{u}^{\mathrm{T}} V^{\mathrm{T}} \in \partial C(\mathbf{x}) \Leftrightarrow \mathbf{u} \in U$, and σ is the index set pointing to the constraints active at $\mathbf{x}$.

(ii) At any point $\mathbf{x}$ let

$$H = \text{span}\{\boldsymbol{\kappa}_i(V), i = 1, 2, \ldots, |\mu|, \mathbf{a}_i, i \in \sigma\}.$$

If $\mathbf{g} \in H$ then tentative multipliers can be computed from the representation

$$\mathbf{g} = -V\mathbf{u} + \sum_{i \in \sigma} w_i \mathbf{a}_i.$$

Deduce that a downhill direction exists if and only if at least one of the conditions $\mathbf{u} \in U$ and $\mathbf{w} \geq 0$ is violated. Generalize Lemma 4.1 to this case.

(iii) Verify that for $\tau > 0$,

$$P(\mathbf{x}, \tau) = \tau C(\mathbf{x}) + \sum_{i=1}^{n} \Psi(\mathbf{a}_i^{\mathrm{T}} \mathbf{x} - b_i, 1, 0)$$

is an appropriate penalty function for the constrained minimization. Clearly $P(\mathbf{x}, \tau)$ is a PCF. What are the structure functionals for this PCF? What is U?

Exercise 4.2 Derive for the general projected gradient algorithm the analogue of Lemma 2.6 of Chapter 3, giving conditions under which the line search procedure is equivalent to relaxing the new structure functional.

Exercise 4.3 Consider the problem in Exercise 2.1. What happens to the inequalities defining U if some of the η_i are the same? What is the implication of this for the line search phase of the descent algorithms?

Exercise 4.4 Reconcile Exercise 4.5 of Chapter 3 with the discussion of scaling given in this section.

4.5 CONTINUATION AND THE PROJECTED GRADIENT ALGORITHM

In this section an alternative approach to the development of the projected gradient algorithm is considered. Let

$$P(\mathbf{x}, \mathbf{x}_0, \gamma) = \tfrac{1}{2}\|\mathbf{x} - \mathbf{x}_0\|_2^2 + \gamma C(\mathbf{x}) \tag{5.1}$$

where the norm is the euclidean norm and C is a PCF. Then $P(\mathbf{x}, \mathbf{x}_0, \gamma)$ is strictly convex in $\mathbf{x}$ for fixed $\mathbf{x}_0$, γ, and so has a unique minimum $\mathbf{x} = \mathbf{x}(\gamma)$ for fixed $\mathbf{x}_0$ and each γ, $0 \leq \gamma < \infty$. Our interest is in the behaviour of $\mathbf{x}(\gamma)$ as γ becomes large.

Theorem 5.1 *Let*

$$\mathbf{x}(\gamma) = \arg\min_{\mathbf{x}} P(\mathbf{x}, \mathbf{x}_0, \gamma) \tag{5.2}$$

for fixed $\mathbf{x}_0$. *Then there exists a* $\gamma_0 < \infty$ *such that* $\mathbf{x}(\gamma)$ *minimizes* $C(\mathbf{x})$ *for* $\gamma \geq \gamma_0$.

Proof If $\mathbf{x}^*$ minimizes $P(\mathbf{x}, \mathbf{x}_0, \gamma)$ with respect to $\mathbf{x}$ then

$$0 \in \partial P(\mathbf{x}^*, \mathbf{x}_0, \gamma) = (\mathbf{x}^* - \mathbf{x}_0)^{\mathrm{T}} + \gamma \partial C(\mathbf{x}^*). \tag{5.3}$$

The theorem is proved by constructing a suitable $\mathbf{x}^*$ from the set of minimizers of $C(\mathbf{x})$ when γ is large enough.

If the minimizer of $C(\mathbf{x})$ is unique then it is strongly unique by Theorem 3.1. If $\hat{\mathbf{x}}$ minimizes $C(\mathbf{x})$ then $0 \in \operatorname{int} \partial C(\hat{\mathbf{x}})$ so that there exists $\lambda > 0$, independent of γ, such that

$$-\lambda (\hat{\mathbf{x}} - \mathbf{x}_0)^{\mathrm{T}} \in \partial C(\hat{\mathbf{x}}). \tag{5.4}$$

If $\lambda\gamma \geq 1$ then (5.3) is an immediate consequence and $\hat{\mathbf{x}} = \mathbf{x}^*$.

If the minimum of C is not unique then the set of minimizers form a polyhedral set S characterized by an index set π such that

$$\mathbf{x} \in S \Rightarrow \exists\, i \in \pi \ni C(\mathbf{x}) = \mathbf{c}_i^{\mathrm{T}}\mathbf{x} - d_i = C(\hat{\mathbf{x}}).$$

It follows that

$$S = \{\mathbf{x}\,;\, \mathbf{c}_i^{\mathrm{T}}\mathbf{x} - d_i \leq C(\hat{\mathbf{x}}),\ i \in \pi\} \tag{5.5}$$

Now select $\mathbf{x}^*$ to minimize $\frac{1}{2}\|\mathbf{x} - \mathbf{x}_0\|_2^2$ for $\mathbf{x} \in S$. The appropriate necessary conditions for an optimum correspond to Remark 5.7 of Chapter 1 and give multipliers $u_i \geq 0$, $i \in \pi$, such that

$$\mathbf{x}^* - \mathbf{x}_0 = -\sum_{i \in \pi} u_i \mathbf{c}_i \tag{5.6}$$

and

$$u_i(\mathbf{c}_i^{\mathrm{T}}\mathbf{x}^* - d_i - C(\hat{\mathbf{x}})) = 0, \qquad i \in \pi. \tag{5.7}$$

Now, by the convex hull form of $\sigma C(\mathbf{x}^*)$

$$-\frac{(\mathbf{x}^* - \mathbf{x}_0)^{\mathrm{T}}}{\sum_{i \in \pi} u_i} \in \partial C(\mathbf{x}^*),$$

so that, by convexity, as $\mathbf{x}^* \in S \Leftrightarrow 0 \in \partial C(\mathbf{x}^*)$,

$$-\xi \frac{(\mathbf{x}^* - \mathbf{x}_0)^{\mathrm{T}}}{\sum_{i \in \pi} u_i} \in \partial C(\mathbf{x}^*), \qquad 0 \leq \xi \leq 1.$$

This implies that $0 \in \partial P(\mathbf{x}^*, \mathbf{x}_0, \gamma)$ provided $\gamma \geq \sum_{i \in \pi} u_i$.

To apply continuation write

$$0=\mathbf{x}-\mathbf{x}_0+\gamma\{\mathbf{g}+V\mathbf{u}\} \tag{5.8}$$

where $\mathbf{x}=\mathbf{x}(\gamma)$, $\mathbf{u}=\mathbf{u}(\gamma)$ are determined by the minimization procedure. Differentiating gives

$$0=\frac{\mathrm{d}\mathbf{x}}{\mathrm{d}\gamma}+\mathbf{g}+V\mathbf{y} \tag{5.9}$$

where

$$\mathbf{y}=\frac{\mathrm{d}}{\mathrm{d}\gamma}(\gamma\mathbf{u}) \tag{5.10}$$

Also for the operation of differentiation to be possible, local structure must not change in the infinitesmal displacement implied. This requires

$$V^{\mathrm{T}}\frac{\mathrm{d}\mathbf{x}}{\mathrm{d}\gamma}=0. \tag{5.11}$$

Equations (5.9), (5.11) imply that $\mathrm{d}\mathbf{x}/\mathrm{d}\gamma$, $\mathbf{y}$ are piecewise constant (changing only when the structure changes). They also correspond to the equations of the projected gradient method (equations (4.8), (4.9)).

Exercise 5.1 Show how to construct the continuation path from $\gamma=0$ to $\gamma=\gamma_0$. How does this approach compare to the projected gradient algorithm? Which approach is preferable?

NOTES ON CHAPTER 4

Polyhedral convex functions are treated in Rockafellar (1970). Duality and conjugacy properties are considered by McLinden (1982). Example 1.5 is discussed in Whittle (1971). Rank regression is introduced in Jaeckel (1972) and further explored by Hettmansperger and co-workers (for example, Hettmansperger and McKean (1977)), who show that it is a fundamental tool of nonparametric statistics – provided the calculations can be carried out. McKean and Shrader (1980) argue for the objective function of Example 1.7. Aspects of the compact form of the subdifferential, the development of descent directions, and the treatment of degeneracy, have been considered in Osborne (1982, 1983) and Clark and Osborne (1983). However, the development here is new, as is the treatment of the reduced and projected gradient algorithms. That the median of the pairwise averages can be found in $\mathrm{O}(n \log n)$ comparisons is demonstrated in Frederickson and Johnson (1982). An alternative approach to the calculation of linear rank statistics is provided by their property of asymptotic linearity (Jaeckel (1972)). This suggests the use of the secant algorithm, and it has been tried with some success by McKean and Ryan (1977). Here the idea is to apply the secant algorithm to find where $C'(\mathbf{x}+\lambda\mathbf{t}:\mathbf{t})$ passes through zero. It does not cost much to move each value of λ generated to the appropriate end point of the corresponding interval of constancy of C', and with this modification the search algorithm is finite. The main result in Section 4.5 is implicit in Rockafellar (1976). Here we follow Clark and Osborne (1983), who use the continuation result to motivate projected gradient algorithms for several problems including the rank regression problem. This work extends the application to the l_1 problem considered in Clark (1981).

As the interest in this chapter is in problems which are unwieldy to pose as linear programming problems, the dual or ascent algorithms have not received much attention. However, the remarks made in Section 3.5 on ascent methods for polyhedral norm approximation problems remain valid, and Anderson has found that it is straightforward to adapt the algorithm of Anderson and Osborne (1976) to solve rank regression problems.

CHAPTER 5

Least squares and related methods

5.1 INTRODUCTION

Problems that arise in data analysis provide the context for the development of the algorithms that are considered in this chapter. These problems are related in that they can be considered as an aspect of a modelling process in which it is assumed that the model is known and it remains to determine the parameters. Thus they have a great deal in common with discrete approximation problems. They differ because here further assumptions are made concerning the manner in which the data departs from the idealized form represented by the model equations.

Least squares methods are introduced by way of their special relationships with the normal distribution and this central position is justified for more general situations using the Gauss–Markov theorem. Least squares problems subject to linear constraints are also considered. They form an important part of the subclass of quadratic programming problems (the subclass that leads to convex minimization problems). The ready availability of the basic least squares algorithms has lead to weighted least squares formulations in other contexts where the weights now depend on the solution. For completeness, this is illustrated here by considering the method of iteratively reweighted least squares which has been proposed to solve these problems.

The method of least squares is used in many situations. However, there are circumstances in which it is less than satisfactory. These are approached here by asking: 'how sensitive is the solution vector $\mathbf{x}$ to perturbations of the data vector $\mathbf{f}$?' This question is made more specific by assuming that the data departs from an idealized form (say exact data plus experimental errors that are independent and identically distributed) as a result of certain gross measurement errors or blunders which would be recognized if the exact parameter vector were known by the presence of large residuals in the corresponding model equations. A method for determining model parameters which is insensitive to a small number of isolated large residuals is called *resistant.* Such a process could be expected to provide a useful method for detecting and removing bad observations (data screening) as a preliminary step before applying a method appropriate to the idealized error structure to complete the estimation of parameters.

An alternative form of the question which leads to the idea of resistance asks whether the estimation process is *robust*, in the sense that the solution is relatively insensitive to the validity of the particular assumption made regarding the probability distribution of the errors in the data. This is a deeper question in the sense that more information is required. However, the resulting algorithms are usually similar, as there is often a close connection between the ideas of resistance and robustness at the empirical distribution.

Resistance offers a method for comparing algorithms, and useful information often can be obtained by considering the characterization results for the particular minimization problem involved. This is worked out here both for least squares and for estimations obtained by minimizing polyhedral convex functions. For the method obtained by considering the objective function in Example 1.7 of Chapter 4 it is possible to control resistance by 'trimming' the scores η_i. The idea here is to de-emphasize the importance of the large residuals in the estimation procedure. The same idea can be applied to least squares problems and leads to the class of M-estimators. The M-estimator need not give a convex minimization problem. However, in one important case it does, and this is the case which is studied further here. In particular, a finite algorithm is presented based on the idea of continuation with respect to a parameter which serves to classify the residuals as large or small (and so also defines a scale for the method). This algorithm appears to lend itself to the data-screening application. However, the trimming parameters for the methods based on rank (Examples 1.6, 1.7 of Chapter 4) can be used in a similar fashion, and effective computational schemes for these methods have been discussed in Chapter 4. Also, the maximally resistant method in both kinds of approach gives the l_1 approximation problem. Thus there is scope for controversy in deciding between the two types of approach.

5.2 LEAST SQUARES METHODS

The central role that least squares methods play in data analysis can be justified in a number of ways. For example, let the observations f_i, $i = 1, 2, \ldots, n$ in the basic model (1.3) of Chapter 3 have the form

$$f_i = \bar{f}_i + \varepsilon_i, \qquad i = 1, 2, \ldots, n \tag{2.1}$$

where $\bar{f}_i$ is predicted by the model, and ε_i is a normally distributed random variable such that

$$\mathscr{E}\{\varepsilon_i\} = 0, \qquad \mathscr{E}\{\varepsilon_i \varepsilon_j\} = J_{ij}, \qquad i, j = 1, 2, \ldots, n \tag{2.2}$$

where $\mathscr{E}$ is the expectation. Unless otherwise stated, M is assumed to have full rank p (J is assumed positive definite), so that computing $\mathbf{x}$ by

$$\begin{aligned}\mathbf{x} &= \arg\min \mathbf{r}^{\mathrm{T}} J^{-1} \mathbf{r} \\ &= (M^{\mathrm{T}} J^{-1} M)^{-1} M^{\mathrm{T}} J^{-1} \mathbf{f}\end{aligned} \tag{2.3}$$

is equivalent to finding $\mathbf{x}$ to maximize the probability of the actual observed $\mathbf{f}$ and corresponds to an application of the principle of maximum likelihood. This approach assumes knowledge of the distribution of the ε_i but this requirement can be weakened. Let $\hat{\mathbf{x}}$ be a linear estimator of the true vector of parameters $\bar{\mathbf{x}}$ defined on the problem data; then

$$\hat{\mathbf{x}} = T\mathbf{f} \tag{2.4}$$

where $T: R^n \to R^p$ is the matrix of this linear operation. This estimator is unbiased if

$$\mathscr{E}\{\hat{\mathbf{x}}\} = \bar{\mathbf{x}} \tag{2.5}$$

is valid for the class of models expressed by equation (1.1) of Chapter 3, and (2.1). A sufficient condition is

$$TM = I_p \tag{2.6}$$

for then

$$\mathscr{E}\{\hat{\mathbf{x}}\} = T\bar{\mathbf{f}} = TM\bar{\mathbf{x}} = \bar{\mathbf{x}}.$$

To determine T we seek the unbiased estimator for which the variance of $\|\hat{\mathbf{x}} - \bar{\mathbf{x}}\|_2$ is minimized (the norm being the Euclidean norm). Now

$$\begin{aligned} \mathscr{E}\{\|\hat{\mathbf{x}} - \bar{\mathbf{x}}\|_2^2\} &= \mathscr{E}\{\|(TM - I_p)\bar{\mathbf{x}} + T\boldsymbol{\varepsilon}\|_2^2\} \\ &= \|(TM - I_p)\bar{\mathbf{x}}\|_2^2 + \operatorname{trace}(TJT^{\mathrm{T}}) \end{aligned} \tag{2.7}$$

using $\mathscr{E}\{\varepsilon\} = 0$. The first term drops out if and only if the estimator is unbiased. Thus the problem of finding T has the equivalent form

$$\min \operatorname{trace}(TJT^{\mathrm{T}}) \tag{2.8}$$

subject to the linear equality constraints (2.6). This minimization problem can be written

$$\min \sum_{i=1}^{p} \mathbf{t}_i^{\mathrm{T}} J \mathbf{t}_i \tag{2.9}$$

subject to

$$\mathbf{t}_i^{\mathrm{T}} M = \mathbf{e}_i^{\mathrm{T}}, \qquad i = 1, 2, \ldots, p, \tag{2.10}$$

where $\mathbf{t}_i$ is the ith row of T. The multiplier conditions for this problem uncouple to give

$$2\mathbf{t}_i^{\mathrm{T}} J = \mathbf{u}_i^{\mathrm{T}} M^{\mathrm{T}}, \qquad i = 1, 2, \ldots, p,$$

where $\mathbf{u}_i$ is the vector of Lagrange multipliers associated with the ith set of constraints, and this is equivalent to the matrix form

$$2TJ = U^{\mathrm{T}} M^{\mathrm{T}} \tag{2.11}$$

where $\mathbf{u}_i = \boldsymbol{\kappa}_i(U)$, $i = 1, 2, \ldots, p$. It follows that

$$I_p = \tfrac{1}{2} U^{\mathrm{T}} (M^{\mathrm{T}} J^{-1} M),$$

and

$$T = \tfrac{1}{2}U^{\mathrm{T}}M^{\mathrm{T}}J^{-1},$$
$$= (M^{\mathrm{T}}J^{-1}M)^{-1}M^{\mathrm{T}}J^{-1}. \tag{2.12}$$

Thus the estimator T minimizing the variance of $\|\hat{\mathbf{x}} - \bar{\mathbf{x}}\|_2$ is the estimator which minimizes the likelihood when the errors are normal. This minimum-variance result can actually be strengthened slightly. To see this, consider the problem of estimating $\mathbf{a}^{\mathrm{T}}\bar{\mathbf{x}}$ by means of $\mathbf{b}^{\mathrm{T}}\mathbf{f}$.

Theorem 2.1 (*Gauss–Markov*) *The minimum variance linear unbiased estimator of* $\mathbf{a}^{\mathrm{T}}\bar{\mathbf{x}}$ *is given by* $\mathbf{b}^{\mathrm{T}}\mathbf{f}$ *where*

$$\mathbf{b}^{\mathrm{T}} = \mathbf{a}^{\mathrm{T}}T. \tag{2.13}$$

Proof The estimator is unbiased if

$$\mathscr{E}\{\mathbf{b}^{\mathrm{T}}\mathbf{f}\} = \mathbf{b}^{\mathrm{T}}M\bar{\mathbf{x}} = \mathbf{a}^{\mathrm{T}}\bar{\mathbf{x}},$$

that is, if

$$\mathbf{b}^{\mathrm{T}}M = \mathbf{a}^{\mathrm{T}}. \tag{2.14}$$

It follows that any unbiased estimator has the form

$$\mathbf{b}^{\mathrm{T}} = \mathbf{a}^{\mathrm{T}}T + \mathbf{c}^{\mathrm{T}}\{I - MT\} \tag{2.15}$$

where $\mathbf{c}$ is arbitrary (note that MT is a projection matrix). Computing the variance of $\mathbf{b}^{\mathrm{T}}\mathbf{f}$ gives

$$\mathscr{E}\{(\mathbf{b}^{\mathrm{T}}\mathbf{f} - \mathbf{a}^{\mathrm{T}}\bar{\mathbf{x}})^2\} = \mathscr{E}\{((\mathbf{b}^{\mathrm{T}}M - \mathbf{a}^{\mathrm{T}}TM)\bar{\mathbf{x}} - \mathbf{b}^{\mathrm{T}}\boldsymbol{\varepsilon})^2\}$$
$$= \mathbf{b}^{\mathrm{T}}J\mathbf{b} \tag{2.16}$$

as the estimator is assumed unbiased. Now, using (2.15),

$$\begin{aligned}\mathbf{b}^{\mathrm{T}}J\mathbf{b} &= \mathbf{a}^{\mathrm{T}}TJT^{\mathrm{T}}\mathbf{a} + 2\mathbf{a}^{\mathrm{T}}TJ\{I - T^{\mathrm{T}}M^{\mathrm{T}}\}\mathbf{c} + \mathbf{c}^{\mathrm{T}}\{I - MT\}J\{I - T^{\mathrm{T}}M^{\mathrm{T}}\}\mathbf{c}\\ &= \mathbf{a}^{\mathrm{T}}TJT^{\mathrm{T}}\mathbf{a} + \mathbf{c}^{\mathrm{T}}\{I - MT\}J\{I - T^{\mathrm{T}}M^{\mathrm{T}}\}\mathbf{c}\\ &\geqslant \mathbf{a}^{\mathrm{T}}TJT^{\mathrm{T}}\mathbf{a}\end{aligned}$$

as direct expansion using (2.11), gives

$$TJ(I - T^{\mathrm{T}}M^{\mathrm{T}}) = U^{\mathrm{T}}M^{\mathrm{T}}(I - T^{\mathrm{T}}M^{\mathrm{T}}) = U^{\mathrm{T}}(M^{\mathrm{T}} - (M^{\mathrm{T}}T^{\mathrm{T}})M^{\mathrm{T}}) = 0.$$

Remark 2.1 Note that choosing $\mathbf{a} = \mathbf{e}_i$, $i = 1, 2, \ldots, p$, and applying the above theorem shows that (2.3) gives minimum-variance linear unbiased estimates for each component of $\bar{\mathbf{x}}$.

Ordinary least squares

The case where $J = \sigma^2 I$ is of particular importance and we refer to the (*ordinary*) *least squares problem*. It is convenient to consider it separately from the more general case.

Methods for solving the least squares problem fall into two classes depending on whether $M^{\mathrm{T}}M$ (the normal matrix) is formed explicitly or not.

Methods which form the normal matrix

It follows from (2.3) that $\hat{\mathbf{x}}$ satisfies the system of linear equations

$$M^{\mathrm{T}}M\mathbf{x} = M^{\mathrm{T}}\mathbf{f} \tag{2.17}$$

Remark 2.2 This is a $p \times p$ system of equations and it is reasonable to consider this system for modest p and possibly rather large n. Thus it is worth noting that this system can be formed provided the data matrix $[M \mid \mathbf{f}]$ is available a row at a time as

$$G = M^{\mathrm{T}}M = \sum_{i=1}^{n} \mathbf{m}_i \mathbf{m}_i^{\mathrm{T}}, \qquad M^{\mathrm{T}}\mathbf{f} = \sum_{i=1}^{n} f_i \mathbf{m}_i \tag{2.18}$$

By symmetry, only the upper triangle of $M^{\mathrm{T}}M$ need be held and this requires approximately $np^2/2$ multiplications in its generation.

Although standard methods can be applied to solve (2.17), G is symmetric and positive definite and advantage can be taken of this fact. Perhaps the most common approach is to make a Choleski decomposition of G:

$$G = LL^{\mathrm{T}} \tag{2.19}$$

where L is lower triangular, followed by a forward and back substitution to complete the calculation of $\mathbf{x}$. To develop this factorization let L_i, G_i be the $(i \times i)$ leading principal submatrices of L and G respectively. Then L_i and G_i can be partitioned such that

$$\left[\begin{array}{c|c} L_{i-1} & 0 \\ \hline \mathbf{l}_i^{\mathrm{T}} & L_{ii} \end{array}\right]\left[\begin{array}{c|c} L_{i-1}^{\mathrm{T}} & \mathbf{l}_i \\ \hline 0 & L_{ii} \end{array}\right] = \left[\begin{array}{c|c} G_{i-1} & \mathbf{g}_i \\ \hline \mathbf{g}_i^{\mathrm{T}} & G_{ii} \end{array}\right] \tag{2.20}$$

giving

$$L_{i-1}\mathbf{l}_i = \mathbf{g}_i \tag{2.21}$$

and

$$\mathbf{l}_i^{\mathrm{T}}\mathbf{l}_i + L_{ii}^2 = G_{ii} \tag{2.22}$$

showing that the factorization can be constructed a row at a time by solving (2.21) for $\mathbf{l}_i$ (this involves a forward substitution) and then determining L_{ii} from (2.22) to complete the specification of L_i when L_{i-1} is known. This latter calculation costs a square root in addition to the algebraic manipulation.

Remark 2.3 The triangular factors can also be developed a column at a time. This is based on the observation that

$$\sum_{j=1}^{i} L_{ij}\boldsymbol{\kappa}_j(L) = \boldsymbol{\kappa}_i(G), \qquad i = 1, 2, \ldots, p \tag{2.23}$$

so that

$$L_{ii}\boldsymbol{\kappa}_i(L)=\boldsymbol{\kappa}_i(G)-\sum_{j=1}^{i-1}L_{ij}\boldsymbol{\kappa}_j(L)$$

showing that $\boldsymbol{\kappa}_i(L)$ can be found once $\boldsymbol{\kappa}_j(L)$, $j=1,2,\ldots,i-1$ are known. The first $(i-1)$ components of the left-hand side vanish as L is lower triangular. The corresponding equations on the right-hand side are equivalent to (2.21) but here add nothing new. The next equation gives L_{ii} and is exactly (2.22). The remaining equations determine L_{ji}, $j=i+1,\ldots,p$.

Remark 2.4 The Choleski algorithm can be organized to compute a factorization having the form

$$G=LDL^{\mathrm{T}} \tag{2.24}$$

where now L has *unit* diagonal and D is a diagonal matrix (in terms of the above factorization $D_{ii}=L_{ii}^2$, $i=1,2,\ldots,p$). The advantage of this form is that it avoids the need for the calculation of square roots.

Remark 2.5 The Choleski factorization costs approximately $p^3/6$ multiplications and is remarkably accurate in the sense that $\|\bar{L}\bar{L}^{\mathrm{T}}-G\|_{\mathrm{F}}$ is very small where $\bar{L}$ is the actual computed factorization. Of course this is not the same thing as $\|\bar{L}-L\|_{\mathrm{F}}$ being very small. Some problems may be encountered when G is only just positive definite and appropriate action may involve modifying the diagonal of G in such a way as to rescue the factorization without increasing $\|\bar{L}\bar{L}^{\mathrm{T}}-G\|_{\mathrm{F}}$ unnecessarily. For references see the notes for this chapter.

Remark 2.6 The Choleski factorization lends itself to modification in situations where observations are either added to or deleted from the problem. To illustrate the basic idea consider the identity

$$\begin{bmatrix} L_1 & 0 \\ \mathbf{v}^{\mathrm{T}} & \gamma \end{bmatrix}\begin{bmatrix} L_1^{\mathrm{T}} & \mathbf{v} \\ 0 & \gamma \end{bmatrix}=\begin{bmatrix} L_2 & \mathbf{u} \\ 0 & \delta \end{bmatrix}\begin{bmatrix} L_2^{\mathrm{T}} & 0 \\ \mathbf{u}^{\mathrm{T}} & \delta \end{bmatrix}$$

which gives, on multiplying out,

$$L_1L_1^{\mathrm{T}}=L_2L_2^{\mathrm{T}}+\mathbf{u}\mathbf{u}^{\mathrm{T}}, \tag{2.25}$$

$$\mathbf{v}^{\mathrm{T}}L_1^{\mathrm{T}}=\delta\mathbf{u}^{\mathrm{T}}, \tag{2.26}$$

$$\mathbf{v}^{\mathrm{T}}\mathbf{v}+\gamma^2=\delta^2. \tag{2.27}$$

Clearly there is no restriction in assuming that $\delta=1$ (this amounts to rescaling $\mathbf{v}$, γ). To add data assume that $G=L_2L_2^{\mathrm{T}}$. Then it follows from (2.18) and (2.25) that L_1 is the appropriate update of L_2 if $\mathbf{u}$ corresponds to the new row of the data matrix. To compute L_1 we seek an orthogonal matrix Q such that

$$\begin{bmatrix} L_1^{\mathrm{T}} & \mathbf{v} \\ 0 & \gamma \end{bmatrix}=Q\begin{bmatrix} L_2^{\mathrm{T}} & 0 \\ \mathbf{u}^{\mathrm{T}} & 1 \end{bmatrix} \tag{2.28}$$

and it is readily seen that Q can be constructed by the sequence of plane rotations

$$Q = \prod_{i=1}^{p} R(i, p+1, (p+1, i)) \tag{2.29}$$

where $R(i, p+1, (p+1, i))$ is the plane rotation that mixes rows i and $p+1$ of the object matrix and zeros the element in the $(p+1, i)$ position.

To delete data (so that $G \to G - \mathbf{u}\mathbf{u}^{\mathrm{T}}$), set

$$G = L_1 L_1^{\mathrm{T}}, \qquad \mathbf{v} = L_1^{-1}\mathbf{u}$$

and seek an orthogonal matrix such that

$$\begin{bmatrix} L_2^{\mathrm{T}} & 0 \\ \mathbf{u}^{\mathrm{T}} & 1 \end{bmatrix} = Q \begin{bmatrix} L_1^{\mathrm{T}} & \mathbf{v} \\ 0 & \gamma \end{bmatrix} \tag{2.30}$$

where $\gamma^2 = 1 - \mathbf{v}^{\mathrm{T}}\mathbf{v}$ by (2.27). Again Q can be built up using plane rotations, and this gives

$$Q = \prod_{i=p}^{1} R(i, p+1, (i, p+1)). \tag{2.31}$$

where the notation indicates that rows p and $p+1$ are mixed first and the subsequent operations correspond to decreasing values of i. Note that deletion of data makes G less positive definite in a sense, so there is some possibility of problems. The process of deleting data is sometimes called *downdating*, and the condition that the downdated matrix is positive definite is $\gamma^2 > 0$ (Exercise 2.3).

Methods which operate on the data matrix directly

Here the idea is to seek an orthogonal matrix Q such that

$$M = [Q_1 \mid Q_2] \begin{bmatrix} U \\ 0 \end{bmatrix} = Q_1 U \tag{2.32}$$

where, in practice, Q is constructed to make U upper triangular. Necessarily,

$$U^{\mathrm{T}} U = L L^{\mathrm{T}}$$

so that U^{T} corresponds to the L in the Choleski factorization aside from the inherent ambiguity of the sign in extracting squareroots. Multiplying (1.1) of Chapter 3 by Q^{T} gives

$$\mathbf{s} = Q^{\mathrm{T}}\mathbf{r} = \begin{bmatrix} U\mathbf{x} - Q_1^{\mathrm{T}}\mathbf{f} \\ Q_2^{\mathrm{T}}\mathbf{f} \end{bmatrix} \tag{2.33}$$

Now $\mathbf{s}^{\mathrm{T}}\mathbf{s} = \mathbf{r}^{\mathrm{T}}\mathbf{r}$ so the solution to the least squares problem is given by

$$\mathbf{x} = U^{-1} Q_1^{\mathrm{T}}\mathbf{f} \tag{2.34}$$

while

$$\mathbf{r}^{\mathrm{T}}\mathbf{r}=\mathbf{f}^{\mathrm{T}}Q_2Q_2^{\mathrm{T}}\mathbf{f}=\mathbf{f}^{\mathrm{T}}\{I-Q_1Q_1^{\mathrm{T}}\}\mathbf{f}. \tag{2.35}$$

This shows that the solution of the least squares problem can be found once the factorization (2.32) is known, and there are two main classes of algorithms. These correspond to methods which compute Q, and those which compute Q_1 only.

Methods which compute Q

Here the most used method would seem to be the one which builds up Q as a sequence of elementary orthogonal or Householder matrices. This calculation has been discussed before in connection with the projected gradient algorithm which actually solves a least squares problem at each step, and which requires the projections onto both the column space of the matrix (determined by Q_1) and its orthogonal complement (determined by Q_2). Explicit formulae are given in Section A1.3 (Appendix 1). However, it should be noted that competitive alternatives do exist (the fast Givens algorithms).

Methods which compute Q_1

It is an immediate consequence of (2.32) that span $(\boldsymbol{\kappa}_i(M),\ i=1,2,\ldots,p)=$ span $(\boldsymbol{\kappa}_i(Q_1),\ i=1,2,\ldots,p)$. Thus the calculation of Q_1 involves the systematic orthogonalization of the columns of M. Also there is a close connection between the Choleski decomposition (in fact any factorization of this kind) and the basic orthogonalization methods. To see this, note that if W is invertible and

$$M^{\mathrm{T}}M=WW^{\mathrm{T}}$$

then

$$(W^{-1}M^{\mathrm{T}})(MW^{-\mathrm{T}})=I_p \tag{2.36}$$

so that the rows of $W^{-1}M^{\mathrm{T}}$ are orthonormal. In particular, W can be chosen as the L of the Choleski factorization and it turns out to be important to remember that L could be generated either by rows or by columns.

Classical Gram–Schmidt orthogonalization. This corresponds to developing L a row at a time and simultaneously carrying out the forward substitution implied by (2.36) on each column of M^{T} to build up Q_1^{T}. To display the calculation it is convenient to let $[\mathbf{z}]_i$ stand for the first i components of $\mathbf{z}$ and to note that

$$\boldsymbol{\kappa}_i(G)=M^{\mathrm{T}}\boldsymbol{\kappa}_i(M)=\sum_{j=1}^{n}M_{ji}\mathbf{m}_j$$

By (2.21), (2.22),

$$\mathbf{l}_i = \sum_{j=1}^{n} M_{ji} L_{i-1}^{-1} [\mathbf{m}_j]_{i-1} = \sum_{j=1}^{n} M_{ji} [\mathbf{q}_j]_{i-1},$$

$$L_{ii}^2 = \sum_{j=1}^{n} M_{ji}^2 - \mathbf{l}_i^{\mathrm{T}} \mathbf{l}_i,$$

and

$$[\mathbf{q}_j]_i = \begin{bmatrix} [\mathbf{q}_j]_{i-1} \\ \{M_{ji} - \mathbf{l}_i^{\mathrm{T}} [\mathbf{q}_j]_{i-1}\}/L_{ii} \end{bmatrix}, \qquad j = 1, 2, \ldots, n. \tag{2.37}$$

Remark 2.7 Consider the least squares problem to determine $\mathbf{y}$ to minimize $\mathbf{r}^{\mathrm{T}}\mathbf{r}$ in

$$\sum_{j=1}^{i-1} y_k \boldsymbol{\kappa}_j(M) = \boldsymbol{\kappa}_i(M) + \mathbf{r} \tag{2.38}$$

The solution to this problem making use of (2.20) is given by

$$L_{i-1} L_{i-1}^{\mathrm{T}} \mathbf{y} = \mathbf{g}_i$$

so that

$$\begin{aligned} \left\| \mathbf{r} - \sum_{j=1}^{i-1} y_j \boldsymbol{\kappa}_j(M) \right\|_2^2 &= \mathbf{r}^{\mathrm{T}}\mathbf{r} + \mathbf{y}^{\mathrm{T}} L_{i-1} L_{i-1}^{\mathrm{T}} \mathbf{y} \\ &= \mathbf{r}^{\mathrm{T}}\mathbf{r} + \mathbf{l}_i^{\mathrm{T}} \mathbf{l}_i \\ &= \sum_{j=1}^{n} M_{ji}^2 \end{aligned}$$

giving

$$L_{ii}^2 = \mathbf{r}^{\mathrm{T}}\mathbf{r} \tag{2.39}$$

Thus L_{ii}^2 is the sum of the squares of residuals in fitting to $\boldsymbol{\kappa}_i(M)$ in the least squares sense by $\boldsymbol{\kappa}_j(M)$, $j = 1, 2, \ldots, i-1$. As the fitting power of this subset of the columns increases with i there can be a tendency for L_{ii} to become small by the cancellation explicit in (2.22) unless the individual columns of M are strongly independent of each other. It follows that there can be cancellation in (2.37).

Modified Gram–Schmidt orthogonalisation. This procedure corresponds to the alternative of building up L a column at a time. This feature is stressed in this development which is due to Bill Gragg. It is convenient to assume that at the ith step the partial factorization satisfies

$$\begin{bmatrix} L_i & 0 \\ R_i & I_{p-i} \end{bmatrix} \begin{bmatrix} Q_i^{\mathrm{T}} \\ S_i^{\mathrm{T}} \end{bmatrix} = M^{\mathrm{T}}, \; Q_i^{\mathrm{T}} Q_i = I_i, \quad Q_i^{\mathrm{T}} S_i = 0. \tag{2.40}$$

The idea is now to set

$$S_i = [\mathbf{s}_i \mid \bar{S}_i], \qquad Q_{i+1} = [Q_i \mid \mathbf{q}]$$

and consider the modification (which preserves M^T)

$$\begin{bmatrix} L_{i+1} & 0 \\ R_{i+1} & I_{p-i-1} \end{bmatrix} = \begin{bmatrix} L_i & 0 \\ R_i & I_{p-i} \end{bmatrix} \begin{bmatrix} I_i & 0 \\ 0 & I_{p-i} - \mathbf{w}\mathbf{e}_1^T \end{bmatrix},$$

$$\left[I_{p-i} + \frac{\mathbf{w}\mathbf{e}_1^T}{1-w_1} \right] \begin{bmatrix} \mathbf{s}_i^T \\ \bar{S}_i^T \end{bmatrix} = \begin{bmatrix} \mathbf{q}^T \\ S_{i+1}^T \end{bmatrix}$$

where $\mathbf{w}$ has dimension $p-i$ and specifies the current orthogonalisation step. In particular

$$\mathbf{q} = \mathbf{s}_i/(1-w_1), \qquad S_{i+1}^T = \bar{S}_i^T + [\mathbf{w}]^2 \mathbf{q}^T$$

so that (2.40) is satisfied with $i := i+1$ provided

$$(L_{i+1})_{(i+1)(i+1)} = 1 - w_1 = \pm \|\mathbf{s}_i\|,$$

$$\boldsymbol{\kappa}_{i+1}(R_{i+1}) = [\mathbf{w}]^2 = -\bar{S}_i^T \mathbf{q}.$$

It is worth noting that this updating step looks like a tableau processing operation. In subsequent sections it will be applied to $[M^T \mid I]$ to generate $[Q_1^T \mid L^{-1}]$ as an initial step in developing tableau forms for several algorithms.

Remark 2.8 There are important differences in the numerical behaviour of the two algorithms. The classical Gram–Schmidt method gives sequence of vectors $\boldsymbol{\kappa}_j(Q_1)$ which rapidly lose orthogonality, and implementations of this algorithm must be modified to include 'reorthogonalisation' steps. On the other hand the modified Gram–Schmidt algorithm has a much better track record. It differs from the classical procedure in that $(L_{i+1})_{(i+1)(i+1)} = \|\mathbf{s}_i\|$ is computed as a sum of squares rather than as a subtraction so that the computed $\boldsymbol{\kappa}_i(Q_1)$ are close to correctly normalised, and in that the partially orthogonalised rows of S_i are used at the ith stage rather than the rows of M so that the potential for cancellation in this step is reduced. However, it appears that Householder transformations provide the most stable method for producing Q_1 with columns that are nearly orthogonal (see the discussion of termination on p108 of Section 2.6).

The choice between the different methods for solving (2.17) depends on considerations of both stability and computational cost. Computing cost favours the Choleski method (see Table 2.1), and as the normal matrix requires less storage than the data matrix M, and as it lends itself to being computed by reading a row of M at a time from backing store, it could well have advantages on machines with small stores.

However, orthogonal transformations can be used in conjunction with methods that modify the factorization of the data matrix when a row is added. These can be developed in a similar manner to the update step defined in (2.28). Larger blocks of data can be added by noting that if

$$M = \begin{bmatrix} M_1 \\ M_2 \end{bmatrix},$$

Table 2.1 Approximate number of multiplications in the basic factorizations.

Choleski	$np^2/2+p^3/6$	Includes cost of forming upper triangle of G
Householder	$np^2-p^3/3$	Gives Q in terms of Householder transformations
Modified Gram–Schmidt	$np^2+p^3/6$	Gives Q_1 and L^{-1} explicitly

and if M_1 possesses the orthogonal factorization

$$M_1=Q\begin{bmatrix}U_1\\0\end{bmatrix},$$

then

$$\left[\begin{array}{c|c}Q^{\mathrm{T}} & \\ \hline & I\end{array}\right]\begin{bmatrix}M_1\\M_2\end{bmatrix}=\begin{bmatrix}U_1\\0\\M_2\end{bmatrix}, \tag{2.41}$$

and the transformed matrix can be reduced to upper triangular form by orthogonal transformations which do not disturb the band of zeros. This procedure is a direct generalization of (2.28).

The error analysis for the least squares problem has been discussed briefly in Section 2.6. That there is a difference between the methods that form the normal matrix and the others can be seen by considering the resulting equations. The solution of

$$G\mathbf{x}=M^{\mathrm{T}}\mathbf{f}$$

must give a perturbation result depending on the sensitivity of G, and by (6.43) of Chapter 2, we have

$$\operatorname{cond}(G)=\operatorname{cond}(M)^2.$$

The orthogonal transformation appears to cancel the factor L analytically in the equivalent form of (2.17),

$$LL^{\mathrm{T}}\mathbf{x}=LQ_1^{\mathrm{T}}\mathbf{f}$$

and so gives hope for a better result. However, the perturbation result (6.44) of Chapter 2 shows that it is not possible to reduce the sensitivity to just a dependence on $\operatorname{cond}(M)$ unless the original system is consistent.

Remark 2.9 The methods based on the use of stable orthogonal transformations do permit error bounds having the form of the optimal perturbation results (6.44) of Chapter 2. As this is significantly more favourable than the result for the normal equations system when the model equations are close to consistent, this has lead to the general use of the orthogonal factorization methods. The following example shows that formation of the normal matrix can lead to problems in certain circumstances. Let

$$M = \begin{bmatrix} 1 & 1 & 1 \\ 1 & \varepsilon & 0 \\ 1 & 0 & \varepsilon \\ 1 & 0 & 0 \end{bmatrix}$$

then

$$G = M^{\mathrm{T}}M = \begin{bmatrix} 4 & 1+\varepsilon & 1+\varepsilon \\ 1+\varepsilon & 1+\varepsilon^2 & 1 \\ 1+\varepsilon & 1 & 1+\varepsilon^2 \end{bmatrix}$$

If $|\varepsilon|$ is so small that $1+\varepsilon^2 = 1$ but $1+\varepsilon \neq 1$ in a finite precision calculation, then G is recorded as singular to working accuracy. However, the orthogonal factorization procedures give a full rank U, and in this sense should be preferred. In the data analysis context there remain questions to be answered (like how did we come to have a data matrix with condition number $O(1/\varepsilon)$?). One possible resolution of this difficulty is sketched in Exercise 2.5.

Generalised least squares

If $J \neq \sigma^2 I$ then (2.3) is referred to as the *generalized least squares problem*. In practice the major problem may well be determining J (the important case in which J is given by a functional form involving a small number of parameters which must be estimated from the data leads to a nonlinear problem which is outside the scope of this work). However, it is assumed here that J is given a priori and is positive definite. This permits the generalized least squares problem to be reduced to the ordinary least squares problem. Let

$$J = SS^{\mathrm{T}} \tag{2.42}$$

where S is the lower triangular factor in the Choleski decomposition of J. Then

$$\mathbf{r}^{\mathrm{T}} J^{-1} \mathbf{r} = \mathbf{s}^{\mathrm{T}} \mathbf{s} \tag{2.43}$$

where

$$\mathbf{s} = S^{-1}\mathbf{r} = S^{-1}\{M\mathbf{x} - \mathbf{f}\} \tag{2.44}$$

Any reasonable method for solving the standard problem now gives a satisfactory algorithm provided J is not nearly singular. This can lead to numerical problems in circumstances in which the generalized least squares problem has a well-determined solution. In fact, it is possible that the generalized least squares problem can be interpreted to have a well-determined solution even when J is singular (the positive semidefinite case) and an alternative approach is preferred. To develop this, note that the generalized least squares problem can be written

$$\min \mathbf{s}^{\mathrm{T}}\mathbf{s} \tag{2.45}$$

subject to

$$S\mathbf{s} = M\mathbf{x} - \mathbf{f} \tag{2.46}$$

which is in the form of an equality-constrained least squares problem. By (6.1), (6.2) of Chapter 3, this problem is exactly equivalent to

$$\min \mathbf{s}^{\mathrm{T}}\mathbf{s}$$

subject to

$$Z^{\mathrm{T}}S\mathbf{s} = -Z^{\mathrm{T}}\mathbf{f} \tag{2.47}$$

where Z is any matrix of maximum rank such that

$$Z^{\mathrm{T}}M = 0 \tag{2.48}$$

To solve this problem let P be the matrix of the orthogonal factorization

$$S^{\mathrm{T}}Z = P\begin{bmatrix} T \\ 0 \end{bmatrix} \tag{2.49}$$

and set

$$P^{\mathrm{T}}\mathbf{s} = \begin{bmatrix} \mathbf{c}_1 \\ \mathbf{c}_2 \end{bmatrix} \tag{2.50}$$

The problem now becomes

$$\min \{\mathbf{c}_1^{\mathrm{T}}\mathbf{c}_1 + \mathbf{c}_2^{\mathrm{T}}\mathbf{c}_2\} \tag{2.51}$$

subject to

$$T^{\mathrm{T}}\mathbf{c}_1 = -Z^{\mathrm{T}}\mathbf{f} \tag{2.52}$$

which has the solution

$$\mathbf{c}_1 = -T^{-\mathrm{T}}Z^{\mathrm{T}}\mathbf{f},$$
$$\mathbf{c}_2 = 0. \tag{2.53}$$

This analysis leads immediately to an existence theorem for the generalizd least squares problem.

Theorem 2.2 *A necessary and sufficient condition for the generalized least squares problem to have a unique solution for every data vector* $\mathbf{f}$ *is that* $Z^{\mathrm{T}}S$ *have its full row rank.*

Remark 2.10 The condition of the theorem is equivalent to the requirement that the matrix $[S \mid M]$ have rank n. This ensures that the equality constraints (2.46) on the minimization problem (2.45) are consistent.

Corollary 2.1 *$Z^{\mathrm{T}}S$ has its full row rank if and only if $Z^{\mathrm{T}}J$ has its full row rank.*

Proof This is equivalent to the proposition

$$\mathbf{y}^{\mathrm{T}}S=0 \Leftrightarrow \mathbf{y}^{\mathrm{T}}J=0,$$

and this follows on writing

$$S=\left[\begin{array}{c|c} S_1 & 0 \\ \hline S_2 & 0 \end{array}\right], \qquad J=\begin{bmatrix} S_1S_1^{\mathrm{T}} & S_1S_2^{\mathrm{T}} \\ S_2S_1^{\mathrm{T}} & S_2S_2^{\mathrm{T}} \end{bmatrix}$$

with S_1 $k\times k$ nonsingular.

Remark 2.11 In Section 3.6 calculation of Z based on an elimination factorization of M is described. Of course, orthogonal techniques are available also. In particular, $Z=Q_2$ in the orthogonal factorization of M will do.

Exercise 2.1 Show that

$$\mathscr{E}\,\|\hat{\mathbf{x}}\|_2^2=\|\mathbf{x}\|_2^2+\text{trace}\,(M^{\mathrm{T}}J^{-1}M)^{-1}.$$

This result shows that 'on average' the length of $\hat{\mathbf{x}}$ overestimates that of the true parameter vector $\mathbf{x}$.

Exercise 2.2 Parallel the development in Remark 2.6 in the case that it is the LDL^{T} factorization of (2.24) that is being modified.

Exercise 2.3 Let G be positive definite. Under what conditions is $G-\mathbf{u}\mathbf{u}^{\mathrm{T}}$ positive definite?

Exercise 2.4 Show how to choose transformations Q and (orthogonal) P such that

$$Q[M\mid SP]=\left[\begin{array}{c|c} 0 & \\ R & W \end{array}\right]$$

where R and W are lower triangular. How does this procedure relate to the method described in the text for solving the generalized least squares problem?

Exercise 2.5 Let M_k consist of k columns of M, say $\mathbf{m}_1,\mathbf{m}_2,\ldots,\mathbf{m}_k$ in some ordering, $G_k=M_k^{\mathrm{T}}M_k$, and $\mathbf{x}_k=\begin{pmatrix}\mathbf{x}_{k-1}^{(k)}\\ x_k\end{pmatrix}$, $\mathbf{r}_k=M_k\mathbf{x}_k-\mathbf{f}$ be the solution of the corresponding least squares problem.

(i) Show that

$$\|\mathbf{r}_k\|_2^2=\|\mathbf{r}_{k-1}\|_2^2-x_k^2\,\|P_{k-1}\mathbf{m}_k\|_2^2$$

and

$$\|P_{k-1}\mathbf{m}_k\|_2^2 = 1/(G_k^{-1})_{kk}$$

where

$$P_k = I - M_k G_k^{-1} M_k^{\mathrm{T}}.$$

(ii) Discuss the use of this result: (a) for deciding which of the uncomitted columns of M to choose to give the maximum reduction in the sum of squares if this reduction is significant; and (b) for deciding if each variable (column of M) is contributing significantly to reducing the sum of squares. The resulting procedure for selecting an efficient subset of variables for solving the least squares problem is called *stepwise regression.* Is it possible for the variables in the selected subset to become nearly linearly dependent?

(iii) Show that the stepwise regression tests can be implemented neatly in conjunction with the use of orthogonal factorization methods for solving linear least squares problems (related applications are given in (3.48) for variable selection and in equations (6.19), (6.20) of Chapter 2 for variable deletion).

(iv) Consider the problem data

$$M = \begin{bmatrix} 0 & 1 & 1.1 \\ 0 & 1 & 0.9 \\ 1 & 0 & 0.9 \\ 1 & 0 & 1.1 \end{bmatrix}, \qquad \mathbf{f} = \begin{bmatrix} 1 \\ 1 \\ 1 \\ 1 \end{bmatrix}.$$

Show that the best subset of two variables does not contain the best subset of one variable.

Exercise 2.6 Useful diagnostic information on the appropriateness of a least squares fit can be obtained by considering the standardized residuals:

$$r_i' = r_i/s\sqrt{1-v_i} \quad \text{where} \quad (n-p)s^2 = \sum_{i=1}^{u} r_i^2, \quad \text{and} \quad v_i = \mathbf{m}_i^{\mathrm{T}}(M^{\mathrm{T}}M)^{-1}\mathbf{m}_i;$$

and the jack-knife residuals:

$$r_i^* = (\mathbf{m}_i^{\mathrm{T}}\mathbf{x}^{(i)} - f_i)/s^{(i)}(1 + \mathbf{m}_i^{\mathrm{T}}(M^{(i)\mathrm{T}}M^{(i)})^{-1}\mathbf{m}_i)^{\frac{1}{2}}$$

where $\mathbf{x}^{(i)}$ solves the least squares problem with the ith observation omitted.

(i) Show that

$$r_i^* = \frac{r_i}{s^{(i)}(1-v_i)^{1/2}} = \frac{r_i'}{\{(n-p)/(n-p-1) - (r_i')^2/(n-p-1)\}^{1/2}}$$

(ii) Show that these quantities can be readily calculated from the orthogonal factorization of M.

5.3 LEAST SQUARES SUBJECT TO LINEAR CONSTRAINTS

The least squares problem subject to linear constraints

$$\min_{\mathbf{x}\in X} \mathbf{r}^{\mathrm{T}} J^{-1}\mathbf{r} : X=\{\mathbf{x};\, A\mathbf{x}\geqslant \mathbf{b}\} \tag{3.1}$$

is a particular case of the general quadratic programming problem (QP):

$$\min_{\mathbf{x}\in X} F(\mathbf{x}) \qquad \text{QP}$$

where

$$F(\mathbf{x}) = -\mathbf{c}^{\mathrm{T}}\mathbf{x}+\tfrac{1}{2}\mathbf{x}^{\mathrm{T}}G\mathbf{x}, \tag{3.2a}$$

$$G=2M^{\mathrm{T}}J^{-1}M, \tag{3.2b}$$

$$\mathbf{c}=2M^{\mathrm{T}}\mathbf{f}. \tag{3.2c}$$

It possesses the important simplifying feature that, necessarily, G is at least positive semidefinite so that $F(\mathbf{x})$ is convex. If G is positive definite, so that $F(\mathbf{x})$ is strictly convex, then even the unconstrained problem has a unique solution. This stands in contrast to the linear problem which is unbounded unless constraints are present. One consequence is that the minimum no longer need be at an extreme point of the feasible region.

The problem dual to (3.1) is (compare Exercise 6.3 of Chapter 1)

$$\max_{\mathbf{x},\mathbf{u}} \mathbf{u}^{\mathrm{T}}\mathbf{b}-\tfrac{1}{2}\mathbf{x}^{\mathrm{T}}G\mathbf{x} \tag{3.3}$$

subject to the constraints

$$G\mathbf{x}-A^{\mathrm{T}}\mathbf{u}-\mathbf{c}=0, \tag{3.4}$$

$$\mathbf{u}\geqslant 0. \tag{3.5}$$

If G is positive definite then $\mathbf{x}$ can be eliminated explicitly to give the dual problem in the form

$$\max_{\mathbf{u}\geqslant 0} \mathbf{u}^{\mathrm{T}}\{\mathbf{b}-AG^{-1}\mathbf{c}\}-\tfrac{1}{2}\mathbf{u}^{\mathrm{T}}AG^{-1}A^{\mathrm{T}}\mathbf{u}. \tag{3.6}$$

This problem is again in the form QP but the constraints have the simpler form of positivity conditions on the variables. In the particular case of the *restricted least squares problem* the constraints on the primal are also positivity constraints ($A=I$, $\mathbf{b}=0$) and the dual program simplifies to

$$\min_{\mathbf{u}\geqslant 0} \mathbf{u}^{\mathrm{T}}G^{-1}\mathbf{c}+\tfrac{1}{2}\mathbf{u}^{\mathrm{T}}G^{-1}\mathbf{u} \tag{3.7}$$

The Kuhn–Tucker conditions for this problem give

$$G\mathbf{x}-\mathbf{u}-\mathbf{c}=0, \tag{3.8a}$$

$$\mathbf{x}\geqslant 0, \qquad \mathbf{u}\geqslant 0, \qquad \mathbf{x}^{\mathrm{T}}\mathbf{u}=0 \tag{3.8b}$$

where $\mathbf{x}$ has been written for the multiplier vector to bring out of the essential identity of primal and dual in this case. This problem is a linear complementarity problem. It corresponds to the simplest reduction of QP to this form.

The problem of minimizing a quadratic form subject to linear equality constraints (EQP),

$$\min_{\mathbf{x}} F(\mathbf{x}) \qquad \text{EQP}$$

subject to

$$D\mathbf{x} = \mathbf{d} \tag{3.9}$$

where $D: R^p \to R^q$, $q < p$, rank $(D) = q$, can be solved directly and for this reason proves to be of importance in developing algorithms for QP. Here the Kuhn–Tucker conditions give directly a system of linear equations for the solution vector and the vector of multipliers. We have

$$G\mathbf{x} - \mathbf{c} = D^{\mathrm{T}}\mathbf{u},$$

$$D\mathbf{x} - \mathbf{d} = 0,$$

or in matrix form,

$$\begin{bmatrix} G & -D^{\mathrm{T}} \\ -D & 0 \end{bmatrix} \begin{bmatrix} \mathbf{x} \\ \mathbf{u} \end{bmatrix} = \begin{bmatrix} \mathbf{c} \\ -\mathbf{d} \end{bmatrix} \tag{3.10}$$

Lemma 3.1 *Let D have the factorization*

$$D^{\mathrm{T}} = [Q_1 \mid Q_2] \begin{bmatrix} U \\ 0 \end{bmatrix} \tag{3.11}$$

where Q is orthogonal and U is $q \times q$ *upper triangular. Then EQP has a unique solution if and only if* $Q_2^{\mathrm{T}} G Q_2$ *is nonsingular.*

Proof Direct calculation gives

$$\left[\begin{array}{c|c} Q^{\mathrm{T}} & \\ \hline & I_q \end{array}\right] \left[\begin{array}{c|c} G & -D^{\mathrm{T}} \\ \hline -D & 0 \end{array}\right] \left[\begin{array}{c|c} Q & \\ \hline & I_q \end{array}\right] = \begin{bmatrix} Q_1^{\mathrm{T}} G Q_1 & Q_1^{\mathrm{T}} G Q_2 & -U \\ Q_2^{\mathrm{T}} G Q_1 & Q_2^{\mathrm{T}} G Q_2 & 0 \\ -U^{\mathrm{T}} & 0 & 0 \end{bmatrix}$$

and the desired result follows by inspection.

Remark 3.1 This result should be compared with Theorem 2.2, to which it is closely related. However, it is assumed here that the matrix of the system of equality constraints D has full rank so that the only possibility for rank deficiency occurs in G. If $G = M^{\mathrm{T}} J^{-1} M$ and J is nonsingular, then the theorem has the equivalent statement that EQP has a unique solution if and only if $\{\mathbf{v}; M\mathbf{v} = 0\} \cap \{\mathbf{v}; D\mathbf{v} = 0\} = \varnothing$.

An active set algorithm for QP

This result forms the basis of an important class of *active set* methods for solving QP problems. The idea is to use the set of constraints active at the current point to define an EQP, and to use the solution of this problem to define a descent direction. By this means we can hope to achieve a finite algorithm in which the combinatorial element enters in the selection of the active constraints defining the EQP. Other choices of descent direction are possible (see Exercise 3.3), but the discussion of the steepest descent algorithm applied to a quadratic form in Section 1.7 shows that the straightforward application of the reduced and projected gradient algorithms is likely to be inefficient. Let the current point be $\mathbf{x}$ (it is assumed to be feasible), and let the index set σ point to the active constraints. It is assumed that $|\sigma| \leq p$ – essentially a nondegeneracy assumption – and that rank $(A_\sigma) = |\sigma|$. We consider the EQP in the form

$$\min_{\mathbf{t}} F(\mathbf{x}+\mathbf{t})$$

subject to

$$A_\sigma \mathbf{t} = 0. \tag{3.12}$$

The Kuhn–Tucker conditions give the system of equations

$$G\mathbf{t} + \mathbf{g} - A_\sigma^{\mathrm{T}} \mathbf{u} = 0, \tag{3.13a}$$

$$A_\sigma \mathbf{t} = 0, \tag{3.13b}$$

where

$$\mathbf{g} = G\mathbf{x} - \mathbf{c} = \nabla F(\mathbf{x})^{\mathrm{T}} \tag{3.14}$$

There are two cases to consider.

(i) $\mathbf{g} \notin H = \operatorname{span}(\mathbf{a}_{\sigma(i)},\ i = 1, 2, \ldots, |\sigma|)$. Here (3.13) determines a unique, nontrivial $\mathbf{t}$ under the conditions specified in Lemma 3.1. As $\mathbf{t}$ gives the step to the minimum it clearly defines a direction of descent. To show this algebraically, note that from (3.13):
(a) $\mathbf{g}^{\mathrm{T}}\mathbf{t} = -\mathbf{t}^{\mathrm{T}} G\mathbf{t}$, and
(b) $\mathbf{t} \in \operatorname{span}(\boldsymbol{\kappa}_i(Q_2),\ i = 1, 2, \ldots, p - |\sigma|)$
where Q_2 spans the null space of A_σ. Then

$$\text{(b)} \Rightarrow \mathbf{t} = Q_2 \mathbf{w}$$

so that

$$\mathbf{g}^{\mathrm{T}}\mathbf{t} = -\mathbf{w}^{\mathrm{T}} Q_2^{\mathrm{T}} G Q_2 \mathbf{w} < 0$$

provided EQP has a unique solution.

(ii) $\mathbf{g} \in H$. This situation is analogous to that encountered, for example, in the reduced and projected gradient algorithms for linear programming. That is, the current EQP gives $\mathbf{t} = 0$ and a binding constraint must be relaxed to generate a downhill direction. That $\mathbf{t} = 0$ follows from $\mathbf{g} \in H$ which implies that $\mathbf{g} = A_\sigma^{\mathrm{T}}\mathbf{u}$, $\mathbf{t} = 0$ satisfies (3.13) and hence is the only solution by uniqueness. If $\mathbf{u} \geq 0$ then $\mathbf{x}$ is optimal for QP. Otherwise let $u_k < 0$ and consider the modified system

$$G\bar{\mathbf{t}} + \mathbf{g} - A_\sigma^{\mathrm{T}}\bar{\mathbf{u}} = 0; \tag{3.15a}$$

$$A_\sigma\bar{\mathbf{t}} = \gamma_k\mathbf{e}_k. \tag{3.15b}$$

That this system has a unique solution for each γ_k follows from the assumption of uniqueness for (3.12). Also $\bar{\mathbf{t}}$ is downhill provided $\gamma_k > 0$, for

$$\mathbf{g}^{\mathrm{T}}\bar{\mathbf{t}} = \mathbf{u}^{\mathrm{T}}A_\sigma\bar{\mathbf{t}} = \gamma_k u_k < 0.$$

However, there is a 'natural' choice for γ_k which is the one for which $\bar{u}_k = 0$ and $\bar{\mathbf{t}}$ is the step to the minimum of the EQP with the kth constraint deleted. The condition that $\gamma_k > 0$ in this case is exactly the condition that the modified EQP has a unique solution. This is shown here under the assumption that G is invertible; the more general case is left to Exercise 3.2. It follows directly from (3.15) that

$$\begin{bmatrix} \bar{\mathbf{t}} \\ \bar{\mathbf{u}} \end{bmatrix} = \begin{bmatrix} \mathbf{t} \\ \mathbf{u} \end{bmatrix} - \gamma_k \begin{bmatrix} G & -A_\sigma^{\mathrm{T}} \\ -A_\sigma & 0 \end{bmatrix}^{-1} \begin{bmatrix} 0 \\ \mathbf{e}_k \end{bmatrix} \tag{3.16}$$

whence

$$0 = \bar{u}_k = u_k + \gamma_K \mathbf{e}_k^{\mathrm{T}}(A_\sigma G^{-1}A_\sigma^{\mathrm{T}})^{-1}\mathbf{e}_k \tag{3.17}$$

as the full rank assumption on A_σ ensures $A_\sigma G^{-1}A_\sigma^{\mathrm{T}}$ is positive definite, so that $u_k < 0$ implies $\gamma_k > 0$.

Selecting k to determine the constraint to relax according to the rule

$$k = \arg\min\,(u_i,\ i = 1, 2, \ldots, |\sigma|) \tag{3.18a}$$

corresponds to a USE test. As in Chapter 2, it is not a scale-invariant test, and the simplest such test corresponds to

$$k = \arg\min\,(u_i\,\|\mathbf{a}_{\sigma(i)}\|_2,\ i = 1, 2, \ldots, |\sigma|) \tag{3.18b}$$

which suggests that it is a good idea to normalize the rows of A to have norm unity as a preliminary step. The corresponding NSE test is easy to develop apparently only in the case in which G is positive definite, and it is sensible to define

$$\|\mathbf{t}\|_G^2 = \mathbf{t}^{\mathrm{T}}G\mathbf{t}. \tag{3.19}$$

Note that the scaling in question here is the relative scaling of the rows of the constraint matrix A. It follows that (3.19) is a reasonable measure for specifying directional derivatives only if all directions are given comparable weighting. In other words, contours of $\|\cdot\|_G$ should not be too strongly elliptical. Apart from this reservation, the development is similar to that given for the linear programming case. The defining equations give (remember $\bar{u}_k = 0$)

$$\mathbf{t}^T G \mathbf{t} = \|\mathbf{t}\|_G^2 = -\mathbf{g}^T\mathbf{t}, \qquad \bar{\mathbf{t}}^T G \bar{\mathbf{t}} = -\mathbf{g}^T\bar{\mathbf{t}}$$

so that, using (3.16),

$$\begin{aligned}\left(\frac{\mathbf{g}^T\bar{\mathbf{t}}}{\|\bar{\mathbf{t}}\|_G}\right)^2 &= \left(\frac{\mathbf{g}^T\mathbf{t}}{\|\mathbf{t}\|_G}\right)^2 - \gamma_k \mathbf{g}^T G^{-1} A_\sigma^T (A_\sigma G^{-1} A_\sigma^T)^{-1}\mathbf{e}_k \\ &= \left(\frac{\mathbf{g}^T\mathbf{t}}{\|\mathbf{t}\|_G}\right)^2 - \gamma_k u_k \\ &= \left(\frac{\mathbf{g}^T\mathbf{t}}{\|\mathbf{t}\|_G}\right)^2 + \frac{u_k^2}{\mathbf{e}_k^T (A_\sigma G^{-1} A_\sigma^T)^{-1}\mathbf{e}_k},\end{aligned} \tag{3.20}$$

where it has again been assumed that G is nonsingular (but here this is a necessary condition for (3.19) to provide a sensible norm), and where (3.13) has been solved for $\mathbf{u}$, giving

$$\mathbf{u} = (A_\sigma G^{-1} A_\sigma^T)^{-1} A_\sigma G^{-1}\mathbf{g}. \tag{3.21}$$

The formula (3.20) should be compared with (6.23) of Chapter 2. In fact, a tableau form for the active set strategy is given below and in this organization of the calculation the NSE test is handled exactly as in the tableau form of the projected gradient algorithm in Chapter 2, Section 2.6.

The second phase of the descent calculation involves a step in the descent direction. Here this can be made until either

(a) a new constraint becomes active, or
(b) the minimum of the current EQP is reached.

This corresponds to

$$\mathbf{x} := \mathbf{x} + \lambda\mathbf{t}$$

where

$$\lambda = \min\left(1, \min_{\substack{j\in\sigma^c \\ \mathbf{a}_j^T\mathbf{t}<0}} \frac{b_j - \mathbf{a}_j^T\mathbf{x}}{\mathbf{a}_j^T\mathbf{t}}\right). \tag{3.22}$$

To compute the reduction achieved in this step we solve (3.13) for $\mathbf{t}$ (equivalently $\bar{\mathbf{t}}$ on the updated constraint set $A_{\bar{\sigma}}$) to obtain

$$\mathbf{t} = -Q_2(Q_2^T G Q_2)^{-1} Q_2^T \mathbf{g}, \tag{3.23}$$

and

$$F(\mathbf{x}+\lambda\mathbf{t}) - F(\mathbf{x}) = \left(\lambda - \frac{\lambda^2}{2}\right)\mathbf{g}^T\mathbf{t}, \tag{3.24}$$

where as usual columns of Q_2 span the null space of A_σ.

Organization of the computation can depend on the need to guard against ill-conditioned G. If this is a possibility then perhaps the best strategy is one which keeps and updates an orthogonal fractorization of A_σ^T. This term disturbs the Choleski factorization of $Q_2^T G Q_2$ which is then restored by systematic use of orthogonal transformations. This approach has the advantage of working only with the matrix which necessarily is well conditioned at the solution if the QP is well behaved.

However, if G is well conditioned then the active set algorithm lends itself to a tableau form of organization. The first step is to make the factorization

$$G = LL^T \tag{3.25}$$

and transform the EQP to obtain

$$-L^T\mathbf{t} = L^{-1}\mathbf{g} - L^{-1}A_\sigma^T\mathbf{u}, \tag{3.26a}$$

and

$$A_\sigma L^{-T}L^T\mathbf{t} = 0 \tag{3.26b}$$

where

$$L^{-1}\mathbf{g} = L^T\mathbf{x} - L^{-1}\mathbf{c} \tag{3.27}$$

This shows that $L^T\mathbf{t}$ satisfies a projected gradient system so that it is no surprise that the organizational details are similar to those developed in Section 2.6. To set up the tableau, let

$$L^{-1}A_\sigma^T = [Q_1 \mid Q_2]\begin{bmatrix} U \\ 0 \end{bmatrix}, \tag{3.28}$$

$$W = Q^T[L^{-1}A^T \mid L^{-1} \mid L^{-1}\mathbf{g}] \tag{3.29}$$

and

$$\mathbf{w}^T = [\mathbf{x}^T A^T - \mathbf{b}^T \mid \mathbf{x}^T]. \tag{3.30}$$

The multiplier vector is given by

$$\mathbf{u} = U^{-T}Q_1^T L^{-1}\mathbf{g} \tag{3.31}$$

and, as $L^T\mathbf{t}$ is in the span of the columns of Q_2,

$$Q_2^T L^T\mathbf{t} = -Q_2^T L^{-1}\mathbf{g}. \tag{3.32}$$

Now, for $i = 1, 2, \ldots, n$,

$$w_i = \mathbf{a}_i^T\mathbf{x} - b_i$$

so that

$$\begin{aligned} \frac{dw_i}{d\lambda} &= \mathbf{a}_i^T\mathbf{t} = \mathbf{a}_i^T L^{-T}L^T\mathbf{t}, \\ &= \mathbf{a}_i^T L^{-T}Q_2 Q_2^T L^T\mathbf{t}, \\ &= (-\mathbf{a}_i^T L^{-T}Q_2)(Q_2^T L^{-1}\mathbf{g}); \end{aligned} \tag{3.33}$$

while for $i = n+1, \ldots, n+p$,

$$w_i = x_i$$

so that

$$\frac{dw_i}{d\lambda} = t_i = \mathbf{e}_i^T L^{-T} L^T \mathbf{t}$$
$$= -(\mathbf{e}_i^T L^{-T} Q_2)(Q_2^T L^{-1} \mathbf{g}). \tag{3.34}$$

Combining (3.33), (3.34) gives

$$\frac{d\mathbf{w}^T}{d\lambda} = -\begin{bmatrix} 0 \\ Q_2^T L^{-1}\mathbf{g} \end{bmatrix}^T Q^T [L^{-1}A^T \mid L^{-1}]. \tag{3.35}$$

Also, at the new point,

$$\mathbf{g} := \mathbf{g} + \lambda G \mathbf{t}$$

so that

$$Q_2^T L^{-1}\mathbf{g} := Q_2^T L^{-1}\mathbf{g} + \lambda Q_2^T L^T \mathbf{t}$$
$$= (1-\lambda) Q_2^T L^{-1}\mathbf{g}$$

whence

$$Q^T L^{-1}\mathbf{g} := \begin{bmatrix} Q_1^T L^{-1}\mathbf{g} \\ (1-\lambda) Q_2^T L^{-1}\mathbf{g} \end{bmatrix}. \tag{3.36}$$

Also, by (3.24),

$$\Delta F = \left(\lambda - \frac{\lambda^2}{2}\right)\mathbf{g}^T \mathbf{t}$$
$$= \left(\lambda - \frac{\lambda^2}{2}\right)\mathbf{g}^T L^{-T} L^T \mathbf{t}$$
$$= \left(\lambda - \frac{\lambda^2}{2}\right)\mathbf{g}^T L^{-T} Q_2 Q_2^T L^T \mathbf{t}$$
$$= -\left(\lambda - \frac{\lambda^2}{2}\right) \|Q_2^T L^{-1}\mathbf{g}\|_2^2. \tag{3.37}$$

This shows that the quantities required in the descent algorithm can be computed readily from the current tableau. Updating the tableau to take account of modifications in $L^{-1}A_\sigma^T$ follows exactly as described in Section 2.6.

Example 3.1 The above procedure has been applied to the problem

$$\min_{\mathbf{x}} \tfrac{1}{2}\mathbf{r}^T\mathbf{r}$$

subject to the constraints

$$\mathbf{r} \geq 0.$$

The particular application occurred in diffuse scattering studies in crystallography. Here

$$G = M^T M = LL^T, \qquad A = M,$$

so that forming $L^{-1}M^T$ corresponds to orthogonalizing the rows of M^T and the modified Gram–Schmidt orthogonalization procedure can be used to initialize the tableau. As a by-product the solution of the unconstrained least squares problem is obtained. If this is infeasible then in this particular application physical considerations are available to give a suitable starting point. The current application has been made using USE tests in determining the descent vector.

In the particular case that

$$\text{rank}\, M = p = n - 1$$

then the approach used in Section 3.6 yields an explicit solution. In this case there is a unique $\mathbf{z}$ (apart from sign) satisfying

$$\mathbf{z}^T M = 0,$$
$$\mathbf{z}^T \mathbf{z} = 1,$$

and we can replace the equality constraints

$$\mathbf{r} = M\mathbf{x} - \mathbf{f}$$

by

$$\mathbf{z}^T \mathbf{r} = \gamma = -\mathbf{z}^T \mathbf{f}.$$

Thus the problem becomes

$$\min \mathbf{r}^T \mathbf{r}$$

subject to

$$\mathbf{z}^T \mathbf{r} = \gamma, \qquad \mathbf{r} \geqslant 0.$$

The Kuhn–Tucker conditions give

$$\mathbf{r} = \mu \mathbf{z} + \boldsymbol{\lambda}, \qquad \boldsymbol{\lambda} \geqslant 0, \qquad \lambda_i r_i = 0, \qquad i = 1, 2, \ldots, n.$$

It follows that

$$0 = \lambda_i \mu z_i + \lambda_i^2, \qquad i = 1, 2, \ldots, n$$

so that

$$\begin{aligned} \lambda_i &= 0, \qquad \text{sgn}\,(\mu z_i) \geqslant 0, \\ &= -\mu z_i, \qquad \text{sgn}\,(\mu z_i) < 0. \end{aligned}$$

Also

$$\begin{aligned} \mathbf{z}^T \mathbf{r} = \gamma &= \mu\, \|\mathbf{z}\|_2^2 - \sum_{\mu z_i < 0} \mu z_i^2 \\ &= \mu \sum_{\mu z_i > 0} z_i^2. \end{aligned}$$

Thus

$$\operatorname{sgn} \gamma = \operatorname{sgn} \mu,$$

$$\mu = \gamma \Big/ \sum_{\gamma z_i > 0} z_i^2,$$

and

$$r_j = \max(\gamma z_j, 0) \Big/ \sum_{\gamma z_i > 0} z_i^2, \qquad j = 1, 2, \ldots, n.$$

Remark 3.2 It is easy to give conditions under which the active set algorithm is finite. Essentially there are only a finite number of possible EQPs and the algorithm either solves an EQP or tangles with an extra constraint at each step. Thus it threads through a finite number of possibilities, always reducing the function value.

Note that a nondegeneracy assumption that at most one new constraint becomes active in each descent step is implicit in the above development. The alternative situation is important, in particular, in developing a descent direction when $\mathbf{g} \in H$ and there are too many constraints in A_σ. In this case it is necessary to find $\mathbf{t}$ such that the direction reduces F. This requires that $\mathbf{t}$ satisfy

$$\mathbf{g}^T\mathbf{t} < 0$$

and is feasible for the set of active constraints so that

$$A_\sigma \mathbf{t} \geq 0.$$

This system is readily identified with the systems considered in the linear programming case. Thus we seek a direction of recession for the linear programming problem

$$\min_{\mathbf{t}} \mathbf{g}^T\mathbf{t} \tag{3.38a}$$

subject to

$$A_\sigma \mathbf{t} \geq 0 \tag{3.38b}$$

and this can be found by considering also a perturbed program which is initially nondegenerate. By the standard recursive argument the desired descent vector is found in a finite number of steps.

The restricted least squares problem

The least squares problem subject to positivity constraints on the variables can be solved effectively by the active set algorithm. However, its special form and relatively frequent occurrence have lead to the development of procedures having the characteristic features that:

(a) a least squares problem for a subset of the variables is solved at each stage (it is assumed without further comment that this is possible); and
(b) some kind of search of the possible subsets is undertaken to find the optimum one.

One justification offered for this approach is that programs for solving least squares problems are readily available. However, there are $2^p - 1$ nontrivial subsets of the p independent variables, so that for all but the smallest problems a simple-minded approach would seem inappropriate. For this reason attention has been directed to ways of improving the efficiency of search of the set of possible subproblems. Also, if the search steps are restricted to adjacent sets of variables (sets obtained from the given set by at most one addition or deletion) then the procedures for adding or deleting a variable from a least squares problem can be employed (this has been discussed in Section 2.6 when considering the projected gradient algorithm which solves a thinly disguised least squares problem at each step), and this results in considerable computing economies. Not surprisingly, there are close connections between the search strategies restricted to adjacent sets at each stage, and methods for solving the linear complementarity form of the restricted least squares problem (3.8).

Basic to this class of algorithms is the manner in which track is kept of the subsets of the independent variables defining the possible subproblems which must be checked for feasibility (independent variables positive) and optimality. The subproblems are conveniently listed using a search tree. An example in the case $p=3$ is given in Figure 3.1.

Here each node determines a subproblem, but also defines the operation of branching on a variable. In this operation a particular variable is fixed for inclusion in all left-descending branches from the given node and excluded from all right-descending branches. In Figure 3.1 the variables for branching are written to the right of the comma in the bracketed list which defines the subproblem, while those that are fixed are written to the left. Note that all nodes on a left-descending branch specify the same subproblem.

An efficient search method is one which visits a comparatively small number of nodes in determining the optimum. An important aspect of this is being able to decide at the current node that no descendant of that node can be optimal. If this decision can be made then the search tree can be 'pruned' at this point. For this problem there is a simple pruning strategy possible. Note that subproblem minima must increase on all right-descending branches of the search tree as at each step a further variable is fixed at zero level. It follows that if the current subproblem minimum is greater than a

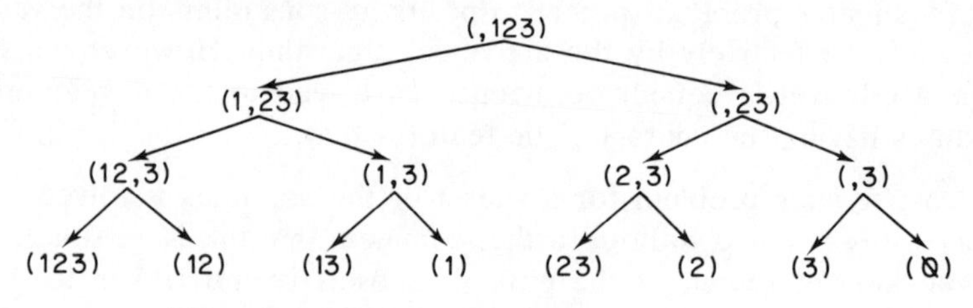

Figure 3.1 Listing of all subsets in case $p=3$

previously encountered minimum for a subproblem with a feasible solution, then there is no point in considering the descendents of this node in the search tree.

This general approach is called *branch and bound* and is a valuable problem-solving device. To make it more formal in the present context we set $\sigma = \{i;\ x_i = 0\}$ and denote by F_σ the restricted sum of squares obtained by setting $x_i = 0$, $i \in \sigma$ in F, and by $\mathbf{x}_\sigma$ the solution of the restricted least squares problem. Note that σ also identifies a node of the search tree. However, there is a possible ambiguity here as $\mathbf{x}_\sigma$ may contain accidental zero components, so that more than one node of the tree can give the same $\mathbf{x}_\sigma$. This ambiguity can be removed by defining the node of the search tree by always taking the largest σ when this problem occurs.

Algorithm

(i) *Set* $\sigma = \{\varnothing\}$, bnd $= +\infty$
(ii) *While there are unprocessed nodes do* (iii)–(vi).
(iii) *Until* $\mathbf{x}_\sigma \geqslant 0$ *or* $F_\sigma(\mathbf{x}_\sigma) \geqslant$ bnd *do*
  min $F_\sigma(\mathbf{x})$
  select variable on which to branch
  (*a usual choice is* $j = \arg\min\ (\mathbf{x}_\sigma)_i,\ i \in \sigma^c$))
  update σ.
(iv) *If* $\mathbf{x}_\sigma \geqslant 0$ *then* bnd := min (bnd, $F_\sigma(\mathbf{x}_\sigma)$).
(v) *Prune at current node.*
(vi) *Back up to unprocessed node.*

There is still scope to improve this procedure. Matters which deserve further attention include:

(a) the process of selecting the variable on which to branch;
(b) the process of selecting the node to back up to after the processing of the current branch is terminated; and
(c) recognizing that a minimum has been obtained other than by exhausting all possibilities.

Let σ_i denote the current node in some enumeration of the search tree, $\mathbf{x}_i$ the corresponding subproblem solution, and $\mathbf{y}_i$ some other vector satisfying $(\mathbf{y}_i)_j = 0$, $j \in \sigma_i$, and $(\mathbf{y}_i)_j > 0$, all j such that $(\mathbf{x}_i)_j < 0$, so that $F(\mathbf{y}_i) > F(\mathbf{x}_i)$ if $\mathbf{x}_i \not\geqslant 0$.

Lemma 3.2 *Let* $\mathbf{y}_i \geqslant 0$. *Then there is some descendant* $\sigma_r \supseteq \sigma_i$ *such that* $\mathbf{x}_r \geqslant 0$ *and* $F_r(\mathbf{x}_r) \leqslant F_i(\mathbf{y}_i)$ *with strict inequality if* $\mathbf{x}_i \not\geqslant 0$.

Proof The result is trivial if $\mathbf{x}_i \geqslant 0$. Otherwise, assuming no ties – this only leads to a larger index set at the next stage – let

$$k = \arg\min \left\{ \frac{(\mathbf{y}_i)_j}{|(\mathbf{x}_i)_j|}, \qquad (\mathbf{x}_i)_j < 0 \right\},$$

$$\sigma_{i+1} = \sigma_i \cup \{k\},$$

and

$$\mathbf{y}_{i+1} = \frac{(\mathbf{y}_i)_k}{(\mathbf{y}_i)_k + |(\mathbf{x}_i)_k|}\,\mathbf{x}_i + \frac{|(\mathbf{x}_i)_k|}{(\mathbf{y}_i)_k + |(\mathbf{x}_i)_k|}\,\mathbf{y}_i.$$

Convexity gives

$$F(\mathbf{x}_{i+1}) \leqslant F(\mathbf{y}_{i+1}) < F(\mathbf{y}_i)$$

and clearly the process can be repeated until $\mathbf{x}_{i+s} \geqslant 0$ ($\sigma_{i+s} = \{1, 2, \ldots, p\}$ is presumably possible) and

$$F(\mathbf{x}_{i+s}) \leqslant F(\mathbf{y}_{i+s}) \leqslant \cdots < F(\mathbf{y}_i).$$

The result which is used to refine the branching phase of the algorithm is as follows.

Theorem 3.1 *At any node σ_i such that $j \in \sigma_i^c \Rightarrow |(\mathbf{x}_i)_j| > 0$ it is only necessary to branch on j such that $(\mathbf{x}_i)_j < 0$ in order to find the best feasible solution among the descendents of σ_i (index sets σ_r such that $\sigma_r \supseteq \sigma_i$).*

Proof Let $\nu_i = \{j;\, (\mathbf{x}_i)_j < 0\}$. Let $\mathbf{x}_r$ be the best feasible descendant of $\mathbf{x}_i$ and assume that it could not be reached by a sequence of branches involving at least one $j \in \nu_i$. This implies that

$$(\mathbf{x}_r)_j > 0, \forall\, j \in \nu_i.$$

Then there exists $\mathbf{y}_i \geqslant 0$ constructed as a convex combination of $\mathbf{x}_r$ and $\mathbf{x}_i$ (and therefore feasible for σ_i), and convexity gives

$$F(\mathbf{y}_i) \leqslant \theta F(\mathbf{x}_i) + (1-\theta) F(\mathbf{x}_r) < F(\mathbf{x}_r).$$

However, by Lemma 3.2 there exists a feasible descendant of σ_i (say σ_s) such that $F(\mathbf{x}_s) \leqslant F(\mathbf{y}_i)$. This contradicts the assumption that σ_r defines the best feasible descendant.

Remark 3.3 This result shows that at any node it is necessary to branch only on j such that $(\mathbf{x}_\sigma)_j < 0$. All other branches can be pruned. For example, consider the tree in Figure 3.1. If $x_1 > 0$, and x_2, $x_3 < 0$, then only the left half of the search tree need be considered and the branches corresponding to deleting 1 can be ignored.

Optimality can be tested using the Kuhn–Tucker conditions. A straightforward application using (3.8) gives the following result.

Theorem 3.2 *If* $\mathbf{x}_\sigma \geq 0$, *and* $M^TM\mathbf{x}_\sigma - M^T\mathbf{f} \geq 0$ *then* $\mathbf{x}_\sigma$ *solves the restricted least squares problem.*

In the branch-and-bound framework it can be difficult to recognize optimality and it is necessary to continue the algorithm until all nodes have either been processed or pruned. In this case much of the work can be involved in confirming that a minimum has been obtained. Theorem 3.2 shows that this can be avoided in the simple application considered here.

The next result relates to carrying out the bracktrack step and again the multiplier conditions are useful.

Theorem 3.3 *Let* $\mathbf{u}_r$ *be the multiplier vector associated with the EQP on* σ_r. *If* $\mathbf{x}_r \geq 0$ *and* $(\mathbf{u}_r)_k < 0$ *then there is a descendant* σ_s *of* $\sigma_{r-1} = \{i;\ (\mathbf{x}_r)_i = 0,\ i \neq k\}$ *which is feasible and for which* $F(\mathbf{x}_s) < F(\mathbf{x}_r)$.

Proof It follows from (3.8a) that

$$\mathbf{u}_r = \nabla F(\mathbf{x}_r)^T$$

so that, using complementarity and the subgradient inequality,

$$\begin{aligned} F(\mathbf{x}_{r-1}) &\geq F(\mathbf{x}_r) + \mathbf{u}_r^T(\mathbf{x}_{r-1} - \mathbf{x}_r) \\ &= F(\mathbf{x}_r) + (\mathbf{u}_r)_k(\mathbf{x}_{r-1})_k. \end{aligned}$$

Also, for $\gamma > 0$ small enough,

$$F(\mathbf{x}_r + \gamma \mathbf{e}_k) = F(\mathbf{x}_r) + \gamma(\mathbf{u}_r)_k + \tfrac{1}{2}\gamma^2 G_{kk} < F(\mathbf{x}_r).$$

Thus

$$F(\mathbf{x}_{r-1}) = F_{r-1}(\mathbf{x}_{r-1}) < F_r(\mathbf{x}_r) = F(\mathbf{x}_r),$$

whence

$$-(\mathbf{u}_r)_k(\mathbf{x}_{r-1})_k > F(\mathbf{x}_r) - F(\mathbf{x}_{r-1}) > 0$$

and, as

$$\begin{aligned} (\mathbf{u}_r)_k &< 0, \\ (\mathbf{x}_{r-1})_k &> 0. \end{aligned}$$

It follows that there is a convex combination $\mathbf{y}_{r-1}$ of $\mathbf{x}_r$, $\mathbf{x}_{r-1}$ which is feasible for σ_{r-1}, and the result now follows from Lemma 3.2.

Remark 3.4 The proof technique used in Lemma 3.2 generated a sequence σ_i for which the objective function cannot increase too rapidly as it is bounded by the $F(\mathbf{y}_i)$ which are decreasing. Thus it could be more difficult to bound than sequences produced by more heuristic means and so provides a systematic way of generating descendants of the node σ_{r-1} produced by a backtrack step.

Example 3.2 The data in Table 3.1 defines M, $\mathbf{f}$ for a restricted least squares problem with $n = 10$, $p = 6$.

Table 3.1 Data for the restricted least squares example.

M						$\mathbf{f}$
1.0	4.70	7.89	7.93	3.47	8.35	6.94
1.0	3.10	3.46	5.35	2.97	7.11	5.77
1.0	8.34	6.68	1.75	8.68	8.90	8.04
1.0	4.62	2.69	9.20	5.39	1.60	5.12
1.0	1.03	6.22	6.25	4.75	3.61	7.82
1.0	3.26	5.64	9.10	6.53	4.70	13.26
1.0	2.27	5.34	5.15	7.27	3.16	13.47
1.0	7.27	3.64	6.65	7.77	3.78	12.49
1.0	5.93	6.65	8.65	9.77	0.92	11.06
1.0	0.47	0.45	1.63	1.90	8.66	14.40

This problem solved by the basic branch-and-bound algorithm required 32 of the possible 64 nodes to be inspected. This is cut to 11 by the use of the Kuhn–Tucker conditions. The details of the computation with Theorem 3.3 used to specify the backtrack node and the procedure of Lemma 3.2 to determine the subsequent descendants give the results reported in Table 3.2. This indicates that a further substantial improvement has been obtained. Other testing confirms the effectiveness of this final form of the algorithm for the restricted least squares problem (Example 3.3).

The formulation of the restricted least squares problem as a complementarity problem underlines the essential interchangeability of the roles of $\mathbf{x}$ and $\mathbf{u}$. This can be exploited to reduce the amount of preliminary computation required before inspecting nodes of the search tree. In this second phase, each step either zeros an x_i freeing the corresponding u_i or vice versa. The basic problem is to solve (3.8a) written in terms of the data of the restricted least squares problem

$$M^{\mathrm{T}}M\mathbf{x} - M^{\mathrm{T}}\mathbf{f} = \mathbf{u}$$

$$\mathbf{u} \geqslant 0, \qquad \mathbf{x} \geqslant 0, \qquad \mathbf{x}^{\mathrm{T}}\mathbf{u} = 0,$$

Table 3.2 Successive steps of the improved algorithm.

i	x_1	x_2	x_3	x_4	x_5	x_6	$F(\mathbf{x})$
1	−7.27	−1.89	−1.34	0.92	2.91	1.70	32.09
2	0	−1.52	−1.05	0.44	2.23	1.08	36.42
3	0	0	−0.95	0.49	1.21	0.84	102.00
4	0	0	0	0.25	0.84	0.62	127.18
5	13.31	0	0	−0.53	0.22	−0.30	93.51
6	7.52	0	0	0	0.33	0.08	103.49

and the first step of the algorithm described above is to set $\mathbf{u}=0$ and solve for $\mathbf{x}$. This is equivalent to writing (3.8b) in the form

$$(M^{\mathrm{T}}M)^{-1}\mathbf{u}+(M^{\mathrm{T}}M)^{-1}M^{\mathrm{T}}\mathbf{f}=\mathbf{x}.$$

Thus there is a set-up cost implicit in this form of the algorithm which can be reduced by interchanging the roles of $\mathbf{x}$ *and* $\mathbf{u}$. This idea can be pushed somewhat further to avoid forming the normal matrix.

The algorithm works by satisfying (3.8b) and the complementarity condition at each stage and seeking a node which makes both $\mathbf{x}$ and $\mathbf{u}$ feasible. Complementarity can be forced by partitioning $\mathbf{x}$ and $\mathbf{u}$ so that

$$\mathbf{x}=\begin{bmatrix}\mathbf{x}_1\\ \mathbf{x}_2\end{bmatrix} \quad \text{with} \quad \mathbf{x}_2=0, \tag{3.39}$$

and

$$\mathbf{u}=\begin{bmatrix}\mathbf{u}_1\\ \mathbf{u}_2\end{bmatrix} \quad \text{with} \quad \mathbf{u}_1=0. \tag{3.40}$$

To derive equations for $\mathbf{x}_1$ and $\mathbf{u}_2$, we introduce the orthogonal factorization of M:

$$M=[Q_1 \mid Q_2]\begin{bmatrix}U\\ 0\end{bmatrix}, \begin{bmatrix}Q_1^{\mathrm{T}}\\ Q_2^{\mathrm{T}}\end{bmatrix}\mathbf{f}=\begin{bmatrix}\mathbf{c}_1\\ \mathbf{c}_2\end{bmatrix}. \tag{3.41}$$

Substituting in (3.8b) gives

$$\mathbf{u}-U^{\mathrm{T}}U\mathbf{x}=-U^{\mathrm{T}}\mathbf{c}_1$$

or

$$U^{-\mathrm{T}}\mathbf{u}-U\mathbf{x}=-\mathbf{c}_1 \tag{3.42}$$

Let $U, \mathbf{c}_1$ be partitioned to conform with (3.39), (3.40), to give

$$U=\begin{bmatrix}U_1 & U_{12}\\ 0 & U_2\end{bmatrix}, \qquad \mathbf{c}_1=\begin{bmatrix}\mathbf{c}_{11}\\ \mathbf{c}_{12}\end{bmatrix} \tag{3.43}$$

Then substituting into (3.42) gives the pair of equations

$$U_1\mathbf{x}_1=\mathbf{c}_{11}, \qquad \mathbf{u}_2=-U_2^{\mathrm{T}}\mathbf{c}_{12} \tag{3.44}$$

This shows that the computation reduces to a sequence of calculations based on a factorization of the data matrix. However, each step of the algorithm adjusts the partitioning of U and this results in the destruction of its upper triangular form in general. This must be restored by the usual procedure of reordering the variables so that the modified U is in suitable form to bring back to upper triangular form by a small number of plane rotations. This is illustrated in Figure 3.2.

When working with $\mathbf{u}$ initially instead of $\mathbf{x}$, it is not even necessary to begin by computing the factorization (3.41) as this can be developed as the algorithm proceeds. The key point is that $\mathbf{u}_2$ can be found once the part of the transformation necessary for the calculation of $\mathbf{x}_1$ has been carried out.

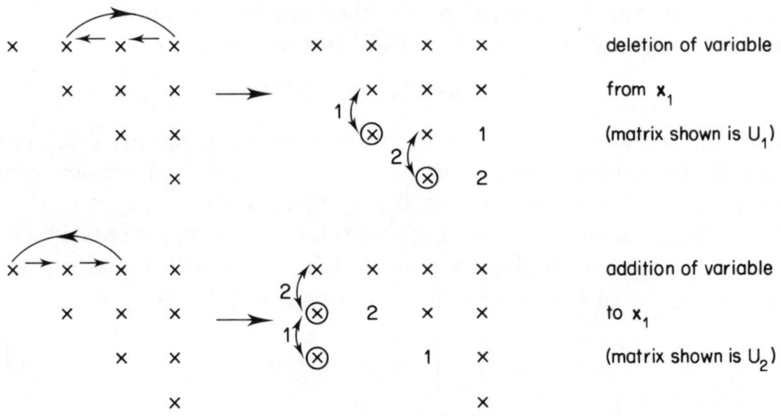

deletion of variable from $\mathbf{x}_1$ (matrix shown is U_1)

addition of variable to $\mathbf{x}_1$ (matrix shown is U_2)

Figure 3.2 Modification of U corresponding to variable exchanges (circles indicate elements zeroed, and numbers correspond to elements introduced by that member of the sequence of plane rotations)

If Q_1, M are partitioned in conformity with $\mathbf{x}$, $\mathbf{u}$, then (3.41) gives

$$Q_{11}^{\mathrm{T}}[M_1 \mid M_2] = [U_1 \mid U_{12}], \tag{3.45}$$

and

$$Q_{12}^{\mathrm{T}}[M_1 \mid M_2] = [0 \mid U_2]. \tag{3.46}$$

Thus

$$-\mathbf{u}_2 = U_2^{\mathrm{T}}\mathbf{c}_{12} = M_2^{\mathrm{T}}Q_{12}Q_{12}^{\mathrm{T}}\mathbf{f}$$

so that by (3.46)

$$-\begin{bmatrix} 0 \\ \mathbf{u}_2 \end{bmatrix} = M^{\mathrm{T}}Q_{12}Q_{12}^{\mathrm{T}}\mathbf{f}$$

$$= M^{\mathrm{T}}(I - Q_{11}Q_{11}^{\mathrm{T}})\mathbf{f}$$

as

$$Q_{12}^{\mathrm{T}}Q_{12} = I - Q_{11}Q_{11}^{\mathrm{T}} - Q_2Q_2^{\mathrm{T}} \qquad \text{and} \qquad M^{\mathrm{T}}Q_2 = 0$$

so that

$$-\mathbf{u}_2 = M_2^{\mathrm{T}}\mathbf{f} - U_{12}^{\mathrm{T}}\mathbf{c}_{11} \tag{3.47}$$

This shows that calculation of $\mathbf{u}_2$ requires only the partial factorization

$$HM = \begin{bmatrix} U_1 & U_{12} \\ 0 & \times\ldots \\ & \ldots \\ & \ldots \\ & \ldots\times \end{bmatrix}, \qquad H\mathbf{f} = \begin{bmatrix} \mathbf{c}_{11} \\ \times \\ \vdots \\ \times \end{bmatrix}.$$

In particular, multipliers can be updated and decisions made on the order in which the remaining columns of M are swept out as the factorization proceeds. Essentially no set-up costs are required for this form of the algorithm.

Remark 3.5 Selecting the variable to fix at zero level by choosing the most negative component is analogous to a USE test and suffers from the same problem of scale dependence. Here there is a simple alternative which corresponds to a test used in stepwise regression (see Exercise 2.5). Consider the first phase of the computation in which components of $\mathbf{u}$ are being fixed at zero in the initial search for a feasible $\mathbf{u}$. If the corresponding matrix factorization is written (at the ith step)

$$M^{(i)} = \begin{bmatrix} U_1 & U_{12} \\ 0 & B \end{bmatrix}, \qquad \mathbf{f}^{(i)} = \begin{bmatrix} \mathbf{c}_{11} \\ \mathbf{d} \end{bmatrix}$$

then the stepwise regression test selects the next column to be processed according to the rule

$$k = \arg\max\,(|\mathbf{d}^{\mathrm{T}}\boldsymbol{\kappa}_j(B)|/\|\boldsymbol{\kappa}_j(B)\|, \qquad j = i+1, \ldots, p).$$

Now

$$-\mathbf{u}_2 = M_2^{\mathrm{T}}\mathbf{f} - U_{12}^{\mathrm{T}}\mathbf{c}_{11} = B^{\mathrm{T}}\mathbf{d}$$

so that the stepwise test is available in the form

$$\mathbf{x}_1 := \begin{bmatrix} \mathbf{x}_1 \\ x_k \end{bmatrix}, \qquad k = \arg\max \frac{-(\mathbf{u}_2)_j}{\|\boldsymbol{\kappa}_j(B)\|_2} \tag{3.48}$$

Note that as the transformation of M is orthogonal it is easy to update the $\|\boldsymbol{\kappa}_j(\mathrm{B})\|^2$ at each step. The similarity between (3.48) and an NSE test should be noted.

Remark 3.6 The discussion up till now has assumed the use of a branch-and-bound strategy so that pruning of the search tree prevents cycling. However, it is tempting to try a stripped-down version which has the form (using a steepest edge selection heuristic):

(i) *Branch on the most negative component of* $\mathbf{u}$ *until* $\mathbf{u} \geqslant 0$.
(ii) *If* $\mathbf{x} \geqslant 0$ *then stop.*
(iii) *Branch on the most negative component of* $\mathbf{x}$.
(iv) *Go to* (i).

The important result that this algorithm cannot cycle follows from Theorem 3.3, *which shows that the successive values obtained at the end of stage* (i) *are strictly decreasing.* In practice it has proved very satisfactory, especially as it is rare for the full factorization (3.41) to be required unless the solution of the ordinary least squares problem is feasible.

The obvious competitive algorithms attack the complementarity problem directly, and these are now considered.

Algorithms for the complementarity problem

The close connection between the branch-and-bound based algorithm and procedures for solving the linear complementarity problem (3.8) is noted above. An alternative approach is to specialize the algorithms designed for more general classes of linear complementarity problem to the restricted least squares problem. Lemke's algorithm and principal pivoting are perhaps the two major methods.

Lemke's algorithm

This has been treated in greater generality than is required here in Section 2.4, where it is shown that if Z is positive definite then the algorithm terminates at a solution of the linear complementarity problem (Remark 4.10 of Chapter 2). In this case it is convenient to treat z_0 in (4.43) of Chapter 2 as a continuation parameter, and it can be shown that the retrogressive behaviour referred to in Remark 4.9 of Chapter 2 does not occur. At the current step let the linear system solved for the basic variables $\mathbf{w}$ have the form

$$\mathbf{w} = \mathbf{q} + Z\mathbf{z} + z_0\mathbf{g} \tag{3.49}$$

where, in particular, $\mathbf{g}$ is the result of applying the pivot transformations to $\mathbf{e}$. Then a necessary condition for w_k to be a blocking variable as z_0 is decreased is

$$\frac{\mathrm{d}w_k}{\mathrm{d}z_0} = g_k > 0. \tag{3.50}$$

We display the exchange of the blocking variable w_k and its complement z_k explicitly, and denote quantities after the exchange by bars. The exchange gives

$$\bar{\mathbf{w}} = \mathbf{w} + \mathbf{e}_k\mathbf{e}_k^{\mathrm{T}}(\mathbf{z} - \mathbf{w}) = E_k\mathbf{w} + \mathbf{e}_k\mathbf{e}_k^{\mathrm{T}}\mathbf{z} \tag{3.51a}$$

and

$$\bar{\mathbf{z}} = \mathbf{z} + \mathbf{e}_k\mathbf{e}_k^{\mathrm{T}}(\mathbf{w} - \mathbf{z}) = E_k\mathbf{z} + \mathbf{e}_k\mathbf{e}_k^{\mathrm{T}}\mathbf{w} \tag{3.51b}$$

where $E_k = I - \mathbf{e}_k\mathbf{e}_k^{\mathrm{T}}$. Now (3.49) can be written

$$E_k\mathbf{w} + \mathbf{e}_k\mathbf{e}_k^{\mathrm{T}}\mathbf{w} = \mathbf{q} + ZE_k\mathbf{z} + \boldsymbol{\kappa}_k(Z)\mathbf{e}_k^{\mathrm{T}}\mathbf{z} + z_0\mathbf{g}$$

and regrouping gives

$$(I - \boldsymbol{\zeta}\mathbf{e}_k^{\mathrm{T}})\bar{\mathbf{w}} = \mathbf{q} + (Z - \boldsymbol{\zeta}\mathbf{e}_k^{\mathrm{T}})\bar{\mathbf{z}} + z_0\mathbf{g} \tag{3.52}$$

where

$$\boldsymbol{\zeta} = \boldsymbol{\kappa}_k(Z) + \mathbf{e}_k.$$

Thus

$$-Z_{kk}\frac{\mathrm{d}\bar{w}_k}{\mathrm{d}z_0} = g_k > 0 \tag{3.53}$$

showing that the complementary variable increases as z_0 decreases, provided $Z_{kk} > 0$. That this is the case is a consequence of Lemma 3.3 below. Note that the diagonal elements of Z are positive if $\mathbf{u}^T Z \mathbf{u} > 0$ for all $\mathbf{u} \neq 0$. This condition is clearly satisfied initially if M has full rank, so that it is necessary to show that it continues to hold after exchanging complementary variables.

Lemma 3.3 *Let $\mathbf{u}^T Z \mathbf{u} \geq 0$, all $\mathbf{u}$, and let $Z \to \bar{Z}$ under the exchange of a complementary pair* (3.51). *Then $\mathbf{u}^T \bar{Z} \mathbf{u} \geq 0$, all $\mathbf{u}$.*

Proof For any $\mathbf{u}$ let

$$\mathbf{v} = Z\mathbf{u}.$$

Then the complementary exchange (3.51) takes this to

$$\bar{\mathbf{v}} = \bar{Z}\bar{\mathbf{u}}$$

and

$$\bar{\mathbf{u}}^T \bar{Z} \bar{\mathbf{u}} = \bar{\mathbf{u}}^T \bar{\mathbf{v}} = \mathbf{u}^T \mathbf{v} = \mathbf{u}^T Z \mathbf{u} \geq 0.$$

Principal pivoting

In Lemke's algorithm, complementarity and feasibility are preserved along the continuation trajectory. By contrast, this algorithm attempts to make the infeasible start $(\mathbf{w}, \mathbf{z}) = (\mathbf{q}, 0)$ feasible while maintaining 'almost complementarity' by permitting $w_i z_i \neq 0$ for at most one i. In outline the algorithm is as follows.

Algorithm

(i) *Select w_i infeasible.*
(ii) *Increase z_i until the process is blocked by $w_k = 0$ or $w_i = 0$.*
(iii) *If $w_i = 0$ then go to (iv);*
else exchange the complementary variables w_k, z_k, go to (ii).
(iv) *Exchange the complementary variables w_i, z_i.*
(v) *If $\bar{\mathbf{w}} \geq 0$ then stop; else go to (i).*

Success of the algorithm requires that the exchange of the blocking variable with its complement is sufficient to preserve feasibility, as z_i is further increased in step (ii). The argument is similar to that used above to justify the continuation form of Lemke's algorithm. Let the linear system be written as

$$\mathbf{w} = \mathbf{q} + Z\mathbf{z} + \lambda\, \boldsymbol{\kappa}_i(Z)$$

so that increasing λ is equivalent to increasing z_i, and $\mathrm{d}w_i/\mathrm{d}\lambda = Z_{ii} > 0$ so that w_i is also increasing. The condition that w_k can be a blocking variable is

$$\frac{\mathrm{d}w_k}{\mathrm{d}\lambda} = Z_{ki} < 0.$$

After the exchange, we have from (3.52) that

$$-Z_{kk}\frac{d\bar{w}_k}{d\lambda}=Z_{ki}$$

showing that $d\bar{w}_k/d\lambda > 0$, as required, as $Z_{kk} > 0$ as a consequence of Lemma 3.3.

Finite termination of the algorithm in the nondegenerate case is now readily demonstrated. First, the minor cycle (steps (ii) and (iii) of the algorithm) cannot repeat partitions of the variables into basic and nonbasic sets, for if a certain set is basic at $\lambda=\lambda^*$ and also for $\lambda=\lambda^{**}>\lambda^*$, then they are basic for $\lambda^*\leqslant\lambda\leqslant\lambda^{**}$ as linear functions of λ. Second, the number of passes through the major cycle is $\leqslant p$ because the number of infeasible variables is reduced by 1 at each pass.

Remark 3.7 The solution trajectory in Lemke's algorithm is essentially determined by the problem data. This is not so in the principal pivoting algorithm when the selection of the infeasible variable to be increased to zero in each major cycle is disposable. Similar considerations apply here as in the branch-and-bound algorithm.

Example 3.3 A generator of restricted least squares problems with known solution can be constructed as follows. Here RND is a uniform pseudo-random number generator for the interval [0, 1], and FNRND is a modified Pareto generator with parameter 1.2 (this generator is discussed in Section 7.4 – the key feature is that the underlying distribution is very long tailed):

(i) Specify n and p. Set the index set σ by a suitable random procedure.
(ii) Specify $\mathbf{x}$, $\mathbf{u}$ and the residual vector $\mathbf{r}$ by
$r_i = \text{RND} - .5$.
if $i\in\sigma$ then $u_i = 10^*$ RND; else $x_i = 10^*$ RND.
(iii) Let $k = \arg\max\,(|r_i|,\ i=1,\ldots,n)$
(iv) $M_{ij} = \text{FNRND}$, $i=1,2,\ldots,n$, $i\neq k$, $j=1,2,\ldots,p$.
(v) Determine $\boldsymbol{\rho}_k(M)$ to satisfy

$$\mathbf{r}^{\mathrm{T}}M=\mathbf{u}^{\mathrm{T}}.$$

(vi) Determine f_i, $i=1,2,\ldots,n$ to satisfy

$$M\mathbf{x}-\mathbf{f}=\mathbf{r}$$

In Table 3.3 the cumulative total for the number of pivot steps for 50 problems solved using the three algorithms Lemke (L), principal pivoting (P), and the cutdown branch-and-bound algorithm (C) are given for a range of values of p and n. FNRND is chosen to generate M because in other contexts it proves to lead to more difficult problems than the other distributions we have considered (see Table 4.1 of Chapter 7). However, in no case did the number of pivot steps exceed $2p$.

Table 3.3 Cumulative totals for the number of pivot steps for 50 randomly generated problems for each p and n value.

p / n	L	P	C
5 / 20	182	200	188
10 / 40	306	388	330
20 / 80	585	647	605

These results show a slight edge for algorithm L, but only the steepest-edge heuristic was used in C and maybe there is scope for improvement here. It is harder to avoid an initial factorization of M or $Z = M^{\mathrm{T}}M$ in the two complementarity algorithms – for example, it is necessary to adjoin a vector to give feasibility in (3.42) to start the Lemke algorithm; and this would seem to be be a definite advantage for algorithm C. As P is also the most complex of the algorithms it does not come out well from the comparison.

Exercise 3.1 If G is only positive semidefinite then QP need not have a bounded solution. Show this for this problem

$$\min_{\mathbf{x} \geqslant 0} -x_1 + \tfrac{1}{2}x_2^2$$

Discuss the associated linear complementarity problem. What result does this example suggest?

Exercise 3.2 Show that if the relaxed EQP in which the kth constraint is deleted from (3.12) has a unique solution then the choice of γ_k which makes $\bar{u}_k = 0$ in (3.15) satisfies $\gamma_k > 0$.

Exercise 3.3 The possibility of difficulty in solving the EQP makes it desirable to have alternate strategies for generating descent directions. Show that both the projected and reduced gradient methods can be used for this purpose. It should be noted that there is scope here for developing these methods further (for example, by special choices of the E_σ matrix in the reduced gradient method). This point is developed further in the next exercise.

Exercise 3.4 Directions $\mathbf{u}, \mathbf{v}$ are said to be conjugate (with respect to the positive definite matrix G) if

$$\mathbf{u}^{\mathrm{T}} G \mathbf{v} = 0.$$

(i) Let $F(\mathbf{x})$ be a positive definite quadratic form with a matrix G, and let $\mathbf{t}_1, \mathbf{t}_2, \ldots, \mathbf{t}_p$ be a set of conjugate directions. Starting from $\mathbf{x}_1$ calculate in sequence

$$\mathbf{x}_{i+1} = \mathbf{x}_i + \lambda_i \mathbf{t}_i, \qquad \lambda_i = \arg\min_{\lambda} F(\mathbf{x}_i + \lambda \mathbf{t}_i), \qquad i = 1, 2, \ldots, n.$$

Show that $\mathbf{x}_{n+1}$ minimizes $F(\mathbf{x})$. What is the corresponding result for the EQP?

(ii) Show that

$$\mathbf{t}_i^{\mathrm{T}}(\mathbf{g}_k - \mathbf{g}_{k-1}) = 0, \qquad \mathbf{g}_{i+1}^{\mathrm{T}} \mathbf{t}_k = 0, \qquad k \leqslant i.$$

The first relation suggests a method for choosing the columns of E_μ in the reduced gradient method to permit the generation of conjugate directions when the line search does not terminate on a new active constraint. Would this procedure be computationally satisfactory?

5.4 ITERATIVELY REWEIGHTED LEAST SQUARES

In this section our concentration on finite algorithms is interrupted to consider a class of iterative methods. These attempt to solve certain estimation problems of fairly general form by means of a sequence of suitably weighted least squares problems in which the weights in the current problem are specified in terms of the solution to the previous one. The justification for considering this approach here is that it has been put forward as a method for solving both the l_1 approximation problem discussed in Chapter 3 and the M-estimation problem considered in Section 5.6. The popularity of iteratively reweighted least squares (IRLS) is due possibly to the ready availability of software for weighted least squares problems. However, the evidence presented here does not provide a recommendation for the technique. It is shown that the convergence rate is only first order in general and that even this cannot be guaranteed for the class of problems considered.

To fix ideas, consider the problem of minimizing

$$F(\mathbf{r}) = \sum_{i=1}^{n} \hat{\rho}(|r_i|) \tag{4.1}$$

where $\hat{\rho}$ is a differentiable function of its argument and positive for positive values. Let $\mathbf{x}^{(k)}$ be the current solution estimate,

$$\mathbf{r}^{(k)} = M\mathbf{x}^{(k)} - \mathbf{f},$$

and assume that M has its maximum rank p. Then a first attempt to turn the problem of minimizing F into a sequence of least squares problems might consider

$$F^{(k)}(\mathbf{r})=\sum_{i=1}^{n}\frac{\hat{\rho}(|r_i^{(k)}|)}{|r_i^{(k)}|^2}\,r_i^2 \tag{4.2}$$

where it is assumed that $r_i^{(k)}\neq 0,\ i=1,2,\ldots,n$, and define

$$\mathbf{x}^{(k+1)}=\arg\min_{\mathbf{x}} F^{(k)}(\mathbf{r}) \tag{4.3}$$

Here IRLS sets out to compute a fixed point by means of the iteration specified in (4.3). However, the form of $F^{(k)}$ given in (4.2) is not satisfactory in general.

Lemma 4.1 *Let* $\mathbf{x}^*$ *minimize* (4.1) *and* $\mathbf{r}^*$ *be the corresponding residual vector. Then* $\mathbf{x}^*$ *is a fixed point of* (4.3) *if* $\hat{\rho}(t)=t^{\alpha}$, $\alpha>1$. *If* $\hat{\rho}(t)$ *does not have this form then* $\mathbf{x}^*$ *is not necessarily a fixed point.*

Proof If $\mathbf{x}^*$ minimizes (4.1) and F is smooth enough, then necessarily

$$0=\nabla F(\mathbf{x}^*)=\sum_{i=1}^{n}\hat{\rho}'(|r_i^*|)\theta_i\mathbf{m}_i^{\mathrm{T}} \tag{4.4}$$

where $\theta_i=\operatorname{sgn}(r_i)$. Now $F^{(k)}$ is strictly convex for all k and thus has a unique minimum. It follows that $\mathbf{x}^*$ is a fixed point of (4.3) provided the least squares condition

$$0=\sum_{i=1}^{n}\frac{\hat{\rho}(|r_i^*|)}{(r_i^*)^2}\,r_i^*\mathbf{m}_i^{\mathrm{T}} \tag{4.5}$$

is satisfied. A sufficient condition for this is

$$\hat{\rho}'(t)=\gamma\hat{\rho}(t)/t$$

for some constant γ. This gives

$$\hat{\rho}=t^{\gamma}.$$

The condition that F be smooth enough gives $\gamma>1$.

To show that if $\hat{\rho}$ is not of this form then minimizers of (4.1) need not be fixed points of (4.3), consider

$$\hat{\rho}(t)=\begin{cases}t^2/2, & |t|\leqslant 1\\ |t|-\frac{1}{2}, & |t|>1\end{cases}$$

and take

$$M^{\mathrm{T}}=[1,1,1,1],\qquad \mathbf{f}^{\mathrm{T}}=[1,5,10,11].$$

Then F is minimized when $x=6$. The IRLS problem corresponding to

$x^{(k)}=6$ requires the minimization of

$$F^{(k)}(x)=\tfrac{1}{2}(x-5)^2+\frac{5-\frac{1}{2}}{25}(x-1)^2+\frac{4-\frac{1}{2}}{16}(x-10)^2+\frac{5-\frac{1}{2}}{25}(x-11)^2$$

and has the solution $x=4398/865\neq 6$.

The restriction on the form of $\hat{\rho}$ introduced in Lemma 4.1 can be avoided by an alternative approach which modifies (4.4) to obtain an IRLS problem. Let

$$\mathbf{g}^{(k)}(\mathbf{x})=\sum_{i=1}^{n}\frac{\hat{\rho}'(|r_i^{(k)}|)}{|r_i^{(k)}|}r_i\mathbf{m}_i \tag{4.6}$$

and define

$$\mathbf{x}^{(k+1)}=\arg\left(\mathbf{g}^{(k)}(\mathbf{x})=0\right). \tag{4.7}$$

Then $\mathbf{x}^{(k+1)}$ solves a weighted least squares problem with a diagonal weighting matrix. The diagonal elements (weights) are given by

$$W_i^{(k)}=\hat{\rho}'(|r_i^{(k)}|)/|r_i^{(k)}|, \qquad i=1,2,\ldots,n. \tag{4.8}$$

By construction, if $\mathbf{x}^*$ minimizes (4.1) then it is a fixed point of (4.7) and any fixed point of (4.7) satisfies the necessary conditions for a minimum of (4.1). However, the stability of the fixed point is not presumed. A sufficient condition for $\mathbf{x}^{(k+1)}$ to be uniquely determined is $W_i^{(k)}>0$, $i=1,2,\ldots,n$. This is implied by $\hat{\rho}(t)$ strictly increasing for $t>0$ and $\hat{\rho}''(0)>0$.

Under certain circumstances it is possible to show the convergence of (4.3).

Lemma 4.2 *Let $\hat{\rho}'(t)/t$ be positive and nonincreasing for $t\geqslant 0$. Then the sequence $\{F(\mathbf{r}^{(k)}\}$ is decreasing (or terminates if $\mathbf{x}^{(k)}$ minimizes (4.1)).*

Proof The idea is to construct a strictly convex dominating function $U^{(k)}(\mathbf{r})$ such that

$$F(\mathbf{r})\leqslant U^{(k)}(\mathbf{r}), \tag{4.9}$$

$$F(\mathbf{r}^{(k)})=U^{(k)}(\mathbf{r}^{(k)}), \tag{4.10}$$

and

$$\arg\min_{\mathbf{x}} U^{(k)}(\mathbf{r})=\mathbf{x}^{(k+1)}. \tag{4.11}$$

For then

$$F(\mathbf{r}^{(k+1)})\leqslant U^{(k)}(\mathbf{r}^{(k+1)})<U^{(k)}(\mathbf{r}^{(k)})=F(\mathbf{r}^{(k)})$$

and this shows that the sequence $\{F(\mathbf{r}^{(k)})\}$ is decreasing provided $\mathbf{x}^{(k+1)}\neq\mathbf{x}^{(k)}$. Let

$$U^{(k)}(\mathbf{r})=\sum_{i=1}^{n}U_i^{(k)}(|r_i|) \tag{4.12}$$

and seek $U_i^{(k)}$ in the form

$$U_i^{(k)} = a_i + \tfrac{1}{2} b_i r_i^2. \tag{4.13}$$

To satisfy (4.10) it is sufficient that

$$\hat{\rho}(|r_i^{(k)}|) = a_i + \tfrac{1}{2} b_i (r_i^{(k)})^2.$$

To fix a_i, b_i we match derivatives at $\mathbf{x}^{(k)}$. This gives

$$U_i^{(k)\prime}(|r_i^{(k)}|) = b_i\, |r_i^{(k)}| = \hat{\rho}'(|r_i^{(k)}|)$$

so that

$$b_i = \hat{\rho}'(|r_i^{(k)}|)/|r_i^{(k)}| = W_i^{(k)},$$

and

$$a_i = \hat{\rho}(|r_i^{(k)}|) - \tfrac{1}{2} W_i^{(k)}\, |r_i^{(k)}|^2.$$

Now

$$\begin{aligned} U_i^{(k)\prime}(|r_i|) - \hat{\rho}'(|r_i|) &= \frac{\hat{\rho}'(|r_i^{(k)}|)}{|r_i^{(k)}|}\, |r_i| - \hat{\rho}'(|r_i|) \\ &= |r_i| \left(\frac{\hat{\rho}'(|r_i^{(k)}|)}{|r_i^{(k)}|} - \frac{\hat{\rho}'(|r_i|)}{|r_i|} \right) \end{aligned}$$

is nonnegative (nonpositive) provided $|r_i| > |r_i^{(k)}|$ ($|r_i| < |r_i^{(k)}|$) by assumption, and this verifies (4.9). The result now follows as

$$U^{(k)}(\mathbf{r}) = \sum_{i=1}^{n} \tfrac{1}{2} W_i^{(k)} r_i^2 + \text{const.} \tag{4.14}$$

is strictly convex as $W_i^{(k)} > 0$ by assumption and is minimized by $\mathbf{x}^{(k+1)}$.

The assumptions of Lemma 4.2 ensure that the set of points $\{\mathbf{x};\, F(\mathbf{r}) \leqslant F(\mathbf{r}^{(k)})\}$ is bounded. A consequence is that the sequence $\{\mathbf{x}^{(k)}\}$ has limit points. These limit points are fixed points of (4.7) as the contrary implies that $U^{(k)}$ would be reduced as it is strictly convex. It follows that the limit points of the IRLS procedure satisfy the necessary conditions for a minimum of (4.1). In summary we have the following result.

Theorem 4.1 *Let $\hat{\rho}(t)$ satisfy the conditions of Lemma 4.2. Then the sequence $\{F(\mathbf{r}^{(k)})\}$ produced by the IRLS procedure converges, and limit points of the sequence $\{\mathbf{x}^{(k)}\}$ minimize $F(\mathbf{r})$.*

This result provides an answer to the convergence question for certain important forms of $\hat{\rho}$. For example, for the M-estimation problem considered in Section 5.6, $\hat{\rho}$ is given by

$$\hat{\rho}(t) = \begin{cases} \tfrac{1}{2} t^2, & |t| \leqslant c \\ c\,|t| - c^2/2, & |t| > c. \end{cases} \tag{4.15}$$

However, the result is certainly not comprehensive, and it does not give rate-of-convergence information. Such information is useful in determining if fixed points are attractive. To investigate the rate of convergence, assume that the iteration has the form

$$\mathbf{x}^{(k+1)} = \mathbf{h}(\mathbf{x}^{(k)}), \tag{4.16}$$

and that $\mathbf{x}^*$ is a fixed point. The following result is well known.

Lemma 4.3 *Let* $\mathbf{h}$ *be sufficiently smooth in a neighbourhood of* $\mathbf{x}^*$ *to permit the expansion*

$$\mathbf{x}^{(k+1)} - \mathbf{x}^* = \nabla\mathbf{h}(\mathbf{x}^*)(\mathbf{x}^{(k)} - \mathbf{x}^*) + o(\|\mathbf{x}^{(k)} - \mathbf{x}^*\|_2) \tag{4.17}$$

for $\|\mathbf{x}^{(k)} - \mathbf{x}^*\|_2$ *small enough. Then a sufficient condition for the convergence of* $\{\mathbf{x}^{(k)}\} \to \mathbf{x}^*$ *is that the spectral radius of* $\nabla\mathbf{h}(\mathbf{x}^*) < 1$.

To calculate $\nabla\mathbf{h}$ let $W(\mathbf{x})$ be the diagonal matrix of the IRLS weights associated with the point $\mathbf{x}$. Then

$$\mathbf{h}(\mathbf{x}) = (M^{\mathrm{T}}WM)^{-1}M^{\mathrm{T}}W\mathbf{f} \tag{4.18}$$

so that

$$\begin{aligned}
\frac{\partial \mathbf{h}}{\partial x_j} &= -(M^{\mathrm{T}}WM)^{-1}M^{\mathrm{T}}\frac{\partial W}{\partial x_j}M(M^{\mathrm{T}}WM)^{-1}M^{\mathrm{T}}W\mathbf{f} + (M^{\mathrm{T}}WM)^{-1}M^{\mathrm{T}}\frac{\partial W}{\partial x_j}\mathbf{f} \\
&= (M^{\mathrm{T}}WM)^{-1}M^{\mathrm{T}}\frac{\partial W}{\partial x_j}\{\mathbf{f} - M(M^{\mathrm{T}}WM)^{-1}M^{\mathrm{T}}W\mathbf{f}\} \\
&= -(M^{\mathrm{T}}WM)^{-1}M^{\mathrm{T}}\frac{\partial W}{\partial x_j}\mathbf{r}^{(k+1)}
\end{aligned} \tag{4.19}$$

where $\mathbf{x}^{(k+1)}$ is obtained by applying one step of the IRLS iteration at the point $\mathbf{x}$. Now

$$\frac{\partial W_i}{\partial x_j} r_i^{(k+1)} = \frac{\mathrm{d}W_i}{\mathrm{d}\,|r_i|}\theta_i r_i^{(k+1)} M_{ij}$$

so that, setting

$$\hat{W} = \operatorname{diag}\left\{\frac{\mathrm{d}W_i}{\mathrm{d}\,|r_i|}\theta_i r_i^{(k+1)}\right\} \tag{4.20}$$

in (4.19), and collecting terms gives

$$\nabla\mathbf{h} = -(M^{\mathrm{T}}WM)^{-1}M^{\mathrm{T}}\hat{W}M. \tag{4.21}$$

By Lemma 4.3 the condition for $\mathbf{x}^*$ to be locally attractive is that all eigenvalues of the symmetric generalized eigenvalue problem with positive definite weighting matrix $M^{\mathrm{T}}WM$,

$$\{-M^{\mathrm{T}}\hat{W}M - \lambda M^{\mathrm{T}}WM\}\mathbf{v} = 0, \tag{4.22}$$

have magnitude <1. Setting $W_i = \hat{\rho}'(|r_i|)/|r_i|$, we obtain

$$\hat{W}_i = \theta_i r_i^{(k+1)} \left\{ \frac{\hat{\rho}''}{|r_i|} - \frac{\hat{\rho}'}{|r_i|^2} \right\}$$

$$= \frac{r_i^{(k+1)}}{r_i} W_i \left\{ \frac{\hat{\rho}''}{\hat{\rho}'} |r_i| - 1 \right\}. \tag{4.23}$$

If now $\mathbf{x} = \mathbf{x}^*$, a fixed point, and $\hat{\rho}(t) = t^\alpha$, then

$$\hat{W}_i = (\alpha - 2) W_i, \qquad \nabla \mathbf{h}(\mathbf{x}^*) = -(\alpha - 2) I. \tag{4.24}$$

Collecting this material together gives the following result.

Theorem 4.2 *The IRLS procedure is at best first-order convergent unless* $\mathrm{d}W_i/\mathrm{d}\,|r_i| = 0$ *(the case of no reweighting). If* $\hat{\rho}(t) = t^\alpha$ *then the multiplier determining the rate of convergence is* $\alpha - 2$ *so that convergence is obtained for* $1 < \alpha < 3$.

Remark 4.1 Convergence can be accelerated easily in the case $\hat{\rho}(t) = t^\alpha$. For example, consider

$$\mathbf{x}^{(k+1)} = \theta \mathbf{x}^{(k)} + (1 - \theta) \mathbf{h}(\mathbf{x}^{(k)})$$

$$= \mathbf{h}^\theta (\mathbf{x}^{(k)}). \tag{4.25}$$

The matrix determining the ultimate rate of convergence is

$$\nabla \mathbf{h}^{(\theta)}(\mathbf{x}^*) = \theta I - (1 - \theta)(\alpha - 2) I$$

$$= 0$$

provided

$$\theta = (\alpha - 2)/(\alpha - 1). \tag{4.26}$$

Remark 4.2 Convergence in the case $|\alpha - 2| = 1$ is not settled by the previous results. It almost is when $\alpha = 1$ but the nondifferentiability of the objective function means the characterization result is wrong. This case is considered further in Theorem 4.3. The case $\alpha = 3$ remains unsettled. However, evidence is given which shows that it cannot always converge.

Theorem 4.3 *If the* l_1 *approximation problem has a unique, nondegenerate solution then the IRLS procedure converges with first order rate, and the multiplier determining the actual speed of convergence is* $\max |\lambda_i|$ *where the* λ_i *are given by equation* (3.1.7).

Proof The characterization condition for the solution of the IRLS problem is

$$\sum_{i=1}^{n} \frac{r_i^{(k+1)}}{|r_i^{(k)}|} \mathbf{m}_i^{\mathrm{T}} = 0. \tag{4.27}$$

Assume now that $\mathbf{x}^*$ solves the l_1 problem and that $\mathbf{x}^{(k)}$ is close to $\mathbf{x}^*$ in the sense that

$$r_i^* = r_i^{(k)} + z_i^{(k)}, \qquad i = 1, 2, \ldots, n \tag{4.28}$$

where the $z_i^{(k)}$ are small. The characterization conditions for the l_1 problem (equations (1.6), (1.7) of Chapter 3) are

$$\sum_{i \in \mu^c} \theta_i \mathbf{m}_i^{\mathrm{T}} + \sum_{i \in \mu} \lambda_i \mathbf{m}_i^{\mathrm{T}} = 0,$$

and

$$-1 \leqslant \lambda_i \leqslant 1, \qquad i \in \mu$$

where $\theta_i = \operatorname{sgn}(r_i^*)$, and $\mu = \{i;\, r_i^* = 0\}$. The uniqueness and nondegeneracy assumptions imply $|\lambda_i| < 1$, $|\mu| = p$, and rank $(M_\mu) = p$. Equation (4.27) can be written

$$\sum_{i \in \mu^c} \frac{r_i^* - z_i^{(k+1)}}{|r_i^*|\, |r_i^{(k)}/r_i^*|} \mathbf{m}_i^{\mathrm{T}} + \sum_{i \in \mu} \frac{z_i^{(k+1)}}{|z_i^{(k)}|} \mathbf{m}_i^{\mathrm{T}} = 0$$

so that

$$\sum_{i \in \mu^c} \theta_i \left(1 - \frac{z_i^{(k+1)}}{r_i^*}\right)\left(1 + \frac{z_i^{(k)}}{r_i^{(k)}}\right) \mathbf{m}_i^{\mathrm{T}} + \sum_{\mu} \frac{z_i^{(k+1)}}{|z_i^{(k)}|} \mathbf{m}_i^{\mathrm{T}} = 0.$$

This simplifies, using the characterization equation (1.6) of Chapter 3, to

$$\sum_{i \in \mu^c} \theta_i \left(-\frac{z_i^{(k+1)}}{r_i^*} + \frac{z_i^{(k)}}{r_i^{(k)}} + \frac{z_i^{(k+1)} z_i^{(k)}}{r_i^* r_i^{(k)}} \right) \mathbf{m}_i^{\mathrm{T}} + \sum_{\mu} \left(\frac{z_i^{(k+1)}}{|z_i^{(k)}|} - \lambda_i \right) \mathbf{m}_i^{\mathrm{T}} = 0 \tag{4.29}$$

Let $\mathbf{z}_\mu^{(k)}$ stand for the vector with components $z_{\mu(i)}^{(k)}$, $i = 1, 2, \ldots, p$. Then

$$\mathbf{z}_\mu^{(k)} = -M_\mu (\mathbf{x}^{(k)} - \mathbf{x}^*) \tag{4.30}$$

and the system is invertible by assumption. Thus each $z_i^{(k)}$, $i \in \mu^c$, is expressible as a linear function of $\mathbf{z}_\mu^{(k)}$ without a constant term. It follows that the system of equations (4.29) can be written in the form

$$\mathbf{z}_\mu^{(k+1)} - \Lambda^{(k)} \mathbf{z}_\mu^{(k)} = \mathbf{s}(\mathbf{z}_\mu^{(k+1)}, \mathbf{z}_\mu^{(k)}) \tag{4.31}$$

where

$$\|\mathbf{s}\|_2 \leqslant K_1 \|\mathbf{z}_\mu^{(k)}\|_2\, \|\mathbf{z}_\mu^{(k+1)}\|_2 + K_2 \|\mathbf{z}_\mu^{(k)}\|_2^2,$$

$$\theta_i^k = \operatorname{sgn}(z_i^{(k)}),$$

and

$$\Lambda^{(k)} = \operatorname{diag}\{\theta_{\mu(i)}^{(k)} \lambda_{\mu(i)}, \qquad i \in \mu\}$$

for constants K_1, K_2 which do not depend on $\mathbf{z}_\mu^{(k)}$. Provided $K_1 \|\mathbf{z}_\mu^{(k)}\|_2 < 1$, then taking norms and rearranging gives

$$(1 - K_1 \|\mathbf{z}_\mu^{(k)}\|_2)\, \|\mathbf{z}_\mu^{(k+1)}\|_2 \leqslant \max_i (|\lambda_i|)\, \|\mathbf{z}_\mu^{(k)}\|_2 + K_2 \|\mathbf{z}_\mu^{(k)}\|_2^2$$

and the required result is a consequence of this inequality.

Example 4.1 This example reports numerical calculations using the IRLS procedure. The problem was generated by forming M by sampling a $[0, 1]$ uniform distribution and then computing $\mathbf{f}$ by assuming that $\mathbf{x} = \mathbf{e}^{(p)}$ and then adding a normally distributed error with standard deviation .1. The values reported in Table 4.1 correspond to $n = 5$, $p = 2$ and $\hat{\rho} = t^\alpha$ for $\alpha = 1$, 1.5, 2.5, 3, 4, and the initial $\mathbf{x}$ is also randomly generated. For each case, values of $F(\mathbf{x}^{(k)})$ and $DX = \|\mathbf{x}^{(k+1)} - \mathbf{x}^{(k)}\|_2$ are given in the table. The predictions of the theory are borne out, in that linear convergence in the case $\alpha = 1$ is clear, as it is in the cases 1.5 and 2.5 where now the common ratio approaches $|\alpha - 2|$ as required by Theorem 4.2. The case $\alpha = 3$ looks as if it might be converging slowly. However, running the program much further gives the successive iterates reported in Table 4.2. These show clearly a two-cycle pattern (no convergence). Similar but stronger behaviour occurs for $\alpha = 4$ (Table 4.3). Further increase in α leads to more complicated behaviour.

The same test problem was used to test the accelerated procedure (4.25). Results are given in Table 4.4. These show clearly enough the eventual convergence acceleration attained. However, the iteration is extremely slow for $\alpha = 1.5$ although eventually giving rapid convergence once $\mathbf{x}$ is close enough to $\mathbf{x}^*$, and it also deteriorates further in the manner suggested by the table for $\alpha > 5$. note that numerical problems must be anticipated when α becomes large as this must exacerbate any differences in the relative scaling of the weights.

Table 4.1 Successive iterations of the IRLS Procedure

i \ α	1		1.5		2.5		3		4	
	DX	*F*	*DX*	*F*	*DX*	*F*	*DX*	*F*	*DX*	*F*
1	.2887	.7604	.2796	.3496	.2443	.0887	.2407	.0467	.2593	.0139
2	.0395	.2549	.0425	.0671	.0932	.0053	.2206	.0023	.3651	.0059
3	.0263	.2443	.0143	.0645	.0442	.0049	.2026	.0017	.4125	.0035
4	.0188	.2374	.0037	.0642	.0231	.0048	.1940	.0020	.4382	.0086
5	.0143	.2327	.0018	.0642	.0113	.0048	.1857	.0016	.4504	.0092
6	.0109	.2292	.0009	.0642	.0057	.0048	.1814	.0019	.4996	.0092
7	.0078	.2267			.0028	.0048	.1772	.0016	.5092	.0032

Table 4.2 Two-cycle in IRLS procedure; $\alpha = 3$.

DX	F	x_1	x_2
.167606	.0015313	.7517241	1.2102947
.167606	.0017512	.8877418	1.1123612
.167606	.0015313	.7517241	1.2102947
.167606	.0017512	.8877418	1.1123612

Table 4.3 Two-cycle in IRLS procedure; $\alpha = 4$.

DX	F	x_1	x_2
.6114533	.000811228	.5836162	1.405599
.6114533	.000825832	1.0389299	.997462
.6114533	.000811228	.5836162	1.405599
.6114533	.000825832	1.0389299	.9974762

Exercise 4.1 In many applications it is necessary to estimate a scale parameter simultaneously with $\mathbf{x}$. Let

$$F(\mathbf{r}, \sigma) = \frac{1}{n} \sum_{i=1}^{n} \left[\hat{\rho}\left(\frac{|r_i|}{\sigma}\right) + a\right]\sigma$$

where $\hat{\rho}(|x|)$, $\hat{\rho}(|x|)/|x|$ are convex, $\hat{\rho}(0) = 0$, and $a > 0$. To determine σ adjoin to (4.4) the extra equation

$$0 = \frac{\partial F}{\partial \sigma} = -\frac{1}{n} \sum_{i=1}^{n} \chi\left(\frac{|r_i|}{\sigma}\right) + a$$

and consider the iterative scheme

$$(\sigma^{(m+1)})^2 = \frac{1}{na} \sum_{i=1}^{n} \chi\left(\frac{|r_i^{(i)}|}{\sigma^{(m)}}\right)(\sigma^{(m)})^2.$$

By constructing the dominating function

$$U(\sigma) = F(\mathbf{r}^{(m)}, \sigma^{(m)}) + a(\sigma - \sigma^{(m)}) + \frac{1}{n} \sum_{i=1}^{n} \chi\left(\frac{|r_i^{(m)}|}{\sigma^{(m)}}\right)\left(\frac{(\sigma^{(m)})^2}{\sigma} - \sigma^{(m)}\right)$$

show that

$$F(\mathbf{r}^{(m)}, \sigma^{(m+1)}) \leq F(\mathbf{r}^{(m)}, \sigma^{(m)}) - \frac{a(\sigma^{(m+1)} - \sigma^{(m)})^2}{\sigma^{(m)}}.$$

Table 4.4 Successive corrections in the accelerated procedure.

i \ α	1.75	2.5	3	3.5	5
1	.232529	.233959	.234851	.235679	.237716
2	.077282	.078215	.117648	.141620	.178451
3	.025064	.026817	.059278	.085328	.134054
4	.005496	.011687	.030614	.051801	.100822
5	.000651	.005253	.018107	.032141	.075982
6	.000040	.000624	.065576	.021420	.057454
7		.000004	.044998	.013461	.043678
8			.010448	.021813	.033477
9			.002655	.0117833	.025915
10			.000099	.001331	.019890
11				.000047	.018182
12					.049825
13					.040415
14					.006262
15					.000049

5.5 SENSITIVITY OF ESTIMATION PROCEDURES

In this section heuristic ideas concerned with developing a basis for the comparison of different estimation and approximation procedures are developed. The basic criterion used is that of *resistance* or insensitivity of the estimate to particular kinds of perturbation in the problem data. It is to be distinguished from the criterion of *robustness* in which the sensitivity of the estimate is considered in relation to variations in the underlying probability model for the data errors.

To motivate the idea of resistance, consider the basic approximation problem with $p=1$, $M=\mathbf{e}^{(n)}$ (the location problem). We consider the following cases.

(i) *Least squares.* Here x is chosen to minimize

$$F(x)=\sum_{i=1}^{n}(x-f_i)^2 \tag{5.1}$$

and is given by

$$x=\frac{1}{n}\sum_{i=1}^{n}f_i. \tag{5.2}$$

The point to notice is that each observed value f_i is equally important in determining x, in the sense that

$$\frac{\partial x}{\partial f_i}=\frac{1}{n}, \qquad i=1,2,\ldots,n. \tag{5.3}$$

Thus the method of least squares provides no obvious discrimination between good and bad observations.

(ii) *Maximum norm approximation.* Here x is chosen to minimize

$$F(x)=\max_i\,(|x-f_i|). \tag{5.4}$$

It is easy to see that there must be two extremal residuals r_u, r_v such that

$$x-f_u=F(x),$$

and

$$x-f_v=-F(x).$$

It follows that

$$x=(f_u+f_v)/2. \tag{5.5}$$

It seems likely that if there are blunders or gross errors in the data then these will be associated with the large residuals. A consequence is that x is determined by the bad data so that this method is certainly not resistant.

(iii) *Approximation in the l_α norm,* $1<\alpha<2$. In this case

$$F(x)=\sum_{i=1}^{n}|x-f_i|^\alpha. \tag{5.6}$$

Proceeding formally, the value of x minimizing F satisfies

$$\sum_{i=1}^{n}|x-f_i|^{\alpha-1}\theta_i=0 \tag{5.7}$$

where $\theta_i=\operatorname{sgn}(r_i)$. Differentiating with respect to f_k gives

$$\left(\sum_{i=1}^{n}|x-f_i|^{\alpha-2}\right)\frac{\partial x}{\partial f_i}-|x-f_k|^{\alpha-2}=0$$

This formula is valid if $r_i\neq 0$ for all i. It follows that

$$\frac{\partial x}{\partial f_k}\leqslant \min_{1\leqslant s\leqslant n}(|r_s|^{2-\alpha})\,|r_k|^{\alpha-2}. \tag{5.8}$$

Now the characterization theorem for l_1 approximation ensures the existence of at least one zero residual. Assuming continuity of the best approximation as a function of α this shows that

$$\frac{\partial x}{\partial f_k} \to 0, \qquad \alpha \to 1$$

provided k is not in the set of zero residuals. On the assumption that blunders show up as large residuals, this suggests that the estimation procedures for $\alpha \to 1$ are increasingly resistant to the effects of blunders.

(iv) *Approximation in the l_1 norm.* The result in this case can be deduced from the previous example. It follows from (5.7) that

$$x = \text{med}\,\{f_i, i = 1, 2, \ldots, n\} \tag{5.9}$$

as this gives as near equal as possible numbers of positive and negative θ_i. As the median is unchanged by perturbations which do not change the sign of r_k it follows that $\partial x/\partial f_k = 0$ for values of k corresponding to the nonzero residuals. This indicates a strong resistance property of l_1 approximation.

(v) *Ranked approximation problems* (Example 1.7 of Chapter 4). Consider the problem of finding x to minimize

$$F(x) = \sum_{i=1}^{n} i\,|r_{\mu(i)}| \tag{5.10}$$

where μ is the ranking set pointing to the residuals in increasing order of magnitude. The argument of Remark 4.3 of Chapter 4 shows that

$$x = \text{med}\left\{\frac{f_i + f_j}{2}, \qquad 1 \leqslant j \leqslant i \leqslant n\right\}. \tag{5.11}$$

This estimator cannot be expected to have similar resistance properties to the l_1 estimator because now $f_i + f_j$ can be small (near the centre of the table of pairwise averages and thus a candidate for x) although both f_i and f_j correspond to extreme values of f and so are candidates for outliers. However, the maximum damage a single bad observation can do in moving from $+\infty$ to $-\infty$ is to move the median n places in the set of $n(n+1)/2$ elements. In this sense the estimator has bounded influence.

(vi) *Trimmed estimates.* Note that the l_1 estimator corresponds to a special case of the ranked estimator in which all the scores are equal, while the l_∞ (maximum norm) estimator correspond to the ranked estimator with all the weight on the residuals of largest magnitude. This suggests that resistance can be increased by deemphasizing the large residuals (in magnitude). One such method chooses the scores according to the rule

$$\eta_i^{(k)} = \begin{cases} \dfrac{i}{nk}, & i < k \\[2ex] \dfrac{1}{n}, & i \geqslant k \end{cases} \tag{5.12}$$

for some integer $k, 1 \leqslant k \leqslant n$. This process is called trimming. The condition for a minimum is now that

$$0 \approx \sum_{i=1}^{k} i \operatorname{sgn}(r_{\mu(i)}) + k \sum_{i=k+1}^{n} \operatorname{sgn}(r_{\mu(i)})$$

$$\approx \sum_{i=1}^{k} \sum_{j=1}^{i} \operatorname{sgn}(r_{\mu(i)} + r_{\mu(j)}) + \sum_{i=k+1}^{k} \sum_{j=1}^{k} \operatorname{sgn}(r_{\mu(i)} + r_{\mu(j)})$$

so that

$$x = \operatorname{med}\left\{\frac{f_{\mu(i)} + f_{\mu(j)}}{2}, i = 1, 2, \ldots, n, j = 1, 2, \ldots, \min(i, k)\right\} \tag{5.13}$$

Now the extreme values of f are not averaged if $k < n$ so the objections raised in the ranked case (v) above increasingly lose their force as k is reduced. Note that $k = 1$ gives the l_1 problem which has been shown already to be resistant.

(vii) *M-estimators.* The same idea of increasing resistance by de-emphasizing the large residuals can be applied in the least squares case. Now we minimize

$$F(x) = \sum_{i=1}^{n} \hat{\rho}(|r_i|) \tag{5.14}$$

where the functional form of $\hat{\rho}$ is chosen to de-emphasize the effect of the large residuals. An appropriate choice for $\hat{\rho}$ is

$$\hat{\rho}(t) = \begin{cases} \frac{1}{2}t^2, & |t| \leqslant c \\ c\,|t| - c^2/2, & |t| > c \end{cases} \tag{5.15}$$

which leads to a convex minimization problem. Here c is a scale parameter which defines the class of large residuals, and the problem is unchanged under the transformation $t \to t/d$, $c \to c/d$. If ν is an index set pointing to the small residuals then the condition for a minimum gives

$$0 = \sum_{i \in \nu} r_i + \sum_{i \in \nu^c} c\theta_i \tag{5.16}$$

showing that x is unaffected by perturbations of the large residuals which do not change the classification of the residuals into large and small.

The basic ideas in these examples can be extended to the multivariate case. It is reasonable to assume that blunders translate into large residuals when measured at the true values of the model parameters. A resistant procedure should then be one for which this is true when the model is evaluated using the computed parameter estimates (if this is not so then surely the parameter estimates are sensitive to the effects of the blunders). Thus such a procedure should give parameters stimates which are insensitive to perturbation of the outlier residuals. However, this criterion is more in the nature of a necessary than of a sufficient condition for resistance.

To extend the argument in the least squares case it appears that some assumption regarding the behaviour of the normal matrix for large n is necessary. For example, assume

$$M_{ij} = m_j(t_i), \qquad i = 1, 2, \ldots, n, \qquad j = 1, 2, \ldots, p \tag{5.17}$$

where the sample points t_i are generated according to the rule

$$t_i = \Psi(i/n), \qquad i = 1, 2, \ldots, n \tag{5.18}$$

where both the $m_j(t)$ and $\Psi(u)$ are sufficiently smooth functions. Then for n large enough,

$$\begin{aligned}(M^{\mathrm{T}}M)_{\alpha\beta} &= \sum_{i=1}^{n} m_\alpha(t_i) m_\beta(t_i) \\ &= n \int_0^1 m_\alpha(\Psi(u)) m_\beta(\Psi(u))\, \mathrm{d}u + \mathrm{O}(1)\end{aligned}$$

Provided the Gram matrix G defined by

$$G_{ij} = \int_0^1 m_i(\Psi(u)) m_j(\Psi(u))\, \mathrm{d}u \tag{5.19}$$

is nonsingular, then

$$\mathbf{x} = \frac{1}{n} G^{-1} M^{\mathrm{T}} \mathbf{f} + \mathrm{o}(1/n) \tag{5.20}$$

so that

$$\frac{\partial \mathbf{x}}{\partial f_i} = \frac{1}{n} G^{-1} \mathbf{m}_i + \mathrm{o}(1/n) \tag{5.21}$$

where $(\mathbf{m}_i)_j = m_j(t_i)$, $j = 1, 2, \ldots, p$. This shows that the sensitivity of the parameter estimates to perturbations in the observations is of the same order in n for each observation. The similarity of the condition on G with the condition needed to show the consistency of the least squares estimates should be noted.

Remark 5.1 This result confirms the earlier suggestion that the method of least squares provides little discrimination between good and bad data. However, the property that $\|\partial \mathbf{x}/\partial f_i\|$ is of the same order in n for each i is not the same as each data value having the same importance. In particular, if $\|\partial \mathbf{x}/\partial f_i\|$ is relatively large then the corresponding equation defining r_i is likely to be important in determining $\mathbf{x}$. Such equations are referred to as *leverage points*. They are likely to be worth checking if gross departures from an assumed M are suspected.

Turning to estimators based on minimizing polyhedral convex functions, we see that the conditions characterizing the minimum give

$$0 = \mathbf{g} + V\mathbf{u}, \qquad \mathbf{u} \in U, \tag{5.22}$$

and that the minimizing $\mathbf{x}$ can be found by solving the system of linear equations

$$\phi_i(\mathbf{r}) = \alpha_i, \qquad i \in \mu \tag{5.23}$$

where the ϕ_i are the structure functionals associated with V. Writing (5.23) in terms of V gives

$$V^{\mathrm{T}}\mathbf{x} - \Phi^{\mathrm{T}}\mathbf{f} = \boldsymbol{\alpha} \tag{5.24}$$

so that

$$V^{\mathrm{T}} \frac{\partial \mathbf{x}}{\partial f_k} = \Phi^{\mathrm{T}}\mathbf{e}_k. \tag{5.25}$$

Clearly resistance requires that the structure functionals do not act on the components of the residual vector corresponding to the gross errors. That is,

$$r_k \quad \text{an outlier} \Rightarrow \phi_i(\mathbf{e}_k) = 0, \qquad i \in \mu. \tag{5.26}$$

In this context trimming can be interpreted as a strategy for attempting to ensure that the resistance condition (5.26) is satisfied.

M-estimation behaves in a similar fashion. The condition for a minimum is

$$\begin{aligned} 0 &= \nabla \sum_{i=1}^{n} \hat{\rho}(|r_i|) \\ &= \sum_{i \in \nu} r_i \mathbf{m}_i^{\mathrm{T}} + c \sum_{i \in \nu^c} \theta_i \mathbf{m}_i^{\mathrm{T}} \end{aligned} \tag{5.27}$$

where $\nu = \{i;\, |r_i| \leqslant c\}$. Thus $\mathbf{x}$ satisfies the system of modified normal equations

$$M_\nu^{\mathrm{T}} M_\nu \mathbf{x} = M_\nu^{\mathrm{T}} \mathbf{f}_\nu - \sum_{i \in \nu^c} \theta_i \mathbf{m}_i \tag{5.28}$$

so that

$$\frac{\partial \mathbf{x}}{\partial f_i} = 0, \qquad i \in \nu^c \tag{5.29}$$

This shows that M-estimation is resistant to perturbations of the large residuals which do not change the classification into large and small specified by the choice of c. As c is decreased it must be expected that the number of residuals classified as large will increase, and in this sense the resistance of the procedure is inversely proportional to c, and c is analogous to a trimming parameter.

That trimming increases resistance is confirmed by looking at the 'fully trimmed' limit in both the estimation procedure based on ranks and the M-estimation procedure.

Lemma 5.1 *Consider the trimmed scores* (5.12) *for the estimation procedure based on ranks* (5.10). *Then setting* $k=1$ *gives the* l_1 *approximation problem.*

Proof Setting $k=1$ gives $\eta_i^{(1)}=1/n,\ i=1,2,\ldots,n.$

Lemma 5.2 *Let* $F(\mathbf{x},c)$ *be the* M*-estimator objective function with its dependence on* c *indicated explicitly. If* $\mathbf{x}_c$ *minimizes* $F(\mathbf{x},c)$, *and* $\mathbf{x}_l$ *minimizes the* l_1 *objective function with corresponding minimum* L_1 *then*

$$\lim_{c\to 0}\frac{1}{c}F(\mathbf{x}_c,c)=L_1 \tag{5.30}$$

Proof Choose c smaller than the smallest in magnitude of the nonzero residuals in the l_1 solution. Then

$$\begin{aligned}\frac{1}{c}F(\mathbf{x}_l,c)&=L_1-|\mu^c|\,c/2\\ &\geqslant\frac{1}{c}F(\mathbf{x}_c,c)\\ &\geqslant\sum_{i\in\nu^c}(|r_i(\mathbf{x}_c)|-c/2)\\ &\geqslant\sum_{i=1}^{n}|r_i(\mathbf{x}_c)|-c\left(\frac{|\nu^c|}{2}+|\nu|\right)\end{aligned}$$

as $|r_i(\mathbf{x}_i)|\leqslant c,\ i\in\nu,$

$$\geqslant L_1-c\left(\frac{|\nu^c|}{2}+|\nu|\right) \tag{5.31}$$

The analogous result can be derived for the rank regression problem (example 4.1.6) where the scores are defined by

$$\eta_i=\eta(i\delta) \tag{5.32}$$

where

$$\eta(t)=\begin{cases}-.5 & 0\leqslant t<\theta,\\ .5\dfrac{t-.5}{.5-\theta}, & \theta\leqslant t\leqslant 1-\theta,\\ .5, & 1-\theta<t\leqslant 1.\end{cases} \tag{5.33}$$

and

$$\delta=1/(n+1).$$

The least resistant case corresponds to $\theta = 0$ and trimming involves increasing θ. In the limit as $\theta \to .5$ the weights become 'sign scores'.

Lemma 5.3 *Rank regression with sign scores is l_1 approximation.*

Proof Note first that rank regression cannot determine an intercept term so it is convenient to separate this out and write the model equations in the form

$$\hat{\mathbf{r}} = [\mathbf{e}^{(n)} \mid M]\begin{bmatrix} x_0 \\ \mathbf{x} \end{bmatrix} - \mathbf{f} \tag{5.34}$$

Let the l_1 solution be $\begin{bmatrix} \bar{x}_0 \\ \mathbf{x}_L \end{bmatrix}$, and the rank regression solution $\mathbf{x}_R$. Then for each $\mathbf{r}(\mathbf{x}) = \hat{\mathbf{r}}\left(\begin{bmatrix} 0 \\ \mathbf{x} \end{bmatrix}\right)$ find γ such that

$$\eta_i(r_{\mu(i)} - \gamma) = |r_{\mu(i)} - \gamma|, \qquad i = 1, 2, \ldots, n$$

where μ is a ranking set pointing to the ordered residuals. In particular,

$$\gamma = \text{med}\,(r_i,\ i = 1, 2, \ldots, n) \tag{5.35}$$

will do. For the l_1 solution this gives

$$\gamma_L = -\bar{x}_0. \tag{5.36}$$

We have the inequalities (writing as before $\Delta(\cdot)$ for the rank regression objective function)

$$\begin{aligned} \Delta(\mathbf{x}_R) \leqslant \Delta(\mathbf{x}_L) &= \sum_{i=1}^{n} \eta_i(r_{\mu(i)}(\mathbf{x}_L) - \gamma_L) \\ &= \min_{\mathbf{x},\gamma} \sum_{i=1}^{n} |r_i - \gamma|, \end{aligned}$$

and

$$\begin{aligned} \Delta(\mathbf{x}_R) &= \sum_{i=1}^{n} \eta_i(r_{\mu(i)}(\mathbf{x}_R) - \gamma_R) = \sum_{i=1}^{n} |r_i(\mathbf{x}_R) - \gamma_R| \\ &\geqslant \min_{\mathbf{x},\gamma} \sum_{i=1}^{n} |r_i - \gamma| \end{aligned}$$

The desired result is an immediate consequence.

Remark 5.2 These results certainly establish the importance of the l_1 approximation problem in parameter estimation calculations. Not only can it be computed efficiently, but it is insensitive to perturbations in the nonzero residuals, and it is approximated by the other methods considered when these are trimmed to improve resistance. However, resistance has been considered in relation to perturbations in **f** and not in M. In this more general situation none of the methods considered (at least in the forms discussed here) appear to have particular virtues. A possible approach to the more general problem is discussed in the next chapter.

Example 5.1 To illustrate the points made in the above discussion and, in particular, the effect of the trimming parameters on the computed parameter estimates, consider the 'stack loss' data set given in Table 5.1.

Table 5.1 The stack loss data set. (Reproduced by permission of John Wiley & Sons Inc. from Daniel and Wood, 1971.)

i	M				**f**
1	1	80	27	89	42
2	1	80	27	88	37
3	1	75	25	90	37
4	1	62	24	87	28
5	1	62	22	87	18
6	1	62	23	87	18
7	1	62	24	93	19
8	1	62	24	93	20
9	1	58	23	87	15
10	1	58	18	80	14
11	1	58	18	89	14
12	1	58	17	88	13
13	1	58	18	82	11
14	1	58	19	93	12
15	1	50	18	89	8
16	1	50	18	86	7
17	1	50	19	72	8
18	1	50	19	79	8
19	1	50	20	80	9
20	1	56	20	82	15
21	1	70	20	91	15

Results of computation using the trimmed rank regression weights (5.33) for $\theta = 0$, .25, and .5 are given in Table 5.2, and the corresponding residual values are given in Table 5.3. This data set has been worked over many times, and there is fairly general agreement that observations 1, 3, 4, and 21 are outliers, and that the remaining data is modelled adequately by least squares. The results presented here tend to confirm this in that:

(i) it is clear that as trimming is increased the suspect observations correspond to the extremes of the ranking set and give residuals that are well separated from the remainder; and

Table 5.2 Parameter values computed using the rank regression procedure with the trimmed weights (5.33). LS is the least squares solution for the full data set, MLS is the least squares solution for the data set omitting observations 1, 3, 4, 21.

	LS	$\theta=0$	$\theta=.25$	$\theta=.5$	MLS
x_2	.72	.791667	.845725	.831884	.80
x_3	1.30	.911111	.661710	.573913	.58
x_4	−.15	−.111111	.104089	−.060870	−.07

(ii) the parameter estimates improve in the sense that they become closer to the parameter estimates given by the least squares problem for the data set with the suspect observations removed.

Note that there is a tendency for the tied residuals to clump in the middle of the ranking set as θ is increased. This is because the discontinuities in derivative can occur only when the tied residuals are associated with distinct scores. In particular, for the sign scores ($\theta=.5$) only one group can be significant and it must contain the zero weight corresponding to $i=11$ as the weights for $i\leqslant 10$ and $i\geqslant 12$ are identical.

The corresponding results using M-estimation are presented in Table 5.4. Here the values of c are the points at which the classification of the residuals into small and large changes. It turns out that the large values of c pick out the generally agreed outliers and these are correctly classified when $c=1.60$. Smaller values of c force other residuals into the large category. For example, choosing $c=.90$ gives eleven large residuals ($|\mu|=10$). Lemmas 5.2 and 5.3 show that the case $c=0$ in Table 5.4 and the case $\theta=.5$ in Table 5.2 should correspond (the fully trimmed limits), and this is readily verified. Both correspond to the solution of the l_1 approximation problem.

Exercise 5.1 Discuss the resistance properties of the norms introduced in Exercise 1.1 of Chapter 3. Can the index sets μ_i used in the definition of the norms be chosen to modify resistance?

5.6 ALGORITHMS FOR M-ESTIMATION

In the previous section the M-estimation problem

$$\min F(\mathbf{x}, c),$$

where

$$F(\mathbf{x}, c)=\sum_{i=1}^{n} \hat{\rho}(|r_i|) \tag{6.1}$$

Table 5.3 Residuals and ranking sets for the values of the trimming parameter. Tied residuals are underlined.

	$\theta = 0$		$\theta = .25$		$\theta = .5$	
i	$\mu(i)$	$r_{\mu(i)}$	$\mu(i)$	$r_{\mu(i)}$	$\mu(i)$	$r_{\mu(i)}$
1	4	33.28	4	31.26	4	32.06
2	3	35.15	3	33.60	3	34.26
3	1	36.04	1	34.26	1	34.63
4	15	38.09	15	36.93	20	38.07
5	11	38.43	20	37.06	15	38.51
6	20	38.44	11	37.70	11	39.16
7	12	38.63	12	38.14	19	39.20
8	16	39.4277	19	38.19	12	39.65
9	10	39.4277	16	38.25	18	39.6899
10	19	39.92	8	38.6357	16	39.6899
11	18	40.12	10	38.6357	8	39.6899
12	8	40.62	18	38.6357	2	39.6899
13	17	40.8944	17	39.3643	10	39.71
14	14	40.8944	2	39.3643	17	40.12
15	2	41.16	7	39.64	7	40.69
16	5	41.46	5	39.937	5	40.91
17	7	41.62	14	39.944	9	41.15
18	9	42.2056	9	40.22	6	41.48
19	13	42.2056	6	40.60	14	41.49
20	6	42.37	13	41.43	13	42.59
21	21	48.53	21	47.96	21	49.17

Table 5.4 The effect of varying c in the M-estimation procedure.

c	$\lvert\nu\rvert$	x_1	x_2	x_3	x_4
3.87	21	−39.92	.716	1.295	−.152
3.21	20	−40.66	.750	1.201	−.143
2.21	19	−41.19	.811	1.010	−.133
1.81	18	−40.99	.834	.906	−.127
1.60	17	−40.19	.825	.827	−.112
.90	10	−38.15	.838	.663	-.106
0	4	−39.69	.832	.574	−.061

and $\hat{\rho}(t)$ is defined in (5.15), was introduced as a way of reducing the influence of the large residuals in the method of least squares. This problem is considered further in this section and efficient finite methods for its solution are developed. Unless otherwise stated it will be assumed that the matrix M has full rank.

Iterative methods have proved popular for finding $\mathbf{x}$ to minimize $F(\mathbf{x}, \mathrm{c})$, the three most often used being:

(i) *Newton's method.* The formal application of Newton's method to find a zero of $\mathbf{g}(\mathbf{x})$ in (4.6) gives the iteration

$$\mathbf{x}^{(k+1)} = \mathbf{x}^{(k)} - [M^{\mathrm{T}} \operatorname{diag}\{\hat{\rho}''(|r_i^{(k)}|),\ i = 1, 2, \ldots, n\}M]^{-1} M^{\mathrm{T}} \boldsymbol{\Psi}^{(k)} \tag{6.2}$$

where

$$\Psi_i^{(k)} = \theta_i^{(k)} \hat{\rho}'(|r_i^{(k)}|),\ i = 1, 2, \ldots, m, \qquad \theta_i^{(k)} = \operatorname{sgn}(r_i^{(k)}). \tag{6.3}$$

The formal nature of this iteration must be indicated as $\hat{\rho}(t)$ in (5.15) is not twice continuously differentiable. However, the method is apparently effective, but it has the disadvantage that the diagonal weighting matrix changes whenever the partitioning of the residuals into large and small changes, and this forces the matrix inverted in (6.2) to be refactorized. However, the iteration terminates in one step if the correct partition into large and small residuals is known (Exercise 6.1(i)) so it has the potential to be rapid. A rigorous analysis of the method does not appear to be available.

(ii) *Modified residual method.* In this case the basic iteration is

$$\mathbf{x}^{(k+1)} = \mathbf{x}^{(k)} - (M^{\mathrm{T}} M)^{-1} M^{\mathrm{T}} \boldsymbol{\Psi}^{(k)} \tag{6.4}$$

This method is convergent (Exercise 6.1(ii)). Although slow, it has the advantage that it can iterate the matrix factorization required in (6.4). For this reason and because of its simplicity it appears to be used frequently.

(iii) *IRLS*. This method gives the iteration

$$\mathbf{x}^{(k+1)} = \mathbf{x}^{(k)} - \left[M^{\mathrm{T}} \operatorname{diag}\left\{\frac{\hat{\rho}'(|r_i^{(k)}|)}{|r_i^{(k)}|},\ i = 1, 2, \ldots, n\right\} M\right]^{-1} M^{\mathrm{T}}\mathbf{r}^{(k)} \tag{6.5}$$

Convergence follows from Theorem 4.1. However, the method suffers from the twin disadvantages of first-order convergence and the need to refactorize in the solution of the linear equations at each iteration.

The algorithm presented here differs from the above methods in being provably finite. It uses continuation with respect to the scale parameter c (thus bypassing the problem of computing an appropriate c *ab initio*), and each step involves a reclassification of the residuals as large and small. This approach has the advantage that it selects automatically a sequence of residuals which can be inspected to determine if they are outliers. Thus it would appear to be suitable for data screening in the context of a believed least squares model and a relatively small number of bad observations. However, our algorithm is not restricted to this kind of exploratory analysis and can be used in conjunction with other methods of scale estimation (for example, that considered in Exercise 4.1). This observation is based on the fact that under change of scale, (5.15) gives

$$\frac{1}{\gamma^2}\hat{\rho}\left(\frac{t}{\sigma}, c\right) = \hat{\rho}\left(\frac{t}{\gamma\sigma}, \frac{c}{\gamma}\right)$$

where the notation has been modified to show explicitly the dependence on a parameter σ (corresponding to the standard deviation of $\mathbf{r}$) as well as on c. It follows that $\mathbf{x}$ minimizing

$$\sum_{i=1}^{n} \hat{\rho}\left(\frac{|r_i|}{\sigma}, c\right)$$

depends only on the product σc so there is no restriction in fixing $\sigma = 1$ in estimating $\mathbf{x}$ as a function of c. Alternatively, $\mathbf{x}$ can be interpreted as a function of σ with c fixed so that we can estimate scale by determining σ to satisfy an auxiliary condition (for example, the condition $\partial F/\partial\sigma = 0$ in Exercise 4.1). In this case it is necessary to fix the values of the parameters c and σ. This is done usually by maximizing the relative efficiency of the estimate of $\mathbf{x}$ and ensuring the consistency of the estimate of σ in the context of an assumed probability model. The quoted references should be consulted for details.

Lemma 6.1 *Let M have its full rank p. Then $F(\mathbf{x}, c)$ is convex in $\mathbf{x}$, and the set of minimizers of $F(\mathbf{x}, c)/c$ is bounded for each fixed $c \geqslant 0$.*

Proof Convexity is an obvious consequence of the convexity of $\hat{\rho}$. To show that the set of minimizers is bounded, it is sufficient to show that $F(\mathbf{x}+\gamma\mathbf{t}, c)\to\infty$, $\gamma\to\infty$ for each $\mathbf{x}$, $\mathbf{t}$. Let

$$k=\arg\max\,(|\mathbf{m}_i^{\mathrm{T}}\mathbf{t}|, \qquad i=1,2,\ldots,n).$$

Then

$$|r_k(\mathbf{x}+\gamma_0\mathbf{t})|\geqslant c$$

provided

$$\gamma_0\,|\mathbf{m}_k^{\mathrm{T}}\mathbf{t}|\geqslant c+|r_k(\mathbf{x})|$$

so that

$$\frac{1}{c}F(\mathbf{x}+\gamma\mathbf{t}, c)\geqslant(\gamma-\gamma_0)\,|\mathbf{m}_k^{\mathrm{T}}\mathbf{t}|\,.$$

Lemma 6.2 *Let $\mathbf{x}^*$ minimize $F(\mathbf{x}, c)$ for fixed $c>0$, and ν be the index set $\nu=\{i; |r_i|\leqslant c\}$. If M has its full rank p then there is no restriction in assuming $|\nu|\geqslant p$.*

Proof If $|\nu|<p$ then there exists nontrivial $\mathbf{t}$ such that

$$M_\nu\mathbf{t}=M_\nu^{\mathrm{T}}M_\nu\mathbf{t}=0.$$

It follows that for $\gamma>0$ small enough so that ν is unchanged,

$$\begin{aligned}\nabla F(\mathbf{x}^*+\gamma\mathbf{t}, c)&=\sum_{i\in\nu}r_i(\mathbf{x}^*+\gamma\mathbf{t})\mathbf{m}_i^{\mathrm{T}}+\sum_{i\in\nu^c}c\theta_i\mathbf{m}_i^{\mathrm{T}}\\&=\nabla F(\mathbf{x}^*, c)+\gamma\mathbf{t}^{\mathrm{T}}M_\nu^{\mathrm{T}}M_\nu\\&=0.\end{aligned}$$

Thus $\mathbf{x}^*+\gamma\mathbf{t}$ also minimizes $F(\mathbf{x}, c)$ for $\gamma>0$ small enough. However, the range of such values of γ is bounded, by Lemma 6.1. Thus there is a first value of γ for which ν must be updated (this requires $|\bar{\nu}|\geqslant|\nu|+1$), and by continuity this point is optimal. The argument can be repeated provided rank $(M_\nu)<p$ – in particular, if $|\nu|<p$.

Corollary 6.1 *If M has full rank p then there is no restriction in assuming*

$$\operatorname{rank}(M_\nu)=p.$$

Remark 6.1 If the correct ν and the partitioning of ν^c into ν^+ and ν^- corresponding to the large positive and large negative residuals respectively is known *ab initio*, then the solution of the M-estimation problem can be found by solving (5.28) for $\mathbf{x}$. This system is

$$M_\nu^{\mathrm{T}}M_\nu\mathbf{x}=M_\nu^{\mathrm{T}}\mathbf{f}_\nu-\sum_{i\in\nu^c}c\theta_i\mathbf{m}_i$$

where $\theta_i=\operatorname{sgn}(r_i)$ as usual, and $(\mathbf{f}_\nu)_i=f_{\nu(i)}$, $i=1,2,\ldots,|\nu|$. This system is uniquely solvable as a consequence of Corollary 6.1.

However, it does not follow necessarily that the solution of the M-estimation problem is unique. To settle this question we compute $F(\mathbf{x}^*+\mathbf{t}, c)-F(\mathbf{x}^*, c)$ where $\mathbf{x}^*$ minimizes $F(\mathbf{x}, c)$ and $\mathbf{t}$ is an arbitrary vector. To simplify this expression set

$$\nu_1=\{i;\, |r_i(\mathbf{x}^*)|<c\},$$

$$\nu_2(\mathbf{t})=\{i;\, |r_i|=c \quad \text{and} \quad \theta_i \mathbf{m}_i^{\mathrm{T}}\mathbf{t}<0\}$$

and

$$\nu_3=\nu\backslash\{\nu_1 \cup \nu_2\}.$$

Provided $\|\mathbf{t}\|$ is sufficiently small not to disturb the classification of residuals which are not *tight* (r_i is tight if $|r_i|=c$), then

$$\begin{aligned} F(\mathbf{x}^*+\mathbf{t}, c)-F(\mathbf{x}^*, c) &= \sum_{\nu_1 \cup \nu_2} \left(\tfrac{1}{2}(r_i+\mathbf{m}_i^{\mathrm{T}}\mathbf{t})^2-\tfrac{1}{2}r_i^2\right) \\ &\quad +\sum_{\nu_3} \left(c\theta_i(r_i+\mathbf{m}_i^{\mathrm{T}}\mathbf{t})-c^2\right)+\sum_{\nu^c} c\theta_i \mathbf{m}_i^{\mathrm{T}}\mathbf{t} \\ &= \sum_{\nu_1 \cup \nu_2} \tfrac{1}{2}(\mathbf{m}_i^{\mathrm{T}}\mathbf{t})^2 \end{aligned} \tag{6.6}$$

where the condition $\nabla F(\mathbf{x}^*, c)=0$ has been used (equation (5.27)). The following result is an immediate consequence.

Lemma 6.3 *There exists a unique minimizer for $F(\mathbf{x}, c)$ if and only if*

$$\dim \{\mathbf{m}_i,\, i \in \nu_1 \cup \nu_2(\mathbf{t})\}=p,\ \forall\, \mathbf{t}.$$

Corollary 6.2 *If $|\nu|=p$, $c>0$, and there are tight residuals, then there exists a direction of non-uniqueness.*

Proof This can be constructed by solving

$$M_\nu \mathbf{t}=\theta_k \mathbf{e}_k$$

where $|r_{\nu(k)}|=c$.

Example 6.1 To illustrate the possibility of a non-unique solution, consider the problem data given in Table 6.1.

Table 6.1 Example of an M-estimation problem with a non-unique solution.

i	M		$\mathbf{f}$
1	1	0	0
2	0	1	1
3	1	1	−5

Provided $6-2c>c$, an optimal solution is $\nu=\{1,2\}$,

$$\mathbf{x}^*=\begin{bmatrix}-c\\1-c\end{bmatrix},\qquad \mathbf{r}=\begin{bmatrix}-c\\-c\\6-2c\end{bmatrix},$$

and both small residuals are tight. Now set

$$\mathbf{x}=\mathbf{x}^*-\begin{bmatrix}\alpha\\\beta\end{bmatrix}$$

with $\alpha\geqslant 0, \beta\geqslant 0$. This gives

$$\mathbf{r}=\begin{bmatrix}-c-\alpha\\-c-\beta\\6-2c-(\alpha+\beta)\end{bmatrix}$$

so that $|r_1|\geqslant c$, $|r_2|\geqslant c$, $\nu_1\cup\nu_2=\varnothing$, and

$$\begin{aligned}F\left(\mathbf{x}^*-\begin{bmatrix}\alpha\\\beta\end{bmatrix},c\right)&=c\{(c+\alpha)+(c+\beta)+(6-2c-(\alpha+\beta))-3c/2\}\\&=F(\mathbf{x}^*,c)\end{aligned}$$

provided $6-2c-(\alpha+\beta)\geqslant c$. The solution to the least squares problem corresponds to $c=2$, so the range of interest for the M-estimator is $0<c<2$. Note that the solution set for $c>0$ tends to the l_1 solution set as $c\to 0$.

To study the behaviour of $F(\mathbf{x},c)$ as c varies let $\mathbf{x}(c)$ minimize F and set $F^*(c)=F(\mathbf{x}(c),c)$.

Lemma 6.4 *$F^*(c)$ is a continuous and increasing function of c on any interval in which $\nu(c)^c\neq\varnothing$.*

Proof Let $c_1>c_2$. Then

$$\begin{aligned}F^*(c_1)&=\sum_{i\in\nu(c_1)}\tfrac{1}{2}r_i^2(\mathbf{x}(c_1))+\sum_{i\in\nu(c_1)^c}\{c_1|r_i(\mathbf{x}(c_1))|-c_1^2/2\}\\&>\sum_{i\in\nu(c_1)}\tfrac{1}{2}r_i^2(\mathbf{x}(c_1))+\sum_{i\in\nu(c_1)^c}\{c_2|r_i(\mathbf{x}(c_1))|-c_2^2/2\},\\&\geqslant F(\mathbf{x}(c_1),c_2),\\&\geqslant F^*(c_2).\end{aligned}$$

This shows that $F^*(c)$ is increasing provided $\nu(c)^c\neq\varnothing$. To show that $F^*(c)$ is continuous, choose $\delta>0$ sufficiently small that

$$c+\delta\leqslant|r_i(\mathbf{x}(c))|,\qquad\forall\, i\in\nu(c)^c.$$

As F^* is an increasing function and as the partitioning of $\mathbf{r}(\mathbf{x}(c))$ does not change when c is changed to $c+\delta$, then

$$F^*(c) \leqslant F^*(c+\delta) \leqslant F(\mathbf{x}(c), c+\delta) = F^*(c) + \delta \sum_{i \in \nu(c)^c} |r_i(\mathbf{x}(c))| - \tfrac{1}{2}|\nu(c)^c|\,((c+\delta)^2 - c^2),$$

and continuity is an immediate consequence (note that the evaluation of $F(\mathbf{x}(c), c+\delta)$ is correct even if there are tight residuals at $c+\delta$).

Corollary 6.3 *Let* $\{c_i\} \to c$. *Then limit points of the sequence* $\{\mathbf{x}(c_i)\}$ *minimize* $F(\mathbf{x}, c)$.

An algorithm for minimizing $F(\mathbf{x}, c)$ assuming that $\mathbf{x} = \mathbf{x}(c)$ defines a unique trajectory is now developed. It is based on the observations:

(i) by Remark 6.1, $\mathbf{x}(c)$ can be found provided the partitioning of the residuals is known; and
(ii) the partitioning of the residuals can be found in the extreme cases corresponding to $c = 0$ (l_1 approximation), and $c \geqslant c_0 = \max(|r_i(\mathbf{x})|, i = 1, 2, \ldots, n)$ where $\mathbf{x}$ is the solution of the least squares problem.

This suggests the use of continuation with respect to c to keep track of the partitioning of the residuals. Assume that at the current c there are no tight residuals. By Corollary 6.1 and Lemma 6.3 the M-estimator is unique. It then follows that there is an interval containing c in which there is a unique continuous solution trajectory $\mathbf{x}(c)$. In fact, expanding the condition (5.27) for a minimum of $F(\mathbf{x}, c)$ and grouping terms gives

$$0 = \nabla F(\mathbf{x}, c) = \Big(\sum_{i \in \nu} \mathbf{m}_i \mathbf{m}_i^{\mathrm{T}}\Big)\mathbf{x} + c \sum_{i \in \nu^c} \theta_i \mathbf{m}_i - \sum_{i \in \nu} f_i \mathbf{m}_i, \tag{6.7}$$

and this equation can be solved for $\mathbf{x}$ as uniqueness forces M_ν to have full rank, as a consequence of Lemma 6.3. It follows that $\mathbf{x} = \mathbf{x}(c)$ *is a linear function of* c *on any interval on which the classification of residuals does not change* (in particular, on any interval on which no residual is tight). Differentiating (6.7) gives $\mathrm{d}\mathbf{x}/\mathrm{d}c$ as the solution of the linear equations

$$M_\nu^{\mathrm{T}} M_\nu \frac{\mathrm{d}\mathbf{x}}{\mathrm{d}c} = -\sum_{i \in \nu^c} \theta_i \mathbf{m}_i \tag{6.8}$$

(and hence $\mathrm{d}r_i/\mathrm{d}c = \mathbf{m}_i^{\mathrm{T}}\,\mathrm{d}\mathbf{x}/\mathrm{d}c$). As might be expected, this is the equation on which to concentrate in order to specify the continuation process.

Given $\mathbf{x}$ and $\mathrm{d}\mathbf{x}/\mathrm{d}c$ for one value of c, we have $\mathbf{x}$ specified for the set of contiguous values of c for which ν defines the optimal partition. This partition must be revised at the first value of c such that a new residual becomes tight (such values are called *critical points*), and the result which makes this straightforward is the following.

Lemma 6.5 *The pointer to an isolated tight residual swaps to the complementary index set when c passes through a critical point.*

Proof This is by direct computation. It is shown that if the pointer moves to the complementary index set and $\mathbf{dx}/dc$ is recomputed, then the tight residual continues to increase (decrease) relative to c. Let

$$G_\nu = \sum_{i \in \nu} \mathbf{m}_i \mathbf{m}_i^{\mathrm{T}} = L_\nu L_\nu^{\mathrm{T}}, \tag{6.9a}$$

where L_ν is invertible (one possibility is the Choleski decomposition but our preferred algorithm will offer an alternative),

$$\mathbf{g}_\nu = \sum_{i \in \nu^c} \theta_i \mathbf{m}_i, \tag{6.9b}$$

and

$$\mathbf{w}_i = L_\nu^{-1} \mathbf{m}_i, \qquad i = 1, 2, \ldots, n. \tag{6.9c}$$

Quantities after the pointer move are indicated by bars. If k is the index of the tight residual then both possibilities are summarized in

$$G_{\bar{\nu}} \frac{\mathrm{d}\bar{\mathbf{x}}}{\mathrm{d}c} = \{G_\nu \pm \mathbf{m}_k \mathbf{m}_k^{\mathrm{T}}\} \frac{\mathrm{d}\bar{\mathbf{x}}}{\mathrm{d}c} = -\mathbf{g}_\nu \pm \theta_k \mathbf{m}_k \tag{6.10}$$

Using (6.9a) gives

$$\{I \pm \mathbf{w}_k \mathbf{w}_k^{\mathrm{T}}\} L_\nu^{\mathrm{T}} \frac{\mathrm{d}\bar{\mathbf{x}}}{\mathrm{d}c} = L_\nu^{\mathrm{T}} \frac{\mathrm{d}\mathbf{x}}{\mathrm{d}c} \pm \theta_k \mathbf{w}_k \tag{6.11}$$

Now

$$\frac{\mathrm{d}\mathbf{r}_k}{\mathrm{d}c} = \mathbf{m}_k^{\mathrm{T}} \frac{\mathrm{d}\mathbf{x}}{\mathrm{d}c} = \mathbf{w}_k^{\mathrm{T}} L_\nu^{\mathrm{T}} \frac{\mathrm{d}\mathbf{x}}{\mathrm{d}c} \tag{6.12}$$

Thus, taking the scalar product of (6.11) with $\mathbf{w}_k$,

$$(1 \pm \|\mathbf{w}_k\|_2^2) \frac{\mathrm{d}\bar{r}_k}{\mathrm{d}c} = \frac{\mathrm{d}r_k}{\mathrm{d}c} \pm \theta_k \|\mathbf{w}_k\|_2^2 \tag{6.13}$$

Setting $\theta_k r_k = |r_k|$ and collecting terms gives

$$\frac{\mathrm{d}\,|\bar{r}_k|}{\mathrm{d}c} - 1 = \frac{\mathrm{d}\,|r_k|/\mathrm{d}c - 1}{1 \pm \|\mathbf{w}_k\|_2^2} \tag{6.14}$$

This shows that $\mathrm{d}\,|\bar{r}_k|/\mathrm{d}c > 1$ or <1 according as $\mathrm{d}\,|r_k|/\mathrm{d}c > 1$ or <1, as $1 - \|\mathbf{w}_k\|_2^2 > 0$ is the condition that $G_{\bar{\nu}}$ be positive definite in the case that the minus sign is appropriate in (6.10).

In other words, an isolated residual which catches up with c passes it, while a residual which is overtaken by c is passed.

In outline the basic form for an algorithm exploiting the idea of continuation is as follows.

Algorithm 6.1

(i) *Solve the l_1 or least squares problem to determine an initial partition.*

(ii) *On the current partition compute* $\mathrm{d}\mathbf{x}(c')/\mathrm{d}c$, $\mathrm{d}\mathbf{r}(c')/\mathrm{d}c$,

$$r_i(c) = r_i(c') + (c - c')\frac{\mathrm{d}r_i}{\mathrm{d}c}(c')$$

where c' is a critical point determining one end of the current interval.

(iii) *Find the next critical point as c is increased/decreased. If c reaches a target value then stop.*

(iv) *Update ν using Lemma* 6.5. *Modify the factorization of G_ν, etc. Repeat* (ii).

If the computation is started by solving the least squares problem then $\nu = \{1, 2, \ldots, n\}$, $\mathrm{d}\mathbf{x}/\mathrm{d}c = \mathrm{d}\mathbf{r}/\mathrm{d}c = 0$, and the first critical point is $c_0 = \max(|r_i|,\ i = 1, 2, \ldots, n)$. The standard methods for solving least squares problems all give the triangular factors of G_ν. However, the modified Gram–Schmidt method also gives

$$L^{-1}M^{\mathrm{T}} = Q_1^{\mathrm{T}} \tag{6.15}$$

where

$$\boldsymbol{\kappa}_i(Q_1^{\mathrm{T}}) = L^{-1}\mathbf{m}_i = \mathbf{w}_i, \qquad i = 1, 2, \ldots, n \tag{6.16}$$

are the quantities appearing in (6.11); so this would appear to be an appropriate method to choose. The other starting point is $c = 0$. Here the solution to the l_1 problem gives $\nu = \mu$, the index set pointing to the zero residuals, and

$$G_\nu = B_\mu B_\mu^{\mathrm{T}}. \tag{6.17}$$

The unusual feature of (6.17) is that the factorization is not specified directly in terms of triangular matrices so there could be some advantage in seeking an update procedure for modifying the factors G_ν at a critical point which is not directly tied to a particular form of solution of the initial problem. To investigate this point further, consider

$$G_{\bar{\nu}} = G_\nu \pm \mathbf{m}_k\mathbf{m}_k^{\mathrm{T}} = L_\nu\{I \pm \mathbf{w}_k\mathbf{w}_k^{\mathrm{T}}\}L_\nu^{\mathrm{T}}$$

where k is the index of the tight residual at the current critical point. Let Q be the orthogonal matrix chosen to satisfy

$$Q^{\mathrm{T}}\mathbf{w}_k = \phi\,\|\mathbf{w}_k\|_2\,\mathbf{e}_\alpha \tag{6.18}$$

where typically Q could be a Householder matrix, $\phi = \pm 1$ is chosen to minimize cancellation in constructing Q, and α is an integer in $1 \leq \alpha \leq p$. Then

$$\begin{aligned} G_{\bar{\nu}} &= L_\nu Q\{I \pm \|\mathbf{w}_k\|_2^2\,\mathbf{e}_\alpha\mathbf{e}_\alpha^{\mathrm{T}}\}Q^{\mathrm{T}}L_\nu^{\mathrm{T}} \\ &= L_{\bar{\nu}}L_{\bar{\nu}}^{\mathrm{T}} \end{aligned} \tag{6.19}$$

where

$$L_{\bar{\nu}} = L_\nu QD \tag{6.20}$$

is not lower triangular in general, even if L_ν is, and

$$D^2 = I \pm \|\mathbf{w}_k\|_2^2 \, \mathbf{e}_\alpha \mathbf{e}_\alpha^{\mathrm{T}} \tag{6.21}$$

is diagonal so that its square root is immediately available.

Remark 6.2 Choice of α does not seem to be important. Methods tested include choosing $\alpha = \arg\max\,(|(\mathbf{w}_k)_i|,\ \ i = 1, 2, \ldots, p)$, cyclically, or at random.

The corresponding transformation of the $\mathbf{w}_i$ is

$$\bar{\mathbf{w}}_i = D^{-1}Q^{\mathrm{T}}\mathbf{w}_i, \qquad i = 1, 2, \ldots, n. \tag{6.22}$$

This is reminiscent of a tableau updating step. Let

$$W = L_\nu^{-1}[M^{\mathrm{T}} \mid I] \tag{6.23}$$

then (6.22) is summarized by

$$\bar{W} = D^{-1}Q^{\mathrm{T}}W. \tag{6.24}$$

This observation provides the basis for a neat implementation of the algorithm in which *a main feature is the cheap updating of* $\mathrm{d}\mathbf{r}/\mathrm{d}c$. The starting point is equation (6.11). Multiplying by $\mathbf{w}_j^{\mathrm{T}}$ gives

$$\frac{\mathrm{d}\bar{r}_j}{\mathrm{d}c} \pm \mathbf{w}_j^{\mathrm{T}}\mathbf{w}_k \frac{\mathrm{d}\bar{r}_k}{\mathrm{d}c} = \frac{\mathrm{d}r_j}{\mathrm{d}c} \pm \theta_k \mathbf{w}_j^{\mathrm{T}}\mathbf{w}_k$$

so that

$$\frac{\mathrm{d}\bar{r}_j}{\mathrm{d}c} = \frac{\mathrm{d}r_j}{\mathrm{d}c} \mp \theta_k \mathbf{w}_j^{\mathrm{T}}\mathbf{w}_k \left(\frac{\mathrm{d}\,|\bar{r}_k|}{\mathrm{d}c} - 1\right) \tag{6.25}$$

This shows that the work involved is in the calculation of the scalar products $\mathbf{w}_j^{\mathrm{T}}\mathbf{w}_k$. To simplify this, note that

$$\begin{aligned}\mathbf{w}_j^{\mathrm{T}}\mathbf{w}_k &= \mathbf{w}_j^{\mathrm{T}}QD^{-1}DQ^{\mathrm{T}}\mathbf{w}_k \\ &= \bar{\mathbf{w}}_j^{\mathrm{T}}D\phi\,\|\mathbf{w}_k\|_2\,\mathbf{e}_\alpha \\ &= \phi D_\alpha\,\|\mathbf{w}_k\|_2\,(\bar{\mathbf{w}}_j)_\alpha\end{aligned} \tag{6.26}$$

Also,

$$\begin{aligned}\bar{\mathbf{w}}_k &= D^{-1}Q^{\mathrm{T}}\mathbf{w}_k \\ &= \frac{\phi\,\|\mathbf{w}_k\|_2}{D_\alpha}\mathbf{e}_\alpha\end{aligned}$$

so that

$$\begin{aligned}\mathbf{w}_j^{\mathrm{T}}\mathbf{w}_k &= D_\alpha^2(\bar{\mathbf{w}}_k)_\alpha(\bar{\mathbf{w}}_j)_\alpha \\ &= (1 \pm \|\mathbf{w}_k\|_2^2)(\bar{\mathbf{w}}_k)_\alpha(\bar{\mathbf{w}}_j)_\alpha\end{aligned} \tag{6.27}$$

Combining this with (6.14) gives

$$\frac{\mathrm{d}\bar{r}_j}{\mathrm{d}c}=\frac{\mathrm{d}r_j}{\mathrm{d}c}\mp\theta_k(\bar{\mathbf{w}}_k)_\alpha(\bar{\mathbf{w}}_j)_\alpha\left(\frac{\mathrm{d}\,|r_k|}{\mathrm{d}c}-1\right). \tag{6.28}$$

There is also a corresponding formula for $\mathrm{d}\bar{x}_j/\mathrm{d}c$:

$$\frac{\mathrm{d}\bar{x}_j}{\mathrm{d}c}=\frac{\mathrm{d}x_j}{\mathrm{d}c}\mp\theta_k(\bar{\mathbf{w}}_{n+j})_\alpha(\bar{\mathbf{w}}_k)_\alpha\left(\frac{\mathrm{d}\,|r_k|}{\mathrm{d}c}-1\right) \tag{6.29}$$

where the $\bar{\mathbf{w}}_{n+j}$ are the columns of the tableau (6.23) set initially to the unit matrix. The derivation is identical.

Remark 6.3 If the algorithm is started with the least squares solution so that continuation reduces c, then it must be expected that the most likely action at a critical point is the reclassification of a tight residual as large. This involves a downdating step in modifying G_ν (compare Remark 2.6). This is the less stable of the two possibilities. However, it is unlikely to cause trouble in the 'usual' situation in which a relatively small number of residuals must be classified as large before the computation terminates (one possibility being that the remaining observations satisfy a test for consistency with a least squares model). On the other hand, if the computation is started with $c=0$ then the usual action is likely to be the classification of a tight residual as small. This requires an updating step on G_ν, and is stable. However, the 'usual' situation requires that most residuals be classified as small before the computation can terminate. In general this would imply many more steps in the case $c=0$.

To prove finiteness of the continuation algorithm it is sufficient to show that the interval $0<c<c_0$ (the critical point determined by the least squares solution) is made up of a finite number of subintervals, on each of which the optimum is characterized by a particular partition $P_\nu=\{\nu, \nu_+^c, \nu_-^c\}$ and hence by particular linear form for $\mathbf{x}(c)$.

Theorem 6.1 *The continuation trajectory on $0<c<c_0$ is made up of a finite number of subintervals on each of which $\mathbf{x}=\mathbf{x}(c)$ is a particular linear function of c.*

Proof Let P_ν determine $\mathbf{x}(c)$ for $c=c_1$ and $c=c_2>c_1$. Then P_ν determines $\mathbf{x}(c)=\mathbf{x}(c_1)+(c-c_1)\,\mathrm{d}\mathbf{x}(c_1)/\mathrm{d}c$ for $c_1\leqslant c\leqslant c_2$, as the corresponding $\mathbf{r}(c)$ is a linear function of c on this interval and hence cannot depart from the classification which applies at c_1 and c_2. It follows that the continuation process cannot cycle between partitions. Thus the continuation trajectory must be made up of a finite number of linear pieces as the total number of P_ν is finite. It is easy to see that it must be continuous.

Remark 6.4 The development of the continuation algorithm assumes that each partition revision is forced by an isolated residual becoming tight. If more than one residual becomes tight simultaneously then a situation occurs which is analogous to degeneracy in the problems considered in the earlier chapters. Theorem 6.1 shows that there must be an appropriate partition on which to compute $d\mathbf{x}/dc$ to reinitiate the continuation process in degenerate situations. In particular, it is necessary only to repartition the tight residuals.

A method which calculates a suitable partition for restarting the continuation process is now developed. Due to David Clark, it does in fact provide an alternative method for calculating the M-estimator for any particular value of c. However, Clark reports that his experiments showed that the method could involve a lot of work if good starting information (in the sense of an accurate classification of most of the residuals as either small or large) is not available.

Definition 6.1 Let $\mu \subseteq \{1, 2, \ldots, n\}$ and set

$$F_\mu(\mathbf{x}, c) = \sum_{i\in\mu} \tfrac{1}{2} r_i^2 + \sum_{i\in\mu^c} c\,|r_i| - c^2/2. \tag{6.30}$$

If $\mathbf{x}_\mu$ minimizes $F_\mu(\mathbf{x}, c)$ then μ is μ^c-*feasible* if

$$|r(\mathbf{x}_\mu)| > c, \qquad i \in \mu^c.$$

If, in addition

$$|r_i(\mathbf{x}_\mu)| \leqslant c, \qquad i \in \mu$$

then the partition is *feasible* and $\mathbf{x}_\mu$ minimizes $F(\mathbf{x}, c)$.

Lemma 6.6 *If μ is μ^c-feasible then*

$$F_\mu(\mathbf{x}_\mu, c) \geqslant F(\mathbf{x}(c), c) \tag{6.31}$$

Proof Let σ point to the subset of the r_i, $i \in \mu$ such that $|r_i| > c$; then

$$\begin{aligned} F_\mu(\mathbf{x}_\mu, c) &= \sum_{i\in\mu} \tfrac{1}{2} r_i^2 + \sum_{i\in\mu^c} c\,|r_i| - c^2/2 \\ &\geqslant \sum_{i\in\mu\backslash\sigma} \tfrac{1}{2} r_i^2 + \sum_{i\in\mu^c\cup\sigma} c\,|r_i| - c^2/2 \\ &= F(\mathbf{x}_\mu, c) \\ &\geqslant F(\mathbf{x}(c), c). \end{aligned}$$

A similar argument gives the following result.

Lemma 6.7 *If $\mu_a = \mu_b \cup \{k\}$ then, in an obvious modification of the notation* *of* (6.30),

$$F_a(\mathbf{x}_a, c) \geqslant F_b(\mathbf{x}_a, c) \geqslant F_b(\mathbf{x}_b, c). \tag{6.32}$$

Equality holds in the first inequality only if $|r_k(\mathbf{x}_a)| = c$.

Algorithm 6.2 *The aim is to find a feasible partition starting from an arbitrary partition and proceeding only by adjacent partition changes so that*

$$\mu := (\mu \cup \{k\} \quad \text{or} \quad \mu \backslash \{k\}).$$

(i) *Starting from any initial partition do until* μ *is* μ^c*-feasible*

$$\mu := \mu \cup \{k\} \quad \textit{where} \quad k \in \mu^c \quad \textit{and} \quad |r_k| \leqslant c.$$

Set $i = 1$, $\mu_1 = \mu$.

(ii) *While* μ_i *is* μ^c*-feasible do*

If μ_i *is feasible then stop; else do*

$$\mu_{i+1} := \mu_i \backslash \{k\} \quad \textit{where} \quad k \in \mu_i \quad \textit{and} \quad |r_k(\mathbf{x}_i)| > c,$$
$$i := i + 1$$

(iii) $\mathbf{y}_{i-1} = \mathbf{x}_{i-1}$ ($\mathbf{x}_{i-1}$ *satisfies* $|r_k(\mathbf{x}_{i-1})| > c$, $k \in \mu_i^c$)

Until μ_i *is* μ^c*-feasible do*

find

$\mathbf{y}_i = \alpha \mathbf{x}_i + (1 - \alpha)\mathbf{y}_{i-1}$, $0 < \alpha \leqslant 1$ *to satisfy*

$|r_j(\mathbf{y}_i)| \geqslant c$, $j \in \mu_i^c$ *and* $|r_k(\mathbf{y}_i)| = c$ *for at least one*

$k \in \mu_i^c$.

$\mu_{i+1} = \mu_i \cup \{k\}$

$i := i + 1$

(iv) *repeat* (ii).

It is clear that each entry of either step (ii) or step (iii) must terminate (for example, if $\mu_i = \varnothing$ so that the problem is effectively one of l_1 minimization, then the l_1 characterization result shows that μ_i cannot be μ^c-feasible for any $c \geqslant 0$, while $\mu_i = \{1, 2, \ldots, n\}$ is trivially μ^c-feasible). Thus to show finite termination of the algorithm it is necessary to show that it cannot cycle.

Theorem 6.2 *Clark's algorithm terminates in a finite number of steps.*

Proof This consists in showing that the sequence of μ^c-feasible solutions generated is strictly decreasing and so cannot repeat the corresponding index sets μ. This shows that the algorithm cannot cycle, and as there are at most a finite number of μ the algorithm must terminate.

First note that in step (ii) of the algorithm, as a consequence of Lemma 6.7,

$$F_{i+1}(\mathbf{x}_{i+1}) \leqslant F_{i+1}(\mathbf{x}_i) < F_i(\mathbf{x}_i) \tag{6.33}$$

as $\mu_{i+1} \cup \{k\} = \mu_i$ and $|r_k(\mathbf{x}_i)| > c$. This shows that the μ^c-feasible solutions produced in step (ii) are strictly decreasing. In step (iii) the first pass gives

$$\begin{aligned} F_{i+1}(\mathbf{x}_{i+1}) &\leqslant F_{i+1}(\mathbf{y}_i) = F_i(\mathbf{y}_i), \\ &\leqslant \alpha F_i(\mathbf{x}_i) + (1-\alpha) F_i(\mathbf{y}_{i-1}), \\ &\leqslant F_i(\mathbf{y}_{i-1}), \\ &= F_i(\mathbf{x}_{i-1}) < F_{i-1}(\mathbf{x}_{i-1}) \end{aligned}$$

where the last inequality is (6.33) above, and where convexity and Lemma 6.7 have been used. If s steps are needed in step (iii) to produce a μ^c-feasible partition, the same argument can be repeated. This gives

$$\begin{aligned} F_{i+s}(\mathbf{x}_{i+s}) &\leqslant F_{i+s}(\mathbf{y}_{i+s-1}) \leqslant \cdots \\ &\leqslant F_i(\mathbf{y}_{i-1}) \\ &< F_{i-1}(\mathbf{x}_{i-1}). \end{aligned}$$

Remark 6.5 The application of this algorithm to resolve degeneracy in the continuation method is rather nice. One obvious approach is to apply it with c incremented by a small amount, the increment being chosen sufficiently small that the manipulations of the algorithm are concerned only with the tight residuals. However, the problem of choosing a suitable value for the increment in c can be avoided by noting that *the classification of the tight residuals on the current partition is determined by* $\mathrm{d}\mathbf{r}/\mathrm{d}c$ *computed on this partition.* Thus all that is required is to work through the algorithm at the critical point with the magnitudes of the tight residuals replaced by $\theta_i \, \mathrm{d}r_i/\mathrm{d}c$ as, for example, $|r_i| > c + \delta$ is equivalent to $\theta_i \, \mathrm{d}r_i/\mathrm{d}c > 1$.

Exercise 6.1 (i) Show that the Newton iteration (6.2) terminates in one step when the partition into large and small residuals is known.
(ii) Discuss the convergence of the modified residual method (6.4).

Exercise 6.2 Consider the problem data

$$M^{\mathrm{T}} = \begin{bmatrix} 1 & 1 & 4 & 2 & 5 \\ 1 & 4 & 1 & 5 & 3 \end{bmatrix},$$

$$\mathbf{f}^{\mathrm{T}} = [1 \quad 7 \quad 2 \quad 12 \quad 1].$$

Let $c = 2$ and minimize F on the assumption that $\nu = \{1, 2, 3, 4\}$. Verify that the partition is feasible. Is it unique? Find the feasible partitions for $0 \leqslant c < \infty$.

Exercise 6.3 Consider the problem data

$$M^{\mathrm{T}}=\begin{bmatrix}1 & 0 & 1\\ 0 & 1 & 1\end{bmatrix},$$
$$\mathbf{f}^{\mathrm{T}}=[9 \quad 3 \quad 3].$$

Let $c=1$ and minimize F for the partitions defined by $\nu=\{1,2,3\}$, $\nu=\{2,3\}$, and $\nu=\{3\}$. Discuss the behaviour of the residuals as a function of changing to an adjacent partition.

Exercise 6.4 Consider the problem data

$$M^{\mathrm{T}}=\begin{bmatrix}1 & 2 & 2 & 0\\ 1 & 3 & 0 & 3\end{bmatrix},$$
$$\mathbf{f}^{\mathrm{T}}=[2 \quad 4 \quad 3 \quad 5].$$

Let $c=1$. Show that $\nu=\{4\}$, $\nu=\{2,3\}$, and $\nu=\{2,3,4\}$ are feasible.

Exercise 6.5 (i) The 'obvious' way to modify G_ν is to use the up-and-down dating formulae developed in Section 6.2 (Remark 2.6). What form does the M-estimation algorithm take in this case? How does its cost compare with the above implementations? (ii) What is the analogue of the product form algorithm of Section 2.4? Compare this with the tableau based algorithm described above.

Exercise 6.6 How must Theorem 6.1 be modified if the uniqueness of $\mathbf{x}(c)$ cannot be assumed?

NOTES ON CHAPTER 5

Clearly it is impossible within the confines of a single chapter to treat adequately even the particular problem of fitting a linear model by least squares. Thus the aim has been to exemplify certain algorithmic ideas in a context which is sufficiently well developed to permit their importance to be assessed.

An appropriate reference on least squares problems is Seber (1977) and an extensive treatment of the main algorithms is given in Hanson and Lawson (1974). The origins of the Choleski method are somewhat obscure, and I am indebted to Allan Miller for the following comments (they derive from a translation by Richard Cottle of a biographic note which appeared in the *Bullétin Géodésique* in 1922 – Choleski was killed in action in August 1918):

> 'Choleski joined the Geodesic Section of the French Geographic Service in 1905. He developed his method of factorization as a way of solving the least squares equations arising in the new cadastral triangulation of France which

began about that time. This was also the principal application of the Gauss–Seidel method which had been in use in German-speaking countries since the 1830s and which Nagel had used to solve the 159 simultaneous equations as part of the survey of Prussia in 1890. We are fortunate that Choleski was probably unaware of the Gauss–Seidel method otherwise he would probably have rejected his new method just as Gauss had rejected Gaussian elimination. The unnamed writer of the 1922 note said that Choleski's method would be extremely useful if it were ever published, indicating that Choleski himself did not do so.'

A stability analysis of the Choleski factorization is given in Wilkinson (1965). The stability of the updating and downdating procedures is analysed in Stewart (1979). The case when G is only just positive definite is considered by Gill, Murray, and Wright (1981). The use of orthogonal transformations in the solution of least squares problems is due to Householder (1958). It was publicized extensively by Golub (1965) and provides a main theme of Golub and Van Loan (1983). The superior stability properties of the orthogonal methods in small residual problems is reported in Golub and Wilkinson (1966), and further aspects such as the growth of errors in intermediate calculations have been considered by Jennings and Osborne (1974) and Stewart (1977). The important result that the modified Gram–Schmidt procedure is stable is due to Bjork (1967). An account of the fast Givens algorithm is given in Hammarling (1973). The generalized least squares problem has been considered by Paige (1979) who developed the factorization given in Exercise 2.3. The general question of the selection of a suitable subset of regression variables, including the stepwise approach sketched in Exercise 2.4, is contained in the forthcoming book by Allan Miller. For information on regression diagnostics see Belsley, Kuh, and Welsch (1980).

Appropriate references on linearly constrained optimization problems, and on quadratic programming problems in particular, are Fletcher (1980), and Gill, Murray, and Wright (1981). Much of the recent algorithmic development is due to Gill and Murray (for example, Gill and Murray (1974)). The special case of the restricted least squares problem is treated by an active set method in Hanson and Lawson (1974). Suitable references on the treatment of the restricted least squares problem as a subset selection problem are Beale (1970) and Armstrong and Frome (1976). The development here follows Clark (1980) and Clark and Osborne (1980). A similar approach to the stripped-down algorithm is attributed to Bard in Bartels, Golub, and Saunders (1970), who also consider the problem of efficient implementation for more general quadratic programming problems. Subset selection has something of the flavour of a dual method such as subopt considered in Section 2.7, especially if the steps of the selection algorithm are considered to be between successive feasible points (for example, such points could be the points at which branching ceases in step (i) of Remark 3.6). An adaptation of this idea of a dual approach to general convex QP has been developed by Goldfarb and Idnani (1983) and appears to be very promising. A systematic pivoting rule which has some of the flavour of a Bard-type scheme is proposed in Murty (1974). The method does not have the freedom of choice in pivot selection of the method described here (cf. Remark 3.5). The presentation of the complementarity algorithms follows Cottle and Dantzig (1968). The use of conjugate directions in solving convex QP was suggested first by Beale (1959). Lemke (1962) suggested this idea should be used in conjunction with (3.7).

The method of iteratively reweighted least squares is usually attributed to Tukey (Beaton and Tukey (1974)). For the l_1 case it was suggested by Schlossmacher (1973). Lemma 4.2 is due to Dutter (1975), while the estimate in Exercise 4.1 comes from Huber (1981). Related applications include the EM algorithm of Dempster, Laird, and Rubin (1977), and the implementation in the programming language GLIM of the algorithm for fitting generalized linear models of Nelder and Wedderburn (1972). Similar rate of convergence estimates to those of Section 4 are obtained also in Section 6.3.

The informal presentation of resistance draws on ideas which have been current for a long time. One reference is Mostellar and Tukey (1977). For a detailed study of robustness see Huber (1981). The trimming of rank regression estimates is considered by Hettmansperger and McKean (1977). The importance of the l_1 estimator is further strengthened by the results of Bassett and Koenker (1978), who show that it gives smaller confidence ellipsoids than least squares whenever the median is more efficient than the mean for the associated location problem. The M in M-estimation is short for maximum likelihood and derives from the resemblance between the objective function (5.14) and a log likelihood. The Huber M-estimator has the important property that it minimizes a measure of maximum loss corresponding to choosing the wrong probability distribution on which to base the estimation from the class of distributions in which a normal distribution is contaminated by an arbitrary (unknown) distribution. The value of c reflects the scale of the contamination. An informal account of this and related material (including the corresponding min-max property for l_1 estimation) is given in Vapnik (1982). Recent developments in the use of M-estimation include the weighting of the individual terms to reduce the influence of high leverage points (see, for example, Krasker and Welsch (1982), Huber (1983)). The stack loss data set was first analysed in Daniel and Wood (1971) (Chapter 5). A recent book which treats the statistical properties of the l_1 estimator is Bloomfield and Steiger (1983).

The summary of iterative algorithms for M-estimation follows Huber (1977). For properties of the M-estimator see Clark (1981) and (1985). For the continuation algorithm see Clark (1981) and Clark and Osborne (1985). The algorithm which attacks the problem of finding a feasible ν directly and its application to resolve degeneracy are given in Clark (1981) and Clark and Osborne (1985).

CHAPTER 6

Some applications to non-convex problems

6.1 INTRODUCTION

In this chapter finite descent algorithms are derived for two problems which have their origins in data analysis, but which differ from those considered previously in being non-convex, in the sense that either the set of feasible points or the objective function is non-convex.

The total approximation problem

Let $Z: R^m \rightarrow R^n$ have rank $m \leqslant n$, and $\|\cdot\|$ be a norm defined on the space of $n \times m$ matrices. Then the total approximation problem is

$$\min_{E \in \mathbb{E}} \|E\| : \mathbb{E} = \{E; \text{rank}\,(Z+E) < m\} \tag{1.1}$$

A form of this problem ocurs in the regression situation considered in the previous chapter when there are errors in recording the matrix of independent variables M in addition to the errors in the dependent variable $\mathbf{f}$. In this case the quantities involved are

$$E = [E_1 \mid \mathbf{r}], \qquad Z = [M \mid \mathbf{f}],$$

and the rank condition is imposed by means of the equation

$$[M + E_1]\mathbf{x} - \mathbf{f} - \mathbf{r} = 0, \tag{1.2}$$

which provides the appropriate extension of (3.1.1), so that the allowable $E \in \mathbb{E}$ are restricted to the open subset of $\mathbb{E}$ satisfying

$$(Z+E)\mathbf{v} = 0, \qquad |v_m| = |v_{p+1}| > 0.$$

For this reason there may exist no bounded solution $\mathbf{x}$ to the errors in the variables problem (1.2) although the corresponding minimization problem (1.1) is well defined. Also, it is frequently the case that certain of the columns of Z are given without error. If Z is partitioned so that

$$Z = [Z_1 \mid Z_2],\, Z_1 : R^{m_1} \rightarrow R^n,\, Z_2 : R^{m_2} \rightarrow R^n,\, m_1 + m_2 = m,$$

where Z_1 are the known columns, then the estimation problem takes the form

$$\min_{E \in \mathbb{E}} \|E\| : \mathbb{E} = \{E;\ \text{rank}\,([Z_1 \mid Z_2 + E]) < m\} \tag{1.3}$$

To see that $\mathbb{E}$ is not convex, consider $Z = I_2$. Then

$$\text{rank}\,(I_2 - \mathbf{e}_1\mathbf{e}_1^{\mathrm{T}}) = \text{rank}\,(I_2 - \mathbf{e}_2\mathbf{e}_2^{\mathrm{T}}) = 1$$

but

$$\text{rank}\,(I_2 - \theta\mathbf{e}_1\mathbf{e}_1^{\mathrm{T}} - (1-\theta)\mathbf{e}_2\mathbf{e}_2^{\mathrm{T}}) = 2, \qquad 0 < \theta < 1.$$

It is shown here that the solution of (1.3) can be reduced to the problem of solving a constrained vector minimization problem in R^n provided $\|\cdot\|$ has the property of separability and that in the particular case of the separable norm

$$\|E\| = \sum_{i=1}^{n} \sum_{j=1}^{m_2} |E_{ij}| \tag{1.4}$$

a finite descent algorithm is available.

Computing centres in cluster analysis

Let $\mathbf{a}_i \in R^p$, $i = 1, 2, \ldots, n$, be data vectors summarizing information concerning certain characteristics of n objects. The problem is to find $k \leqslant n$ and index sets $\pi_1, \pi_2, \ldots, \pi_k$ pointing to subgroups $C_1, C_2, \ldots, C_k$ of the objects such that every vector is in some subgroup and the elements of each subgroup have similar characteristics but differ significantly from the elements of the other groups. The problem of finding
the centre of C_i is that of finding

$$\mathbf{z}_i = \arg\min \left(\sum_{j \in \pi_i} d(\mathbf{z}, \mathbf{a}_j) \right) \tag{1.5}$$

where $d(\cdot, \cdot)$ is an appropriate measure of similarity. One approach to cluster analysis makes use of this calculation and seeks k small, and subgroups $C_1, C_2, \ldots, C_k$ such that

$$\min_{C_1, C_2, \ldots, C_k} \sum_{i=1}^{k} \sum_{j \in \pi_i} d(\mathbf{z}_i, \mathbf{a}_j) \tag{1.6}$$

is small. In this sense $C_1, C_2, \ldots, C_k$ explain the variability in the sample $\{\mathbf{a}_i, i = 1, 2, \ldots, n\}$.

Clearly, efficient calculation of the $\mathbf{z}_i$ is essential if the calculation (1.6) is to be feasible for data sets of any size. If $d(\cdot, \cdot)$ is the Euclidean norm then the centre corresponds to the centroid of the subgroup. If $d(\cdot, \cdot)$ is the l_1 norm then the problem becomes

$$\min_{\mathbf{z}} \sum_{j \in \pi_i} \sum_{k=1}^{p} |z_k - (\mathbf{a}_j)_k| \tag{1.7}$$

so that

$$(\mathbf{z}_i)_k = \arg\min \left(\sum_{j \in \pi_i} |t - (\mathbf{a}_j)_k| \right) \tag{1.8}$$

and the solution is obtained by computing medians. This implies that each component of $\mathbf{z}_i$ is equal to a corresponding component of one of the data vectors $\mathbf{a}_j$, $j \in \pi_i$, a fact that is useful subsequently. An important measure of similarity for *nonnegative* vectors is provided by the Jaccard metric. In this case

$$d(\mathbf{z}, \mathbf{a}) = 1 - \frac{\sum_{i=1}^{p} \min(z_i, a_i)}{\sum_{i=1}^{p} \max(z_i, a_i)} = \frac{2\|\mathbf{z}-\mathbf{a}\|_1}{\|\mathbf{z}\|_1 + \|\mathbf{a}\|_1 + \|\mathbf{z}-\mathbf{a}\|_1} \tag{1.9}$$

where the norm is again the l_1 norm. In the scalar case

$$d(z, a) = \begin{cases} 1 - \dfrac{z}{a}, & 0 \leqslant z < a \\[2ex] 1 - \dfrac{a}{z}, & z \leqslant a \end{cases}$$

so that $d(\mathbf{z}, \mathbf{a})$ is not convex. However, $d(\mathbf{z}, \mathbf{a})$ does satisfy the triangle inequality. For example, in the scalar case, if $0 < z < y < a$ then

$$\begin{aligned} d(z, y) + d(y, a) &= 1 - \frac{z}{y} + 1 - \frac{y}{a} \\ &= 1 - \frac{z}{a} + \frac{z}{a} - \frac{z}{y} + \frac{y}{y} - \frac{y}{a} \\ &= 1 - \frac{z}{a} + (y - z)\left(\frac{1}{y} - \frac{1}{a}\right) \\ &> 1 - \frac{z}{a} = d(z, a). \end{aligned}$$

Thus $d(\mathbf{z}, \mathbf{a})$ satisfies the axioms for a metric on the set of nonnegative vectors. For this example it is also possible to give a finite algorithm.

Non-convex problems differ from those considered already in two important ways:

(i) It is no longer true that a local minimum is a global minimum. In general, multiple local minima occur in non-convex problems.

(ii) Under rather minimal regularity assumptions it is true that multiplier relations or first-order necessary conditions hold at local minima. However, the correspondence between local minima and stationary points (points which satisfy the first order necessary conditions) is no longer one-to-one. This corresponds to the existence of stationary points which are neither maxima or minima. Thus further investigation is necessary to classify a stationary point in the non-convex case. Second derivative information can be used for this purpose if it is available.

The two examples considered in this section permit the above points to be illustrated. The approach used is to adapt results from previous chapters where possible, and this proves to be reasonably straightforward. Such new theory as is needed is developed in the next section, and is largely based on a form of the implicit function theorem which does not require differentiability. Finiteness follows because it is possible to characterize the candidates for stationary points in a manner similar to extreme points in linear programming, and then to use the polyhedral nature of the objective functions to show that there are at most a finite number of these. Descent algorithms are then used to generate a new candidate point with decreased function value if the current extreme point is not a stationary point. Thus termination and finiteness are immediate consequences.

6.2 CONDITIONS FOR A STATIONARY POINT

The problems considered here are put together from convex functions and inherit certain smoothness properties as a consequence. However, the convexity of the problem is all too easily lost. For example, the set of points such that $g(\mathbf{x})=0$ is in general not convex if g is convex. One exception is the case g affine considered in Lemma 1.5.4, and to generalize this result to develop first order necessary conditions a suitable tool is an implicit function theorem.

Theorem 2.1 *Let*

$$F\begin{bmatrix}\mathbf{x}\\ \mathbf{y}\end{bmatrix}=\begin{bmatrix}\mathbf{f}(\mathbf{x},\mathbf{y})\\ \mathbf{y}\end{bmatrix}:(A\subseteq R^n)\times(B\subseteq R^m)\rightarrow(C\subseteq R^n)\times(B\subseteq R^m)$$

be a homeomorphism which maps neighbourhoods of the origin into neighbourhoods of the origin (so that $\mathbf{f}(0,0)=0$). Then there exist open neighbourhoods $A_0\subseteq A$, $B_0\subseteq B$ such that for all $\mathbf{y}\in B_0$ the equation

$$\mathbf{f}(\mathbf{x},\mathbf{y})=0 \tag{2.1}$$

has a continuous solution $\mathbf{x}=\boldsymbol{\psi}(\mathbf{y})$.

Proof Because F is a homeomorphism there is a continuous inverse mapping

$$G\begin{bmatrix}\mathbf{x}\\ \mathbf{y}\end{bmatrix}=\begin{bmatrix}\mathbf{g}_1(\mathbf{x},\mathbf{y})\\ \mathbf{g}_2(\mathbf{x},\mathbf{y})\end{bmatrix}.$$

Thus for all $\mathbf{y}$ with $\|\mathbf{y}\|$ small enough,

$$\begin{bmatrix} 0 \\ \mathbf{y} \end{bmatrix} = F \circ G \begin{bmatrix} 0 \\ \mathbf{y} \end{bmatrix} = \begin{bmatrix} \mathbf{f}(\mathbf{g}_1(0, \mathbf{y}), \mathbf{g}_2(0, \mathbf{y})) \\ \mathbf{g}_2(0, \mathbf{y}) \end{bmatrix}$$

It follows that $\mathbf{g}_2(0, \mathbf{y}) = \mathbf{y}$, so that

$$\mathbf{f}(\mathbf{g}_1(0, \mathbf{y}), \mathbf{y}) = 0$$

showing that

$$\mathbf{x} = \boldsymbol{\psi}(\mathbf{y}) = \mathbf{g}_1(0, \mathbf{y}) \tag{2.2}$$

is the desired solution.

Remark 2.1 To apply this result it is necessary to verify that F is a homeomorphism with the required properties. Let $\mathbf{w} = \begin{bmatrix} \mathbf{x} \\ \mathbf{y} \end{bmatrix}$. Then an important sufficient condition is that there exists a neighbourhood N of the origin in $R^n \times R^m$ and constants $K_1 \geq K_2 > 0$ depending on the choice of norm such that, for all $\mathbf{w}_1, \mathbf{w}_2 \in N$,

$$K_1 \|\mathbf{w}_1 - \mathbf{w}_2\| \geq \|F\mathbf{w}_1 - F\mathbf{w}_2\| \geq K_2 \|\mathbf{w}_1 - \mathbf{w}_2\| \tag{2.3}$$

It is an immediate consequence of (2.3) that (2.2) satisfies a Lipschitz condition.

Lemma 2.1 *Let* $\mathbf{f}(\mathbf{w})$ *be continuously differentiable in a neighbourhood* N *of* $\mathbf{w} = \mathbf{w}_0$, *and let*

$$\text{rank}\,(\nabla \mathbf{f}(\mathbf{w}_0)) = n. \tag{2.4}$$

Then there exists a neighbourhood $N_1 \subseteq N$ *containing* $\mathbf{w}_0$ *such that* (2.3) *is satisfied.*

Proof By assumption there exists an ordering of the components of $\mathbf{w} = \begin{bmatrix} \mathbf{x} \\ \mathbf{y} \end{bmatrix}$ such that $\nabla_x \mathbf{f}(\mathbf{w}_0)$ has full rank. This implies that

$$\nabla F(\mathbf{w}_0) = \begin{bmatrix} \nabla_x \mathbf{f} & \nabla_y \mathbf{f} \\ 0 & I_m \end{bmatrix} \tag{2.5}$$

has full rank, and hence that

$$\min_{\|\mathbf{t}\|=1} \|\nabla F(\mathbf{w}_0)\mathbf{t}\| \geq \delta > 0 \tag{2.6}$$

The result now follows from this as

$$\|F\mathbf{w}_1 - F\mathbf{w}_2\| = \|\nabla F(\mathbf{w}_0)(\mathbf{w}_1 - \mathbf{w}_2) + (\nabla F(\mathbf{w}_2) - \nabla F(\mathbf{w}_0))(\mathbf{w}_1 - \mathbf{w}_2) + o(\|\mathbf{w}_1 - \mathbf{w}_2\|)\|$$

provided N_1 is chosen such that $\|\nabla F(\mathbf{w}_2) - \nabla F(\mathbf{w}_0)\|$ is sufficiently small in a consistent matrix norm.

An immediate consequence is the usual form of the implicit function theorem (it is only necessary to apply a translation that takes $\mathbf{w}_0 \to 0$). However, here it is required that the differentiability condition be relaxed and be replaced by the convexity of some or all of the components of $\mathbf{f}(\mathbf{w})$. We treat the convex case, and it is assumed that $\mathbf{w}_0 \in \operatorname{int} \bigcap_{i=1}^{n} \operatorname{dom}(f_i)$.

Definition 2.1 The *generalized Jacobian* at $\mathbf{w}_0$ is the set

$$\mathscr{J}(\mathbf{f}, \mathbf{w}_0) = \{J; \boldsymbol{\rho}_i(J) \in \partial f_i(\mathbf{w}_0), \qquad i = 1, 2, \ldots, n\}. \tag{2.7}$$

The rank of $\mathscr{J}$ is n if the rank of each $J \in \mathscr{J}$ is n. If $\mathbf{f}$ is differentiable then $\mathscr{J}(\mathbf{f}, \mathbf{w}_0) = \nabla \mathbf{f}(\mathbf{w}_0)$.

Lemma 2.2 *Let*

(i) $\mathbf{w}_0 \in \operatorname{int} \bigcap_{i=1}^{n} \operatorname{dom}(f_i) = N$, *and*

(ii) $\operatorname{rank}(\mathscr{J}(\mathbf{f}, \mathbf{w}_0)) = n$.

Then there exists a neighbourhood $N_1 \subseteq N$ *of* $\mathbf{w}_0$ *such that* (2.3) *is satisfied.*

Proof Let $\mathbf{w}_1, \mathbf{w}_2 \in N$. Then the subgradient inequality gives

$$\mathbf{f}(\mathbf{w}_1) - \mathbf{f}(\mathbf{w}_2) \geqslant J_2(\mathbf{w}_1 - \mathbf{w}_2), \ \forall\, J_2 \in \mathscr{J}(\mathbf{f}, \mathbf{w}_2),$$

and

$$\mathbf{f}(\mathbf{w}_2) - \mathbf{f}(\mathbf{w}_1) \geqslant J_1(\mathbf{w}_2 - \mathbf{w}_1), \ \forall\, J_1 \in \mathscr{J}(\mathbf{f}, \mathbf{w}_1). \tag{2.8}$$

Now (specializing the sequence if necessary) select J_2 as a limit of the sequences J_1 as $\mathbf{w}_1 \to \mathbf{w}_2$. Provided $\|\mathbf{w}_2 - \mathbf{w}_0\|$ is small enough, then $\operatorname{rank}(J_2) = n$, as the contrary implies that there is a sequence $\{\mathbf{w}_i\} \to \mathbf{w}_0$ such that $\operatorname{rank}(\mathscr{J}(\mathbf{f}, \mathbf{w}_i)) < n$, contradicting the assumption on $\mathscr{J}(\mathbf{f}, \mathbf{w}_0)$. The desired result can now be deduced by noting that if

$$a \geqslant b, \qquad -a \geqslant -b + \varepsilon \tag{2.9}$$

then necessarily $\varepsilon \leqslant 0$ and the only possibilities are displayed in Table 2.1.

Table 2.1 Possible inequalities from (2.9).

	$b > 0$	$b < 0$
$a > 0$	$\lvert a\rvert \geqslant \lvert b\rvert$	$\lvert a\rvert \leqslant \lvert\varepsilon\rvert, \lvert b\rvert \leqslant \lvert\varepsilon\rvert$
$a < 0$	—	$\lvert a\rvert \geqslant \lvert b\rvert - \lvert\varepsilon\rvert$

Identifying the inequalities (2.8) componentwise with (2.9) and choosing the norm to be the maximum norm gives

$$\|F\mathbf{w}_1 - F\mathbf{w}_2\|_\infty \geq \left\| \begin{bmatrix} J_2(\mathbf{w}_1 - \mathbf{w}_2) \\ \mathbf{y}_1 - \mathbf{y}_2 \end{bmatrix} \right\|_\infty - \left\| \begin{bmatrix} (J_1 - J_2)(\mathbf{w}_1 - \mathbf{w}_2) \\ 0 \end{bmatrix} \right\|_\infty$$

where the partitioning of $\mathbf{w}$ into $\mathbf{x}$ and $\mathbf{y}$ is treated as in Lemma 2.1. This gives the right-hand inequality in (2.3) for $\mathbf{w}_1, \mathbf{w}_2$ in some small enough neighbourhood N_1 of $\mathbf{w}_0$ as $\|J_1 - J_2\| = o(1)$ by construction. The left-hand inequality follows from convexity (for example, (4.4) of Chapter 1).

To show that these results can be used to derive multiplier conditions, consider the problem

$$\min_{\mathbf{z} \in Z} h(\mathbf{z}) : Z = \{\mathbf{z}; \mathbf{g}(\mathbf{z}) = 0\} \tag{2.10}$$

where h, g_i, $i = 1, 2, \ldots, s$ are convex functions on R^p and $s < p$. Let $\mathbf{z}^*$ be a local minimum so that $h(\mathbf{z}^*) \leq h(\mathbf{z})$ for all $\mathbf{z}$ sufficiently close to $\mathbf{z}^*$,

$$\mathbf{f}(\mathbf{w}) = \begin{bmatrix} h(\mathbf{z}) - h(\mathbf{z}^*) + t \\ \mathbf{g}(\mathbf{z}) \end{bmatrix}, \qquad \mathbf{w} = \begin{bmatrix} \mathbf{z} \\ t \end{bmatrix},$$

and assume that there is no relation of linear dependence between the elements of $\partial h(\mathbf{z}^*)$ and $\partial g_i(\mathbf{z}^*)$, $i = 1, 2, \ldots, s$. Then $\mathcal{J}\left(\mathbf{f}, \begin{bmatrix} \mathbf{z}^* \\ 0 \end{bmatrix}\right)$ has full rank $s+1$ so that the implicit function theorem can be used to solve for $\mathbf{x} = \boldsymbol{\psi}(\mathbf{y})$ where t can be chosen among the variables in $\mathbf{y}$ corresponding to the partitioning $x_i = z_i$, $i = 1, 2, \ldots, s+1$, $y_i = z_{s+i+1}$, $i = 1, 2, \ldots, p-s-1 = q$, $y_{p+1} = t$,

$$\left[\begin{array}{c|c} \multicolumn{2}{c}{J_w} \\ \hline 0 & I_{q+1} \end{array}\right] = \left[\begin{array}{c|c|c} \multicolumn{2}{c|}{J_z} & \mathbf{e}_1 \\ \hline 0 & I_q & 0 \\ \hline \multicolumn{2}{c|}{0} & 1 \end{array}\right], \qquad J_w \in \mathcal{J}\left(\mathbf{f}, \begin{bmatrix} \mathbf{z}^* \\ 0 \end{bmatrix}\right), \qquad J_z \in \mathcal{J}\left(\begin{bmatrix} h \\ \mathbf{g} \end{bmatrix}, \mathbf{z}^*\right).$$

But then for $t > 0$,

$$h(\mathbf{z}) = h(\mathbf{z}^*) - t < h(\mathbf{z}^*)$$

which is a contradiction. Thus there exists a relation of the form

$$\lambda_0 \mathbf{v} = \sum_{i=1}^{s} \lambda_i \mathbf{v}_i \tag{2.11}$$

where $\mathbf{v}^T \in \partial h(\mathbf{z}^*)$, $\mathbf{v}_i^T \in \partial g_i(\mathbf{z}^*)$, $i = 1, 2, \ldots, s$. Also, $\lambda_0 = 0$ is possible only if rank $(\mathcal{J}(\mathbf{g}, \mathbf{z}^*)) < s$. This proves the result.

Theorem 2.2 *Let* $h, g_i, i = 1, 2, \ldots, s$ *be convex with* $\operatorname{dom} h \cap \left\{\bigcap_{i=1}^{s} \operatorname{dom} g_i\right\} \neq \emptyset$, *and* $\mathbf{z}^*$ *a local minimum of* (2.10). *Then first-order neces-*

sary conditions are given by (2.11). *As* rank $(\mathcal{J}(\mathbf{g}, \mathbf{z}^*)) = s$ *then* λ_0 *can be chosen to be* 1.

Example 2.1 (i) Let $Z: R^m \to R^n$, $n \geq m$, have rank m and consider the problem

$$\min_{\mathbf{v} \in V} \|Z\mathbf{v}\|_R : V = \{\mathbf{v}; \|\mathbf{v}\|_D = 1\} \tag{2.12}$$

This problem involves minimizing a continuous function on a compact set, so the minimum exists. However, the problem is not convex because the minimization problem is equivalent to one subject to the inequality constraint $\|\mathbf{v}\|_D \geq 1$ with the feasible region given by the outside of the unit ball. Let $\mathbf{v}^*$ be a local minimum of (2.12). The problem functions are convex and $0 \notin \partial \|\mathbf{v}^*\|_D$ so that Theorem 2.2 applies. The multiplier condition (2.11) gives

$$\mathbf{y}^T Z = \lambda \mathbf{w}^T \tag{2.13}$$

where $\mathbf{y}^T \in \partial \|Z\mathbf{v}^*\|_R$ and $\mathbf{w}^T \in \partial \|\mathbf{v}^*\|_D$. To calculate λ multiply by $\mathbf{v}^*$ to obtain

$$\|Z\mathbf{v}^*\|_R = \lambda \|\mathbf{v}^*\|_D. \tag{2.14}$$

For this example the multiplier result can be refined to show that for *every* $\mathbf{w}^T \in \partial \|\mathbf{v}^*\|_D$ there exists $\mathbf{y}^T \in \partial \|Z\mathbf{v}^*\|_R$ such that (2.13) holds. For the alternative is that there exists $\hat{\mathbf{w}}$ such that $\lambda \hat{\mathbf{w}}^T \notin \partial \|Z\mathbf{v}^*\|_R Z$. As $\lambda > 0$ the standard separation argument ensures the existence of $\mathbf{t}$ such that

$$\mathbf{y}^T Z\mathbf{t} < 0, \forall\, \mathbf{y}^T \in \partial \|Z\mathbf{v}^*\|_R$$

and

$$\hat{\mathbf{w}}^T \mathbf{t} \geq 0.$$

This $\mathbf{t}$ can be used to obtain a contradiction. Choose $\mathbf{y}_\gamma^T \in \partial \|Z(\mathbf{v}^* + \gamma \mathbf{t})\|_R$ such that $\mathbf{y}_\gamma^T \to \mathbf{y}^T \in \partial \|Z\mathbf{v}^*\|_R$ as $\gamma \to 0$. Then for $\gamma > 0$ small enough to ensure $\mathbf{y}_\gamma^T Z\mathbf{t} < 0$, the subgradient inequality gives

$$\begin{aligned}\|Z\mathbf{v}^*\|_R &\geq \|Z(\mathbf{v}^* + \gamma \mathbf{t})\|_R - \gamma \mathbf{y}_\gamma^T Z\mathbf{t} \\ &> \|Z(\mathbf{v}^* + \gamma \mathbf{t})\|_R\end{aligned}$$

while

$$\|\mathbf{v}^* + \gamma \mathbf{t}\|_D \geq \|\mathbf{v}^*\|_D + \gamma \hat{\mathbf{w}}^T \mathbf{t} \geq \|\mathbf{v}^*\|_D.$$

Thus

$$\|Z\mathbf{v}^*\|_R > \frac{\|Z(\mathbf{v}^* + \gamma \mathbf{t})\|_R}{\|\mathbf{v}^* + \gamma \mathbf{t}\|_D}$$

showing that the objective function is reduced at the neighbouring feasible point $\dfrac{\mathbf{v}^* + \gamma \mathbf{t}}{\|\mathbf{v}^* + \gamma \mathbf{t}\|_D}$, and this is a contradiction.

(ii) Consider the problem

$$\max_{\mathbf{v} \in V} \|Z\mathbf{v}\|_R : V = \{\mathbf{v}; \|\mathbf{v}\|_D = 1\} \tag{2.15}$$

By the argument used above, the maximum is attained. However, maximizing a convex function does not fit the theoretical framework (nor does minimizing the concave function $-\|Z\mathbf{v}\|_R$), although the modifications required are straightforward. The difficulty can be avoided by using the linear homogeneity of the norm function to write the problem in the alternate form

$$\min_{\mathbf{v}\in V} \|\mathbf{v}\|_D : V = \{\mathbf{v} : \|Z\mathbf{v}\|_R = 1\} \tag{2.16}$$

and V is compact as Z has full rank. If $\mathbf{v}^*$ is a local minimum then the multiplier condition becomes

$$\mathbf{w}^T = \lambda \mathbf{y}^T Z \tag{2.17}$$

where $\mathbf{w}^T \in \partial \|\mathbf{v}^*\|_D$, $\mathbf{y}^T \in \partial \|Z\mathbf{v}^*\|_R$, and $\lambda = \|\mathbf{v}^*\|_D / \|Z\mathbf{v}^*\|_R$. This time the argument used in the first part of this example shows that for every $\mathbf{y}^T \in \partial \|Z\mathbf{v}^*\|_R$ there exists $\mathbf{w}^T \in \partial \|\mathbf{v}^*\|_D$ such that (2.17) holds.

Remark 2.2 What is being computed in Example 2.1(ii) is a norm for the rectangular matrix Z. Such a norm is called an operator norm or subordinate norm. If $\|\cdot\|_R$ and $\|\cdot\|_D$ are Euclidean vector norms, then the solutions to Example 2.1(i) and (ii) are the smallest and largest singular values respectively.

Here *stationary points* are taken to be points at which (2.11) holds. A stationary point is a *local minimum* if $h(\mathbf{z}) \geq h(\mathbf{z}^*)$ for all feasible $\mathbf{z}$ close enough to $\mathbf{z}^*$. It is an *isolated local minimum* if $h(\mathbf{z}) > h(\mathbf{z}^*)$. This classification makes it necessary to consider function values at feasible points in the immediate vicinity of $\mathbf{z}^*$. To do this it is convenient to consider sequences of points converging directionally to $\mathbf{z}^*$.

Definition 2.2 The sequence $\{\mathbf{t}_i\} \to 0$ has the *directional limit* $\mathbf{t}$ if $\{\mathbf{t}_i/\|\mathbf{t}_i\|_2\} \to \mathbf{t}$. In this case $\{\mathbf{t}_i\}$ is a *directional sequence.*
The next result characterizes the directional limits of sequences $\{\mathbf{z}_i\} \to \mathbf{z}^*$ ($\{\mathbf{z}_i - \mathbf{z}^*\} \to 0$). As usual, $\mathbf{z}_i$ is feasible if $\mathbf{g}(\mathbf{z}_i) = 0$.

Lemma 2.3 *Let* rank $(\mathcal{J}(\mathbf{g}, \mathbf{z}^*)) = s$. *Then* $\mathbf{t}$ *is the directional limit of the sequence of feasible points* $\{\mathbf{z}^* + \mathbf{t}_i\}$ *where* $\{\mathbf{t}_i\} \to 0$ *if and only if*

$$g_i'(\mathbf{z}^* : \mathbf{t}) = 0, \qquad i = 1, 2, \ldots, s. \tag{2.18}$$

Proof (i) Let $\mathbf{t}$ be the directional limit of the sequence $\{\mathbf{t}_i\}$. The subgradient inequality applied at $\mathbf{z}_i = \mathbf{z}^* + \|\mathbf{t}_i\|_2 \mathbf{t}$ with $J_i \in \mathcal{J}(\mathbf{g}, \mathbf{z}_i)$ gives (as $\mathbf{g}(\mathbf{z}^* + \mathbf{t}_i) = 0$ by assumption):

$$\frac{\mathbf{g}(\mathbf{z}_i) - \mathbf{g}(\mathbf{z}^* + \mathbf{t}_i)}{\|\mathbf{t}_i\|_2} \leq J_i\left(\mathbf{t} - \frac{\mathbf{t}_i}{\|\mathbf{t}_i\|_2}\right) \tag{2.19}$$

so that $\mathbf{g}'(\mathbf{z}^* : \mathbf{t}) \leq 0$. However, if $\hat{J}_i \in \mathcal{J}(\mathbf{g}, \mathbf{z}^* + \mathbf{t}_i)$ is matched to J_i under a limiting operation which takes $\mathbf{z}_i$ to $\mathbf{z}^*+\mathbf{t}_i$ for each i, then

$$\frac{\mathbf{g}(\mathbf{z}_i)-\mathbf{g}(\mathbf{z}^*+\mathbf{t}_i)}{\|\mathbf{t}_i\|_2} \geqslant \hat{J}_i\left(\mathbf{t}-\frac{\mathbf{t}_i}{\|\mathbf{t}_i\|_2}\right) \tag{2.20}$$

giving $\mathbf{g}'(\mathbf{z}^*:\mathbf{t}) \geqslant 0$ as the right-hand sides of (2.19) and (2.20) tend to zero.

(ii) Now let $\mathbf{t}$ be such that $\mathbf{g}'(\mathbf{z}^*:\mathbf{t})=0$. To construct a directional sequence prescribe $\{\lambda_i>0\} \to 0$ and set

$$\mathbf{z}=\begin{bmatrix}\mathbf{x}\\ \mathbf{y}\end{bmatrix}, \qquad \mathbf{t}=\begin{bmatrix}\mathbf{t}_1\\ \mathbf{t}_2\end{bmatrix}, \qquad J=[J_1 \mid J_2]$$

where the partitioning of $J \in \mathcal{J}(\mathbf{g}, \mathbf{z}^*)$ is made so that

$$\operatorname{rank}(J_1)=s \quad \text{and} \quad J=\lim_{i\to\infty} J_i=\lim_{i\to\infty} \hat{J}_i.$$

Now apply the implicit function theorem to obtain

$$\mathbf{x}_i=\mathbf{x}^*+\boldsymbol{\psi}(\lambda_i\mathbf{t}_2),$$
$$\mathbf{y}_i=\mathbf{y}^*+\lambda_i\mathbf{t}_2.$$

It follows from (2.3) that

$$\left\|\begin{bmatrix}0\\ \lambda_i\mathbf{t}_2\end{bmatrix}\right\| \geqslant K_2\left\|\begin{bmatrix}\boldsymbol{\psi}(\lambda_i\mathbf{t}_2)\\ \lambda_i\mathbf{t}_2\end{bmatrix}\right\|$$

so that $\frac{1}{\lambda_i}\boldsymbol{\psi}(\lambda_i\mathbf{t}_2)$ is bounded as $i\to\infty$; so there is no restriction in assuming that a convergent subsequence is chosen and i specialized accordingly. By (2.19), (2.20) it follows that

$$0=J\begin{bmatrix}\mathbf{t}_1-\lim_{i\to\infty}\dfrac{\boldsymbol{\psi}(\lambda_i\mathbf{t}_2)}{\lambda_i}\\ 0\end{bmatrix}=J_1\left(\mathbf{t}_1-\lim_{i\to\infty}\frac{\boldsymbol{\psi}(\lambda_i\mathbf{t}_2)}{\lambda_i}\right)$$

But J_1 has full rank, so that

$$\{\mathbf{t}_i\}=\left\{\begin{bmatrix}\boldsymbol{\psi}(\lambda_i\mathbf{t}_2)\\ \lambda_i\mathbf{t}_2\end{bmatrix}\right\}$$

is an appropriate directional sequence.

To specify the second-order conditions, assume that h, g_i, $i=1,2,\ldots,s$, are at least twice continuously differentiable (although this can be weakened as consideration is restricted to directional sequences), and that $\nabla\mathbf{g}(\mathbf{z}^*)$ has full rank. This permits choosing $\lambda_0=1$ in (2.11). If the other multipliers are λ_i, $i=1,2,\ldots,s$, then the Lagrangian

$$\mathcal{L}(\mathbf{z})=h(\mathbf{z})-\sum_{i=1}^{s}\lambda_i g_i(\mathbf{z}) \tag{2.21}$$

satisfies

$$\nabla \mathscr{L}(\mathbf{z}^*) = 0. \tag{2.22}$$

Theorem 2.3 (*Second-order necessary conditions*) *If* $\mathbf{z}^*$ *is a local minimum then*

$$\mathbf{t}^T \nabla^2 \mathscr{L}(\mathbf{z}^*)\mathbf{t} \geq 0 \tag{2.23}$$

for every $\mathbf{t}$ *satisfying* $\mathbf{g}'(\mathbf{z}^* : \mathbf{t}) = 0$.

Proof Let $\{\mathbf{t}_i\}$ be a feasible directional sequence with limit $\mathbf{t}$. Then

$$h(\mathbf{z}^* + \mathbf{t}_i) = \mathscr{L}(\mathbf{z}^* + \mathbf{t}_i) = h(\mathbf{z}^*) + \tfrac{1}{2}\mathbf{t}_i^T \nabla^2 \mathscr{L}(\mathbf{z}^*)\mathbf{t}_i + o(\|\mathbf{t}_i\|_2^2) \geq h(\mathbf{z}^*)$$

provided $\|\mathbf{t}_i\|_2$ is small enough, and (2.23) is an immediate consequence.

Theorem 2.4 (*Second-order sufficiency conditions*) *If*

$$\mathbf{t}^T \nabla^2 \mathscr{L}(\mathbf{z}^*)\mathbf{t} > 0 \tag{2.24}$$

for every $\mathbf{t}$ *such that* $\mathbf{g}'(\mathbf{z}^* : \mathbf{t}) = 0$, *then there exists* $\gamma > 0$ *and an open neighbourhood* N *of* $\mathbf{z}^*$ *such that if* $\mathbf{z}^* + \mathbf{t}_i \in N$ *then*

$$h(\mathbf{z}^* + \mathbf{t}_i) \geq h(\mathbf{z}^*) + \gamma \|\mathbf{t}_i\|_2^2 \tag{2.25}$$

holds for every feasible directional sequence $\{\mathbf{t}_i\}$.

Proof Assume the contrary. Then there exists a directional sequence and corresponding $\{\gamma_i > 0\} \to 0$ such that

$$\mathscr{L}(\mathbf{z}^* + \mathbf{t}_i) \leq h(\mathbf{z}^*) + \gamma_i \|\mathbf{t}_i\|_2^2.$$

But this implies

$$\mathbf{t}^T \nabla^2(\mathbf{z}^*)\mathbf{t} = \lim_{i \to \infty} \frac{\mathscr{L}(\mathbf{z}^* + \mathbf{t}_i) - h(\mathbf{z}^*)}{\|\mathbf{t}_i\|_2^2} \leq 0,$$

and this gives a contradiction.

Example 2.2 As in Example 2.1(i), let $h = \|Z\mathbf{v}\|_R$, and $g = \|\mathbf{v}\|_D - 1$. Then

$$\mathscr{L} = \|Z\mathbf{v}\|_R - \|Z\mathbf{v}^*\|_R(\|\mathbf{v}\|_D - 1).$$

If the norms are Euclidean then

$$\nabla^2 \mathscr{L} = \frac{Z^T Z}{\{\mathbf{v}^T Z^T Z \mathbf{v}\}^{1/2}} - \frac{Z^T Z \mathbf{v}\mathbf{v}^T Z^T Z}{\{\mathbf{v}^T Z^T Z \mathbf{v}\}^{3/2}} - \frac{\|Z\mathbf{v}^*\|_2}{\{\mathbf{v}^T \mathbf{v}\}^{1/2}} \left\{ I - \frac{\mathbf{v}\mathbf{v}^T}{\mathbf{v}^T \mathbf{v}} \right\}$$

Here $\mathbf{v}^*$ is the singular vector associated with the smallest singular value of Z (the singular values are assumed to be distinct). Then the condition for $\mathbf{t}$ to be the limit of a directional sequence is

$$\mathbf{t}^T \mathbf{v}^* = 0 = \mathbf{t}^T Z^T Z \mathbf{v}^*$$

so that

$$\mathbf{t}^{\mathrm{T}}\nabla^2\mathcal{L}(\mathbf{v}^*)\mathbf{t}=\frac{1}{\|Z\mathbf{v}^*\|_2}\{\mathbf{t}^{\mathrm{T}}Z^{\mathrm{T}}Z\mathbf{t}-\|Z\mathbf{v}^*\|_2^2\}$$

$$\geq\frac{\sigma_2^2-\sigma_1^2}{\sigma_1}>0$$

where $0<\sigma_1<\sigma_2<\ldots$ are the singular values of Z in increasing order.

Exercise 2.1 (i) Let $\mathbf{x}^*$ minimize $f(\mathbf{g}(\mathbf{x}))$ where f is convex and the components of $\mathbf{g}$ are either differentiable or convex or concave. Use Theorem 2.2 to derive necessary conditions characterizing $\mathbf{x}^*$.

(ii) Apply this result to the problem of minimizing $f=\sum_{i=1}^{m}|f_i-\max(z_i,\mathbf{m}_i^{\mathrm{T}}\mathbf{x})|$. For $f=|1-\max(0,x)|+|-1-\max(0,x)|$, interpret the cases $x=0$ and $x=1$. Also show that if $f_i>z_i$, $i=1,2,\ldots,n$, then only minor changes need be made to an l_1 algorithm to minimize f. This problem corresponds to l_1 estimation of a censored tobit model.

Exercise 2.2 Let $\|\mathbf{v}\|_\beta=\left\{\sum_{i=1}^{m}|v_i|^\beta\right\}^{1/\beta}$, and $\beta>1$. For the constraint $\|\mathbf{v}\|_\beta=1$ show that $\mathbf{t}$ is the limit of a directional sequence at $\mathbf{v}$ if and only if

$$\mathbf{t}^{\mathrm{T}}D_v\mathbf{t}=0$$

where

$$D_v=\operatorname{diag}\{|v_i|^{\beta-2}\text{ if }v_i\neq0,\text{ else }0,\ i=1,2,\ldots,m\}.$$

6.3 THE TOTAL APPROXIMATION PROBLEM IN SEPARABLE NORMS

If a matrix mapping $R^m\to R^n$ is considered as a particular organization of a vector in $T^{n\times m}$ then a matrix norm can be defined by adapting the corresponding vector norm. For example, the matrix norm corresponding to the Euclidean vector norm under this association is the Euclidean or Frobenius norm of the matrix. The tools required for manipulating these matrix norms are straightforward modifications of their vector counterparts. For example, the subdifferential of $\|E\|$ is

$$\partial\|E\|=\{G:R^m\to R^n;\|S\|\geq\|E\|+\operatorname{trace}((S-E)^{\mathrm{T}}G),\forall S:R^m\to R^n\},\tag{3.1}$$

and (compare (4.13), (4.14) of Chapter 1) $G\in\partial\|E\|$ is equivalent to the statements:

(i) $\|E\|=\operatorname{trace}(G^{\mathrm{T}}E)$, and
(ii) $\|G\|^*\leq1$,

where

$$\|G\|^* = \max_{\|S\| \leq 1} \text{trace}\,(S^T G). \tag{3.2}$$

and $\|\cdot\|^*$ is the polar or dual norm to $\|\cdot\|$. The roles of the norm and its dual can be interchanged in this definition.

Here attention is restricted to norms which are *separable.*

Definition 3.1 A $\|\cdot\|$ on the set of $n \times m$ matrices is separable if there exist vector norms $\|\cdot\|_R$ and $\|\cdot\|_D$ on R^n and R^m, the range and domain spaces respectively, such that for all $\mathbf{u} \in R^n$, $\mathbf{v} \in R^m$,

(i) $$\|\mathbf{u}\mathbf{v}^T\| = \|\mathbf{u}\|_R \, \|\mathbf{v}\|_D^* \tag{3.3a}$$

and

(ii) $$\|\mathbf{u}\mathbf{v}^T\|^* = \|\mathbf{u}\|_R^* \, \|\mathbf{v}\|_D \tag{3.3b}$$

The second part of the definition is needed (or at least makes life easier) in proving the following lemma, which shows that certain elements in the subdifferential of $\|\mathbf{u}\mathbf{v}^T\|$ have rank 1 and are known when the subdifferentials of $\|\mathbf{u}\|_R$, $\|\mathbf{v}\|_D^*$ are known.

Lemma 3.1 *Let $\|\cdot\|$ be separable. Then*

$$E = \mathbf{u}\mathbf{v}^T \Rightarrow G = \mathbf{s}\mathbf{t}^T \in \partial\,\|E\| \tag{3.4}$$

where $\mathbf{s}^T \in \partial\,\|\mathbf{u}\|_R$, $\mathbf{t}^T \in \partial\,\|\mathbf{v}\|_D^*$.

Proof It follows from the definition of $\mathbf{s}$ and $\mathbf{t}$ and (3.3a) that

$$\begin{aligned}\text{trace}\,(G^T\mathbf{u}\mathbf{v}^T) &= \mathbf{v}^T G^T \mathbf{u} \\ &= (\mathbf{v}^T\mathbf{t})(\mathbf{s}^T\mathbf{u}) \\ &= \|\mathbf{u}\|_R \, \|\mathbf{v}\|_D^* = \|\mathbf{u}\mathbf{v}^T\|.\end{aligned}$$

Thus $G \in \partial\,\|\mathbf{u}\mathbf{v}^T\|$ provided $\|G\|^* \leq 1$. But this is an immediate consequence of (3.3b) as $\|\mathbf{s}\|_R^* \leq 1$ and $\|\mathbf{t}\|_D \leq 1$ follow from the assumptions as in Example 4.4 of Chapter 1.

The next theorem generalizes the well-known result that the Euclidean or Frobenius norm of a matrix is not a subordinate or operator norm for any vector norm pair (the case of equality in (3.5) below).

Theorem 3.1 *Let $\|\cdot\|$ be separable. Then for any $E : R^m \to R^n$*

$$\|E\| \geq \max_{\|\mathbf{v}\|_D \leq 1} \|E\mathbf{v}\|_R \tag{3.5}$$

Proof The optimization problem in (3.5) is discussed in Example 2.1(ii). There it is shown that the maximum is attained for $\mathbf{v}=\mathbf{v}^*$, and that there exist $\mathbf{w}^T \in \partial \|\mathbf{v}^*\|_D^*$, $\mathbf{y}^T \in \partial \|E\mathbf{v}^*\|_R$, $\lambda = \|E\mathbf{v}^*\|_R / \|\mathbf{v}^*\|_D$ such that

$$\mathbf{y}^T E = \lambda \mathbf{w}^T \tag{3.6}$$

Let $G=\mathbf{y}\mathbf{v}^{*T}$. Then $\|G\|^* \leqslant 1$ by (3.3b). It now follows that

$$\begin{aligned}\|E\| &\geqslant \text{trace}\,(G^T E)\\ &= \text{trace}\,(\mathbf{v}^*\mathbf{y}^T E)\\ &= \mathbf{y}^T E \mathbf{v}^*\\ &= \|E\mathbf{v}^*\|_R\end{aligned}$$

Example 3.1 (i) If equality holds in (3.5) for all $E: R^m \to R^n$ then the norm is a subordinate or operator norm. In this case

$$\|E\| = \max_{\|\mathbf{v}\|_D \leqslant 1} \|E\mathbf{v}\| = \max_{\substack{\|\mathbf{y}\|_R^* \leqslant 1\\ \|\mathbf{v}\|_D \leqslant 1}} \mathbf{y}^T E \mathbf{v} \tag{3.7}$$

so that $G=\mathbf{y}\mathbf{v}^T \in \partial \|Z\|$. Equation (3.3a) is an immediate consequence. Equation (3.3b) requires that

$$\|\mathbf{s}\|_R^* \|\mathbf{t}\|_D = \max_{\|E\| \leqslant 1} \text{trace}\,(E^T \mathbf{s}\mathbf{t}^T)$$

and follows on substituting $1 \geqslant \|E\|$ in (3.7).

(ii) Let T_1, T_2 be positive diagonal matrices and consider the generalized Frobenius norm

$$\begin{aligned}\|\mathbf{u}\mathbf{v}^T\|^2 &= \|T_1\mathbf{u}\mathbf{v}^T T_2\|_F^2\\ &= \sum_{i=1}^{m} v_i^2 (T_2)_i^2 \|T_1\mathbf{u}\|_2^2\\ &= \|T_1\mathbf{u}\|_2^2 \|T_2\mathbf{v}\|_2^2.\end{aligned}$$

It follows that (3.3) holds with $\|\mathbf{u}\|_R = \|T_1\mathbf{u}\|_2$, $\|\mathbf{v}\|_D = \|T_2^{-1}\mathbf{v}\|_2$, and $\|E\|^* = \|T_1^{-1} E T_2^{-1}\|_F$. Theorem 3.1 gives

$$\begin{aligned}\|E\| &\geqslant \max_{\mathbf{v}^T T_2^{-2}\mathbf{v} \leqslant 1} \|T_1 E \mathbf{v}\|_2\\ &= \max_{\mathbf{w}^T\mathbf{w} \leqslant 1} \|T_1 E T_2 \mathbf{w}\|_2.\end{aligned}$$

Thus $\|E\|$ is bounded below by the largest singular value of $T_1 E T_2$.

(iii) In the l_α norm

$$\|E\| = \left\{\sum_{i=1}^{n}\sum_{j=1}^{m} |E_{ij}|^\alpha\right\}^{1/\alpha}.$$

It follows that

$$\|\mathbf{u}\mathbf{v}^T\|^\alpha = \sum_{j=1}^{m} |v_j|^\alpha \sum_{i=1}^{n} |u_i|^\alpha$$

so that

$$\|\mathbf{u}\mathbf{v}^{\mathrm{T}}\| = \|\mathbf{u}\|_{\alpha}\ \|\mathbf{v}\|_{\alpha}.$$

Thus $\|\cdot\|_{\mathrm{R}}$ is the l_α norm on R^n and $\|\cdot\|_{\mathrm{D}}$ is the l_β norm on R^m where $1/\alpha + 1/\beta = 1$. This verifies that (3.3a) and (3.3b) are similar with α and β interchanged. The extreme cases $\alpha = 1$, $\beta = \infty$ and $\alpha = \infty$, $\beta = 1$ are included in this result.

(iv) It is not difficult to find meaningful norms which are not separable. Consider the case where the E_{ij} in (1.1) correspond to measurement errors with expectation given by

$$\mathscr{E}\{\boldsymbol{\kappa}_i(E)\boldsymbol{\kappa}_j(E)^{\mathrm{T}}\} = W_i\,\delta_{ij}, \qquad i, j = 1, 2, \ldots, m$$

where the W_i are known variance-covariance matrices. If the E_{ij} are normally distributed then likelihood considerations lead to the minimization of

$$\|E\|^2 = \sum_{i=1}^{m} \boldsymbol{\kappa}_i(E)^{\mathrm{T}} W_i^{-1} \boldsymbol{\kappa}_i(E).$$

Here

$$\|\mathbf{u}\mathbf{v}^{\mathrm{T}}\| = \sum_{i=1}^{m} v_i^2 \mathbf{u}^{\mathrm{T}} W_i^{-1} \mathbf{u}$$

so the norm is separable only if $\mathbf{u}^{\mathrm{T}} W_i^{-1} \mathbf{u} = k_i\ \|\mathbf{u}\|_{\mathrm{R}}^2$.

The characteristic feature of the total approximation problem in a separable norm is that the solution has rank 1. We consider here the more general problem (1.3).

Theorem 3.2 *Let $\|\cdot\|$ be any separable norm on $R^{m_2} \to R^n$. If*

$$E = -Z\begin{bmatrix} \hat{\mathbf{v}}_1 \\ \hat{\mathbf{v}}_2 \end{bmatrix} \mathbf{w}_2^{\mathrm{T}} \tag{3.8}$$

where $\hat{\mathbf{v}}$ solves the vector minimization problem

$$\min_{\|\mathbf{v}_2\|_{\mathrm{D}}=1} \|Z\hat{\mathbf{v}}\|_{\mathrm{R}} \tag{3.9}$$

and $\hat{\mathbf{w}}_2^{\mathrm{T}} \in \partial \|\hat{\mathbf{v}}_2\|_{\mathrm{D}}$, then E solves (1.3).

Proof If $E \in \mathbb{E}$ then it follows from the constraint on (1.3) that there exists $\bar{\mathbf{v}}$, $\|\bar{\mathbf{v}}_2\|_{\mathrm{D}} = 1$, such that

$$[Z_1 \mid Z_2]\begin{bmatrix} \bar{\mathbf{v}}_1 \\ \bar{\mathbf{v}}_2 \end{bmatrix} = -E\bar{\mathbf{v}}_2$$

Thus for each feasible E, by Theorem 3.1,

$$\min_{\|\mathbf{v}_2\|_{\mathrm{D}}=1} \|Z\mathbf{v}\|_{\mathrm{R}} \leqslant \|Z\bar{\mathbf{v}}\|_{\mathrm{R}} = \|E\bar{\mathbf{v}}_2\|_{\mathrm{R}} \leqslant \|E\|. \tag{3.10}$$

However, $E=-Z\hat{\mathbf{v}}\hat{\mathbf{w}}_2^T$ satisfies the rank condition and gives equality in (3.10) as a consequence of (3.3a).

The above argument provides the hint for setting up the appropriate Lagrangian function for (1.3). This is

$$\mathscr{L}=\|E\|+\boldsymbol{\lambda}^T[Z_1 \mid Z_2+E]\mathbf{v}+\gamma(\|\mathbf{v}_2\|_D-1) \tag{3.11}$$

where the first n equality constraints in

$$\mathbf{g}(E, \mathbf{v})=\begin{bmatrix}[Z_1 \mid Z_2+E]\mathbf{v} \\ \|\mathbf{v}_2\|_D-1\end{bmatrix}=0 \tag{3.12}$$

ensure that rank $([Z_1 \mid Z_2+E])<m$, while the last constraint serves as a normalizing condition. In this case,

$$\mathscr{J}(\mathbf{g}, E, \mathbf{v})=\left\{\left[\begin{array}{c|c|c|c}[Z_1 \mid Z_2+E] & (\mathbf{v}_2)_1 I_n & \cdots & (\mathbf{v}_2)_{m_2} I_n \\ \hline [0 \mid \mathbf{w}_2^T] & 0 & & 0\end{array}\right], \mathbf{w}_2^T \in \partial\|\mathbf{v}_2\|_D\right\} \tag{3.13}$$

has rank $n+1$ as $\mathbf{v}_2, \mathbf{w}_2$ necessarily $\neq 0$ when rank $(Z)=m$. Thus the first-order necessary conditions (2.11) characterize the stationary points of (1.3).

Lemma 3.2 *The necessary conditions for a solution of* (1.3) *are that there exist* $\boldsymbol{\lambda} \in R^n$, $G \in \partial\|E\|$ *such that*

$$\boldsymbol{\lambda}^T[Z_1 \mid Z_2+E]=0 \tag{3.14}$$

and

$$G+\boldsymbol{\lambda}\mathbf{v}_2^T=0. \tag{3.15}$$

Proof The necessary conditions that follow from considering (3.11) are

$$\boldsymbol{\lambda}^T[Z_1 \mid Z_2+E]+\gamma[0 \mid \mathbf{w}_2^T]=0 \tag{3.16}$$

for some $\mathbf{w}_2^T \in \partial\|\mathbf{v}_2\|_D$, and

$$G_{ij}+\lambda_i(\mathbf{v}_2)_j=0, i=1,2, \ldots, n, \qquad j=1,2, \ldots, m_2$$

for some $G \in \partial\|E\|$. Taking the scalar product of (3.16) with $\mathbf{v}$ gives

$$\boldsymbol{\lambda}^T[Z_1 \mid Z_2+E]\mathbf{v}+\gamma \mathbf{w}_2^T \mathbf{v}_2=0$$

so that $\gamma=0$ by (3.12) and the result follows.

Remark 3.1 Separability of the norm is not assumed in Lemma 3.2, so that it is interesting that (3.15) shows that $G \in \partial\|E\|$ is of rank 1.

Theorem 3.3 *For a given separable norm let* $\hat{\mathbf{v}}$ *be a local minimizer of* (3.9). *If* $\mathbf{w}_2^T \in \partial\|\hat{\mathbf{v}}_2\|_D$ *then*

$$\hat{E}=-Z\hat{\mathbf{v}}\mathbf{w}_2^T$$

is a stationary point of the total approximation problem.

Proof The argument used in Example 2.1(i) shows that for *each* $\mathbf{w}_2^T \in \partial \|\hat{\mathbf{v}}_2\|_D$, there is a $\mathbf{y}^T \in \partial \|Z\hat{\mathbf{v}}\|_R$ such that

$$\mathbf{y}^T Z = \|Z\hat{\mathbf{v}}\|_R [0 \mid \mathbf{w}_2^T]. \tag{3.17}$$

$$G = -\mathbf{y}\hat{\mathbf{v}}_2^T.$$

By Lemma 2.1, $G \in \partial \|E\|$. Thus it remains to show that $\boldsymbol{\lambda} = \mathbf{y}$ satisfies (3.14). Here

$$\begin{aligned}\mathbf{y}^T[Z_1 \mid Z_2 + E] &= \mathbf{y}^T[Z_1 \mid Z_2 - Z\hat{\mathbf{v}}\mathbf{w}_2^T] \\ &= \mathbf{y}^T Z_2 - \|Z\hat{\mathbf{v}}\|_R \mathbf{w}_2^T \\ &= 0.\end{aligned}$$

In the case when the norm is smooth it is natural to consider iterative methods for solving (3.9). For example, if $\|\cdot\|_R$ is l_α and $\|\cdot\|_D$ is l_β corresponding to the separable norm treated in Example 3.1(iii) with $\alpha, \beta > 1$, then stationary points are found by solving $\nabla \mathscr{L} = 0$ where

$$\mathscr{L} = \|Z\mathbf{v}\|_\alpha - \gamma(\|\mathbf{v}_2\|_\beta - 1). \tag{3.18}$$

It is convenient to set $Z\mathbf{v} = \mathbf{r}$. Then a direct calculation gives the first-order necessary conditions in the form

$$0 = \frac{\mathbf{r}^T D_r Z}{\|Z\mathbf{v}\|_\alpha^{\alpha-1}} - \gamma \frac{\mathbf{v}^T D_v}{\|\mathbf{v}_2\|_\beta^{\beta-1}} \tag{3.19}$$

where

$$\begin{aligned} D_r &= \operatorname{diag}\{|r_i|^{\alpha-2} \text{ if } r_i \neq 0 \text{ else } 0,\ i = 1, 2, \ldots, n\}, \\ D_v &= \operatorname{diag}\{0,\ i = 1, 2, \ldots, m_1,\ |(\mathbf{v}_2)_j|^{\beta-2} \text{ if } (\mathbf{v}_2)_j \\ &\qquad \neq 0 \text{ else } 0,\ i = m_1 + j,\ j = 1, 2, \ldots, m_2\}, \end{aligned}$$

and

$$\gamma = \|Z\mathbf{v}\|_\alpha / \|\mathbf{v}_2\|_\beta.$$

One possible way of attempting to satisfy (3.19) is by the iteration

$$\mathbf{v}^{(i+1)} = \mathbf{h}(\mathbf{v}^{(i)}) / \|\mathbf{h}(\mathbf{v}^{(i)})_2\|_\beta \tag{3.20}$$

where

$$\mathbf{h}(\mathbf{v}) = (Z^T D_r Z)^{-1} D_v \mathbf{v} \tag{3.21}$$

It turns out that the iteration has similar convergence properties to the IRLS procedures considered in Section 5.4. Here second-order sufficiency is assumed and used explicitly in the following result.

Theorem 3.4 *Let $\hat{\mathbf{v}}$ be an isolated local minimum of* (3.9). *Then $\hat{\mathbf{v}}$ is a local point of attraction for the iteration* (3.20) *and the convergence rate is first order provided:*

(i) $\mathbf{r} = Z\hat{\mathbf{v}}$ *has no zero components (this condition ensures that* $\mathbf{h}$ *is well defined and $Z^T D_r Z$ is positive definite)*;

(ii) $|\alpha-2|<1$; *and*
(iii) *the second-order sufficiency conditions hold.*

The proof involves computing $\nabla\{\mathbf{h}(\hat{\mathbf{v}})/\|\mathbf{h}(\hat{\mathbf{v}})_2\|_\beta\}$ and showing that its eigenvalues have magnitude <1 provided the stated conditions hold. Several intermediate steps are required and these are summarized below, before returning to the proof of the theorem.

Lemma 3.3

$$\nabla\{\mathbf{h}(\mathbf{v})/\|\mathbf{h}(\mathbf{v})_2\|_\beta\}=\frac{1}{\|\mathbf{h}(\mathbf{v})_2\|_\beta}\left(I-\frac{\mathbf{h}\mathbf{h}^{\mathrm{T}}D_h}{\|\mathbf{h}_2\|_\beta^\beta}\right)\nabla\mathbf{h}(\mathbf{v}) \tag{3.22}$$

where

$$D_h=\operatorname{diag}\{0,\ i=1,2,\ldots,m_1,\ |(\mathbf{h}_2)_j|^{\beta-2}\ \textit{if}\ (\mathbf{h}_2)\neq 0\ \textit{else}\ 0,\ i=m_1+j,\ j=1,2,\ldots,m_2\}.$$

Lemma 3.4

$$\nabla\mathbf{h}(\mathbf{v})=-(\alpha-2)\,\|\mathbf{h}_2\|_\beta\,I+(\beta-1)(Z^{\mathrm{T}}D_rZ)^{-1}D_v$$

and

$$\nabla\{\mathbf{h}(\hat{\mathbf{v}})/\|\mathbf{h}(\hat{\mathbf{v}})_2\|_\beta\}=\{I-\hat{\mathbf{v}}\hat{\mathbf{v}}^{\mathrm{T}}D_v\}\{-(\alpha-2)I+\frac{\beta-1}{\|\mathbf{h}_2\|_\beta}(Z^{\mathrm{T}}D_rZ)^{-1}D_v\} \tag{3.23}$$

Lemma 3.5 *Consider the genralized eigenvalue problem*

$$[Z^{\mathrm{T}}D_rZ-\mu_iD_v]\mathbf{u}_i=0$$

where the matrices are evaluated at $\mathbf{v}=\hat{\mathbf{v}}$. *Then the solutions fall into three groups:*

(i) $$\mathbf{u}_1=\hat{\mathbf{v}},\qquad \mu_1=\gamma\frac{\|Z\hat{\mathbf{v}}\|_\alpha^{\alpha-1}}{\|\hat{\mathbf{v}}_2\|_\beta^{\beta-1}}=\frac{\|Z\hat{\mathbf{v}}\|_\alpha^\alpha}{\|\hat{\mathbf{v}}\|_\beta^\beta}=\frac{1}{\|\mathbf{h}_2\|_\beta},$$

(ii) *vectors* $\mathbf{u}_i$, *and values* $\mu_i>0$, $i=2,3,\ldots,k\leq m$ *where the eigenvalues* μ_i *are finite; and*
(iii) *the vectors associated with the infinite eigenvalues – that is* $\mathbf{u}_{k+1},\ldots,\mathbf{u}_m$ *forming a basis for the set of vectors satisfying*

$$Z^{\mathrm{T}}D_rZ\mathbf{u}\neq 0,\qquad D_v\mathbf{u}=0.$$

It is convenient to choose these vectors to be orthogonal with respect to $Z^{\mathrm{T}}D_rZ$ *as weighting matrix.*

Note that it is the vectors in categories (ii) and (iii) which span the set of directional limits for the constraint $\|\mathbf{v}_2\|_\beta=1$ at $\hat{\mathbf{v}}$ (Exercise 2.1).

Lemma 3.6 *For $\mathscr{L}$ given by (3.18),*

$$\nabla^2\mathscr{L}(\hat{\mathbf{v}}) = \frac{\alpha-1}{\|Z\hat{\mathbf{v}}\|_\alpha^{\alpha-1}}\left\{Z^T D_r Z - \frac{Z^T D_r \mathbf{r}\mathbf{r}^T D_r Z}{\|Z\hat{\mathbf{v}}\|_\alpha^{\alpha-2}}\right\}$$
$$-\frac{\gamma(\beta-1)}{\|\hat{\mathbf{v}}_2\|_\beta^{\beta-1}}\left\{D_v - \frac{D_v \hat{\mathbf{v}}\hat{\mathbf{v}}^T D_v}{\|\hat{\mathbf{v}}_2\|_\beta^{\beta-2}}\right\}. \tag{3.24}$$

The second-order sufficiency conditions can be satisfied only if

$$\mu_i > \frac{\beta-1}{\alpha-1}\mu_1, \qquad i=2,\ldots,k. \tag{3.25}$$

Proof By using the orthogonality condition $\mathbf{u}_i^T D_v \mathbf{u}_j = 0$, $i \neq j$, checking the second-order second-order sufficiency conditions reduces to verifying that $\mathbf{u}_i^T \nabla^2\mathscr{L}(\hat{\mathbf{v}})\mathbf{u}_i > 0$, $i=1,2,\ldots,m$. This condition is trivially satisfied for $i > k$, and the remaining cases give (3.25).

Proof of Theorem 3.4 For the matrix defined in (3.23) it follows that
(i) $\hat{\mathbf{v}}$ is an eigenvector associated with the eigenvalue zero, and
(ii) $D_v\mathbf{u} = 0 \Rightarrow \mathbf{u}$ is an eigenvector with eigenvalue $-(\alpha-2)$.
Thus $|\alpha-2| < 1$ is a necessary condition for convergence if $k > m$. Direct calculation shows that $\mathbf{u}_i$ is an eigenvector of (3.23) with corresponding eigenvalue

$$-(\alpha-2) + (\beta-1)\frac{\mu_1}{\mu_i}, \qquad i=2,\ldots,k.$$

Now Lemma 3.6 gives

$$\alpha - 1 > (\beta-1)\frac{\mu_1}{\mu_i}$$

so that

$$\alpha - 2 - (\beta-1)\frac{\mu_1}{\mu_i} > -1.$$

Thus to verify convergence under the stated conditions it remains to show that

$$-1 < -(\alpha-2) + (\beta-1)\frac{\mu_1}{\mu_i}.$$

But $-1 < -(\alpha-2)$ is assumed.

Exercise 3.1 (i) Discuss in detail the solution of the total approximation problem in the case $\alpha=\beta=2$. Are there any problems in extending this discussion to cover the case considered in Example 3.1(ii)?

(ii) Show that the distance of the point $\mathbf{z}_i^T = \boldsymbol{\rho}_i(Z)$ from the plane $\mathbf{z}^T\mathbf{v} = 0$ is $|r_i|/\|\mathbf{v}\|_2$ where $r_i = \mathbf{z}_i^T\mathbf{v}$. Use this result to interpret the total approximation problem when $p=q=2$. Why does it follow from this formulation that the solution of the total least squares problem has rank 1? In this form the total approximation problem has been called orthogonal least squares.

6.4 THE TOTAL l_1 PROBLEM

In Example 3.1(iii) it was shown that the l_1 matrix norm

$$\|S\| = \sum_{i=1}^{n} \sum_{j=1}^{m} |S_{ij}|$$

is separable with

$$\|\mathbf{u}\|_R = \|\mathbf{u}\|_1 = \sum_{i=1}^{n} |u_i|$$

and

$$\|\mathbf{v}\|_D = \|\mathbf{v}\|_\infty = \max_{1 \leq i \leq m} |v_i|.$$

This case is considered further in this section, and it will be shown that the polyhedral nature of the quantities involved permits more to be said about the nature of the solution. This can be constructed using Theorem 3.2, given the solution of the vector minimization problem

$$\min_{\|\mathbf{v}_2\|_\infty = 1} \left\| [Z_1 \mid Z_2] \begin{bmatrix} \mathbf{v}_1 \\ \mathbf{v}_2 \end{bmatrix} \right\|_1 \tag{4.1}$$

Theorem 4.1 *A necessary and sufficient condition for* $\hat{\mathbf{v}}$ *to be a local minimum of* (4.1) *is that for each* $\mathbf{w}_2^T \in \partial \|\hat{\mathbf{v}}_2\|_\infty$ *there exists* $\mathbf{y}^T \in \partial \|Z\hat{\mathbf{v}}\|_1$ *satisfying*

$$\mathbf{y}^T Z = \|Z\hat{\mathbf{v}}\|_1 [0 \mid \mathbf{w}_2^T] \tag{4.2}$$

Proof The necessity follows by the argument used in Example 2.1(i). To prove sufficiency note that if $\mathbf{t} = \begin{bmatrix} \mathbf{t}_1 \\ \mathbf{t}_2 \end{bmatrix}$ is the limit of a feasible directional sequence at $\hat{\mathbf{v}}$, then because $\|\mathbf{v}_2\|_\infty$ is polyhedral $\hat{\mathbf{v}} + \lambda \mathbf{t}$ is feasible for $\lambda > 0$ small enough. Let $\mathbf{w}_2^T \in \partial \|\hat{\mathbf{v}}_2\|_\infty$ satisfy $\mathbf{w}_2^T \mathbf{t}_2 = 0$ ($\mathbf{w}_2$ exists by Lemma 2.3), and let $\mathbf{w}_2$ select $\mathbf{y}^T \in \partial \|Z\hat{\mathbf{v}}\|_1$ by (4.2). Then

$$\begin{aligned} \|Z(\hat{\mathbf{v}} + \lambda \mathbf{t})\|_1 &\geq \mathbf{y}^T Z \mathbf{v} + \lambda \mathbf{y}^T Z \mathbf{t} \\ &\geq \|Z\hat{\mathbf{v}}\|_1 + \lambda \|Z\hat{\mathbf{v}}\|_1 \mathbf{t}_2^T \mathbf{w}_2 \\ &\geq \|Z\hat{\mathbf{v}}\|_1. \end{aligned}$$

This result can be interpreted in terms of the number of zero components in $Z\hat{\mathbf{v}}$ and extremal components in $\hat{\mathbf{v}}$. Let

$$\mu = \{i; \boldsymbol{\rho}_i(Z)\hat{\mathbf{v}} = 0\}, \tag{4.3}$$

and

$$\nu = \{i; |(\hat{\mathbf{v}}_2)_i| = \|\hat{\mathbf{v}}_2\|_\infty\} \tag{4.4}$$

be the index sets describing these quantities.

Theorem 4.2 *In the neighbourhood of a local solution either there exists a solution* $\hat{\mathbf{v}}$ *to* (4.1) *such that* $|\mu| + |\nu| \geq m$, *or there is a nearby point at which* *there is a downhill direction and which can be reached without increasing the objective function.*

Proof Let $\hat{\mathbf{v}}$ be a solution of (4.1) with $|\mu|+|\nu|\leq m-1$. Then there always exists a non-trivial $\mathbf{s}\in R^m$ such that

$$\mathbf{z}_i^T\mathbf{s}=0, \qquad i\in\mu,$$

$$s_j=0, \qquad j\in\nu,$$

and

$$\mathbf{g}^T\mathbf{s}\leq 0,$$

where, as usual, $\mathbf{z}_i^T=\boldsymbol{\rho}_i(Z)$, $\theta_i=\operatorname{sgn}(\mathbf{z}_i^T\hat{\mathbf{v}})$, and

$$\mathbf{g}=\sum_{i\in\mu^c}\theta_i\mathbf{z}_i.$$

It follows that for $\tau>0$ small enough

$$\|Z(\hat{\mathbf{v}}+\tau\mathbf{s})\|_1=\|Z\hat{\mathbf{v}}\|_1+\tau\mathbf{g}^T\mathbf{s}\leq\|Z\hat{\mathbf{v}}\|_1$$

and

$$\|\hat{\mathbf{v}}_2+\tau\mathbf{s}_2\|_\infty=\|\hat{\mathbf{v}}_2\|_\infty=1.$$

Thus $\hat{\mathbf{v}}+\tau\mathbf{s}$ is a solution for all small enough τ. However, there must be a first value τ_0 of τ for which the slope of one of the piecewise linear functions $\|Z(\hat{\mathbf{v}}+\tau\mathbf{s})\|_1$ and $\|\hat{\mathbf{v}}_2+\tau\mathbf{s}_2\|_\infty$ changes as both $\to\infty$ as $\tau\to\infty$. This must correspond either to at least one new zero residual ($|\mu(\tau_0)|\geq|\mu|+1$) or at least one new extremal component in $\hat{\mathbf{v}}_2+\tau\mathbf{s}_2$ ($|\nu(\tau_0)|\geq|\nu|+1$), and the argument can be repeated if there is not a downhill direction at $\hat{\mathbf{v}}+\tau_0\mathbf{s}$ – in particular, if just $|\mu|$ is increased. However, if $\nu(\tau_0)\geq|\nu|+1$ and $|\mu(\tau_0)|=|\mu|$ then $\mathbf{s}$ is downhill for minimizing $\|Z\mathbf{v}\|_1/\|\mathbf{v}_2\|_\infty$ at $\mathbf{v}=\hat{\mathbf{v}}+\tau_0\mathbf{s}$, because at this point $\|\hat{\mathbf{v}}_2+\tau\mathbf{s}_2\|_\infty$ begins to increase.

Remark 4.1 If $\hat{\mathbf{v}}$ is an isolated minimum then only the first option of the theorem is possible. The existence of a downhill direction at a nearby point does not conflict with $\hat{\mathbf{v}}$ being a local minimum, as this requires only that there is no downhill direction at $\hat{\mathbf{v}}$. However, if $\mathbf{s}$ is downhill at $\hat{\mathbf{v}}+\tau_0\mathbf{s}$ then $\nu\cap\nu(\tau_0+\delta)=\varnothing$ for all $\delta>0$ sufficiently small. The implication is that while it is possible to build up μ systematically, $|\nu|>1$ appears to happen only by chance, and it is argued that this corresponds to a degenerate situation.

Example 4.1 Let

$$[Z_1\,|\,Z_2]=\left[\begin{array}{c|cc}1&-1&0\\1&0&-1\\0&1&0\\0&0&1\end{array}\right].$$

The minimum of (4.1) is $\|Z\mathbf{v}\|_1 = 2$, and optimal solutions include

$$v_1 = 1, v_2 = 1, v_3 = 0, |\mu| = 2, |\nu| = 1,$$

and

$$v_1 = 1, v_2 = 0, v_3 = 1, |\mu| = 2, |\nu| = 1.$$

These are the $\eta = 0$ cases of the solutions

$$v_1 = 1, v_2 = 1, v_3 = \eta, |\mu| = 1, |\nu| = 1,$$

and

$$v_1 = 1, v_2 = \eta, v_3 = 1, |\mu| = 1, |\nu| = 1,$$

valid for $0 < \eta < 1$. When $\eta = 1$ we obtain the degenerate solution

$$v_1 = 1, v_2 = 1, v_3 = 1, |\mu| = 2, |\nu| = 2.$$

Sufficient conditions for an isolated solution when $|\nu| = 1$ are given in the next theorem, which should be compared with Exercise 1.1 of Chapter 3. The argument does not extend when $|\nu| > 1$ (this is the subject of Exercise 4.1). In fact it then suggests how it may be possible to construct directions of nonuniqueness. This reinforces Remark 4.1 that $|\nu| > 1$ corresponds to degenerate behaviour.

Theorem 4.3 *Let $|\nu| = 1$ and $\hat{\mathbf{v}}$ be a local solution of* (4.1). *Then $\hat{\mathbf{v}}$ is an isolated solution provided*

$$\operatorname{rank}\left((\mathbf{z}_i, i \in \mu, |y_i| < 1), \begin{bmatrix} 0 \\ \mathbf{w}_2 \end{bmatrix}\right) = m \tag{4.5}$$

where $\mathbf{w}_2^{\mathrm{T}} \in \partial \|\hat{\mathbf{v}}_2\|_\infty = \phi_k \mathbf{e}_k^{\mathrm{T}}$ corresponding to $\nu = \{k\}$ and $\phi_k = \operatorname{sgn}((\hat{\mathbf{v}}_2)_k)$, and $\mathbf{y}$ is aligned with $\mathbf{w}_2$ by (4.2).

Proof Let $\mathbf{s} \neq 0$ satisfy $(\mathbf{s}_2)_k = 0$ so that $\tau\mathbf{s}$ is a feasible displacement at $\hat{\mathbf{v}}$ for $\tau > 0$ small enough, but is otherwise arbitrary. In this displacement

$$\|Z(\hat{\mathbf{v}} + \tau\mathbf{s})\|_1 - \|Z\hat{\mathbf{v}}\|_1 = \tau\left(\mathbf{g}^{\mathrm{T}}\mathbf{s} + \sum_{i \in \mu} |\mathbf{z}_i^{\mathrm{T}}\mathbf{s}|\right).$$

Equation (4.2) gives

$$\mathbf{g}^{\mathrm{T}}\mathbf{s} + \sum_{i \in \mu} y_i \mathbf{z}_i^{\mathrm{T}}\mathbf{s} = \|Z\hat{\mathbf{v}}\|_1 \, \mathbf{w}_2^{\mathrm{T}}\mathbf{s}_2 = 0$$

Thus

$$\begin{aligned} \mathbf{g}^{\mathrm{T}}\mathbf{s} + \sum_{i \in \mu} |\mathbf{z}_i^{\mathrm{T}}\mathbf{s}| &= \sum_{i \in \mu} \{|\mathbf{z}_i^{\mathrm{T}}\mathbf{s}| - y_i \mathbf{z}_i^{\mathrm{T}}\mathbf{s}\} \\ &\geqslant \sum_{i \in \mu} \{1 - |y_i|\} \, |\mathbf{z}_i^{\mathrm{T}}\mathbf{s}| > 0 \end{aligned}$$

as the alternative is $\mathbf{z}_i^{\mathrm{T}}\mathbf{s} = 0$ for $i \in \mu$ and $|y_i| < 1$. But then the rank condition (4.5) gives $\mathbf{s} = 0$, as $(\mathbf{s}_2)_k = 0$ is assumed. This shows that $\mathbf{s}$ is a direction of ascent.

Definition 4.1 The point $\mathbf{v}$ is *nondegenerate* if $|\mu| = k \leq m-1$, $|\nu| = 1$, and

$$\operatorname{rank}\left((\mathbf{z}_i,\ i \in \mu),\ \left(\begin{bmatrix} 0 \\ \mathbf{w}_2 \end{bmatrix},\ \mathbf{w}_2 \in \partial \|\mathbf{v}_2\|_\infty\right)\right) = k+1 \tag{4.6}$$

At a nondegenerate local minimum $\hat{\mathbf{v}}$ only one column of E is different from zero. In the sense argued above, this represents the usual case. It provides the basic structural assumption for the descent algorithm that is developed here.

The descent algorithm

Here consideration is restricted to the nondegenerate situation, and an algorithm can then be given as a simple extension of a standard descent algorithm for the l_1 discrete approximation problem. The algorithm has two main parts:

(i) At the current point $\mathbf{v}$ a descent step is performed on the problem

$$\min_{(\mathbf{v}_2)_k = \phi_k} F(\mathbf{v}) \tag{4.7}$$

where $F(\mathbf{v}) = \|Z\mathbf{v}\|_1$. This step is essentially a standard l_1 descent step and consists of two components:

(a) A descent vector $\mathbf{t}$ is computed satisfying

$$F'(\mathbf{v}:\mathbf{t}) < 0, \qquad (\mathbf{t}_2)_k = 0. \tag{4.8}$$

If a suitable $\mathbf{t}$ cannot be found then the current point is optimal. The argument is standard (Exercise 4.2).

(b) Given $\mathbf{t}$ find λ to minimize $F(\mathbf{v}+\lambda\mathbf{t})$ and set $\tilde{\mathbf{v}} = \mathbf{v}+\lambda\mathbf{t}$.

(ii) In the line search no attempt is made to impose the constraint $\|\mathbf{v}_2\|_\infty = 1$ although it holds for small enough λ because $(\mathbf{t}_2)_k = 0$. For larger values of λ this only ensures that $\|\mathbf{v}_2 + \lambda\mathbf{t}_2\|_\infty \geq 1$, and it is frequently the case that the norm is increased. But $\tilde{\mathbf{v}}$ can be made feasible by renormalization to give

$$\bar{\mathbf{v}} = \tilde{\mathbf{v}}/\|\tilde{\mathbf{v}}_2\|_\infty, \quad \text{and} \quad F(\bar{\mathbf{v}}) \leq F(\tilde{\mathbf{v}})$$

If renormalization is necessary then the objective function is strictly decreased, and the index of the component of maximum modulus in $\bar{\mathbf{v}}_2$ is $\bar{k} \neq k$.

The reduced gradient method is readily adapted to generate a suitable descent direction. Let

$$B_\mu = [Z_\mu \mid J_\mu]$$

where

$$\boldsymbol{\kappa}_i(Z_\mu) = \mathbf{z}_{\mu(i)}, \qquad i = 1, 2, \ldots, |\mu|,$$

and J_μ consists of columns of the unit matrix *including* $\mathbf{e}_k$ chosen to make $\operatorname{rank}(B_\mu) = m$. It is convenient to arrange that $\boldsymbol{\kappa}_m(B_\mu) = \mathbf{e}_k$. Define $\mathbf{u} = \begin{bmatrix} \mathbf{u}_1 \\ \mathbf{u}_2 \end{bmatrix}$

by

$$B_\mu \mathbf{u} = -\mathbf{g}$$

Then the argument of Section 3.2 gives

$$\mathbf{t} = \operatorname{sgn}(u_j)\boldsymbol{\kappa}_j(B_\mu^{-T})$$

is a descent direction provided
(a) $j \leq |\mu|$ and $|u_j| > 1$, or
(b) $|\mu| < j < m$ and $|u_j| \neq 0$.
It is possible for both alternatives to give adequate descent directions. In this case, and assuming that scale differences are not important so that steepest-edge considerations suffice, a possible criterion is

$$j = \arg\max(|u_j| - 1, j = 1, 2, \ldots, |\mu|, \zeta\,|u_j|, j = |\mu| + 1, \ldots, m - 1)$$

where $\zeta > 0$ is chosen to favour deleting columns of J_μ unless there is a relatively large violation of the multiplier bound. The advantages of this test in rank-deficient problems have been noted already. But here this advantage is diagnostic only – if rank $(Z) < m$ then $\hat{E} = 0$.

The line search does not vary from the l_1 discrete approximation case, nor does the subsequent updating of μ and B_μ. The new feature comes in the renormalizing steps when the components of $\mathbf{v}_2$ must be scanned for the one of maximum modulus and the corresponding unit vector entered into B_μ. Here the policy of making $\boldsymbol{\kappa}_m(B_\mu) = \mathbf{e}_k$ pays off. It is only necessary to modify the method of Bartels and Golub so that when B_μ is updated after the line search the new entry is made into column $m - 1$. Note that neither this update nor the update to column m to take account of renormalisation alters the upper Hessenberg form of the modified U.

Remark 4.2 If $\mathbf{v}$ is chosen to be $\mathbf{e}_{\nu(1)}$ initially, then at each subsequent step of the algorithm

$$\begin{aligned}\mathbf{v}^T B_\mu \mathbf{u} &= v_{\nu(1)} u_m \\ &= \phi_s u_m, \qquad s = \nu(1),\end{aligned}$$

as $\mathbf{v}^T\boldsymbol{\kappa}_i(Z_\mu) = 0$, $i = 1, 2, \ldots, |\mu|$, by the definition of μ, and $\mathbf{v}^T\boldsymbol{\kappa}_i(J_\mu) = 0$, $i = 1, 2, \ldots, m - |\mu| - 1$, by the choice of initial conditions. However,

$$\begin{aligned}\mathbf{v}^T B_\mu \mathbf{u} &= -\mathbf{v}^T\mathbf{g} \\ &= -F(\mathbf{v}).\end{aligned}$$

Thus, at each step, the correct multiplier estimate is given for the scaling constraint. This has value as a numerical check, but it also shows that information is not available from this source concerning the appropriateness of the scaling constraint.

Remark 4.3 Finiteness of the algorithm in the nondegenerate case follows from the usual arguments. If the initial vector satisfies $\mathbf{v} = \mathbf{e}_{\nu(1)}$ then at each

Table 4.1 Values of F, $\mathbf{v}$, $\mathbf{u}$ for successive iterations of the reduced gradient algorithm applied to the data given in Table 5.1 of Chapter 5 (the stack loss data set).

i	1	2	3	4	5	6	7	8	9	10
F	368.0	138.3	64.86	64.64	1.359	.9891	.9022	.8840	.8043	.7840
v_1	0	0	0	0	1	1	1	1	1	1
v_2	0	0	−1	−1	.0269	−.01522	−.00494	−.00878	−.00923	−.00845
v_3	0	0	0	.0934	.00034	−.02174	−.02612	−.02239	−.00848	−.01079
v_4	0	−.1724	.4982	.5201	.00190	0	−.00435	−.00321	−.00541	−.00524
v_5	1	1	.9774	.9933	.02688	.02174	.01306	.01592	.01200	.01031
u_1	−21.00	.3103	.7641	.7590	−1.370	−1.815	−2.204	−2.013	−1.439	−.4647
u_2	−1269.	−89.00	−2.328	4.834	.9542	−2.	−3.891	3.101	−.5001	.8594
u_3	−443.0	−40.86	5.760	−3.279	5.667	4.076	3.997	−1.601	1.214	.7616
u_4	−1812.	−.3103	−3.524	−.3145	7.434	−.2727	.2001	.3972	.5295	.6277
u_5	−368.0	−138.3	64.86	64.64	−1.359	−.9891	−.9022	−.8840	−.8043	−.7840

subsequent step $\mathbf{v}$ is uniquely determined by the linear equations

$$\mathbf{v}^T B_\mu = \phi_{\nu(1)}\mathbf{e}_m^T,$$

and each possible B_μ is associated with a particular value of F. As the algorithm reduces F at each step, it follows that the configuration specified by B_μ cannot repeat. As there are only a finite number of such configurations, finiteness is an immediate consequence.

To illustrate the use of the algorithm, it was applied to $Z: R^5 \to R^{21}$ given in Table 5.1 of Chapter 5. In Table 4.1, values of $F, \mathbf{v}, \mathbf{u}$ are given for the successive steps of the algorithm starting from $\mathbf{v} = \mathbf{e}_5$. In this case the same solution is found for each of the starting points $\mathbf{v} = \mathbf{e}_j$, $j = 1, 2, \ldots, 5$, and the optimal scaling could be predicted by noting that the elements in $\boldsymbol{\kappa}_1(Z)$ are small compared with those in the other columns so that choosing $\mathbf{v} = \mathbf{e}_j$, $j > 1$, would be likely to make v_1 large. However, the first column corresponds to an intercept term and is known exactly. If this column is not included ($\mathbf{v}_2^T = (v_2, v_3, v_4, v_5)$) then the results are rather different and several isolated local minima are found by varying the initial conditions so that $\mathbf{v} = \mathbf{e}_j$, $j = 2, 3, \ldots, 5$. Table 4.2 gives the values of $F, \mathbf{v}, \mathbf{u}$ obtained corres-

Table 4.2 Solutions produced by starting with different coordinate vectors as initial conditions when the intercept is not constrained. The solutions corresponding to $\mathbf{e}_3, \mathbf{e}_4$ are identical. The permutation of the components of $\mathbf{u}$ is an implementation artefact.

$\mathbf{v}_0$	$\mathbf{e}_2$	$\mathbf{e}_3$	$\mathbf{e}_4$	$\mathbf{e}_5$
F	43.75	24.20	24.20	42.08
v_1	−34.83	−20.50	−20.50	39.69
v_2	1	.00334	.00334	−.8319
v_3	.3013	1	1	−.5739
v_4	−.1949	.05071	.05071	.06087
v_5	−.8558	−.2900	−.2900	1
u_1	−.1464	.5527	.04604	.1899
u_2	−.06796	.04604	.4184	−.5580
u_3	.4696	.4184	.5527	−.7290
u_4	.7447	−.0171	−.0171	.6391
u_5	−43.75	−24.20	−24.20	−42.08

ponding to each of these initial conditions. It is readily checked that isolated local solutions are obtained. Also the solution with $v_5 = 1$ gives the solution to the l_1 regression problem (the results should be compared with Table 5.2 of Chapter 5).

To conclude this section, the resistance properties of the total l_1 problem solution are considered. Recall that in the l_1 discrete approximation problem it is the f_i corresponding to the nonzero residuals that can be allowed to vary substantially (one interpretation of resistance in this case is that the solution does correspond to data which yields small residuals for the given model). However, we have seen that in the usual case in the total l_1 problem only one column of E contains nonzero elements. To examine resistance it is necessary to evaluate the rate of change of the solution quantities with respect to the elements of Z. The defining relations are

$$\mathbf{g} + \sum_{i \in \mu} y_i \mathbf{z}_i - \gamma \begin{bmatrix} 0 \\ \mathbf{w}_2 \end{bmatrix} = 0$$

where

$$|\mu| = m - 1, \ -1 \leq y_i \leq 1, \ i \in \mu, \ \gamma = \|Z\hat{\mathbf{v}}\|_1, \ \mathbf{w}_2 = \phi_k \mathbf{e}_k,$$

$$\mathbf{z}_i^{\mathrm{T}} \hat{\mathbf{v}} = 0, \ i \in \mu,$$

and

$$\|\hat{\mathbf{v}}_2\|_\infty = 1$$

Let Z_{pq} be a general element of Z subject to $p \notin \mu$. Differentiating the above conditions with respect to Z_{pq} gives

$$\theta_p \mathbf{e}_q + \sum_{i \in \mu} \frac{\mathrm{d}y_i}{\mathrm{d}Z_{pq}} \mathbf{z}_i - \frac{\mathrm{d}\gamma}{\mathrm{d}Z_{pq}} \begin{bmatrix} 0 \\ \phi_k \mathbf{e}_k \end{bmatrix} = 0$$

$$\mathbf{z}_i^{\mathrm{T}} \frac{\mathrm{d}\hat{\mathbf{v}}}{\mathrm{d}Z_{pq}} = 0, \ i \in \mu,$$

and

$$\frac{\mathrm{d}(\hat{\mathbf{v}}_2)_k}{\mathrm{d}Z_{pq}} = 0.$$

In particular, $\mathrm{d}\mathbf{v}/\mathrm{d}Z_{pq} = 0$ (this corresponds to the l_1 regression problem result obtained in Section 5.5). However, the new feature emerges that

$$\frac{\mathrm{d}}{\mathrm{d}Z_{pq}} \begin{bmatrix} \mathbf{y} \\ \gamma \end{bmatrix} \neq 0$$

and is independent of Z_{pq}. The conditions that the perturbation

$$Z_{pq} \to Z_{pq} + \Delta_{pq}$$

leaves $\hat{\mathbf{v}}$ unchanged now become:

(i) that the signs of the nonzero residuals are not changed, so that

$$\theta_p (\mathbf{z}_p^{\mathrm{T}} \hat{\mathbf{v}} + \Delta_{pq} \hat{v}_q) > 0, \ p \in \mu^c;$$

and
(ii) that the multipliers remain bounded by ± 1 so that

$$1-\left|y_i+\frac{dy_i}{dZ_{pq}}\Delta_{pq}\right|\geq 0,\ i\in\mu.$$

This second condition puts strict limits on the size of perturbations that leave $\hat{\mathbf{v}}$ a solution. *In this sense the total l_1 problem is significantly less resistant than the corresponding discrete approximation problem.*

Exercise 4.1 The analogue of the rank condition in Theorem 4.3 when $|\nu|>1$ is that rank $\{(\mathbf{z}_i,\ i\in\mu,\ |y_i|<1),\ (\mathbf{e}_j,\ j\in\nu)\}=m$ for some $\mathbf{y}$ such that the pair $(\mathbf{w}_2\in\partial\|\hat{\mathbf{v}}_2\|_\infty,\ \mathbf{y}\in\partial\|Z\hat{\mathbf{v}}\|_1)$ are aligned by (4.2). Show that this does not suffice to guarantee that $\hat{\mathbf{v}}$ is an isolated minimum when $|\nu|>1$.

Exercise 4.2 If $\mathbf{v}$ is nondegenerate, show that if there does not exist $\mathbf{t}$ satisfying (4.8) then $\mathbf{v}$ is optimal.

Exercise 4.3 Parallel the discussion of this section for the case $\|\cdot\|_R=\|\cdot\|_\infty$, $\|\cdot\|_D=\|\cdot\|_1$.

Exercise 4.4 In order to improve the resistance of orthogonal least squares (Exercise 3.1) it has been suggested to consider the problem

$$\min_{\|\mathbf{v}\|_2=1}\|Z\mathbf{v}\|_1.$$

(i) Derive the multiplier conditions for this problem and use these to characterize the conditions under which a descent direction is possible (show that a descent direction is possible in general if any multiplier associated with a zero residual is at its bounds).

(ii) Develop the descent algorithms for this problem in the usual way. Are they finite?

(iii) How does the resistance of this method compare with that of the total l_1 problem?

6.5 FINDING CENTRES IN THE JACCARD METRIC

The problem considered in this section is that of finding $\mathbf{z}\in R^p$ to minimize

$$h(\mathbf{z})=\sum_{i=1}^{n}f_i(\mathbf{z})/g_i(\mathbf{z}) \tag{5.1}$$

where

$$f_i(\mathbf{z})=\|\mathbf{z}-\mathbf{a}_i\|_1,$$
$$g_i(\mathbf{z})=\|\mathbf{z}\|_1+\|\mathbf{a}_i\|_1+\|\mathbf{z}-\mathbf{a}_i\|_1,$$

and $\mathbf{a}_i,\ \mathbf{z}\geq 0$. In particular, both f_i and g_i are convex functions of $\mathbf{z}$. The key result requires a minimizer of (5.1) to interpolate the data in the sense expressed in Definition 5.1, and this result depends in turn on our ability to compute directional derivatives of h of all orders so that arguments based on first- and second-order conditions are available.

Lemma 5.1 *Let $\mathbf{z} \geq 0$, and $\mathbf{t}$ be a feasible displacement at $\mathbf{z}$. Then directional derivatives of all orders and a one-sided Taylor series exist for h. In particular,*

$$h'(\mathbf{z}:\mathbf{t}) = \sum_{i=1}^{n} \frac{g_i(\mathbf{z})f_i'(\mathbf{z}:\mathbf{t}) - f_i(\mathbf{z})g_i'(\mathbf{z}:\mathbf{t})}{g_i^2} \tag{5.2}$$

and

$$h''(\mathbf{z}:\mathbf{t}) = 2\sum_{i=1}^{n} \frac{f_i(\mathbf{z})g_i'(\mathbf{z}:\mathbf{t})^2 - g_i(\mathbf{z})f_i'(\mathbf{z}:\mathbf{t})g_i'(\mathbf{z}:\mathbf{t})}{g_i^3} \tag{5.3}$$

Proof This follows on noting that a direct calculation yields, for $\gamma > 0$ small enough,

$$h(\mathbf{z}+\gamma\mathbf{t}) = \sum_{i=1}^{n} \frac{f_i(\mathbf{z}) + \gamma f_i'(\mathbf{z}:\mathbf{t})}{g_i(\mathbf{z}) + \gamma g_i'(\mathbf{z}:\mathbf{t})} \tag{5.4}$$

Definition 5.1 $\mathbf{z}$ *interpolates* $\{\mathbf{a}_i, i = 1, 2, \ldots, n\}$ if there exists an index set σ such that

$$\sigma = \{i; \forall j \ \exists \ i \in \sigma \ni (\mathbf{a}_i)_j = z_j\} \tag{5.5}$$

Theorem 5.1 *Let $\hat{\mathbf{z}}$ be a local minimum of (5.1). Then $\hat{\mathbf{z}}$ interpolates $\{\mathbf{a}_i, i = 1, 2, \ldots, n\}$.*

Proof Assume the result is not true. Then feasibility implies that there exists k such that

$$0 \leq \hat{z}_k \neq (\mathbf{a}_i)_k, \ i = 1, 2, \ldots, n.$$

Let $\xi = \pm 1$. Then, whenever the direction is feasible,

$$f_i'(\hat{\mathbf{z}}:\xi\mathbf{e}_k) = \theta_{ik}\xi,$$

and

$$g_i'(\hat{\mathbf{z}}:\xi\mathbf{e}_k) = \xi(1 + f_i'(\hat{\mathbf{z}}:\xi\mathbf{e}_k))$$

where

$$\theta_{ik} = \operatorname{sgn}(z_k - (\mathbf{a}_i)_k), \quad i = 1, 2, \ldots, n$$

so that

$$h'(\hat{\mathbf{z}}:\xi\mathbf{e}_k) = \xi\left\{\sum_{\theta_{ik}>0} \frac{g_i - 2f_i}{g_i^2} - \sum_{\theta_{ik}<0} \frac{1}{g_i}\right\}. \tag{5.6}$$

In particular, if $\hat{z}_k < \min_i ((\mathbf{a}_i)_k)$, then

$$h'(\hat{\mathbf{z}}:\mathbf{e}_k) = -\sum_{i=1}^{n} \frac{1}{g_i} < 0,$$

while if $\hat{z}_k > \max_i ((\mathbf{a}_i)_k)$, then

$$h'(\hat{\mathbf{z}}:-\mathbf{e}_k) = -\sum_{i=1}^{n} \frac{g_i - 2f_i}{g_i^3} = -\sum_{i=1}^{n} \frac{\|\hat{\mathbf{z}}\|_1 + \|\mathbf{a}_i\|_1 - \|\hat{\mathbf{z}} - \mathbf{a}_i\|_1}{g_i^3} < 0.$$

Thus it suffices to restrict consideration to $\min_i ((\mathbf{a}_i)_k) < \hat{z}_k < \max_i ((\mathbf{a}_i)_k)$. In this case $h'(\hat{\mathbf{z}}:\xi\mathbf{e}_k) = \xi h'(\hat{\mathbf{z}}:\mathbf{e}_k)$, and this is consistent with no downhill direction only if $h'(\hat{\mathbf{z}}:\xi\mathbf{e}_k) = 0$. Evaluating (5.3) gives

$$h''(\mathbf{z}:\xi\mathbf{e}_k) = 4 \sum_{\theta_{ik}>0} \frac{2f_i - g_i}{g_i^3} = 4 \sum_{\theta_{ik}>0} \frac{\|\hat{\mathbf{z}} - \mathbf{a}_i\|_1 - \|\hat{\mathbf{z}}\|_1 - \|\mathbf{a}_i\|_1}{g_i^3} < 0$$

This indicates that $\xi\mathbf{e}_k$ is a downhill direction at $\hat{\mathbf{z}}$, which gives a contradiction.

The method of proof gives the following result, which is of independent interest.

Corollary 5.1 *Let $z_k \neq (\mathbf{a}_i)_k$, $i = 1, 2, \ldots, n$. Then $h'(\mathbf{z}:\mathbf{e}_k) = 0$ is possible only if $\mathbf{z}$ is a local maximum of h in the direction $\mathbf{e}_k$.*

Theorem 5.1 shows that the minimum of $h(\mathbf{z})$ can be found by the exhaustive search of a finite number of possibilities so that a finite algorithm for minimizing (5.1) exists. To systematize the search it is natural to consider a descent calculation in which the possible descent directions at the current point $\mathbf{z}$ are restricted to the coordinate directions $\pm\mathbf{e}_k$, $j = 1, 2, \ldots, p$. There is no restriction in assuming that $\mathbf{z}$ interpolates the data in the sense made explicit in (5.4), and a suitable initial point is given by the l_1 centre (1.8). If $h'(\mathbf{z}:\mathbf{e}_k) < 0$ so that $\mathbf{e}_k$ is a descent direction and $z_k = (\mathbf{a}_i)_k$, then the adjacent interpolation point is

$$\bar{\mathbf{z}} = \mathbf{z} + \bar{\delta}\mathbf{e}_k \tag{5.7}$$

where

$$\bar{\delta} = \min_{s,\, (\mathbf{a}_s)_k > (\mathbf{a}_i)_k} (\mathbf{a}_s)_k - (\mathbf{a}_i)_k$$

Lemma 5.2

$$h(\bar{\mathbf{z}}) < h(\mathbf{z})$$

Proof This is a consequence of $h'(\mathbf{z} + \delta\mathbf{e}_k : \mathbf{e}_k) < 0$, $0 \leq \delta < \bar{\delta}$. This follows from the continuity of $h'(\mathbf{z} + \delta\mathbf{e}_k : \mathbf{e}_k)$ because $h'(\mathbf{z} + \delta\mathbf{e}_k : \mathbf{e}_k) = 0$ is possible for $0 < \delta < \bar{\delta}$ only at a local maximum in the direction $\mathbf{e}_k$, by Corollary 5.1. But $\mathbf{e}_k$ is a descent direction at $\mathbf{z}$.

To specify the coordinate directions that are downhill at $\mathbf{z}$ it is convenient to partition σ to give

$$\delta = \bigcup_{j=1}^{p} \sigma[j] : \sigma[j] = \{i; (\mathbf{a}_i)_j = z_j, i = 1, 2, \ldots, n\}.$$

Then the condition that $\xi \mathbf{e}_j$ $(\xi = \pm 1)$ is downhill is that

$$h'(\mathbf{z}:\xi\mathbf{e}_j) = \sum_{k\in\sigma[j]^c} \frac{(g_k - f_k)\theta_{kj} - f_k}{g_k^2}\xi + \sum_{k\in\sigma[j]} \frac{(g_k - f_k) - \xi f_k}{g_k^2} < 0 \qquad (5.8)$$

This condition can always be applied to check for a downhill directon or local optimality. However, if $|\sigma[j]| = 1$, $j = 1, 2, \ldots, p$, then the calculation can be simplified slightly. Set

$$h_j' = \sum_{k\in\sigma[j]^c} \frac{(g_k - f_k)\theta_{kj} - f_k}{g_k^2}, \qquad \{l\} = \sigma[j],$$

and compute λ_j by

$$\lambda_j \frac{g_l - f_l}{g_l^2} - \frac{f_l}{g_l} + h_j' = 0. \qquad (5.9)$$

Lemma 5.3 sgn $(\lambda_j)\mathbf{e}_j$ *is a descent direction at* $\mathbf{z}$ *if and only if* $|\lambda_j| > 1$.

Proof The two cases $\xi = \pm 1$ are computed separately.

(i)
$$\begin{aligned} h'(\mathbf{z}:\mathbf{e}_j) &= h_j' - \frac{f_l}{g_l^2} + \lambda_j \frac{g_l - f_l}{g_l^2} + (1-\lambda_j)\frac{g_l - f_l}{g_l^2} \\ &= (1-\lambda_j)\frac{g_l - f_l}{g_l^2} \end{aligned}$$

$$\begin{cases} \geq 0 & \text{if } |\lambda_j| \leq 1 \\ < 0, & \lambda_j > 1. \end{cases}$$

(ii)
$$\begin{aligned} h'(\mathbf{z}:-\mathbf{e}_j) &= -h_j' + \frac{g_l}{g_l^2} \\ &= -h_j' + \frac{f_l}{g_l^2} - \lambda_j \frac{g_l - f_l}{g_l^2} + (1+\lambda_j)\frac{g_l - f_l}{g_l^2} \\ &= (1+\lambda_j)\frac{g_l - f_l}{g_l^2} \end{aligned}$$

$$\begin{cases} \geq 0 & \text{if } |\lambda_j| \leq 1 \\ < 0, & \lambda_j < -1. \end{cases}$$

Remark 5.1 Because the searching is done in the coordinate directions, only single components of $\mathbf{z}, \mathbf{z} - \mathbf{a}_i$ change. This simplifies updating f_i, g_i in performing the computations necessary in (5.7) or (5.8) to select a descent direction, and this is where the bulk of the work lies. Also, the analogue of a line search is possible, for at the adjacent point $\bar{\mathbf{z}}$ it may be the case that $\mathbf{e}_j$ is still downhill and further progress can be made without the need to search over the remaining coordinate vectors.

Example 5.1 Let the $\mathbf{a}_i$ be the columns of the 4×8 matrix

$$\begin{bmatrix} 1 & 1 & 3 & 4 & 2 & 6 & 5 & 7 \\ 5 & 4 & 7 & 7 & 1 & 4 & 1 & 7 \\ 3 & 6 & 11 & 2 & 5 & 1 & 1 & 5 \\ 7 & 5 & 2 & 5 & 8 & 2 & 9 & 3 \end{bmatrix}$$

Then $h(\mathbf{z})$ has at least two local minima corresponding to

$$\mathbf{z}^{\mathrm{T}}=[4,5,5,7] \quad \text{and} \quad \mathbf{z}^{\mathrm{T}}=[4,7,5,5].$$

This shows that a unique minimum of (5.1) cannot be expected. The l_1 centre is

$$\mathbf{z}^{\mathrm{T}}=((3 \text{ or } 4),(4 \text{ or } 5),(3 \text{ or } 5),5).$$

In Figure 5.1 a graph of $h(\mathbf{z})$ is given for $\mathbf{z}^{\mathrm{T}}=[4,4,5,t]$, $2\leqslant t\leqslant 9$. This suggest the two local minima, and the graph is very flat (little discriminatory power) for $5\leqslant t\leqslant 7$.

Exercise 5.1 (i) Show that if $\mathbf{z}>0$ is a local minimum then there exist $\mathbf{y}_i^{\mathrm{T}}\in\partial f_i(\mathbf{z})$ such that

$$\sum_{i=1}^{n}\frac{(g_i-f_i)\mathbf{y}_i-f_i\mathbf{e}}{g_i^2}=0.$$

How must the equation be modified if it is only true that $\mathbf{z}\geqslant 0$?

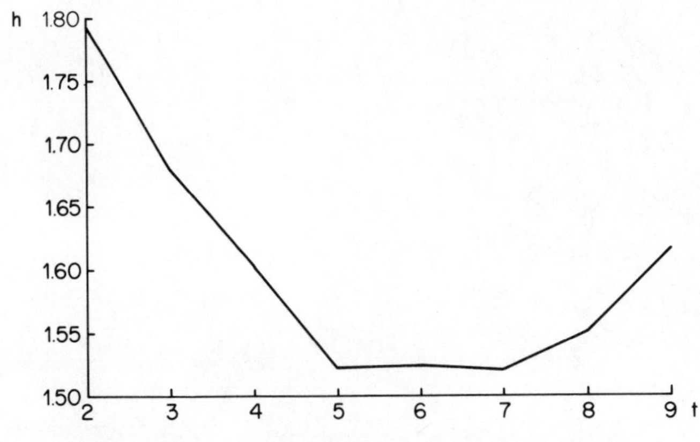

Figure 5.1 Graph of $h(4,4,t,t)$, $2\leqslant t\leqslant 9$

(ii) If $|\lambda_j| \leq 1$, $j = 1, 2, \ldots, p$, then the above equation is satisfied by

$$(\mathbf{y}_i)_k = \begin{cases} \theta_{ik}, & i \notin \sigma[k], \\ \lambda_k, & i \in \sigma[k], \qquad k = 1, 2, \ldots, p. \end{cases}$$

for $i = 1, 2, \ldots, n$.

NOTES ON CHAPTER 6

The systematic treatment of nondifferentiable non-convex problems is comparatively recent and owes a lot to the theory of generalized derivatives developed by Clarke (1975) (see also Clarke (1983)). This theory is not needed here as the convex origins of the problems imply sufficient regularity, but concepts such as the generalized Jacobian emerge. The implicit function theorem follows Jittorntrum (1978) and has similarities with Clarke (1976), and the derivation of the first-order necessary conditions follows Hiriart-Urruty (1979). The treatment of the second-order conditions is standard (for example Hestenes (1966)). The problem in Exercise 2.1(ii) is taken from Powell (1984). A finite algorithm is given in Womersely (1985). The term 'total least squares' is introduced in Golub and Van Loan (1980). Prior to this important paper, this particular application had appeared under many different names in a wide range of contexts. The application to operator norms is given in Watson (1983). The treatment here follows Osborne and Watson (1985), who introduce the concept of separability and develop the finite algorithm in the l_1 case. The convergence result in Theorem 3.4 is given in Watson (1984). Exercise 4.4 is suggested by Späth (1982). Related problems are considered in Watson (1982a). The discussion of the algorithm for finding a centre in the Jaccard metric is due to Watson (1982b).

CHAPTER 7

Some questions of complexity and performance

7.1 INTRODUCTION

In this chapter certain questions which pertain to the effectiveness of the class of methods suggested for solving linear programming problems and minimizing polyhedral convex functions are considered. The aim is to provide more specific information than can be conveyed by a finiteness proof (a finite bound can be totally unacceptable when interpreted in terms of available resources), and to attempt to suggest what the expected performance measured in terms of computing cost might be in certain circumstances.

Remark 1.1 The estimates given in earlier chapters for the work involved in updating the tableau show that the expected computing cost Σ will have the general form

$$\Sigma \sim f(n, p)np, \qquad \frac{p}{n} \to 0, \tag{1.1}$$

where np is the number of elements in the data matrix, and reflects the property that in each iteration each tableau element enters into at most a small fixed number of operations; and $f(n, p)$ is, up to a method-dependent constant multiple, the expected number of iterations.

Thus to provide complexity estimates it is necessary to be able to estimate $f(n, p)$. It is clear that $f(n, p)$ grows at least as fast as p which is the usual number of descent steps required to establish a first extreme point. Also, there are really two interrelated questions involved. These are:

(i) Can we give problem classes for which $f(n, p)$ can be estimated?

and

(ii) For what range of values of $f(n, p)$ do the class of descent algorithms lead to practical computing algorithms?

For present purposes three cases are distinguished on the basis of the apparent or likely behaviour of $f(n, p)$. To do this we need a concept of slow growth, and by this we will mean that the observed (or likely, or actual) growth rate is less rapid than any positive power of n. The three cases are:

(i) $\dfrac{f(n, p)}{p}$ a function of slow growth (with n);

(ii) $\dfrac{f(n, p)}{n}$ a function of slow growth; and

(iii) $\dfrac{f(n, p)}{np}$ a function of rapid growth.

In the first case the complexity estimate obtained is broadly in line with that for such precisely specified algorithms as the methods for the linear least squares problem discussed in Section 5.2. This is a very satisfactory result; and there is a generally held opinion that this case describes the usual behaviour of the simplex algorithm. This point is explored further in Section 7.4. An example of the second kind of behaviour is provided by the problem considered in Example 5.1 of Chapter 3. Here the dependence of the number of iterations on n comes about because most iterations correspond to moves to an immediately adjacent configuration, with the result that it takes $|j-i|$ steps to replace $\mathbf{m}_i$ by $\mathbf{m}_j$ in one column of B_μ. That this example was given to emphasize the point that a better method is available for this problem perhaps suggests that problems with this second class of estimates are marginal for the descent algorithms and that an alternative approach should be sought. It is interesting that in the example under consideration the appropriate trick is just to consider a dual formulation.

In the third case, which corresponds to even faster growth with n, the descent algorithms rapidly become impractical, and there exist problems for which particular implementations in the set of possible descent algorithms (the set includes the particular implementations of the simplex method) have $f(n, p)$ of fast growth. The particular case in which $f(n, p) = \gamma^p$, $\gamma > 1$ when n and p are increased systematically corresponds to the case of exponential complexity, and in the next section examples with this complexity are considered. These examples are special not only to the particular descent algorithm but even to the implementation details. This raises the question: is the worst-case complexity for the problem class exponential in the sense that examples with exponential complexity will exist for whatever kind of algorithm might conceivably be developed? It is a very recent and significant result that the answer to this question is 'no'. It was settled by the explicit derivation of an algorithm with polynomial worst-case complexity for linear programs with integer coefficients.* A brief sketch of this material is given in Section 7.3. We note here, but do not pursue further, that examples of

* The actual requirement is that the number of digits needed to code or store the problem be finite. Some such condition is necessary for any realization of an algorithm by computer and in this sense is not a restriction. However, the condition is also necessary in the sense that the result is not true in the hypothetical case of exact real arithmetic.

exponential worst-case behaviour are known for algorithms which fall within the general class of methods developed here for quadratic programming and for the linear complementarity problem.

There is an alternative aspect to the question of finding the average number of iterations needed to solve problems within particular problem classes, and this is the activity of generating representative examples of the problem class. Ability to do this is needed not only to permit the gathering of statistics on the number of iterations in using Monte Carlo methods to estimate expected behaviour, but also in order to carry out the tests needed to be able to decide between the possible implementation choices such as whether to use USE or NSE tests, whether to relax off active constraints with infeasible multipliers when a downhill direction is already available, and so on. This question is considered in Section 7.4, and applications to the problems of deciding between various implementation strategies are considered in Section 7.5.

7.2 WORST-CASE BEHAVIOUR OF THE BASIC ALGORITHMS

The feature in common to all members of the class of descent algorithms for linear programming and for minimizing polyhedral convex functions is that after an initial phase, consideration is restricted to points which satisfy a condition of the form $\mathbf{g} \in H$ (compare (5.8), (6.7) of Chapter 2, (2.10) of Chapter 3, and (4.1) of Chapter 4 for example). It is a convenient shorthand to refer to such points as extreme points, and the number of such points provides an obvious upper bound for the complexity of any descent algorithm. In the particular case of LP1, a bound for the number of extreme points is $\binom{n}{p} = O(n^p)$ as $n \to \infty$ and is clearly exponential in p. However, it can be a gross overestimate. For example, if $p = 2$ then each edge of the feasible region can contain at most two extreme points so the total number N_v must satisfy the inequalities valid for $n \geq 4$

$$N_v \leq n < \binom{n}{2} = \tfrac{1}{2}(n^2 - n).$$

But polyhedral sets with exponential numbers of extreme points are easy to construct, and the unit cube in p dimensions provides an interesting example. In this case we have the $2p$ constraint inequalities

$$x_j \geq 0, \qquad -x_j \geq -1, \qquad j = 1, 2, \ldots, p,$$

and there are 2^p extreme points corresponding to the 2^p different ways of putting 0 or 1 into each of p locations.

Having shown that the number of extreme points can be exponential, the next question is: how accessible are these points under a particular implementation of a descent algorithm and choice of starting point? Here an interesting example is obtained by perturbing the unit cube in p dimensions

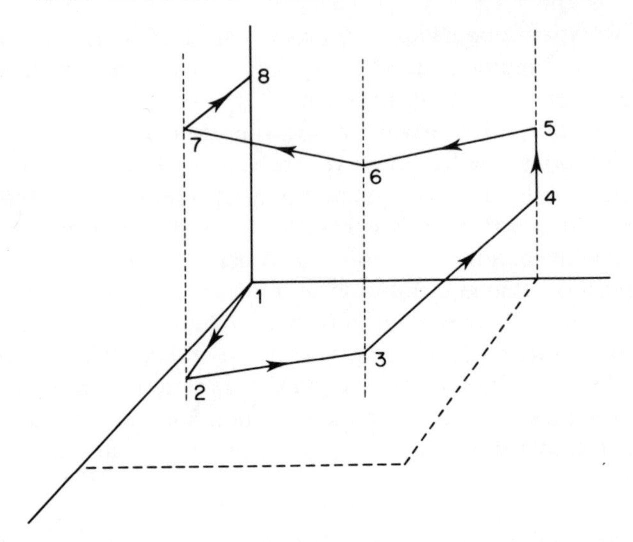

Figure 2.1 Path on which x_3 increases and which visits every vertex of the perturbed cube in R^3

in such a way that it is possible to visit all vertices by a path on which one coordinate is always increasing. This is illustrated in Figure 2.1 for $p=3$. The distorted figure can be generated by rotating faces $(1, 2, 7, 8)$ about $(1, 8)$ and $(3, 4, 5, 6)$ about $(4, 5)$ and then faces $(5, 6, 7, 8)$ about $(7, 8)$ and $(1, 2, 3, 4)$ about $(1, 2)$, the numbering corresponding to faces and edges of the unperturbed figure.

If the vertices are labelled so that the labelling increases on the increasing path, then it is possible to describe an algorithm following this trajectory by characterizing the constraints by the vertices they contain and giving explicitly the constraint deleted at each step. This is done in Table 2.1, which

Table 2.1 Patterns of deletions generating the path in Figure 2.1. Note that as the problem is to maximize x_3, it is the constraints with positive multipliers that are candidates for deletion.

Vertex	Constraint deleted	Remaining constraints and multipliers
1	(1, 4, 5, 8)	(1, 2, 3, 4) (+), (1, 2, 7, 8) (+)
2	(1, 2, 7, 8)	(1, 2, 3, 4) (+), (2, 3, 6, 7) (−)
3	(2, 3, 6, 7)	(1, 2, 3, 4) (+), (3, 4, 5, 6) (−)
4	(1, 2, 3, 4)	(3, 4, 5, 6) (−), (1, 4, 5, 8) (−)
5	(1, 4, 5, 8)	(3, 4, 5, 6) (+), (5, 6, 7, 8) (−)
6	(3, 4, 5, 6)	(5, 6, 7, 8) (−), (2, 3, 6, 7) (−)
7	(2, 3, 6, 7)	(5, 6, 7, 8) (−), (1, 2, 7, 8) (−)

also gives the signs of the multipliers in the relation $\mathbf{g} \in H$ at each extreme point to indicate which constraints can be relaxed to give a trajectory on which x_3 increases.
Now, by (5.45), (5.46) of Chapter 2, for example, the active constraints at each vertex can be scaled so that the USE test selects the corresponding edge in Figure 2.1. This scaling is easily seen to be consistent. The most difficult case corresponds to the constraints at vertex 1. Here it is necessary to scale $(1, 4, 5, 8)$ relative to $(1, 2, 3, 4)$ and $(1, 2, 7, 8)$ so that edge $(1, 2)$ is preferred to edges $(1, 4)$ and $(1, 8)$, and then to scale $(1, 2, 7, 8)$ relative to $(1, 2, 3, 4)$ so that edge $(2, 3)$ is preferred to edge $(2, 7)$. Constraint $(1, 2, 3, 4)$ is relaxed at vertex 4, and here the edge $(4, 5)$ is selected unambiguously. The only other point of interest is at vertex 5, where constraint $(1, 4, 5, 8)$ is deleted for the second time. The scale of this constraint was specified at vertex 1, but that of $(3, 4, 5, 6)$ which is the only competitor is still disposable and can be adjusted so that the edge $(5, 6)$ is selected.

It is not too difficult to extend the above argument from the deformed cube in R^k to the corresponding problem in $R^{k+1} = R^k \times [0, 1]$ (the case $k = 4$ is suggested in Exercise 2.1). The conclusion is that there are problems with $2p$ constraints in R^p which have solution trajectories passing through 2^p vertices. It follows that the worst-case behaviour of the descent algorithms for LP1 under USE tests is exponential.

It is the corresponding question concerning the complexity of the simplex algorithm which has been most studied, and a similar result is available. Consider the problem defined by

$$A^{\mathrm{T}} = [A_1 \mid I]$$

where

$$(A_1)_{ij} = \begin{cases} 2 * 10^{i-j}, & j < i \\ 1, & j = i \\ 0, & j > i, \end{cases} \qquad i = 1, 2, \ldots, p,$$

$$\mathbf{b}^{\mathrm{T}} = [\mathbf{b}_1^{\mathrm{T}} \mid 0]$$

where

$$b_i = 10^{p-i}, \qquad i = 1, 2, \ldots, p,$$

and

$$c_i = 10^{i-1}, \qquad i = 1, 2, \ldots, p.$$

This problem is worked in detail in Example 4.1 of Chapter 2 for $p = 3$. It can be shown that the pattern of behaviour revealed there holds in general and that the number of steps in the simplex algorithm with a USE edge selection criterion is $O(2^p)$. Thus, this example demonstrates worst-case exponential behaviour of the simplex algorithm. The extent to which this worst-case result depends on the USE test has been investigated also, and examples of exponential behaviour have been constructed both for the NSE test, and for the edge-selection rule in which the actual increase in the objective function is maximized at each step.

Exercise 2.1 Discuss the deformation of the unit cube in R^4 to provide a trajectory made up of edges which visits every vertex and on which x_4 is increasing. Show that the scaling of the constraint equations can be chosen so that this trajectory is realized.

Exercise 2.2 Use the simplex method to solve the linear programming problem given in this section when $p=4$ and using the obvious starting point (the aim is to determine the sequence of steps with a minimum of work).

7.3 THE ELLIPSOID METHOD

The discovery that the worst-case complexity of the simplex method increased exponentially with the dimension lead immediately to the question: is such behaviour characteristic of all possible algorithms for the linear programming problem? This question was answered in the negative in 1979 when Khachiyan proved that his ellipsoid method is polynomial (that is, there is a fixed α such that the complexity is dominated by n^{α}). Thus far the ellipsoid method has proved unsatisfactory for explicit computations.

The basic problem addressed by the ellipsoid method is that of finding a feasible solution to the system of strict inequalities (3.1) or demonstrating that no solution is possible. Let $A: R^p \to R^n$, $\mathbf{c} \in R^p$ have *integer* components. The system considered is

$$A^T\mathbf{u} < \mathbf{c}, \qquad \mathbf{u} > 0, \tag{3.1}$$

and it is readily put into the form associated with the simplex algorithm by adding slack variables. The restriction to bounded integer components is essential to the subsequent argument. It means that the constraint hyperplanes are either parallel or intersect at angles which cannot be vanishingly small. The discussion given here assumes exact arithmetic. It can be modified to take account of finite precision. The first step is to determine bounds for the magnitude of extreme points of (3.1) and for the volume of the feasible region. These bounds are given in terms of the quantity L defined by

$$L = \sum_{i=1}^{n} \sum_{j=1}^{p} \log_2 (|A_{ij}|+1) + \sum_{i=1}^{n} \log_2 (|c_i|+1) + \log_2 np + 2 \tag{3.2}$$

which gives for each particular instance of (3.1) the length of the corresponding binary encoding of the data.

Lemma 3.1 *Every vertex of* $U = \{\mathbf{u} : A^T\mathbf{u} \leqslant \mathbf{c}, \mathbf{u} \geqslant 0\}$ *satisfies* $\|\mathbf{u}\|_\infty \leqslant 2^L/np$. *The components of* $\mathbf{u}$ *are rational numbers with denominators less than* $2^L/np$.

Proof Let σ be an index set pointing to the active constraints at a particular vertex. Then Cramer's rule gives

$$u_j = \frac{\det(B_\sigma - (\boldsymbol{\kappa}_j(B_\sigma) - \mathbf{c})\mathbf{e}_j^T)}{\det(B_\sigma)} \tag{3.3}$$

Now $\det(B_\sigma) \geq 1$ as it is necessarily an integer. Applying Hadamard's inequality to the numerator (and writing this as $\det(D_j)$ for convenience) gives

$$\begin{aligned} |\det(D_j)| &\leq \prod_{i=1}^{p} \|\boldsymbol{\kappa}_i(D_j)\|_2 \\ &\leq \prod_{i=1}^{p} \prod_{m=1}^{p} (|(D_j)_{im}| + 1) \\ &\leq 2^{L - \log_2 np - 2} \leq 2^L/np \end{aligned} \tag{3.4}$$

Here the second inequality uses a crude upper bound while the third notes that the quantities in brackets constitute a subset of the problem data. The same argument serves to bound $|\det(B_\sigma)|$ which is the denominator in u_j in unreduced form.

A further application of the argument gives the following result.

Corollary 3.1 *Every vertex of the polytope* $\hat{U} = U \cap \left\{\mathbf{u} < \left\lfloor \frac{2^L}{n} \right\rfloor \mathbf{e}\right\}$ *has coordinates which are rational numbers with denominators that are less than* $2^L/np$ *in absolute value.*

Lemma 3.2 *If (3.1) has a solution then the volume of its solution space inside the sphere* $\|\mathbf{u}\|_2 < 2^L$ *is at least* $2^{-(n+1)L}$.

Proof The solution set of (3.1) is restricted to the positive orthant and as a consequence contains no lines. It follows that if the solution set is nonempty then it contains a vertex $\mathbf{u}$ with $\|\mathbf{u}\|_\infty < \left\lfloor \frac{2^L}{n} \right\rfloor$. In this case $\hat{U}$ has a nontrivial interior. As it is bounded it must have $n+1$ vertices $\mathbf{u}_1, \mathbf{u}_2, \ldots, \mathbf{u}_{n+1}$ not on a hyperplane and contained within the indicated sphere. Let V be the volume of this simplex. Then, as $\hat{U} \subseteq U$, it provides the required lower bound. Now

$$V = \frac{1}{n!} \left| \det \begin{bmatrix} 1 & & 1 \\ \mathbf{u}_1 & \ldots & \mathbf{u}_{n+1} \end{bmatrix} \right|$$

For each i, $i = 1, 2, \ldots, n+1$, let γ_i be an integer such that $\gamma_i \mathbf{u}_i = \mathbf{v}_i$ has integral components. By Lemma 3.1, $\gamma_i < \lfloor 2^L/n \rfloor$. Thus

$$V = \frac{1}{n!} \frac{1}{\gamma_1 \gamma_2 \ldots \gamma_{n+1}} \left| \det \begin{bmatrix} \gamma_1 \cdots \gamma_{n+1} \\ \mathbf{v}_1 \quad \mathbf{v}_{n+1} \end{bmatrix} \right| \geqslant \frac{1}{n!} \left(\frac{2^L}{n} \right)^{-(n+1)}$$
$$\leqslant 2^{-(n+1)L}. \tag{3.5}$$

We are now in a position to describe the ellipsoid method. By Lemma 3.1, an n-sphere of radius 2^L is large enough to contain all extreme points. The ellipsoid method now constructs a sequence of ellipsoids of strictly decreasing volume, each of which contains the set of feasible points in the original sphere, until either the centre of the current ellipsoid is feasible or until its volume is reduced below $2^{-(n+1)L}$. In the latter case an appeal to Lemma 3.2 shows that the feasible region is empty. The construction is indicated in Figure 3.1.

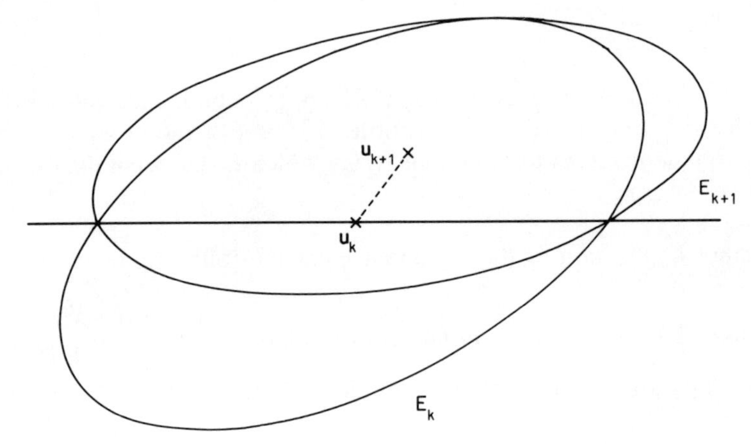

Figure 3.1 The basic step of the Khachiyan algorithm

Let the current ellipsoid E_k with centre $\mathbf{u}_k$ be

$$E_k = \{\mathbf{u};\, (\mathbf{u}-\mathbf{u}_k)^{\mathrm{T}} G_k^{-1} (\mathbf{u}-\mathbf{u}_k) \leqslant 1\}. \tag{3.6}$$

Then the steps of the algorithm are:

(i) find the most violated constraint at $\mathbf{u}_k$ – say

$$\mathbf{a}^{\mathrm{T}}\mathbf{u}_k > c;$$

and

(ii) set

$$\mathbf{u}_{k+1} = \mathbf{u}_k - \tau\{G_k\mathbf{a}/(\mathbf{a}^{\mathrm{T}}G_k\mathbf{a})^{1/2}\}, \tag{3.7}$$
$$G_{k+1} = \delta\{G_k - \rho(G_k\mathbf{a}\mathbf{a}^{\mathrm{T}}G_k)/\mathbf{a}^{\mathrm{T}}G_k\mathbf{a}\} \tag{3.8}$$

where

$$\tau = 1/(n+1), \qquad \rho = 2/(n+1), \quad \text{and} \quad \delta = n^2/(n^2-1).$$

To verify that the second step has the properties claimed, it is convenient to make an affine transformation taking E_k to the unit sphere centred at the origin. If T is the required transformation then

$$\mathbf{u}-\mathbf{u}_k \rightarrow T\mathbf{v} \tag{3.9}$$

and constraints are preserved provided

$$\mathbf{a} \rightarrow T^{-\mathrm{T}}\bar{\mathbf{a}} \tag{3.10}$$

Under this transformation

$$E_k = \{\mathbf{v};\, \mathbf{v}^{\mathrm{T}} T^{\mathrm{T}} G_k^{-1} T\mathbf{v} \leqslant 1\}$$

so that T must be chosen to satisfy

$$T^{\mathrm{T}} G_k^{-1} T = I \tag{3.11}$$

Also,

$$T^{-1} G_k \mathbf{a} = T^{-1} G_k T^{-\mathrm{T}} T^{\mathrm{T}} \mathbf{a} = \bar{\mathbf{a}} \tag{3.12}$$

so that

$$T^{-1} G_{k+1} T^{-\mathrm{T}} = \delta\left\{I - \rho \frac{\bar{\mathbf{a}}\bar{\mathbf{a}}^{\mathrm{T}}}{\bar{\mathbf{a}}^{\mathrm{T}}\bar{\mathbf{a}}}\right\}, \tag{3.13}$$

and

$$\mathbf{v}_{k+1} = -\tau\bar{\mathbf{a}}/\{\bar{\mathbf{a}}^{\mathrm{T}}\bar{\mathbf{a}}\}^{1/2}. \tag{3.14}$$

In particular, there is no restriction in assuming $\|\bar{\mathbf{a}}\|_2 = 1$.

We show that E_{k+1} contains the feasible region.

Lemma 3.3 $E_k \cap \{\mathbf{v}^{\mathrm{T}}\bar{\mathbf{a}} < 0\} \subset E_{k+1}$.

Proof The most important points to check lie on the surface of E_k, and these have the general form

$$\mathbf{v} = -\alpha\bar{\mathbf{a}} + \beta\mathbf{t}$$

where

$$\bar{\mathbf{a}}^{\mathrm{T}}\mathbf{t} = 0, \qquad \alpha^2 + \beta^2 = 1, \qquad \alpha \geqslant 0, \qquad \beta \geqslant 0.$$

Let

$$\begin{aligned} S_{k+1} &= (\mathbf{u}-\mathbf{u}_{k+1})^{\mathrm{T}} G_{k+1}^{-1} (\mathbf{u}-\mathbf{u}_{k+1}) \\ &= \delta^{-1}\left\{\|\mathbf{v}-\mathbf{v}_{k+1}\|_2^2 + \frac{\rho}{1-\rho}(\bar{\mathbf{a}}^{\mathrm{T}}(\mathbf{v}-\mathbf{v}_{k+1}))^2\right\} \\ &= \delta^{-1}\left((\alpha-\tau)^2 + \beta^2 + \frac{\rho}{1-\rho}(\alpha-\tau)^2\right) \end{aligned} \tag{3.15}$$

The problem reduces to maximizing S_{k+1} subject to the constraints on α, β. The necessary conditions give

$$2\delta^{-1}(1+\rho/(1-\rho))(\alpha-\tau) = 2\mu\alpha - \lambda,$$
$$2\delta^{-1}\beta = 2\mu\beta - \nu,$$

where μ, λ, ν are the multipliers. If $\alpha > 0$, $\beta > 0$ then $\lambda = \nu = 0$. In this case it is straightforward to calculate the stationary value and it gives $S_{k+1} < 1$. Alternatively, either $\alpha = 0$ or $\beta = 0$. We have

$$\alpha = 0 \Rightarrow S_{k+1} = \delta^{-1}\left(1 + \frac{\tau^2}{1-\rho}\right) = 1,$$

and

$$\beta = 0 \Rightarrow S_{k+1} = \frac{\delta^{-1}}{1-\rho}(1-\tau)^2 = 1.$$

Thus E_{k+1} contains the intersection of E_k with $\bar{\mathbf{a}}^T\mathbf{v} = 0$ and touches E_k at $\mathbf{v} = -\bar{\mathbf{a}}$ (in particular, we verify the correctness of Figure 3.1).

The next result shows that E_{k+1} has a smaller volume than E_k.

Lemma 3.4 *Let V_k denote the volume of E_k. Then*

$$V_{k+1} = \Gamma(n) V_k \tag{3.16}$$

where

$$\Gamma(n) = \frac{n}{n+1}\left(\frac{n^2}{n^2-1}\right)^{(n-1/2)} < 2^{-1/2(n+1)} \tag{3.17}$$

Proof To evaluate V_{k+1} it is convenient to make the transformation $\mathbf{v} - \mathbf{v}_{k+1} = L\mathbf{t}$, where $LL^T = \delta(I - \rho\bar{\mathbf{a}}\bar{\mathbf{a}}^T)$. Then

$$V_{k+1} = \det L V_k \tag{3.18}$$

and

$$\begin{aligned} \det L &= \det\,(\delta(I - \rho\bar{\mathbf{a}}\bar{\mathbf{a}}^T))^{1/2} \\ &= \delta^{n/2}(1-\rho)^{1/2}. \end{aligned} \tag{3.19}$$

The result now follows on substituting the values of δ and ρ. To obtain the inequality for $\Gamma(n)$, note that

$$\frac{n}{n+1} = 1 - \frac{1}{n+1} < e^{-1/(n+1)}, \qquad \frac{n^2}{n^2-1} = 1 + \frac{1}{n^2-1} < e^{1/(n^2-1)}.$$

This estimate of $\Gamma(n)$ makes it possible to give a bound for the number of steps of the ellipsoid method necessary to reduce V_k below the lower bound for the volume of the feasible region derived in Lemma 3.2.

Lemma 3.1 *In at most $4(n+1)^2L$ steps the ellipsoid method either returns a feasible point of (3.1) or shows that the feasible region is empty.*

Proof It follows from Lemma 3.4 that after k steps,

$$V_k < 2^{-k/2(n+1)} V_0 < 2^{-k/2(n+1)} 2^{nL+n} \tag{3.20}$$

where V_0 is bounded by the cube with side 2×2^L centred at 0. Thus an upper bound for k is obtained by satisfying the inequality (note $L>n$)

$$2^{-k/2(n+1)}2^{(n+1)L}<2^{-(n+1)L}.$$

To apply this result to the linear programming problem it is convenient to start with the symmetric formulation (Exercise 2.2 of Chapter 2). This shows that the primal and dual problems have a solution if and only if there exists a feasible point of the system of inequalities

$$\begin{aligned}\mathbf{c}^{\mathrm{T}}\mathbf{x}&\geqslant \mathbf{b}^{\mathrm{T}}\mathbf{u},\\ A\mathbf{x}&\geqslant \mathbf{b},\\ \mathbf{x}&\geqslant 0,\\ A^{\mathrm{T}}\mathbf{u}&\leqslant \mathbf{c},\\ \mathbf{u}&\geqslant 0.\end{aligned} \tag{3.21}$$

This system is not in the form of strict inequalities required by the ellipsoid method. This problem can be circumvented, but again the essentially integral nature of the original constraints is required.

Theorem 3.2 *The system of linear inequalities*

$$\mathbf{a}_i^{\mathrm{T}}\mathbf{u}\leqslant c_i, \qquad i=1,2,\ldots,p \tag{3.22}$$

has a solution if and only if the system of strict linear inequalities

$$\mathbf{a}_i^{\mathrm{T}}\mathbf{u}<c_i+2^{-L}, \qquad i=1,2,\ldots,p \tag{3.23}$$

has a solution.

Remark 3.1 This result is proved using an intermediate result which is derived by an argument which is essentially similar to that employed in the proof of Theorem 3.1 of Chapter 2. Let $r_i(\mathbf{u})=\mathbf{a}_i^{\mathrm{T}}\mathbf{u}-c_i$ and $\mathbf{u}_0$ arbitrary. Then there exists $\mathbf{u}_1$ satisfying

(i) $r_i(\mathbf{u}_1)\leqslant \max(0, r_i(\mathbf{u}_0)), \qquad i=1,2,\ldots,p,$ and

(ii) $\mathbf{a}_j\in \operatorname{span}(\mathbf{a}_i;\, r_i(\mathbf{u}_1)\geqslant 0), \qquad j=1,2,\ldots,p.$

Proof Let $\mathbf{u}_0$ satisfy (3.23) so that $r_i(\mathbf{u}_0)<2^{-L}$. It is necessary to construct a solution to (3.22). By the above remark, there is no restriction in assuming that

$$\nu=\{i;\, r_i(\mathbf{u}_0)\geqslant 0\}=\nu_1\cup\nu_2 \tag{3.24}$$

where ν_1 points to a basis for the $\mathbf{a}_i$. Let $\mathbf{z}$ be any solution of the system

$$\mathbf{a}_i^{\mathrm{T}}\mathbf{z}=c_i, \qquad i\in\nu_1. \tag{3.25}$$

It will be shown that $\mathbf{z}$ satisfies (3.22). By Cramer's rule applied to a nonsingular submatrix of the $\mathbf{a}_i$, $i \in \nu$, we obtain the representation

$$\mathbf{a}_j = \sum_{i \in \nu_1} \frac{D_{ij}}{D} \mathbf{a}_i, \qquad j = 1, 2, \ldots, p, \tag{3.26}$$

where, by Lemma 3.1, D_{ij}, D are integers $<2^L/n\,|\nu_1|$ in magnitude, and D can be taken positive. Now

$$\begin{aligned}\mathbf{a}_j^{\mathrm{T}}\mathbf{z} - c_j &= \sum_{i \in \nu_1} \frac{D_{ij}}{D}(\mathbf{a}_i^{\mathrm{T}}\mathbf{z} - c_i) + \sum_{i \in \nu_1} \frac{D_{ij}}{D} c_i - c_j \\ &= \sum_{i \in \nu_1} \frac{D_{ij}}{D}(\mathbf{a}_i^{\mathrm{T}}\mathbf{u}_0 - r_i(\mathbf{u}_0)) - (\mathbf{a}_j^{\mathrm{T}}\mathbf{u}_0 - r_j(\mathbf{u}_0)) \\ &= r_j(\mathbf{u}_0) - \sum_{i \in \nu_1} \frac{D_{ij}}{D} r_i(\mathbf{u}_0) \end{aligned} \tag{3.27}$$

Thus

$$D(\mathbf{a}_j^{\mathrm{T}}\mathbf{z} - c_j) \leq \frac{2^L}{n\,|\nu_1|} 2^{-L} + |\nu_1| \frac{2^L}{n\,|\nu_1|} 2^{-L} < 1 \tag{3.28}$$

But the left-hand side is an integer (it is equal to $\sum_{\nu_1} D_{ij}c_i - Dc_j$), so that

$$\mathbf{a}_j^{\mathrm{T}}\mathbf{z} - \mathbf{c}_j \leq 0, \qquad j = 1, 2, \ldots, p.$$

Thus $\mathbf{z}$ satisfies (3.22). As any $\mathbf{u}$ satisfying (3.22) satisfies (3.23), the reverse implication is trivial.

Remark 3.2 Finding $\mathbf{z}$ given $\mathbf{u}_0$ requires constructing ν containing ν_1 pointing to a basis for the $\mathbf{a}_i$. The basic operation in this, as in the subsequent computation of $\mathbf{z}$, is the solution of linear equations. It is a simple consequence that the complexity of the steps which go from the solution produced by the ellipsoid method to the solution of the linear program is polynomial in n.

Remark 3.3 In the above discussion of the ellipsoid method, no effort has been made either to refine the inequalities used or to examine the possibility of making more effective cuts in order to reduce the size of the set containing feasible points more rapidly, but certainly something can be done on both counts. However, the volume of the solution space of (3.21) is zero so the volume of the solution set of the perturbed problem is going to be small, and there does not seem to be convincing evidence that the order of the estimate of $4(n+1)^2L$ is much reduced by these refinements of the method. While there may be more chance of reducing the estimate by using problem information to reduce L, it seems that this approach to the linear programming problem is going to compare badly with the descent methods in any case in which $f(n, p)/p$ is of slow growth.

This suggests that it could be more favourable to look for alternative ways of attacking the linear programming problem. One possibility is as follows. Assume that upper and lower bounds to the optimum (say ζ_u, ζ_l) are known. Then the feasibility of the system

$$A^T\mathbf{u} < \mathbf{c}, \qquad \mathbf{u} > 0, \qquad \mathbf{b}^T\mathbf{u} > (\zeta_l + \zeta_u)/2$$

can be tested. If this system is feasible then ζ_l can be updated, otherwise ζ_u, and this permits the ellipsoid method to be used in conjunction with a bisection process to successively refine estimates of the optimum. By this means an estimate of the optimum to within any specified tolerance ε can be found by a process which is polynomial in p, n, and $\log_2 1/\varepsilon$. To determine the bounds ζ_l, ζ_u two cases are relevant:

(i) If the linear program is known to be bounded then the ellipsoid method can be used to compute a feasible point. If this is found at the kth step, then suitable values are

$$\zeta_l = \mathbf{b}^T\mathbf{u}_k, \qquad \zeta_u = \mathbf{b}^T\mathbf{u}_k + (\mathbf{b}^T G_k \mathbf{b})^{1/2}.$$

The second bound is found by considering the points on E_k where the normal to the tangent plane is parallel to $\mathbf{b}$.

(ii) If the linear program is not known to be bounded a priori, then a suitable upper bound can be found by applying the ellipsoid method to the dual problem.

This process has the disadvantage that the system of inequalities considered can be infeasible, causing the ellipsoid method to run its full number of steps; but it can be modified to overcome this. Also, the integral nature of the data can be exploited in this case to obtain an exact solution from a sufficiently good approximate solution in a polynomially bounded number of steps.

Remark 3.4 There is nothing particularly special about the form of (3.1). It is just that the ellipsoid method has been developed with the problems solved by the simplex algorithm in mind. There is no trouble in adjusting it to solve LP1 subject to the proviso that the feasible region contain no lines (Exercise 3.1).

The ellipsoid method can be extended to show that the minimum of a polyhedral convex function can be found in polynomial time. This does not follow directly from the formulation of these problems as linear programs, because the number of constraints can be exponentially large with n, so that L is also exponentially large. However, this problem is readily circumvented by noting that L is never used except in the context of bounding extreme points, and all that is necessary is to use problem structure to obtain suitable inequalities.

Example 3.1 In the l_1 problem the constraint set has the form (1.22) of Chapter 3:

$$P\{M\mathbf{x}-\mathbf{f}\} \leq h\mathbf{e}^{(N)}, \qquad h \geq 0,$$

where the rows of P correspond to the $N=2^n$ different ways of putting ± 1 into n locations. Any extreme point will satisfy a set of linear equations of the form

$$[B_pM \mid \mathbf{e}^{(p+1)}]\begin{bmatrix} \mathbf{x} \\ -h \end{bmatrix} = B_p\mathbf{f} \tag{3.29}$$

where the rows of B_p are a selection of $p+1$ rows from the N rows of P. Proceeding as in Lemma 3.1, and using Hadamard's inequality to bound the determinants which appear in Cramer's rule, leads to inequalities of the form

$$D \leq (p+1)^{(p+1)/2} \max \left\{ \prod_{j=1}^{p} \sum_{i=1}^{n} |M_{ij}| \sum_{i=1}^{n} |f_i|, \prod_{j=1}^{p} \sum_{i=1}^{n} |M_{ij}|, \right.$$
$$\left. \prod_{\substack{j \neq k \\ j=1}}^{p} \sum_{i=1}^{n} |M_{ij}| \sum_{i=1}^{n} |f_i|, \qquad k=1,2,\ldots,p \right\} \tag{3.30}$$

where use has been made of the obvious estimates

$$\boldsymbol{\rho}_i(B_p)\boldsymbol{\kappa}_j(M) \leq \sum_{i=1}^{n} |M_{ij}|,$$

$$\boldsymbol{\rho}_i(B_p)\mathbf{f} \leq \sum_{i=1}^{n} |f_i|.$$

It follows that $\log D$ can be bounded by a low-degree polynomial in the problem dimensions. Other useful information is that the feasible region contains no lines as a consequence of Theorem 1.1 of Chapter 3, and that row generation (see Section 3.5) can be used to find the most violated constraint at each step.

There is no difficulty in developing similar bounds to (3.30) for the other PCFs considered in Chapter 4 (see Exercise 3.2 below).

Exercise 3.1 Adapt the ellipsoid method to solve LP1.

Exercise 3.2 Establish the analogue of the bound (3.30) for the problem considered in Example 1.7 of Chapter 4.

7.4 DETERMINING EXPECTED BEHAVIOUR

Experience tends to suggest that the simplex algorithm is really rather efficient in practice, and that the more extreme types of worst-case behaviour are seldom if ever encountered outside of contrived situations. It even seems that these extreme cases are critically dependent on particular heuristic choices within particular algorithm specifications, so there might even be a sense in which they are unstable. These considerations have lead to efforts being directed towards finding a wider justification for the use of the descent algorithms by considering whether it is possible to categorize average performance relative to a particular problem class in some sense. The theoretical difficulties in the way of computing the expected number of iterations for problems considered as samples from a specified probability distribution which serves to define a particular class seem considerable, but some progress has been made, and there is a dearth of alternative suggestions.

Perhaps the most successful attempt to compute this expected behaviour is due to Smale. He considers the linear programming problem formulated as a complementarity problem (Exercise 2.3(iii) of Chapter 2), and shows that the expected number of steps in the Lemke algorithm grows more slowly than any fixed power of n for fixed p as $n \to \infty$. The argument uses the structure of the Lemke algorithm as a continuation method explicitly, so that the other continuation methods may be amenable to similar treatment.

To specify the density determining the class of problems it is noted that the number of iterations is invariant under the operation of multiplying the data matrix by $\lambda > 0$, and is similarly invariant under multiplication of the right-hand side $\mathbf{b}$ by $\mu > 0$ (both operations can be absorbed in $\mathbf{x}$ and do not change the decision process). Thus, for the purpose of studying numbers of iterations, there is no restriction in assuming that A and $\mathbf{b}$ are each distributed on the surface of a sphere (in R^{np} and R^n respectively). Without further a priori assumptions, the appropriate probability density in both cases would seem to be the uniform density on the surface of the sphere, with the random variables whose realizations determine A and $\mathbf{b}$ being independent.

An alternative to the theoretical calculation of the expected numbers of iterations is the estimation of this quantity on the basis of explicit solution of sequences of randomly generated problems. This is expensive as the problem dimensions and the number of trials get large; but it is easy to check different algorithms. In this context there are some results for the simplex algorithm based on sampling from spherical uniform distributions. These suggest that the expected number of iterations appears to be $O(p \log p \log n/p)$, and the $O(\log n)$ growth is completely compatible with the predictions of Smale's theory.

However, there are good reasons for looking at rather more simple-minded approaches in which the problem data for the Monte Carlo calculations is obtained by repeatedly sampling a univariate random number generator. In particular, such problems are easy to construct and serve such practical ends as providing a ready means for comparing implementation heuristics provided that they satisfy the requirement of being in some sense representative. But this is the weak point in any scheme of Monte Carlo sampling for testing algorithms – applications are unlikely to have a random origin.

Remark 4.1 This point is well illustrated by Example 5.1 of Chapter 3. Here the descent methods for l_∞ approximation clearly take $O(n)$ steps when the problem data is *smoothly varying*. However, for problems generated from random data obtained by repeated univariate sampling the descent methods have proved perfectly satisfactory and certainly do not show anything like an $O(n)$ growth in the number of iterations. Thus the Monte Carlo approach gives misleading information on the performance of the descent algorithms for an important class of applications.

It is important in generating random problems for the purpose of evaluating the performance of numerical algorithms, to be able to assess the accuracy of the solution process, and the best way to do this is to generate problems with known solutions. In outline, a process for generating a PCF with a known minimum is as follows:

(i) Define U by selecting the structure functionals ϕ_i, $i = 1, 2, \ldots, p$ characterizing the minimum by an appropriate random process.
(ii) Select $\mathbf{u} \in U$ to define the optimum multiplier vector, again by an appropriate random process.
(iii) Specify $M, \mathbf{f}$ by random sampling from a suitable probability distribution.
(iv) $\mathbf{g}$ is known once M and U are given. But we also know $\mathbf{g}$ from $0 = \mathbf{g} + V\mathbf{u}$. This requires that M be adjusted so that the two specifications are consistent.
(v) The solution $\mathbf{x}$ must satisfy the structure conditions

$$\phi_i(\mathbf{r}) = \alpha_i, \qquad i = 1, 2, \ldots, p.$$

It is usual to determine $\mathbf{x}$ by random sampling and these equations must then be used to modify $\mathbf{f}$ to ensure consistency.

Remark 4.2 Basic to the random sampling procedures is a uniform random number generator on $[0, 1]$ called RND. In applications a second generator FNRND is also used. This second generator serves to define the problem class and the effect of varying FNRND is considered subsequently.

Example 4.1 Linear programming. This is perhaps the simplest application of the general approach sketched above but it serves to illustrate the main features. The notation is based on the BASIC language used in most of the implementations.

(i) In this step the constraints active at the minimum are selected. The following procedure generates σ in the array IR by selecting P numbers from (1, . . . , N) without replacement.

```
    MAT Z=(1)
    FOR I=1 TO P
L1: K=MAX (1,INT(N*RND+.999))
    IF Z(K)=0 THEN GOTO L1
    Z(K)=0
    IR(I)=K
    NEXT I
```

(ii) Any set of P positive numbers will do as multipliers. Here S1 is a scale factor.

```
FOR I=1 TO P
U(I)=S1*RND
NEXT I
```

(iii) Random sampling is used to generate A.

```
FOR I=1 TO N
FOR J=1 TO P
A(I,J)=FNRND
NEXT J
NEXT I
```

(iv) In this case the optimality conditions give **c**.

```
MAT C=(0)
FOR I=1 TO P
K=IR(I)
FOR J=1 TO P
C(J)=C(J)+U(I)*A(K,J)
NEXT J
NEXT I
```

(v) In this step we generate a random **x** and then determine **b** so that **x** is feasible and σ points to the active constraints. Here S2 is a scale factor.

```
FOR I=1 TO P
X(I)=FNRND
NEXT I
FOR I=1 TO N
B(I)=-Z(I)*S2*RND
FOR J=1 TO P
B(I)=B(I)+A(I,J)*X(J)
NEXT J
NEXT I
```

Example 4.2 l_1 approximation. In this case it is easy enough to follow the general program outlined above.

(i) Selection of the residuals zero at the optimum can use the same code as in the previous example.

(ii) To select the optimum multipliers an appropriate sequence is

```
FOR I=1 TO P
U(I)=2*(RND-.5)
NEXT I
```

(iii) Can use the same code as in the previous example.

(iv) At this point we need the signs of the residuals at the optimum in order to specify **g**. The residuals (stored in Z) and **g** can be generated as follows.

```
  MAT G=(0)
  FOR I=1 TO N
  IF Z(I)=0 THEN GOTO L2
  Z(I)=S2*(RND-.5)
  S=SGN(Z(I))
  FOR J=1 TO P
  G(J)=G(J)+S*A(I,J)
  NEXT J
L2: NEXT I
```

Now the consistency which must be established between the two ways of defining **g** can be used to give the row of A corresponding to the multiplier of largest magnitude. To specify this multiplier let K = ARG (MAX (|U(I)|, I = 1, P)). Then we can proceed as follows.

```
  FOR I=1 TO P
  S=G(I)
  FOR J=1 TO P
  IF K=J THEN GOTO L3
  S=S+U(J)*A(IR(J),I)
L3: NEXT J
  A(IR(K),I)=-S/U(K)
  NEXT I
```

(v) Again it is convenient to specify **x** and determine **f** (which here is held in the array B).

```
MAT B=Z
FOR I=1 TO N
FOR J=1 TO P
B(I)=B(I)+A(I,J)*X(J)
NEXT J
NEXT I
```

Example 4.3 l_∞ approximation. In this case the structure functionals are more complex but this need not intrude in the problem generator.

(i) Select P+1 points to be the points of extremal deviation.

(ii) Determine multipliers by selecting P+1 positive random numbers and normalizing them to sum to 1.

(iii) As before.

(iv) Non-extremal residuals can be fixed as in the l_1 case. But now the extremal residuals must be specified also.

```
    FOR I=1 TO N
    IF Z(I)=1 THEN GOTO L3
    Z(I)=S2*SGN(RND-.5)
    GOTO L4
L3: Z(I)=2*S2*(RND-.5)
L4: NEXT I
```

Again it is convenient to achieve consistency in the multiplier conditions by using these to determine the row of A corresponding to the multiplier of largest magnitude. The procedure is the same as in the previous example but here the equation solved is $\sum_{i=1}^{p+1} u_i \boldsymbol{\rho}_{\sigma(i)}(A)=0$.

(v) The actual residuals are defined in Z so that B can be obtained exactly as in the previous example.

Example 4.4 Rank regression. This example is of interest because now steps (i) and (ii) become much more involved.

(i) Now the steps required are
   (a) allocate the m_i, $i=1,2,\ldots,n_g$, $\sum_{i=1}^{n_g} m_i = p$,
   (b) allocate m_i+1 residuals to the ith subgroup,
   (c) allocate an ordering and values consistent with that ordering to all residuals.

(ii) It is now necessary to determine $\mathbf{u} \in U$. That is, a feasible point must be found to the systems of linear inequalities (1.33) of Chapter 4. These are not quite so intimidating as they appear because an order can be imposed on the components of $\mathbf{u}^{(i)}$ by adjoining extra constraints and the randomizing can be carried out subsequently. However, the most obvious approach to the calculation of a feasible point for a system of linear inequalities is by linear programming.

Once these hurdles have been crossed the rest of the exercise is relatively plain-sailing. Also the problems in stage (ii) can be reduced if $h(\cdot)$ is not fixed but can be selected within a restricted class of similar entities as in Exercise 4.1.

Assuming that the above procedure settles the mechanics of problem generation, it remains to settle the choice of the random number generator FNRND. Most reported work seems to have used uniform generators suitably shifted and scaled, maybe because they are readily available. Normal generators have also proved popular, presumably for similar reasons. However, some disquiet has been voiced. For example, it has been noted that provided the distributions possess bounded fourth moments then the scalar products of the normals to the constraint hyperplanes converge in probability to a constant characteristic of the distribution, while the standard deviation of the scalar product converges in probability to zero as $p \to \infty$. The worry is that this appears to be indicative of rather special behaviour and reinforces the question of how representative these problems can be considered to be.

Example 4.5 It is clear that the choice of distribution can be important. Taking FNRND = RND, $p = 5$, $n = 20$, and using the procedure in Example 4.1 to generate problems which were solved using the reduced gradient algorithm with the penalty procedure to provide a feasible starting point and $1{\cdot}1 \sum u_i$ as penalty parameter produced a remarkably regular results. In 99 out of 100 trials just 5 iterations were required, with the remaining trial taking 6. With $p = 10$, $n = 40$ the pattern was essentially repeated, the problems taking 10 iterations in almost every case. In contrast taking FNRND = 10*(RND − .5) gave much more interesting results. The starting point chosen has been given by (5.20) of Chapter 2 in all cases.

Three random number generators have been singled out for further exploration. These are:

(i) FNRND = 10*(RND − .5);
(ii) FNRND = $(1-2\alpha)/(\alpha-1)+(1-\text{RND})**(-1/\alpha)$, $\alpha = 1{\cdot}2$; and
(iii) FNRND = TAN (π*(RND − .5)).

The second generator is based on a Pareto distribution and has finite first moment. The third generator is based on the Cauchy distribution. The corresponding functions are plotted in Figure 4.1. The reason for comparing the results obtained with these generators is to see to what extent the existence of moments is important in problem generation. In Table 4.1, median and maximum numbers of iterations for 100 trials in which LP1 was solved by the reduced gradient method are given for different problem sizes. The results appear to indicate that the Pareto distribution generates marginally more difficult problems although the general behaviour is rather similar. The most pleasing feature is that the hoped-for pattern of behaviour with $f(n, p)/p$ growing only slowly appears to be confirmed (about the only statement that the limited amount of data will support is that the rate of growth appears somewhat faster than $\log n$ in all cases). That the number of bounded moments does not figure as a major factor casts doubt on validity of the worries stemming from the estimates of the angles between the constraint normals. It could well be skewness rather than the number of finite moments which is the major contributor to problem difficulty.

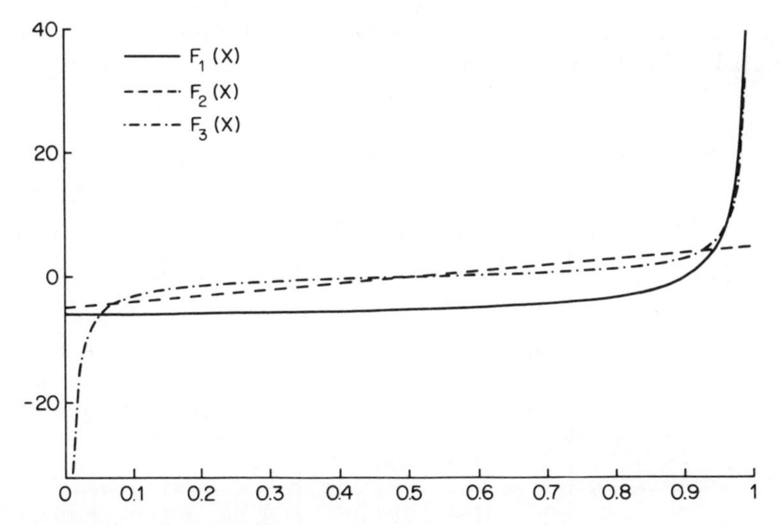

Figure 4.1 Plot of FNRND. $F1 = (1-2\alpha)/(\alpha-1)+(1-X)**(-1/\alpha)$, $\alpha = 1.2$, $F2 = 10*(X-.5)$, $F3 = TAN(\pi*(X-.5))$

Rather more information about performance can be obtained from Table 4.2. This gives the median number of iterations required to solve 100 problems for both the reduced and projected gradient algorithms with NSE tests for each of the range of values $p = 5\times 2^{i-1}$, $i = 1(1)5$, $n = p\times 2^j$, $i+j \leq 6$. Here logarithmic growth with n for fixed p corresponds to constant differences between adjacent entries in each row, and this would appear to be confirmed by the table. The two algorithms reveal very similar performances although there is some evidence that the number of iterations taken by the reduced gradient method increases more rapidly with p. For example, fitting to the iteration data using the model $y = p + \alpha p \log p \log(n/p)$ gives $\alpha = .371$ for the reduced gradient algorithm and $\alpha = .359$ for the projected gradient algorithm. The corresponding F values for the fits are 67 and 365, indicating that the model appears satisfactory.

Table 4.1 Comparison of algorithm performance under different random number generators

Dimensions	$p=5, n=20$		$p=10, n=40$		$p=20, n=80$	
Distribution	Med	Max	Med	Max	Med	Max
Pareto	9	15	22	33	53	70
$10*(RND-.5)$	8	13	18	25	42	57
Cauchy	8	15	19	29	46	65

Table 4.2 Median numbers of iterations for 100 problems solved by the projected and reduced gradient algorithms for the specified values of p and n

	PRGD						REGD					
p \ n	10	20	40	80	160	320	10	20	40	80	160	320
5	7	10	11	13	14	15	6	8	10	12	13	14
10		16	21	26	32	37		14	20	25	32	36
20			34	49	65	81			34	49	67	86
40				73	118	164				77	131	182
80					169	284					195	366

Table 4.3 Study of the convergence of mean and median as measures of the number of iterations.

Dimension	$p=5, n=20$		$p=10, n=40$	
No. of runs	Mean	Median	Mean	Median
10	9.8	10/11	23.2	23
20	10.4	10	22.7	21/22
30	10.0	9/10	22.0	21/22
40	9.7	9	21.8	21
50	9.6	9	21.7	21
60	9.7	9/10	21.8	21
70	9.6	9	21.8	21
80	9.6	9	21.5	21
90	9.5	9	21.6	21
100	9.6	9	21.3	21

The algorithms are stable enough to handle the $p=40$, $n=160$ cases in single precision on a Univac 1100–80 (7–8 decimals), and give between 2 and 5 significant figures in the solution $\mathbf{x}$. At this level of precision both algorithms require several reinversions per problem in the NSE recurrence. No reinversions are required in double precision, and the projected gradient results show a slight but consistent edge in accuracy, which is high in both cases.

The median has been selected to measure the number of iterations because it is more resistant to outliers than the obvious candidate, the mean (see Section 5.5). Thus it might be hoped that it would settle down faster for samples of modest size. An attempt to gauge this is presented in Table 4.3, where the mean and median are compared for samples increasing in increments of 10 from 10 to 100 for the particular case of the Pareto distribution. The median may have an edge, but both measures would seem to converge rapidly enough to provide the information required from samples of this size.

Exercise 4.1 The generation of rank regression problems can be simplified by noting that the structure functionals are independent of the scores η_i. Thus these are available also. Show that they can be chosen to simplify the problem of satisfying (1.33) of Chapter 4.

7.5 SOME IMPLEMENTATION CONSIDERATIONS

In this section the techniques for the Monte Carlo generation of problems with known solutions are used to provide source data concerning implementation strategies. This approach seems adequate for this purpose because, apart from the exceptional cases of exponential complexity, the decisions are between different options in what amounts to fine tuning the basic process. Topics considered here include:

(i) Is it a good idea to relax an active constraint if the multiplier is negative before an extreme point has been established, and how large a tolerance should be used to decide this question?
(ii) Is the extra work involved in implementing NSE tests justified?
(iii) How should the linear search be carried out in cases in which it is relevant?

When should constraints be relaxed?

The conclusion in this case seems, unambiguously, to be that there is no value in this strategy when $\mathbf{c} \notin H$ unless a very large relative violation occurs in the multiplier. In the reduced gradient algorithm the NSE test which determines the column of B_ν to be deleted to determine the descent direction is specified in (5.22) of Chapter 2:

$$s = \arg\max_i \left(-\frac{u_i}{\|\mathbf{t}_i\|}, 1 \leq i \leq k, \tau \frac{|u_i|}{\|\mathbf{t}_i\|}, k+1 \leq i \leq p \right)$$

This is considered for the following values of τ:

(i) $\tau = 10^{-3}$, which implies that the constraint with the largest negative multiplier will almost always be relaxed;
(ii) $\tau = 1$, which implies a compromise; and
(iii) $\tau = 10^3$ which implies that the action will almost always be to delete a column of E_ν when $\mathbf{c} \notin H$.

In the projected gradient algorithm a similar test is provided by deciding to relax the constraint associated with a negative multiplier if

$$\max_{i, u_i < 0} \frac{u_i^2}{(U^T U)_{ii}^{-1}} > \tau \|\mathbf{t}\|^2 \tag{5.1}$$

where $\mathbf{t}$ is the descent vector in the case that no constraint is relaxed. Again, the same values of τ are considered, and the interpretation is very much as before (if the test fails $\mathbf{t}$ is used as the descent direction). Results are presented in Table 5.1 and give the cumulative number of iterations for the solution of 10 problems of type LP1 for the different values of τ and for each algorithm. The problem dimensions are $p = 10$, $n = 40$.

Table 5.1 Relaxing constraints is not a good idea. Cumulative results for 10 problem replications.

τ	10^{-3}	1	10^3
REDG	367	209	204
PRGD	245	218	204

USE vs. NSE tests

This comparison is relevant only to the linear programming and l_1 approximation problems for the reasons explained in Section 4.4. For these problems the advantage of NSE tests is that they are scale independent, and any difficulties lie in the additional work involved. This is asymptotically small for large n and fixed p, but does add an $O(p^2)$ term which can be significant when p/n is not negligible. Thus it is important that the NSE test show improved performance. In Table 5.2 results of comparisons are given for both tests, for both reduced and projected gradient algorithms applied to LP1, and for several different choices of p and n. For each case the median

Table 5.2 USE and NSE tests compared.

Dimension		$p=5$	$n=20$	$p=10$	$n=40$	$p=20$	$n=80$
		Med	Max	Med	Max	Med	Max
REGD	USE	9	15	22	33	53	70
	NSE	8	13	20	28	50	64
PRGD	USE	10	15	22	31	54	82
	NSE	10	14	20	30	49	67

and maximum number of iterations for 10 replications are given, and an initial feasible point has been found using the penalty procedure of Section 2.5 with penalty parameter given by $1{\cdot}1 \sum_{i=1}^{p} u_i$ (note that the optimum multipliers are generated as part of the problem set-up procedure described in the previous section). On the basis of these figures we conclude that the NSE test has about a 10% edge on the USE test and significantly reduces the maximum number of iterations needed. These observations are consistent with other reports but perhaps less conclusive. One reason for this could very likely be that the generated problems are usually pretty well scaled (consider Figure 4.1).

It is interesting that the iteration counts for the reduced and projected gradient methods are very close especially when NSE tests are used.

Line searches

With the exception of the linear programming and l_∞ approximation problems, line searching in the direction of the current descent direction is an important feature of our descent algorithms for minimizing polyhedral convex functions. This is illustrated in Table 5.3, which gives the median

Table 5.3 Median number of iterations and cummulative line search steps for ten problems with $p=10$.

Dimension	$n=20$		$n=40$		$n=80$		$n=160$		$n=320$	
	ITS	CUM	ITS	CUM	ITS	CUM	ITS	CUM	ITS	CUM
RGLAD	16	49 / 52	22 / 23	103 / 105	30	220 / 230	32 / 33	406 / 418	31 / 33	813 / 815
PGLAD	17	50	22 / 25	91 / 93	24 / 25	142 / 153	29 / 31	299 / 315	31 / 32	688

Table 5.4 Progress of one problem with $p=10$, $n=40$ solved by the projected gradient algorithm. POS is the position of the optimum λ in the sorted list.

I	$\lvert\Lambda\rvert$	POS	CUM
1	40	10	10
2	35	18	28
3	31	17	42
4	33	21	63
5	22	6	69
6	22	1	70
7	18	3	73
8	16	1	74
9	15	1	75
10	21	1	76
11	12	2	78
12	25	3	81
13	19	1	82
14	11	2	84
15	22	1	85
16	22	2	87
17	16	2	89
18	18	1	90
19	18	1	91

numbers of iterations and cumulative line search steps for ten l_1 approximation problems, for both reduced and projected gradient algorithms, and for a range of problem dimensions. It will be seen that the number of line search steps clearly increases with n and is large enough to justify careful consideration. An interesting feature is the somewhat better performance of the projected gradient method (the problems solved being identical in both cases). NSE tests were used. In Table 5.4 the number of steps taken in the line search in each iteration of a particular problem is given together with the number of elements in Λ (Lemma 2.4 of Chapter 3). It will be seen that $|\Lambda|$ remains large with m throughout the calculations, but the position of the optimum λ_i in the sorted list changes significantly – from being somewhere in the middle of the list in the initial steps (large changes) to being always near the head of the list from about one-third of the way through the calculation in the example given (smaller changes and a slower rate of progress).

These results show that determining the optimum λ^* requires a nontrivial amount of sorting, at least early in the calculation. The following scheme for the l_1 problem makes use of the observation that Λ need not be completely sorted. Rather it is sufficient to partition it into two sets Λ_L, Λ_R such that

$$\lambda \in \Lambda_L \Rightarrow \lambda < \lambda^*, \qquad \lambda \in \Lambda_R \Rightarrow \lambda \geqslant \lambda^*$$

This optimum value λ^* is computed recursively:

(i) Set the directional derivative Dx, Λ.
(ii) If $|\Lambda| \leq 3$ complete by an explicit calculation. Otherwise set the partition bound $\hat{\lambda}$ as the median of the first, last, and middle elements of Λ.
(iii) Partition Λ into Λ_L, Λ_R such that

$$\Lambda_L = \{\lambda_k;\ \lambda_k < \hat{\lambda}\},$$
$$\Lambda_R = \{\lambda_k;\ \lambda_k \geq \hat{\lambda}\}$$

(iv) Compute $G = Dx + \sum_{\lambda_k \in \Lambda_L} 2\,|\mathbf{m}_k^T \mathbf{t}|$
(v) If $G > 0$ go to (vii).
(vi) $Dx := G$, $\Lambda := \Lambda_R$, go to (ii).
(vii) $\Lambda := \Lambda_L$, go to (ii).

The hard work is done in step (iii) which computes the set partition. If no prior information is available on the location of λ^* in the sorted order, then reasons of efficiency dictate that $\hat{\lambda}$ should be chosen as the median of the $\lambda_i \in \Lambda$ to minimize the maximum size of the set to be inspected after the partitioning step. The strategy of picking $\hat{\lambda}$ as the median of first, last, and middle members of Λ is a heuristic motivated by this observation. The following algorithm achieves the desired partitioning in $|\Lambda|$ comparisons. It is due to Hoare and is perhaps best known as the initial step of the standard sorting routine QUICKSORT.

Algorithm

```
Lb = 1, Ub = |Λ| = N, D = λ̂, B = λ(N)
WHILE Lb < Ub DO
BEGIN
LOOP: WHILE λ(Lb) < D DO Lb = Lb + 1
      λ(Ub) = λ(Lb), Ub = Ub − 1
      WHILE λ(Ub) > D DO Ub = Ub − 1
      λ(Lb) = λ(Ub), Lb = Lb + 1
      GO TO LOOP
END
λ(Ub) = B
```

If $\hat{\lambda}$ is actually the median of Λ at each stage then this process takes $< 2N = N + N/2 + N/2^2 + \ldots$ comparisons to find λ^*. This figure will be increased in practice due to errors in the heuristic determining $\hat{\lambda}$. An actual figure of 4N corresponding to a 100% error in the estimated number of comparisons just about makes it worth while to look at alternatives for the tail of the computation when λ^* is always in the first few elements of the sorted list. In this case a simple comparison sort would be effective.

However, it is also important to get the sorting cost into perspective. The total cost using the above partitioning algorithm can be estimated by ITS$*4n$ comparisons (say) which must be compared with ITS$*np$ multiplications, which gives the order of work in tableau processing (here ITS is the number of descent steps in the computation). Thus it is worth taking steps to avoid $O(n^2)$ comparisons on any descent step, but it is perhaps not justified to introduce a second method to try to minimize the number of comparisons in the tails of the computation where, as illustrated in Table 5.4, it is always going to be a small multiple of n for either of the two possibilities.

The conclusion from the above discussion is that the line search problem is in good shape for the piecewise linear class of PCFs. The difficulties for the more elaborate examples 1.6, 1.7 of Chapter 4 have been indicated already in Remark 4.3 of Chapter 4. To put this problem into context $O(n^2)$ structure functionals imply up to $O(n^4)$ comparisons *per iteration* in a simple comparison sort. Although the ordering of the residuals can be used to cut down the number of comparisons to $O(n^2)$ at most per iteration by noting that ties must occur first between residuals which are adjacent in the ranking list, the optimum use of the structure in the definition of the data to be sorted is an important and presently unsolved problem, and the need to maintain an ordered list is a further constraint not present in the piecewise linear problems.

NOTES ON CHAPTER 7

There is a very substantial literature on the complexity of the simplex algorithm, and only a few items are noted here. The example showing worst-case exponential complexity of the simplex algorithm is due to Klee and Minty (1972). Their result was modified in Jeroslow (1973) to show exponential complexity for the maximum objective function improvement selection rule. An example showing exponential complexity with an NSE selection rule has been given by Goldfarb and Sit (1979). The explicity formulated problem with exponential complexity is discussed by Avis and Chvatal (1976). The treatment of Khachiyan's ellipsoid method is based on the report of Aspvall and Stone (1979) and the survey article of Bland, Goldfarb, and Todd (1981). That the restriction to integer arithmetic is essential is discussed by Traub and Wozniakowski (1982). The estimate of average numbers of iterations referenced is due to Smale (1983). The simulation results using spherical uniform distributions are due to Dunham, Kelly, and Tolle (1977). The concern over the use of randomly generated data for algorithm evaluation is expressed by van Dam, Frenk, and Telgen (1983). The use of Monte Carlo techniques goes back at least to Quandt and Kuhn (1964). Our procedures follow Bartels (1980). The particular form of the Pareto generator is copied from Bloomfield and Steiger (1980), who also suggest the use of Hoare's partitioning algorithm in the line search. Apart from this reference, the need to carry out the line search efficiently has received little explicit acknowledgement in the literature. However, the distributed program based on Bartels, Conn, and Sinclair (1978) uses a heap-sorting technique and indicates they are aware of the problem. Perhaps the secant algorithm applied to the directional derivative offers the best hope of carrying out the line search in the ranked residuals problems (see notes on Chapter 4).

APPENDIX 1

Some basic linear algebra

A1.1 NOTATION

Many of the formal aspects of this work are concerned with manipulating structured components of matrices and require a working notation to describe these operations. This is developed in this appendix and applied to the construction of certain of the basic transformations needed. Matrices are written in uppercase letters and described by the associated mapping. Thus $A: R^p \rightarrow R^n$ defines a real matrix A of n rows and p columns which gives an explicit representation of a mapping from a p-dimensional domain space to an n-dimensional range space. Note that the dimensions are not indicated explicitly and should be clear from the context. However, it will be convenient on occasion to denote the k-dimensional unit matrix by I_k. The operation of defining a submatrix by selecting certain rows (or columns depending on context) is represented by specifying an index set which contains the row indices, for example $\nu = \{i, j, k, \ldots\}$. The name is always written as a lowercase Greek letter. The indices can also be enumerated ($\nu = \{\nu(1), \nu(2), \ldots, \nu(k)\}$ (where $k = |\nu|$ is the number of elements in ν), and the complement ν^c contains the row indices not in ν (so that $\nu \cup \nu^c = \{1, 2, \ldots, n\}$). The submatrix specified by reference to ν is written A_ν, and this subscript will be used to refer to other subset quantities contingent on ν – for example, if $[A \mid \mathbf{b}]$ is a partitioned matrix then $\mathbf{b}_\nu$ would be the subvector of $\mathbf{b}$ pointed to by the elements of ν.

Vectors are written in lowercase letters in bold type. Except in situations where a natural definition dictates otherwise, roman script is used. To associate vectors with the rows or columns of a matrix, two special operators are used. These are

$$\boldsymbol{\rho}_i(A) = \mathbf{e}_i^{\mathrm{T}} A \qquad \text{(selecting the } i\text{th row of } A\text{), and}$$

$$\boldsymbol{\kappa}_j(A) = A\mathbf{e}_j \qquad \text{(selecting the } j\text{th column of } A\text{).}$$

Here $\mathbf{e}_i$ is the ith coordinate vector in R^n and, in general, the dimension will be clear from the context. If not, it will be indicated by the means of a superscript. For example, let $\mathbf{e}_i \in R^k$ (alternatively this could be written $\mathbf{e}_i^{(k)}$);

then

$$\mathbf{e}^{(k)} = \sum_{i=1}^{k} \mathbf{e}_i$$

is the column of ones in R^k. As usual, superscript T denotes transpose.

On occasions it will be convenient to adopt the convention that $\mathbf{a}_i^{\mathrm{T}}$ is the ith row of A. For example,

$$\boldsymbol{\rho}_i(A_\nu) = \mathbf{a}_{\nu(i)}^{\mathrm{T}}, \qquad i = 1, 2, \ldots, |\nu|.$$

To select consecutive elements from a vector the notation used is

$$[\mathbf{x}]_j^i = \begin{bmatrix} x_i \\ x_{i+1} \\ \vdots \\ x_j \end{bmatrix}.$$

If $i = 1$ then the superscript is suppressed and if $j = \dim \mathbf{x}$ the subscript is suppressed. These correspond to the usual cases.

Example (i) Example of the use of the operation $\times$

$$\mathbf{x} = [\mathbf{x}]_i \times [\mathbf{x}]^{i+1}.$$

A1.2 ELEMENTARY MATRICES

Much of numerical algebra can be developed in a convenient and constructive manner using elementary matrices as basic building blocks. An *elementary matrix* $E(\mathbf{u}, \mathbf{v}, \alpha): R^n \to R^n$ is a matrix of the specific form identity I plus a rank-one correction. Thus

$$E(\mathbf{u}, \mathbf{v}, \alpha) = I - \alpha \mathbf{u}\mathbf{v}^{\mathrm{T}}. \tag{2.1}$$

A major property is a composition rule

$$\begin{aligned} E(\mathbf{u}, \mathbf{v}, \alpha)E(\mathbf{u}, \mathbf{v}, \beta) &= I - (\alpha + \beta - \alpha\beta\mathbf{v}^{\mathrm{T}}\mathbf{u})\mathbf{u}\mathbf{v}^{\mathrm{T}} \\ &= E(\mathbf{u}, \mathbf{v}, \alpha + \beta - \alpha\beta\mathbf{v}^{\mathrm{T}}\mathbf{u}). \end{aligned} \tag{2.2}$$

In particular, the choice

$$\beta = -\alpha/(1 - \alpha\mathbf{v}^{\mathrm{T}}\mathbf{u}) \tag{2.3}$$

gives

$$E(\mathbf{u}, \mathbf{v}, \beta) = E(\mathbf{u}, \mathbf{v}, \alpha)^{-1}.$$

Also $E(\mathbf{u}, \mathbf{v}, \alpha)$ has an $(n-1)$-fold eigenvalue 1 with invariant (column) subspace orthogonal to $\mathbf{v}$, and a simple eigenvalue $1 - \alpha\mathbf{v}^{\mathrm{T}}\mathbf{u}$ and corresponding (right) eigenvector $\mathbf{u}$. It follows that

$$\det E = 1 - \alpha\mathbf{v}^{\mathrm{T}}\mathbf{u}. \tag{2.4}$$

Example (i) Elementary permutation matrices Let P_{ij} be the matrix which when applied to the general vector $\mathbf{x}$ has the property that it exchanges the i and j components of $\mathbf{x}$. That is,

$$\begin{aligned} P_{ij}\mathbf{x} &= \mathbf{x} + (x_j - x_i)\mathbf{e}_i + (x_i - x_j)\mathbf{e}_j \\ &= \mathbf{x} - (x_i - x_j)(\mathbf{e}_i - \mathbf{e}_j) \\ &= [I - (\mathbf{e}_i - \mathbf{e}_j)(\mathbf{e}_i - \mathbf{e}_j)^{\mathrm{T}}]\mathbf{x}. \end{aligned} \tag{2.5}$$

This shows that the operation can be effected with P_{ij} an elementary matrix. The resulting P_{ij} is called an elementary permutation matrix. Note that

$$P_{ij}P_{ij} = I. \tag{2.6}$$

Example (ii) Elementary lower triangular matrices Consider the matrix $L(i, \mathbf{x})$ specified by the conditions

(a) $$L(i, \mathbf{x})\mathbf{x} = \begin{bmatrix} [\mathbf{x}]_i \\ 0 \end{bmatrix}, \tag{2.7}$$

and

(b) $$\text{if} \quad \mathbf{w} = \begin{bmatrix} [\mathbf{w}]_{i-1} \\ 0 \end{bmatrix}$$

then

$$L(i, \mathbf{x})\mathbf{w} = \mathbf{w}. \tag{2.8}$$

These conditions can be satisfied by an elementary matrix. Let

$$L = I - \alpha\mathbf{u}\mathbf{v}^{\mathrm{T}},$$

then (2.8) requires

$$[\mathbf{v}]_{i-1}^{\mathrm{T}}[\mathbf{w}]_{i-1} = 0$$

so that $[\mathbf{v}]_{i-1} = 0$. Thus L can be lower triangular only if $\mathbf{v} = \mathbf{e}_i$. Condition (a) now gives

$$L\mathbf{x} = \mathbf{x} - \alpha\mathbf{x}_i\mathbf{u} = \begin{bmatrix} [\mathbf{x}]_i \\ 0 \end{bmatrix}$$

and this can be satisfied by choosing

$$\alpha = \frac{1}{x_i}, \qquad \mathbf{u} = \begin{bmatrix} 0 \\ [\mathbf{x}]^{i+1} \end{bmatrix}.$$

Note that the calculation $L(i, \mathbf{x})\mathbf{t}$ given L requires $n - i$ multiplications or one multiplication for each zero introduced in $\mathbf{x}$.

Example (iii) Elementary orthogonal matrices The condition for the elementary matrix E to be orthogonal is

$$I=(I-\alpha \mathbf{v}\mathbf{u}^{\mathrm{T}})(I-\alpha \mathbf{u}\mathbf{v}^{\mathrm{T}})=I-\alpha(\mathbf{v}\mathbf{u}^{\mathrm{T}}+\mathbf{u}\mathbf{v}^{\mathrm{T}})+\alpha^{2}(\mathbf{u}^{\mathrm{T}}\mathbf{u})\mathbf{v}\mathbf{v}^{\mathrm{T}}.$$

There is no restriction in assuming $\|\mathbf{u}\|_2=\|\mathbf{v}\|_2=1$ (norms are Euclidean vector norms), and it is readily seen that the only solution is given by $\mathbf{u}=\mathbf{v}$, $\alpha=2$. In particular, E is symmetric and $\det(E)=-1$ so that reflection as well as rotation is involved in the transformation.

Here we construct the particular elementary orthogonal matrix (or Householder transformation) $H(i,\mathbf{x})$ defined by the conditions:

(a)
$$H(i,\mathbf{x})\mathbf{x}=\begin{bmatrix}[\mathbf{x}]_{i-1}\\ \theta\,\|[\mathbf{x}]^{i}\|_2\,[\mathbf{e}_i]^{i}\end{bmatrix} \tag{2.9}$$

where the second component of the right-hand side is determined by the condition that length is unchanged in an orthogonal transformation, and $\theta=\pm1$ so that there is a degree of freedom in the choice of sign; and

(b)
$$H(i,\mathbf{x})\mathbf{w}=\mathbf{w} \tag{2.10}$$

whenever

$$\mathbf{w}=\begin{bmatrix}[\mathbf{w}]_{i-1}\\ 0\end{bmatrix}.$$

Substituting the explicit form for an elementary orthogonal matrix in (a) gives

$$[I-2\mathbf{u}\mathbf{u}^{\mathrm{T}}]\mathbf{x}=\mathbf{x}-2(\mathbf{u}^{\mathrm{T}}\mathbf{x})\mathbf{u}=\begin{bmatrix}[\mathbf{x}]_{i-1}\\ \theta\,\|[\mathbf{x}]^{i}\|_2\,[\mathbf{e}_i]^{i}\end{bmatrix}$$

so that $\mathbf{u}$ is parallel to $\mathbf{v}$, where

$$\mathbf{v}=\begin{bmatrix}0\\ [\mathbf{x}]^{i}-\theta\,\|[\mathbf{x}]^{i}\|_2\,[\mathbf{e}_i]^{i}\end{bmatrix}.$$

Thus $\mathbf{u}$ is obtained by normalizing $\mathbf{v}$. This gives

$$\mathbf{u}=\frac{\mathbf{v}}{\sqrt{2\,\|[\mathbf{x}]^{i}\|_2\,(\|[\mathbf{x}]^{i}\|_2-\theta x_i)}}, \tag{2.11}$$

and it is usual to choose $\theta=-\operatorname{sgn}(x_i)$ in order to avoid the possibility of cancellation in computing the denominator. It is verified readily that the zero elements in $\mathbf{v}$ suffice to ensure that (2.10) is always satisfied.

Note that the computation of $H(i,\mathbf{x})\mathbf{w}$ for a general $\mathbf{w}$ involves:

(i) $(n-i+2)$ multiplications to compute $\mathbf{v}^{\mathrm{T}}\mathbf{w}$, $2\mathbf{v}^{\mathrm{T}}\mathbf{w}/\|\mathbf{v}\|^2$, and
(ii) $(n-i+1)$ multiplications to compute $\mathbf{w}-2(\mathbf{v}^{\mathrm{T}}\mathbf{w}/\|\mathbf{v}\|_2^2)\mathbf{v}$.

In particular, if $n=2$, $i=1$, then the cost is 5 multiplications. This corresponds to an important special case in which it is possible to do significantly better by a somewhat different approach. However, if $n-i$ is significantly bigger than 1 then the cost approaches 2 multiplications per element reduced to zero. This is twice the cost of the elimination procedure based on elementary lower triangular matrices.

Example (iv) Elementary rotation matrices The simplest nontrivial orthogonal transformation is provided by the realization of a rotation through an angle θ in a two-dimensional subspace. The matrix representation can be made symmetric if it is combined with a reflection, and then it can be displayed as an elementary orthogonal matrix. If the matrix of this combined transformation is R_{ij} then

$$R_{ij}=I-\{(1-\cos\theta)\mathbf{e}_i\mathbf{e}_i^{\mathrm{T}}-\sin\theta\mathbf{e}_i\mathbf{e}_j^{\mathrm{T}}-\sin\theta\mathbf{e}_j\mathbf{e}_i^{\mathrm{T}}+(1+\cos\theta)\mathbf{e}_j\mathbf{e}_j^{\mathrm{T}}\}, \tag{2.12}$$

and it is a straightforward calculation that this is an elementary orthogonal matrix with

$$\mathbf{u}=\sin(\theta/2)\mathbf{e}_i+\cos(\theta/2)\mathbf{e}_j. \tag{2.13}$$

The application that is of importance here corresponds to the use of the R_{ij} to reduce systematically to zero particular elements of a matrix. Operating on A (premultiplying) by R_{ij} combines (mixes) rows i and j of A and leaves the other rows unchanged. In particular,

$$\begin{aligned}\mathbf{e}_i^{\mathrm{T}}R_{ij}A&=\cos\theta\boldsymbol{\rho}_i(A)+\sin\theta\boldsymbol{\rho}_j(A),\\ \mathbf{e}_j^{\mathrm{T}}R_{ij}A&=\sin\theta\boldsymbol{\rho}_i(A)-\cos\theta\boldsymbol{\rho}_j(A).\end{aligned} \tag{2.14}$$

If the element in the (j, s) position is reduced to zero, then

$$\sin\theta A_{is}-\cos\theta A_{js}=0$$

so that

$$\begin{aligned}\cos\theta&=A_{is}/d,\\ \sin\theta&=A_{js}/d,\end{aligned} \tag{2.15}$$

where

$$d^2=A_{is}^2+A_{js}^2.$$

Similar formulae hold if the zeroed element is in the (i, s) position and it should be noted that the calculation involves a square root. It will be convenient to denote the resulting matrix by $R(i, j, (j, s))$, as this shows explicitly the rows that are mixed and the element which is reduced to zero.

It appears from (2.14) that the cost of applying R to A is four multiplications for each column for each zero generated (that is, about twice the cost of a Householder transformation if a number of zeros have to be produced in any column). However, the cost can easily be reduced to three multiplications. Let sn = sin θ, cs = cos θ. Then a suitable sequence is

```
mu = sn/(1+cs)
FOR k = 1 TO p
w = cs*A(i,k) + sn*A(j,k)
A(j,k) = mu*(A(i,k)+w) - A(j,k)
A(i,k) = w
NEXT k
```

In certain circumstances it is possible to reduce the number of multiplications per column of A to two and to avoid the need to compute square roots, and these features serve to make the resulting method competitive with the use of Householder transformations in such problems as solving systems of linear equations. For our purposes, methods based on these fast Given's transformations are attractive for tableau processing using the projected gradient form of the basic descent algorithms in the case that the factorization of the matrix of active constraints has to be updated to take account of deletion of a constraint from the active set (see Section 2.6, for example). However, they complicate the description of the algorithm because of the need to keep a diagonal scaling matrix which adds a further level of detail. For this reason the discussion in the text is presented in terms of the standard form of plane rotation.

A1.3 MATRIX FACTORIZATIONS BASED ON ELEMENTARY MATRICES

An immediate application of elementary matrices is to the transformation of systems of linear equations into a form which can be solved readily. We first consider a basic elimination algorithm which makes use of elementary lower triangular matrices. Here

$$L(i, \mathbf{x}) = I - \frac{1}{x_i}\begin{bmatrix} 0 \\ [\mathbf{x}]^{i+1} \end{bmatrix}\mathbf{e}_i^T$$

so that control of the size of the elements in L (and hence of the potential of the transformation to cause cancellation) requires control of the size of x_i relative to the size of the elements in $[\mathbf{x}]^{i+1}$. Clearly the situation is favourable if x_i is the largest element in magnitude in $[\mathbf{x}]^i$, and this can be achieved by permuting the elements of $\mathbf{x}$ using an elementary permutation matrix if necessary. The standard error analyses confirm the importance of this strategy.

The first step of the elimination is typical and proceeds as follows:

(i) the rows of A are permuted to bring the element of largest modulus in the first column into the first or pivotal position

$$A = A^{(1)} = P_{1i_1}P_{1i_1}A^{(1)} = P_{1i_1}\bar{A}^{(1)} \tag{3.1}$$

where permutation exchanges rows 1 and i_1, and

(ii) zeros are introduced into the first column of $\bar{A}^{(1)}$ using an elementary lower triangular matrix

$$\begin{aligned}\bar{A}^{(1)} &= L(1, \boldsymbol{\kappa}_1(\bar{A}^{(1)}))^{-1}L(1, \boldsymbol{\kappa}_1(\bar{A}^{(1)}))\bar{A}^{(1)} \\ &= L(1, \boldsymbol{\kappa}_1(\bar{A}^{(1)}))^{-1}A^{(2)}.\end{aligned} \tag{3.2}$$

The next step is to eliminate elements in the second column using $L(2, \cdot)$. The properties of elementary lower triangular matrices ensure that the first column is unchanged. Thus the zeros can be introduced column by column, and after $p-1$ steps the factorization obtained is

$$A = \prod_{k=1}^{p-1} P_{ki_k} L(k, \boldsymbol{\kappa}_k(\bar{A}^{(k)}))^{-1} A^{(p)} \tag{3.3}$$

where $A^{(p)}$ is upper triangular. This factorization can be represented as the product of permutation matrices times a lower triangular matrix times $A^{(p)}$. The basic manipulation can be illustrated by considering the product $L(i, \mathbf{x})P_{jk}$ where $i<j\leq k$. This gives

$$\begin{aligned}\left(I - \frac{1}{x_i}\begin{bmatrix} 0 \\ [\mathbf{x}]^{i+1}\end{bmatrix}\mathbf{e}_i^{\mathrm{T}}\right)P_{jk} &= P_{jk}\left(I - \frac{1}{x_i}P_{jk}\begin{bmatrix} 0 \\ [\mathbf{x}]^{i+1}\end{bmatrix}\mathbf{e}_i^{\mathrm{T}}P_{jk}\right) \\ &= P_{jk}\left(I - \frac{1}{x_i}\begin{bmatrix} 0 \\ [P_{jk}\mathbf{x}]^{i+1}\end{bmatrix}\mathbf{e}_i^{\mathrm{T}}\right) \\ &= P_{jk}L(i, P_{jk}\mathbf{x}).\end{aligned} \tag{3.4}$$

Applying this result to (3.3) gives

$$A = \prod_{k=1}^{p-1} P_{ki_k} \prod_{k=1}^{p-1} L\left(k, \prod_{j=k+1}^{p-1} P_{ji_j}\boldsymbol{\kappa}_k(\bar{A}^{(k)})\right)^{-1} A^{(p)}, \tag{3.5}$$

and if the $\alpha_i\mathbf{u}_i$ defining the transformations are stored on the zeros introduced at the current stage then the interchanges in (3.5) are carried out automatically when the rows of A are interchanged.

Remark 3.1 This procedure is called LU factorization of A with partial pivoting. In complete pivoting, row and column interchanges are performed to bring the element of largest modulus in the remaining submatrix into the pivotal position.

If Householder transformations are used to factorize A then the corresponding formulae are

$$\begin{aligned}A &= A^{(1)}, \\ A^{(i+1)} &= H(i, \boldsymbol{\kappa}_i(A^{(i)}))A^{(i)}, \qquad i = 1, 2, \ldots, p-1, \\ A &= \prod_{i=1}^{p-1} H(i, \boldsymbol{\kappa}_i(A^{(i)}))A^{(p)}\end{aligned} \tag{3.6}$$

where $A^{(p)}$ is upper triangular. Thus this sequence realizes the factorization of A into the product of an orthogonal and an upper triangular matrix. This factorization can also be realized using plane rotations, but in this case it is necessary to spell out the sequence in which the individual elements in the subdiagonal of A are reduced to zero. If the zeros are introduced column by column as in the case of the Householder transformations above, then the generation of zeros within a particular column can be described by

$$A^{(i,k)} = \prod_{s=k}^{n} R(i, s, (s, i))A^{(i)},$$

and the column-by-column sequence reduces to

$$A^{(i+1)} = A^{(i,i+1)}.$$

The final form of the transformation is

$$A = \prod_{i=1}^{p-1} \prod_{k=n}^{i+1} R(i, k, (k, i))A^{(p)}. \tag{3.7}$$

To carry out these factorization sequences in each case requires only that n (column dimension) is greater than p (row dimension), and they terminate after $(p-1)$ steps or at the ith stage if $[\boldsymbol{\kappa}_i(A^{(i)})]^i = 0$. If the second alternative obtains then it may be that $[\boldsymbol{\kappa}_k(A^{(i)})]^i \neq 0$ for some $k > i$, and in this case a column interchange $(A^{(i)} := A^{(i)}P_{ik})$ permits further progress to be made. Otherwise the rank of A is $(i-1)$. However, this deduction assumes exact arithmetic and in practice a more sophisticated approach will be required. The related problem of determining if a given vector lies in a particular subspace occurs in the problem of terminating the descent algorithms when it is suspected that the matrix of constraints does not have full rank. This problem is discussed in Sections 2.5 and 2.6.

NOTES ON APPENDIX 1

The basic references are Householder (1964), Stewart (1973),and Wilkinson (1965). The stable two-multiplication modifications of plane rotations are discussed in detail in Hammarling (1974).

APPENDIX 2

Algorithms for continuous approximation problems

A2.1 INTRODUCTION

Several of the algorithms for discrete approximation problems possess fairly natural analogues for solving the corresponding continuous approximation problems. The discrete algorithms of interest are the descent method for l_1 approximation and the ascent or dual method for l_∞ approximation. The key features of the continuous methods which make them relevant are:

(i) Attention is fastened on the characterization equations for the solution of the problem. These give a system of nonlinear equations for the set of zeros or extrema characterizing the solution which are solved by Newton's method, and each iteration results in these sets of tentative critical points being updated (exchanged) for sets which hopefully better satisfy the characterization equations.

(ii) Close to the solution the algorithm uncouples the problem to first order into a set of one-dimensional subproblems in which a particular critical point (zero or extremum) is adjusted in a manner analogous to the discrete algorithm.

It is no longer reasonable to expect a finite termination property. However, in certain circumstances which correspond closely to the case in which decomposition into subproblems occurs, a second-order rate of convergence can be proved, and this implies that a number of steps proportional to the log of the problem tolerance is required. Arguing by analogy, this suggests a dependence on p in the number of iterations required in the corresponding discrete problem. This contrasts with the observed poor behaviour of the wrong algorithm for discrete problems with smoothly varying data (the exchange algorithm for l_1 problems and the descent algorithm for l_∞ problems). Thus the performance of the algorithms in the continuous case gives insight into circumstances in which certain of the discrete algorithms are likely to perform well.

A2.2 THE L_1 APPROXIMATION PROBLEM

The problem considered in this section is that of finding a vector of parameters $\mathbf{x}$ to solve the problem

$$\min_{\mathbf{x}} \|r(\tau, \mathbf{x})\|_1 = \min_{\mathbf{x}} \int_0^1 \left| \sum_{i=1}^{p} x_i m_i(\tau) - f(\tau) \right| d\tau \tag{2.1}$$

where the data functions $f, m_i, i=1,2,\ldots,p$ are assumed to be sufficiently smooth functions (for example $f, m_i \in C^\alpha[0,1]$ with $\alpha>2$), and where the problem has a unique solution $\mathbf{x}^*$ characterized by conditions analogous to the particular form of (1.6) of Chapter 3 given explicitly in (1.7) of that chapter. These conditions are

$$\int_0^1 \operatorname{sgn}(r(\tau, \mathbf{x}^*))\mathbf{m}_i(\tau)\,\mathrm{d}\tau = 0, \qquad i=1,2,\ldots,p, \tag{2.2}$$

or, equivalently,

$$\int_0^1 \operatorname{sgn}(r(\tau, \mathbf{x}^*))\,\nabla_x r\,\mathrm{d}\tau = \int_0^1 \operatorname{sgn}(r(\tau, \mathbf{x}^*))\mathbf{m}(\tau)^{\mathrm{T}}\,\mathrm{d}\tau = 0, \tag{2.3}$$

Uniqueness is a consequence of the Haar condition (Remark 1.4 of Chapter 3), which here takes the form that the matrix M,

$$M_{ij} = m_i(t_j), \qquad i, j = 1,2,\ldots,p, \tag{2.4}$$

has rank p, whenever $t_1, t_2, \ldots, t_p$ are distinct points in $[0,1]$, and we make the further assumption that there exists an open, connected neighbourhood X of $\mathbf{x}^*$ in which $r(\tau, \mathbf{x})$ has exactly p *simple* zeros $t_1, t_2, \ldots, t_p$ in $(0,1)$ for all $\mathbf{x} \in X$.

Remark 2.1 It follows from the above assumptions by an application of the implicit function theorem (Section 6.2) that the system of equations

$$r(t_i, \mathbf{x}) = 0, \qquad i=1,2,\ldots,p \tag{2.5}$$

can be solved for $t_i = t_i(\mathbf{x}) \in C^{\alpha-1}(X)$ for $i=1,2,\ldots,p$.

The Harr condition is sufficient to ensure that $\mathbf{t}=\mathbf{t}(\mathbf{x})$ can be inverted to give $\mathbf{x}=\mathbf{x}(\mathbf{t})$ where $\mathbf{t} \in T \subset [0,1]^p$ is again open and connected. Differentiating (2.5) gives for $\mathbf{x} \in X$,

$$\frac{\partial r}{\partial \tau}(t_i, \mathbf{x})\,\nabla_x t_i = -\mathbf{m}^{\mathrm{T}}(t_i)$$

so that

$$R\,\nabla_x \mathbf{t} = -\Theta M^{\mathrm{T}} \tag{2.6}$$

where R is nonsingular,

$$R = \operatorname{diag}\left\{\left|\frac{\partial r}{\partial \tau}(t_i, \mathbf{x})\right|, \qquad i=1,2,\ldots,p\right\} \tag{2.7}$$

and

$$\begin{aligned}\Theta &= \operatorname{diag}\{\theta_i, \qquad i=1,2,\ldots,p\} \\ &= \operatorname{diag}\left\{\operatorname{sgn}\left(\frac{\partial r}{\partial \tau}(t_i, \mathbf{x})\right) = \operatorname{sgn}(r(t_i+0, \mathbf{x})), \qquad i=1,2,\ldots,p\right\}.\end{aligned} \tag{2.8}$$

The system of equations (2.2) can be written in the equivalent form

$$0=\mathbf{g}(\mathbf{t})=\int_0^1 Q(\tau,\mathbf{t})\mathbf{m}(\tau)\,d\tau \tag{2.9}$$

where

$$Q(\tau,\mathbf{t})=\sum_{i=1}^{p+1}\theta_{i-1}H(\tau-t_{i-1})H(t_i-\tau) \tag{2.10}$$

where $H(\cdot)$ is the Heavyside unit function, $t_0=0$, and $t_{p+1}=1$. Newton's method applied to (2.9) to find $\mathbf{t}^*=\mathbf{t}(\mathbf{x}^*)$ gives the correction

$$\Delta\mathbf{t}=-(\nabla_t\mathbf{g})^{-1}\mathbf{g} \tag{2.11}$$

where

$$\begin{aligned}\nabla_t\mathbf{g}&=\int_0^1 \mathbf{m}(\tau)\,\nabla_t Q\,d\tau,\\ &=2\int_0^1 \mathbf{m}(\tau)\left(-\sum_{i=1}^{p}\theta_i\,\delta(\tau-t_i)\mathbf{e}_i^{\mathrm{T}}\right)d\tau,\\ &=-2M\Theta,\end{aligned} \tag{2.12}$$

and $\delta(\cdot)$ is the Dirac delta function. In particular, this shows that the Haar condition is sufficient for the iteration to be defined.

This approach uses the characterization equations explicitly but does not connect Newton's method directly with a minimization procedure. To do this we attack the minimization problem directly. Let

$$F(\mathbf{t})=\|r(\tau,\mathbf{x}(\mathbf{t}))\|_1=\int_0^1 Q(\tau,\mathbf{t})r(\tau,\mathbf{x}(\mathbf{t}))\,d\tau. \tag{2.13}$$

The necessary conditions for a minimum give

$$\begin{aligned}0=\nabla_t F&=\left(\int_0^1 Q(\tau,\mathbf{t})\mathbf{m}(\tau)^{\mathrm{T}}\,d\tau\right)\nabla_t\mathbf{x}-2\sum_{i=1}^{p}|r(t_i,\mathbf{x}(\mathbf{t}))|\\ &=\mathbf{g}(\tau)^{\mathrm{T}}\nabla_t\mathbf{x}\end{aligned} \tag{2.14}$$

as the second term vanishes as a consequence of (2.5). This is equivalent to (2.2) as $\nabla_t\mathbf{x}$ is nonsingular.

Newton's method applied to (2.14) gives

$$\Delta\bar{\mathbf{t}}=-(\nabla_t^2F)^{-1}(\nabla_tF)^{\mathrm{T}} \tag{2.15}$$

where

$$\nabla_t^2F=(\nabla_t\mathbf{x})^{\mathrm{T}}\nabla_t\mathbf{g}+\sum_{i=1}^{p}g_i\,\nabla_t^2x_i. \tag{2.16}$$

By comparing (2.14), (2.15), and (2.16) with (2.11) it will be seen that

$$\|\Delta\bar{\mathbf{t}}-\Delta\mathbf{t}\|_2=O(\|\mathbf{g}\|_2^2)$$

as $\|\mathbf{g}\|_2\to 0$. However, perhaps the most interesting result follows from

(2.16) on putting $\mathbf{x}=\mathbf{x}^*$:

$$\begin{aligned}\nabla_t^2 F(\mathbf{t}(\mathbf{x}^*)) &= (\nabla_t \mathbf{x})^{\mathrm{T}} \nabla_t \mathbf{g} \\ &= -2(\nabla_t \mathbf{x})^{\mathrm{T}} M\Theta \\ &= 2(\nabla_t \mathbf{x})^{\mathrm{T}} (\nabla_x \mathbf{t})^{\mathrm{T}} R \\ &= 2R. \end{aligned} \tag{2.17}$$

This shows that $\nabla_t^2 F(\mathbf{t}(\mathbf{x}))$ tends to a diagonal matrix, so that the corrections to each t_i are determined independently to first order in $\|\mathbf{g}\|_2$. However, the Newton correction (2.15) is equivalent to stepping to the minimum of a second-order exact approximation to F. This shows that the process of independent minimization in each variable is not far away from the Newton iteration. Thus a method which minimizes F with respect to each of the components of $\mathbf{t}$ in sequence (more or less) could be expected to have excellent convergence properties. But this is essentially what happens in the descent methods for the discrete l_1 problem. Thus it should not be surprising that the algorithm gives a good account of itself.

A2.3 APPROXIMATION IN THE MAXIMUM NORM

In the above considerations the correction between the continuous and discrete problems is at best approximate so that the argument leans to some extent on analogy. However, the connection between maximum norm approximation and the discrete l_∞ problem is much closer, the continuous case algorithms actually solving discrete problems at each step. The maximum norm problem is

$$\min_{\mathbf{x}} \max_{0 \leqslant \tau \leqslant 1} \left| \sum_{i=1}^{p} x_i m_i(\tau) - f(\tau) \right| = \min_{\mathbf{x}} \|r(\tau, \mathbf{x})\|_\infty \tag{3.1}$$

The basis for the algorithmic developments is the result that the solution of (3.1) is characterized by:

(i) a critical set of points $t_1, t_2, \ldots, t_k \in [0, 1]$; and
(ii) multipliers λ_i, $i=1, 2, \ldots, k$, satisfying

(a) $\sum_{i=1}^{k} \lambda_i m_i(t_j) = 0, \quad j=1, 2, \ldots, k,$ and

(b) $\lambda_i \theta_i \geqslant 0,$ and $\sum_{i=1}^{k} |\lambda_i| = 1,$

where $\theta_i = \operatorname{sgn}(r(t_i, \mathbf{x}))$. It follows that if M satisfies the Haar condition then necessarily $k \geqslant p+1$. The characterization results should be compared with the condition for an optimum solution of the discrete l_∞ problem given in Theorem 1.3 of Chapter 3. Two algorithms for the maximum norm problem are considered, both associated with the name of the Russian mathematician Remes.

Algorithm 1 The first algorithm of Remes tries to fill out [0, 1] with a set of points which will eventually have a critical set as limit points. Let the current set of points be $T_s = \{t_1, \ldots, t_s\}$, $\mathbf{x}_s$ solve the l_∞ problem

$$\min_{\mathbf{x}} \max_{t \in T_s} |r(t, \mathbf{x})| \tag{3.2}$$

and h_s be the corresponding optimal value in (3.2). The algorithm now finds a point τ_s such that

$$|r(\tau_s, \mathbf{x}_s)| = \|r(\tau, \mathbf{x}_s)\|_\infty, \tag{3.3}$$

and checks that

$$h_s < |r(\tau_s, \mathbf{x}_s)| .$$

If this inequality is satisfied, T_{s+1} is defined by

$$T_{s+1} = T_s \cup \{\tau_s\}, \tag{3.4}$$

and the process repeated iteratively. It is sufficient for the success of this algorithm that the initial matrix M_0 have rank p, where

$$(M_0)_{ij} = m_j(t_i), \qquad j = 1, 2, \ldots, p, \qquad t_i \in T_0,$$

and T_0 is the initial set of points.

If this condition holds then limit points of the iteration solve (3.1). There is considerable generality and robustness in this procedure.

Algorithm 2 The second algorithm of Remes works with ordered sets of exactly $p+1$ points which in the simplest case are assumed to be approximations to a critical set in which the residuals at successive points alternate in sign. At the sth stage let this be $U_s = \{t_1^s, \ldots, t_{p+1}^s\}$. Then an estimate of the best approximation can be found by solving for h_s, $\mathbf{x}_j$ the system of linear equations

$$r(t_i^s, \mathbf{x}_s) = (-1)^{i-1} h_s, \qquad i = 1, 2, \ldots, p+1 \tag{3.5}$$

The points of extremal deviation of the error curve $r(\tau, \mathbf{x}_s)$ are now calculated, and the points in U_{s+1} are selected from among these extrema (Figure 3.1) so that

$$\frac{\mathrm{d}r}{\mathrm{d}\tau}(t_i^{s+1}, \mathbf{x}_s) = 0.$$

This procedure is more fragile than the first algorithm, and convergence proofs depend heavily on the restrictive assumptions. However, local convergence and rate of convergence results can be developed in the situation in which (3.1) has a solution which is strongly unique.

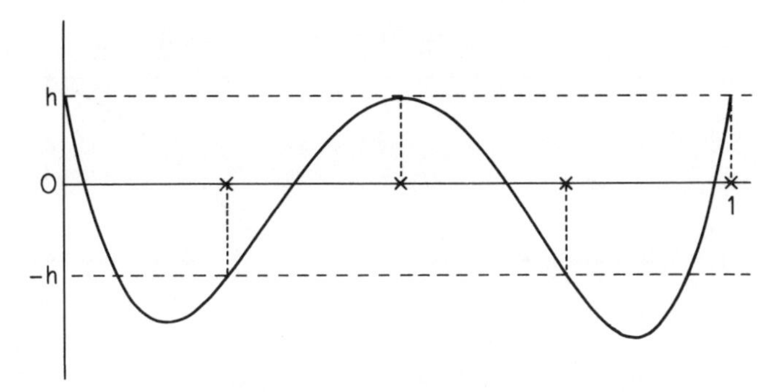

Figure 3.1 Selection of points in the second algorithm of Remes. Extrema of the error curve determine the next U_s

If the Haar condition does not hold then there may be fewer than $p+1$ points of extremal deviation. To see that this is awkward, note that the first algorithm leads to a sequence of linear programming problems which it is appropriate to solve using the exchange algorithm of Section 3.5, and this automatically forces $p+1$ points of extremal deviation if rank $(M_s)=p$. Thus the limiting set can contain fewer than $p+1$ points only if certain of the points defining the successive basis matrices coalesce as the algorithm proceeds. It follows that the successive basis matrices become increasingly *singular* and this term is used to describe such problems.

Example 3.1 Consider

$$r(\tau, \mathbf{x}) = \tau^2 - x_1\tau - x_2 e^{\tau}.$$

The error curve has just two points of extremal deviation in $(0, 2)$, at $t_1 = .406\,375\,74\ldots$ and $t_2 = 2$, and the extremal deviation is $h = .538\,245\,32\ldots$. A sketch of a typical residual curve produced by the first algorithm is given in Figure 3.2. Progress of the algorithm is summarized in Table 3.1. This can be compared with the usual case by noting that as $t=0$ is not a point of extremal deviation the solution is a best approximation on the larger interval $[\bar{t}, 2]$ where $\bar{t} = -.768\,600\,78\ \ldots$, and where there are now three points of extremal deviation. The result of applying the exchange algorithm in this case is summarized in Table 3.2.

This example suggests:

(a) that the first algorithm is capable of excellent performance when the problem is not singular, and
(b) that in the singular case the numerical difficulties appear to be compounded by slow convergence.

To conclude this discussion, an indication is given of how the rate of

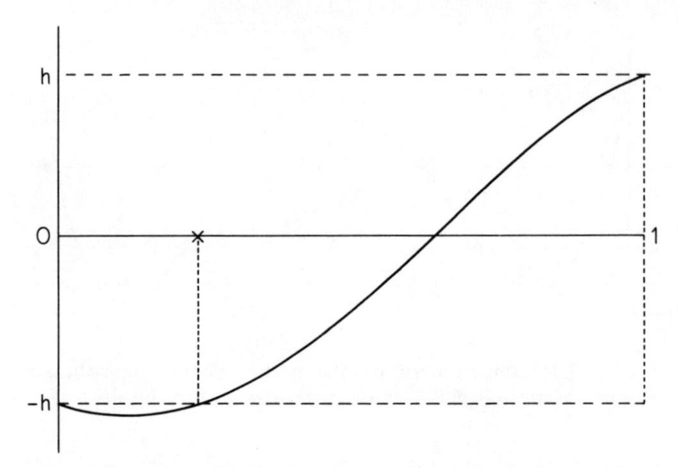

Figure 3.2 A typical residual curve for the singular problem

Table 3.1 Progress of the first algorithm in the singular case.

Iteration	0	1	2	9	15
t_1	0	0	.2347	.4055	.406356
t_2	1	.4789	.4789	.4074	.406395
t_3	2	2	2	2	2
h	.4083	.5224	.5313	.5382448	.53824532

Table 3.2 Progress of the first algorithm, the well-behaved case.

Iteration	0	1	2
t_1	$\bar{t}$	$\bar{t}$	$\bar{t}$
t_2	0	.3796	.4061
t_3	2	2	2
h	.453954	.537840	.53824526

convergence of the exchange algorithms can be analysed. Two methods are presented which have different advantages. Both concentrate on the second algorithm. The first shows the uncoupling into subproblems which occurred also in the L_1 algorithm, and this permits conclusions to be drawn also about the convergence of the first algorithm. Interestingly, it does not use linearity in the argument. The second is an application of Newton's method.

The key assumptions are:

(i) There exists an open connected neighbourhood X containing $\mathbf{x}^*$ such that for each $\mathbf{x} \in X$ there exists a critical set consisting of $p+1$ distinct points t_i, $i=1, 2, \ldots, p+1$, with $t_1 = 0$ and $t_{p+1} = 1$, such that

$$r(t_i, \mathbf{x}) = \theta_i h, \qquad i = 1, 2, \ldots, p+1 \tag{3.6}$$

and

$$\frac{\partial r}{\partial \tau}(t_i, \mathbf{x}) = 0, \qquad i = 2, 3, \ldots, p \tag{3.7}$$

where $\theta_i = \operatorname{sgn}(r(t_i^*, \mathbf{x}^*))$.

(ii) The matrix $[M^{\mathrm{T}} \mid \boldsymbol{\theta}]$ is nonsingular.

(iii) If $\dfrac{\partial}{\partial \tau}(t_i, \mathbf{x}) = 0$ so that t_i is not, in general, an end point of $[0, 1]$, then $\theta_i \dfrac{\partial^2 r}{\partial \tau^2}(\tau, \mathbf{x}) > 0$ in a neighbourhood of t_i. This ensures that the conditions $\dfrac{\partial r}{\partial \tau}(t_i, \mathbf{x}) = 0$, $i = 2, 3, \ldots, p$, determine $t_i = t_i(\mathbf{x})$, $i = 2, 3, \ldots, p$, for $\mathbf{x} \in X$, by the implicit function theorem.

The result hinges on the following lemma which follows by differentiating selected components of (3.6) with respect to t_j, $j = 2, \ldots, p$.

Lemma 3.1 *Let* $\dfrac{\partial}{\partial \tau} r(t_i^*, \mathbf{x}^*) = 0$, $i = 2, \ldots, p$. *Then*

(i)
$$-[M^{\mathrm{T}} \mid -\boldsymbol{\theta}] \frac{\partial}{\partial t_i} \begin{bmatrix} \mathbf{x} \\ h \end{bmatrix}_{\mathbf{x}^*, h^*} = \frac{\partial r}{\partial \tau}(t_i^*, \mathbf{x}^*) \mathbf{e}_i = 0,$$

and

(ii)
$$-[M^{\mathrm{T}} \mid -\boldsymbol{\theta}] \frac{\partial^2}{\partial t_i \, \partial t_q} \begin{bmatrix} \mathbf{x} \\ h \end{bmatrix}_{\mathbf{x}^*, h^*} = \delta_{iq} \frac{\partial^2 r}{\partial \tau^2}(t_i^*, \mathbf{x}^*) \mathbf{e}_i.$$

In particular, the second derivative matrices of the components of $\mathbf{x}^*$, h^* *are diagonal.*

To apply these results let t_j^{s+1} be an extremum of an approximate error curve,

$$\frac{\partial r}{\partial \tau}(t_j^{s+1}, \mathbf{x}^s) = \frac{\partial r}{\partial \tau}(t_j^*, \mathbf{x}^*) = 0.$$

Then

$$0 = \frac{\partial r}{\partial \tau}(t_j^{s+1}, \mathbf{x}^s)$$

$$= \frac{\partial r}{\partial \tau}(t_j^{s+1}, \mathbf{x}^*) - \frac{\partial r}{\partial \tau}(t_j^*, \mathbf{x}^*) + \frac{\partial r}{\partial \tau}(t_j^{s+1}, \mathbf{x}^s) - \frac{\partial r}{\partial \tau}(t_j^{s+1}, \mathbf{x}^*)$$

$$= \frac{\partial^2 r}{\partial \tau^2}(\bar{t}, \mathbf{x}^*)(t_j^{s+1} - t_j^*) + \frac{\mathrm{d}\mathbf{m}}{\mathrm{d}\tau}(t_j^{s+1})^{\mathrm{T}}(\mathbf{x}^s - \mathbf{x}^*)$$

$$= \frac{\partial^2 r}{\partial \tau^2}(\bar{t}, \mathbf{x}^*)(t_j^{s+1} - t_j^*) + \sum_{i=1}^{p} \frac{\mathrm{d}m_i}{\mathrm{d}\tau}(t_j^{s+1}) \left\{ \frac{1}{2} \sum_{q \in \sigma} \frac{\partial^2 x_i}{\partial t_q^2}(t_q^s - t_q^*)^2 + \mathrm{O}\left(\max_q |t_q^s - t_q^*|^3 \right) \right\}$$

where σ points to the subset of the critical points for which $\frac{\partial r}{\partial \tau}(t_j^*, \mathbf{x}^*) = 0$, and $\bar{t}$ is a mean value, as a consequence of Lemma 3.1. The second-order convergence result follows immediately as a consequence of assumption (iii). The p step second-order convergence of the first algorithm also follows provided the estimates of the critical points are replaced more or less cyclically.

An alternative proof proceeds by noting that given $\mathbf{x}^s$ we determine $\mathbf{t}^{s+1}$ by solving $\frac{\partial r}{\partial \tau}(t_j^{s+1}, \mathbf{x}^s) = 0$, $j = 2, \ldots, p$. Given $\mathbf{t}^{s+1}$, $\mathbf{x}^{s+1}$ is then determined by solving (3.6). This is a system of nonlinear equations having the form

$$r(t_i(\mathbf{x}), \mathbf{x}) = \theta_i h, \qquad i = 1, 2, \ldots, p+1$$

Applying Newton's method gives

$$r(t_i, \mathbf{x}^s) + \left(\nabla_x r + \frac{\partial r}{\partial \tau} \nabla_x t_i \right) \Delta \mathbf{x} = \theta_i (h^s + \Delta h^s),$$

and this reduces to

$$r(t_i(\mathbf{x}^s), \mathbf{x}^{s+1}) = \theta_i h^{s+1}, \qquad i = 1, 2, \ldots, p+1$$

by the condition $\frac{\partial r}{\partial \tau}(t_i(\mathbf{x}^s), \mathbf{x}^s) = 0$. But this is the system of equations determining $\mathbf{x}^{s+1}$, h^{s+1} in the second algorithm.

Remark 3.1 Note that the one-dimensional minimization step appears naturally in the L_1 algorithm, but there is nothing comparable in the maximum norm case. One possible conclusion is that the multiple pivot sequence is likely to be less valuable in l_∞ problems that can be interpreted as discretizations of maximum norm problems, as here the test that determines t_s is directly comparable with the standard USE test. This test is already adequate to give superlinear convergence, suggesting that it will perform well also on the corresponding discretized problem.

NOTES ON APPENDIX 2

The L_1 exchange algorithm is due to Watson (1981). That the Hessian is diagonal in this case is an observation due to Powell. The Remes algorithms are treated in Cheney (1966), Watson (1980), and Powell (1981). The first algorithm has been the subject of much research. Mention is made here of Laurent (1973), who studies among other things the need to retain the points in the successive T_i, and Osborne (1980) who explores the (very considerable) breadth of application of the algorithm. A consequence is that it can be defined for L_1 problems (compare the treatment of polyhedral norms in Section 3.5), but the resulting problem is singular, there being essentially a unique element in the subdifferential. The concept of a singular problem is introduced in Osborne and Watson (1969). The second-order convergence of the second algorithm is due to Veidinger (1960). The treatment here follows Curtis and Powell (1966). It has been generalized in Smith (1979). Powell (1981) gives a full rate of convergence analysis for the first algorithm. The application of Newton's method is due to Meinardus (1967).

References

Abdelmalek, N. N. (1975). An efficient method for the discrete linear L_1 approximation problem, *Math. Comp.*, **29,** 844–850.

Akaike, H. (1959). On successive transformations of a probability distribution and its application to the analysis of the optimum gradient method, *Ann. Inst. Statist. Math., Tokyo,* **11,** 1–16.

Anderson, D. H., and Osborne, M. R. (1976). Discrete linear approximation problems in polyhedral norms, *Num. Math.*, **26,** 179–186.

Armstrong, R. D., and Frome, E. L. (1976). A branch and bound solution of restricted least squares problem, *Technometrics*, **18,** 447–450.

Armstrong, R. D., and Godfrey, J. (1979). Two linear programming algorithms for the linear discrete L_1 norm problem, *Math. Comp.*, **33,** 289–300.

Armstrong, R. D., Frome, E. L., and Kung, D. S. (1979). A revised simplex algorithm for the absolute deviation curve fitting problem, *Commun. Statist. B*, **8,** 175–190.

Armstrong, R. D., and Kung, D. S. (1980). A dual method for discrete Chebyshev curve fitting, *Math. Prog.*, **19,** 186–199.

Aspvall, B., and Stone, R. E. (1979). Khachiyan's linear programming algorithm, *Report CS79-776*, Computer Science Dept. Stanford University.

Avis, D., and Chvatal, V. (1976). Notes on Bland's pivoting rule, *Math. Prog. Study*, **8,** 24–34.

Barrodale, I., and Phillips, C. (1975). Solution of an overdetermined system of linear equations in the Chebyshev norm, *ACM Trans. Math. Software,* **1,** 264–270.

Barrodale, I., and Roberts, F. D. K. (1973). An improved algorithm for discrete l_1 linear approximation, *SIAM J. Numer. Anal.*, **10,** 839–848.

Barrodale, I., and Roberts, F. D. K. (1978). An efficient algorithm for discrete L_1 linear approximation with linear constraints, *SIAM J. Numer. Anal.*, **15,** 603–611.

Barrodale, I., and Young, A. (1966). Algorithms for best L_1 and L_∞ linear approximation on a discrete set, *Numer. Math.*, **8,** 295–306.

Bartels, R. H. (1971). A stabilization of the simplex method, *Numer. Math.*, **16,** 414–434.

Bartels, R. H. (1980). A penalty linear programming method using reduced gradient basis-exchange techniques, *Linear Algebra and Applic.*, **29,** 17–32.

Bartels, R. H., Conn, A. R., and Charalambous, C. (1978). On Clines' direct method for solving overdetermined linear systems in the l_∞ sense. *SIAM J. Numer. Anal.*, **15,** 255–270.

Bartels, R. H., Conn, A. R., and Sinclair, J. W. (1978). Minimization techniques for piecewise differentiable functions: the l_1 solution to an overdetermined linear system. *SIAM J. Numer. Anal.*, **15,** 224–241.

Bartels, R. H., Golub, G. H., and Saunders, M. A. (1970). Numerical techniques in Mathematical Programming. In *Nonlinear Programming* (editors Rosen, J. B., Mangasarian, O. L., and Ritter, K.), 123–176, Academic Press.

Bartels, R. H., and Joe, B. (1983). An exact penalty method for constrained, discrete, l_∞ data fitting, *SIAM Sci. and Stat. Comp.*, **4,** 69–84.

Bassett, G., and Koenker, R. (1978). Asymptotic theory of least absolute error regression, *J. Amer. Stat. Assoc.*, **73,** 618–622.

Beale, E. M. L. (1955). Cycling in the dual simplex algorithm, *Naval Research Logistics Quarterly*, **2,** 269–275.

Beale, E. M. L. (1959). On quadratic programming, *Naval Research Logistics Quarterly*, **6,** 227–244.

Beale, E. M. L. (1970). Selecting an optimum subset. In Abadie, J. (Ed.), *Integer and Nonlinear Programming*, 451–462, North Holland.

Beaton, A. E., and Tukey, J. W. (1974). The fitting of power series, meaning polynomials, illustrated on band-spectrographic data, *Technometrics*, **16,** 147–185.

Belsley, D. A., Kuh, E., and Welsch, R. E. (1980). *Regression Diagnostics: Identifying Influential Data and Sources of Collinearity*, John Wiley, New York.

Bertsekas, D. P., and Mitter, S. K. (1973). A descent method for optimization problems with nondifferentiable cost functions, *SIAM J. Control*, **11,** 637–652.

Bittner, L. (1961). Das Austauschverfahren der linearen Tschebyscheff-Approximation hei nicht erfullter Haarscher Bedingung, *ZAMM*, **41,** 238–256.

Björk, A. (1967). Solving linear least squares problems by Gram–Schmidt orthogonalization. *BIT*, **7,** 1–2.

Bland, R. G. (1977). New finite pivoting rules for the simplex method, *Math. of Operations Res.*, **2,** 103–107.

Bland, R. G., Goldfarb, D., and Todd, M. J. (1981). The ellipsoid method: a survey, *Oper. Res.*, **29,** 1039–1091.

Bloomfield, P., and Steiger, W. L. (1980). Least absolute deviations curve-fitting, *SIAM J. Sci. and Stat. Comp.*, **1,** 290–301.

Bloomfield, P., and Steiger, W. L. (1983). *Least Absolute Deviations*, Birkhauser, Boston.

Charnes, A. (1952). Optimality and degeneracy in linear programming, *Econometrica*, **20,** 160–170.

Cheney, E. W. (1966). *Introduction to Approximation Theory*, McGraw-Hill, New York.

Chvatal, V. (1983). *Linear Programming*, W. H. Freeman and Co.

Clark, D. I. (1980). An algorithm for solving the restricted least squares problem, *J. Austral. Math. Soc. B*, **21,** 345–356.

Clark, D. I. (1981). Finite algorithms for linear optimization problems, Ph.D. Thesis, Australian National University.

Clark, D. I. (1984). The mathematical structure of Huber's *M*-estimator, *SIAM J. Sci. and Stat. Comp.* **6,** 209–219.

Clark, D. I., and Osborne, M. R. (1980). On the implementation of a subset selection algorithm for the restricted least squares problem, *J. Austral. Math. Soc. B*, **22,** 2–11.

Clark, D. I., and Osborne, M. R. (1983). A descent algorithm for minimizing polyhedral convex functions, *SIAM J. Sci. and Stat. Comp.*, **4,** 757–786.

Clark, D. I., and Osborne, M. R. (1985). Finite algorithms for Huber's *M*-estimator, *SIAM J. Sci. and Stat. Comp.* (to appear).

Clarke, F. H. (1975). Generalised gradients and applications, *Trans. Ann. Math. Soc.*, **205,** 247–262.

Clarke, F. H. (1976). On the inverse function theorem, *Pac. J. Math.*, **64,** 97–102.

Clarke, F. H. (1983). *Optimization and Nonsmooth Analysis*, Wiley, New York.

Cline, A. K. (1976). A descent method for the uniform solution to overdetermined systems of linear equations, *SIAM J. Numer. Anal.*, **13,** 293–309.

Conn, A. R. (1976). Linear programming via a nondifferentiable penalty function, *SIAM J. Numer. Anal.*, **13,** 145–154.

Cottle, R. W., and Dantzig, G. B. (1968). Complementary pivot theory of mathematical programming, *J. Linear Algebra Applics*, **1,** 103–125.

Craven, B. D. (1978). *Mathematical Programming and Control Theory*, Chapman and Hall, London.

Curtis, A. R., and Powell, M. J. D. (1966). On the convergence of exchange algorithms for calculating minimax approximations, *Computer J.*, **9,** 78–80.

Daniel, C., and Wood, F. S. (1971). *Fitting Equations to Data*, Wiley, New York.

Dantzig, G. B. (1963). *Linear Programming and Extensions*, Princeton University Press.

Dantzig, G. B. (1982). Reminiscences about the origins of linear programming, *Op. Res. Letters*, **1,** 43–48.

Dantzig, G. B., Orden, A., and Wolfe, P. (1955). The generalised simplex method for minimizing a linear form under linear inequality constraints, *Pacific J. Math.*, **5,** 183–195.

Dempster, A. P., Laird, N. M., and Rubin, D. B. (1977). Maximum likelihood from incomplete data via the EM algorithm, *J. R. Stat. Soc. B*, **39,** 1–38.

Dunham, J. R., Kelly, D. G., and Tolle, J. W. (1977). Some experimental results concerning the expected number of pivots for solving randomly generated linear programs, *Report 77–16*, Operations Research Department, U. North Carolina.

Dutter, R. (1975). Robust regression: different approaches to numerical solutions and algorithms, *Res. Rep. no. 6*, Fachgruppe für Statistic, Eidgen. Technische Hochschule, Zurich.

Fiacco, A. V., and McCormick, G. P. (1968). *Nonlinear Programming*, Wiley, New York.

Fletcher, R. (1980). *Practical Methods of Optimization*, Vol. 1: *Unconstrained Optimization*, Vol. 2: *Constrained Optimization*, Wiley, Chichester.

Frederickson, G. N., and Johnson, D. B. (1982). The complexity of selection and ranking in $X+Y$ and matrices with sorted columns, *J. Comp. Sys. Sci.*, **24,** 194–208.

Gale, D. (1967). A geometric duality theorem with economic applications, *Rev. Econ. Studies*, **34,** 19–24.

Geoffrion, A. M. (Ed.) (1972). *Perspectives on Optimization*, Addison-Wesley, Reading, Mass.

Gill, P. E., Golub, G. H., Murray, W., and Saunders, M. A. (1974). Methods for modifying matrix factorizations, *Math. of Comp.*, **28,** 505–535.

Gill, P. E., and Murray, W. (1973). A numerically stable form of the simplex algorithm, *J. Linear Algebr. and Applic.*, **7,** 99–138.

Gill, P. E., and Murray, W. (Eds), (1974). *Numerical Methods for Constrained Optimization*, Academic Press, London.

Gill, P. E., Murray, W., and Wright, M. H. (1981). *Practical Optimization*, Academic Press, London.

Goldfarb, D., and Reid, J. K. (1977). A practicable steepest edge simplex algorithm, *Math. Prog.*, **12,** 361–371.

Goldfarb, D., and Sit, W. Y. (1979). Worst case behaviour of the steepest edge simplex method, *Discrete Appl. Math.*, **1,** 277–285.

Goldfarb, D., and Idnani, A. (1983). A numerically stable dual method for solving strictly quadratic programs, *Math. Prog.*, **27,** 1–33.

Golub, G. H. (1965). Numerical methods for solving linear least squares problems, *Num. Math.*, **7,** 206–216.

Golub, G. H., and Van Loan, C. F. (1980). An analysis of the total least squares problem, *SIAM J. Numer. Anal.*, **17,** 883–893.

Golub, G. H., and Van Loan, C. F. (1983). *Matrix Computations*, Johns Hopkins University Press, Baltimore.

Golub, G. H., and Wilkinson, J. H. (1966). Iterative refinement of least squares solutions, *Numer. Math.*, **9,** 189–148.

Hadley, G. (1962). *Linear Programming*, Addison-Wesley, Reading, Mass.

Hammarling, S. (1974). A note on modifications to the Givens plane rotation, *J. Inst. Math. Applics.*, **13,** 215–218.

Hanson, R. J., and Lawson, C. L. (1974). *Solving Least Squares Problems*, Prentice-Hall, Englewood Cliffs, New Jersey.

Harris, P. M. J. (1973). Pivot selection methods of the Devex LP code, *Math. Prog.*, **5,** 1–28.

Hestenes, M. R. (1966). *Calculus of Variations and Optimal Control Theory*, Wiley, New York.

Hettmansperger, T. P., and McKean, J. W. (1977). A robust alternative based on ranks to least squares in analyzing linear models, *Technometrics*, **19,** 275–284.

Hiriart-Urruty, J. B. (1979). Tangent cones, generalized gradients and mathematical programming in Banach spaces, *Math. Oper. Res.*, **4,** 79–97.

Hoffman, A. J. (1953). Cycling in the simplex algorithm, National Bureau of Standards, *Report No. 2974.*

Hopper, M. J., and Powell, M. J. D. (1977). A technique that gains speed and accuracy in the minimax solution of overdetermined linear equations. In J. R. Rice (Ed.), *Mathematical Software III*, 15–34, Academic Press, New York.

Householder, A. S. (1958). Unitary triangularization of a nonsymmetric matrix, *J. ACM*, **6,** 339–342.

Householder, A. S. (1964). *The Theory of Matrices in Numerical Analysis*, Blaisdell Pub. Co., New York.

Huber, P. J. (1977). Robust Statistical Procedures, *SIAM Regional Conference Series in Appl. Math.*, No. 27.

Huber, P. J. (1981). *Robust Statistics*, Wiley, New York.

Huber, P. J. (1983). Minimax aspects of bounded-influence regression, *J. Am. Stat. Assoc.*, **78,** 66–80.

Jaeckel, L. A. (1972). Estimating regression coefficients by minimizing the dispersion of the residuals, *Ann. Math. Stats*, **43,** 1449–1458.

Jennings, L. S., and Osborne, M. R. (1974). A direct error analysis for least squares, *Numer. Math.*, **22,** 325–332.

Jeroslow, R. (1973). The simplex algorithm with the pivot rule of maximizing criterion improvement, *Discrete Math.*, **4,** 367–377.

Jittorntrum, K. (1978). Sequential algorithms in nonlinear programming, Ph.D. thesis, Australian National University.

Jittorntrum, K., and Osborne, M. R. (1980). Strong uniqueness and second order convergence in nonlinear discrete approximation, *Numer. Math.*, **34,** 439–455.

John, F. (1948). Extremum problems with inequalities as subsidiary conditions. In *Studies and Essays presented to R. Courant on his 60th Birthday*, 187–204, Interscience, New York.

Kelley, J. E. (Jr.). (1958). An application of linear programming to curve fitting, *SIAM J.*, **6,** 15–22.

Klee, V. L. (1957). Extremal structure of convex sets, *Arch. Math.*, **8,** 234–240.

Klee, V. L., and Minty, G. L. (1972). How good is the simplex algorithm? In Shisha, O. (Ed.), *Inequalities III*, 159–175, Academic Press, New York.

Koopmans, T. C. (Ed.) (1951). *Activity Analysis of Production and Allocation*, Cowles Commission Monograph No. 13, Wiley, New York.

Krasker, W. S., and Welsch, R. E. (1982). Efficient bounded-influence regression estimation, *J. Am. Stat. Assoc.*, **77,** 595–604.

Kuhn, H. W., and Tucker, A. W. (1951). Nonlinear programming. In Neyman, J. (Ed.), *Proceedings of the Second Berkeley Symposium in Mathematical Statistics and Probability*, 481–492, University of California Press, Berkeley.

Laurent, P. J. (1973). Exchange algorithms in convex analysis. In Lorentz, G. G. (Ed.), *Approximation Theory*, 403–409, Academic Press, New York.

Lemke, C. E. (1954). The dual method of solving the linear programming problem, *Naval Research Logistics Quarterly*, **1,** 48–54.

Lemke, C. E. (1962). A method of solution for quadratic programs, *Management Sci.*, **8,** 442–453.

Luenberger, D. G. (1969). *Optimization by Vector Spare Methods*, Wiley, New York.

Luenberger, D. G. (1973). *Introduction to Linear and Nonlinear Programming*, Addison-Wesley, Reading, Mass.

McKean, J. W. and Ryan, T. A. Jr. (1977). An algorithm for obtaining confidence intervals and point estimates based on ranks in the two sample location problem, *Trans. Math. Software*, **3,** 183–185.

McKean, J. W., and Shrader, R. M. (1980). The geometry of robust procedures in linear models, *J. Roy. Stat. Soc. B*, **42,** 366–371.

McLewin, W. (1980). *Linear Programming and Applications*, Input-Output Publishing Co., London.

McLinden, L. (1982). Polyhedral extensions of some theorems on linear programming, *Math. Prog.*, **24,** 162–176.

Meinardus, G. (1967). *Approximation of Functions: Theory and Numerical Methods*, Springer-Verlag, Berlin.

Mostellar, F., and Tukey, J. W. (1977). *Data Analysis and Regression*, Addison-Wesley, Reading, Mass.

Müller-Merbach, H. (1970). *On Round-off Errors in Linear Programming*, Lecture Notes in Operations Research and Linear Systems, No. 37, Springer-Verlag, Berlin.

Murty, K. G. (1974). Note on a Bard-type scheme for solving the complementarity problem, *Opsearch*, **11,** 123–130.

Murty, K. G. (1976). *Linear and Combinatorial Programming*, Wiley, New York.

Nelder, J. A., and Wedderburn, R. M. W. (1972). Generalized linear models, *J. Roy. Stat. Soc.*, **135,** 370–384.

Osborne, M. R. (1978). Implementation of exchange algorithms for polyhedral norm problems. In *Very Informal Proceedings of Gatlinburg VII Meeting on Numerical Linear Algebra*, Computer Science Dept., Stanford University.

Osborne, M. R. (1980). An algorithmic approach to nonlinear approximation problems. In Cheney, E. W. (Ed.), *Approximation Theory III*, 705–710, Academic Press, New York.

Osborne, M. R. (1982). A finite algorithm for the rank regression problem. In Gani, J., and Hannan, E. J. (Eds.), *Essays in Statistical Science*, 241–252, *J. App. Probability*, special volume **19A**.

Osborne, M. R. (1983). An algorithm for minimizing polyhedral convex functions. In Fiacco, A. V., and Kortanek, K. O. (Eds.), *Semi-Infinite Programming and Applications*, Lecture Notes in Economics and Mathematical Systems, No. 215, Springer-Verlag, Berlin.

Osborne, M. R., and Watson, G. A. (1967). On the best linear Chebyshev approximation, *Comp. J.*, **10,** 172–177.

Osborne, M. R., and Watson, G. A. (1969). A note on singular minimax approximation problems, *J. Math. Anal. and Applic.*, **25,** 692–700.

Osborne, M. R., and Watson, G. A. (1985). An analysis of the total approximation problem in separable norms, and an algorithm for the total l_1 problem, *SIAM J. Sci. and Stat. Comp.* **6,** 410–424.

Paige, C. C. (1973). An error analysis of a method for solving matrix equations, *Math. of Comp.*, **27,** 355–359.

Paige, C. C. (1979). Computer solution and perturbation analysis of generalised linear least squares problems, *Math. of Comp.*, **33,** 171–183.

Perold, A. F. (1980). A degeneracy exploiting LU factorization for the simplex method, *Math. Prog.*, **19,** 239–254.

Powell, J. L. (1984). Least absolute deviations estimation for the censored regression models, *J. Econometrics* **25,** 303–325.

Powell, M. J. D. (1981). *Approximation Theory and Methods*, Cambridge University Press, Cambridge.

Quandt, R. E., and Kuhn, H. W. (1964). On upper bounds for the number of iterations in solving linear programs, *Operations Res.*, **12,** 161–165.

Reid, J. K. (1976). Fortran subroutines for handling sparse linear programming bases, *Report AERE-R8269*, Atomic Energy Research Establishment, Harwell.

Reid, J. K. (1982). A sparsity-exploiting variant of the Bartels–Golub decomposition for linear programming bases, *Math. Prog.*, **24,** 55–69.

Remes, E. Y. (1934). Sur un procédé convergent d'approximations successives pour determiner les polynômes d'approximation, *Comptes Rendues*, **198,** 2063–2065.

Robers, P. D., and Ben-Israel, A. (1969). An interval programming algorithm for discrete linear L_1 approximation, *J. Approx. Theory*, **1,** 323–336.

Robers, P. D., and Ben-Israel, A. (1970). A suboptimization method for interval linear programming, *Lin. Alg. and Applic.*, **3,** 383–405.

Robinson, S. M. (1975). Stability theory for systems of inequalities, Part 1: Linear systems, *SIAM J. Numer. Anal.*, **12,** 754–769.

Rockafellar, R. T. (1970). *Convex Analysis*, Princeton University Press, Princeton.

Rockafellar, R. T. (1976). Monotone operators and the proximal point algorithm, *SIAM J. Control and Optimization*, **14,** 877–898.

Rosen, J. B. (1960). The gradient projection method for nonlinear programming, Part 1: Linear constraints, *SIAM J.*, **8,** 181–217.

Saunders, M. A. (1972). Robust form of the Choleski factorization for large-scale linear programming, *Stanford University Report STAN-C5-72-301*.

Schlossmacher, E. J. (1973). An iterative technique for absolute deviations curve fitting, *J. Am. Stat. Assoc.*, **68,** 857–865.

Seber, G. A. F. (1977). *Linear Regression Analysis*, Wiley, New York.

Seneta, E., and Steiger, W. L. (1984). A new LAD curve fitting algorithm: slightly overdetermined systems in L_1, *Discrete Appl. Math.* **7,** 79-91.

Smale, S. (1983). On the average number of steps of the simplex method of linear programming, *Math. Prog.*, **27,** 241–262.

Smith, S. R. (1979). Ph.D. Thesis, Department of Mathematics, U. California Riverside.

Späth, H. (1981). The minisum location problem for the Jaccard metric, *Operations Res. Spektrum*, **3,** 91–94.

Späth, H. (1982). On discrete linear orthogonal L_p-approximation, *ZAMM*, **62,** 354–355.

Spyropoulos, K., Kiountouzis, E., and Young, A. (1973). Discrete approximation in the L_1 norm, *Comp. J.*, **16,** 180–186.

Stewart, G. W. (1973). *Introduction to Matrix Computations*, Academic Press, New YorK.

Stewart, G. W. (1977). On the perturbation of pseudo-inverses, projections and linear least squares, *SIAM Rev.*, **19,** 634–662.

Stewart, G. W. (1979). The effect of rounding error on an algorithm for downdating a Choleski factorization, *J. Inst. Maths. Applic.*, **23,** 203–213.

Stiefel, E. L. (1959). Uber diskrete und lineaire Tschebyscheff-Approximationen, *Numer. Math.*, **1,** 1–28.

Taylor, G. D., and Winter, M. J. (1970). Calculation of best retricted approximations, *SIAM J. Numer. Anal.*, **7,** 248–255.

Traub, J. F., and Wozniakowski, H. (1982). Complexity of linear programming, *Op. Res. Letters,* **1,** 59–62.

Vajda, S. (1958). *Readings in Linear Programming,* Wiley, New York.

van Dam, W. B., Frenk, J. B. G., and Telgen, J. (1983). Randomly generated polytopes for testing mathematical programming algorithms, *Math. Prog.*, **26,** 172–181.

Vapnik, V. (1982). *Estimation of Dependences Based on Empirical Data,* Springer-Verlag, Berlin.

Veidinger, L. (1960). On the numerical determination of the best approximations in the Chebyshev sense, *Numer. Math.*, **2,** 99–105.

Wagner, H. M. (1959). Linear programming techniques for regression analysis, *J. Am. Stat. Assoc.*, **54,** 206–212.

Watson, G. A. (1973). On the best linear one-sided Chebyshev approximation, *J. Approx. Theory,* **7,** 48–58.

Watson, G. A. (1974). The calculation of best restricted approximations, *SIAM J. Numer. Anal.*, **11,** 693–699.

Watson, G. A. (1980). *Approximation Theory and Numerical Methods,* Wiley, Chichester.

Watson, G. A. (1981). An algorithm for linear L_1 approximation of continuous functions, *IMA J. Numer. Anal.*, **1,** 157–168.

Watson, G. A. (1982a). Numerical methods for linear orthogonal L_p approximation, *IMA J. Numer. Anal.*, **2,** 275–288.

Watson, G. A. (1982b). An algorithm for the single facility location problem using the Jaccard metric, *Numerical Analysis Report,* NA/57, University of Dundee.

Watson, G. A. (1983). The total approximation problem. In Schumaker, L. (Ed), *Approximation Theory IV,* 723–728, Academic Press, New York.

Watson, G. A. (1984). The numerical solution of total l_p approximation problems. In Griffiths, D. F. (Ed), *Numerical Analysis,* Dundee 1983, 221–238, *Lecture Notes in Mathematics,* No. 1066, Springer-Verlag, Berlin.

Whittle, P. (1971). *Optimization under Constraints,* Wiley, London.

Wilkinson, J. H. (1965). *The Algebraic Eigenvalue Problem,* Oxford University Press.

Wolfe, P. (1963). A technique for resolving degeneracy in linear programming, *SIAM J.*, **11,** 205–211.

Wolfe, P. (1967). Methods of nonlinear programming. In Abadie, J. (Ed.), *Nonlinear Programming,* 97–131, North-Holland, Amsterdam.

Wolfe, P. (1975). A method of conjugate subgradients for minimizing nondifferentiable functions. In Balinski, M. L., and Wolfe, P. (Eds.), *Nondifferentiable Optimization,* 145–173, Math. Prog. Study No. 3, North-Holland, Amsterdam.

Womersley, R. S. (1985). Censored discrete L_1 approximation, *SIAM J. Sci. and Stat. Comp.* (to appear).

Index